INTERNATIONAL PLANT PROTECTION CENTER
OREGON STATE UNIVERSITY

Properties and Management
of Soils in the Tropics

Properties and Management
of Soils in the Tropics

PEDRO A. SANCHEZ

Department of Soil Science
North Carolina State University

A WILEY-INTERSCIENCE PUBLICATION

JOHN WILEY AND SONS, New York • London • Sydney • Toronto

Library of Congress Cataloging in Publication Data

Sanchez, Pedro A   1940–
   Properties and management of soils in the tropics.

  "A Wiley-Interscience publication."
  Includes bibliographical references and index.
   1. Soils—Tropics. 2. Plant-soil relationships. 3. Soil science. 4. Agriculture—
Tropics. I. Title.

S599.9.T76S26   631.4'913   76-22761
ISBN 0-471-75200-2

Printed in the United States of America

10 9 8 7 6 5 4 3 2 1

PREFACE

The purpose of this book is to apply the principles of soil science to tropical conditions, with emphasis on ways to increase food production in developing countries. Although these principles are universal, their application is site-specific. The proper management of soils in the tropics is considered one of the critical components in the worldwide race between food production and population growth. Much valuable work has been conducted throughout the tropics. This book attempts to compile the available information on the properties and management of tropical soils collected from the literature and from ongoing work at several institutions.

The need for such a synthesis first became apparent to me while teaching an undergraduate course in soil fertility at the University of the Philippines 11 years ago. Students experienced difficulties in relating to examples applying the principles of soil science from textbooks devoted almost exclusively to temperate-region agriculture. A rough outline was prepared at that time, but involvement in a rice research project in Peru prevented further work until I was transferred to Raleigh in a teaching and research position on tropical soils in 1971.

The book is based on a course with the same title given at North Carolina State University during the past 4 years for advanced undergraduates and graduate students. The level of presentation assumes a knowledge of elementary soil science. The presentation can be divided into three sections. The first section (Chapters 1 and 2) defines the tropical environment in physical and human terms and the geographical distribution and classification of soils in the tropics. The second section (Chapters 3 to 9) focuses on specific soil–plant relationships related to physical and chemical soil properties, organic matter, plant nutrients, and methods of soil fertility evaluation. The third section (Chapters 10 and 13) integrates the previous concepts in terms of four principal soil management systems encountered in the tropics: shifting cultivation, rice culture, multiple cropping, and pasture production. Limitation of coverage to these topics does not imply that they are the only important ones, but rather reflects time limitations.

The format is aimed to serve as a textbook for courses on tropical soils and as a reference for agricultural scientists and development workers interested in the tropics. The list of references at the end of each chapter includes those cited in the text and additional ones relevant for further study for interested readers.

I wish to acknowledge the help of several individuals who have aided in the preparation of this book. To my students in Soil Science 501 I am grateful for their incisive questions and criticisms of the concepts and examples presented. For reviewing the various chapters I am indebted to my colleagues E. J. Kamprath, R. B. Cate, Jr., S. W. Buol, W. L. Johnson, and C. K. Hiebsch of North Carolina State University; G. Uehara of the University of Hawaii; S. K. DeDatta of the International Rice Research Institute; and José Toledo of the Instituto Veterinario de Investigación del Trópico y Altura, Peru. For typing and revising the manuscript I am grateful to Miss Bertha Monar, Mrs. Dawn Silsbee, and Mrs. Patrice Hill. To R. J. McCracken and C. B. McCants, former and present heads of the Soil Science Department, I am grateful for their encouragement to write this book.

For permission to use copyright materials I wish to express my appreciation to Academic Press, American Association for the Advancement of Science, American Society of Agronomy, Armand Colin Publishers, *Australian Journal of Agricultural Research, Biometeorology,* Cambridge University Press, Centre for Agricultural Publications and Documentation of Wageningen, CSIRO Australia, *Experimental Agriculture, Journal of Agricultural Science, Journal of Soil Science,* IRI Research Institute, Mouton and Company, *Plant and Soil, Soil Science,* Soil Science Society of America, and the University of Queensland Press.

To my wife, Wendy, and my children Jennifer and Evan, I am grateful for their patience and understanding, and dedicate this work to them.

PEDRO A. SANCHEZ

Raleigh, North Carolina
June 1976

CONTENTS

Properties and Management
of Soils in the Tropics

1
THE TROPICAL ENVIRONMENT

The purpose of this chapter is to provide a framework of the physical and social environment of the tropics. It is this writer's contention that there is nothing special or unique about tropical soils that cannot be understood by analysis. What is special and unique is how the soils behave and are managed within the tropical environment. For this reason an overview of the tropical environment is in order.

The geographic definition of the tropics is "that part of the world located between 23.5 degrees north and south of the Equator." Because of the tilt of the earth's axis, this latitude is the limit of the sun's apparent migration to the north or south of the zenith. Consequently, the tropics is the only part of the world where the sun passes directly overhead.

The tropics comprise 38 percent to the earth's land surface (approximately 5 billion hectares) and 45 percent of the world's population (about 1.8 billion people in 1975). About 72 countries and territories lie wholly or mostly in the tropics. They include most of the "developing" countries except China, Pakistan, Afghanistan, Iran, the Middle East, North Africa, Argentina, Chile, and Uruguay. Only tropical Australia and the states of Hawaii (the United States) and São Paulo (Brazil) are considered "developed" areas in the tropics. Economically, therefore, tropical areas are essentially equivalent to "developing" areas, although the reverse is not true.

1

The literature is full of attempts to quantify precisely the tropical versus extratropical parts of the world. The concept used in this book is the strictly geographical one, which includes the cold tropical highlands as well as the hot lowlands. Like any quantitative definition, it loses meaning at the boundaries because these changes are gradual. The tabular data presented follow the latitudinal definition but are based on entire countries. Thus parts of northern India, northern Bangladesh, northern Mexico, and southern Brazil outside of the tropics are included, whereas parts of southern China within the tropics are excluded.

TEMPERATURE AND SOLAR RADIATION

Temperature

The reason for choosing the latitudinal definition is its ease of quantification in terms of air temperatures. The tropics can be defined as that part of the world where the mean monthly temperature variation is 5°C (9°F) or less between the average of the three warmest and the three coldest months. Table 1.1 shows mean monthly temperatures for June and July. The daily variation is also within this range.

This definition includes the tropical highlands, the difference being in their lower overall temperatures. Mean annual temperatures generally decrease by 0.6°C for every 100 m increase in elevation in the tropics (3.56°F for every 1000 ft). If, for example, at sea level the mean annual

Table 1.1   Mean Monthly Air
Temperatures at Tropical Latitudes
(Sea Level) (°C)

Latitude	January	July	Mean Annual
20°N	22	28	26
15°N	24	28	26
10°N	26	27	26
5°N	26	24	26
0	26	25	26
5°S	26	25	26
10°S	26	24	25
15°S	26	23	24
20°S	25	20	23

temperature is 26°C, then at 1000 m it will be 20°C and at 2000 m it will be 14°C. Local variation in topography, rainfall, and other factors often alters these parameters.

The low but constant temperatures of the tropical highlands constitute one reason why certain temperate crops such as apples and pears, which require winter chilling for high production, perform poorly in these areas. During the growing season temperatures are higher in the temperate region than in the tropical highlands. It takes 5 months to grow a crop of corn in North Carolina and about the same time in the lowlands of Colombia, but in Bogotá, Colombia, at an altitude of 2800 m, corn requires 11 months to mature.

The least temperature variation occurs within 6 degrees of the Equator; as latitude increases, temperature variation also increases, reaching maximum values in the desert areas near the Tropic of Cancer. The widest temperature variation is found in the areas of least rainfall and in those of pronounced but unevenly distributed rainfall.

Soil temperatures in the tropics, as defined in the U.S. Soil Taxonomy System, fall in the categories of the isotemperature regimes, that is, "less than 5°C difference between the mean summer and mean winter temperatures at 50 cm or to a lithic contact if shallower" (Soil Survey Staff, 1970). Mean annual air temperatures closely approximate mean annual soil temperatures in the tropics, according to Smith et al. (1964). The following soil temperature regimes can be estimated from mean annual temperature and elevation data:

Regime	Mean Annual Temperature (°C)	Elevation (m)
Isohyperthermic	> 22	0–600
Isothermic	15–22	600–1800
Isomesic	8–15	1800–3000
Isofrigid	< 8	> 3000

The above definition does not exclude soil temperature variation in the topsoil. This is illustrated in Table 1.2, where the monthly and daily air and soil temperature variations are shown for a soil from Indonesia. Very high soil temperatures have been registered on the surface of bare soils during dry periods. Mohr et al. (1972) reported a record of 86°C at the surface of a bare soil in Zaïre. At a depth of 10 cm the same bare soil had a nearly normal temperature of 30°C. At the same site the surface soil temperature was 34°C under grassland and 25°C under forest. Unless exposed, soil temperatures even at the surface do not seriously exceed air temperatures.

Table 1.2   Mean Monthly and Daily Soil
Temperature Variations at Djakarta,
Indonesia (°C)

Soil Depth (cm)	Highest Month	Lowest Month	Daily Variation
Air	26.6	22.5	6.9
3	29.9	28.3	5.2
5	29.9	28.7	5.0
10	29.9	28.9	3.1
15	30.0	28.7	1.5
30	30.0	28.5	0.3
60	30.8	28.5	0.05
90	29.8	28.7	0.04
110	29.7	28.8	0.04

Source: Mohr et al. (1972).

This is due partly to the relatively low heat capacity of soils, about 0.2 g-cal/cc, one-fifth of the heat capacity of water. Any excess heat is reradiated to the atmosphere. This is the reason why one cannot fry an egg even on the hottest soil, whereas it is possible to do so on asphalt.

People unfamiliar with the tropical region generally consider it oppressively hot and humid. Although this condition certainly exists, it is as broad a generalization as considering the temperate region oppressively cold and dry. I have experienced more oppressively hot and humid conditions in Washington, D.C., during the summer than in the heart of the Amazon Basin, where one can almost always count on cool night breezes. The main point to remember about temperatures—soil or air— in the tropics is their constancy, rather than any absolute value.

Solar Radiation

The tropics receive more annual solar radiation available for photosynthesis than the temperate region because of three factors. (1) The tilt of the earth's axis exposes the tropics to more annual solar radiation at the outer atmosphere than the temperate region. (2) The passage of the sun's rays through a thinner atmosphere (because of a more perpendicular angle) in the tropics decreases the amount of radiation absorbed by the atmosphere. From 56 to 59 percent of the sun's radiation at the rim of the atmosphere reaches the earth's surface in the tropics. In the temperate region 46 percent of the radiation reaches the surface at 40° latitude and only 33 percent at 60° latitude. More ultraviolet and blue–violet rays reach the soil surface in the tropics than in the temperate region. (3) The growing season, as limited by temperature, is longer in the tropics (except at the highest elevations).

A global map of annual solar radiation reaching the earth's surface appears in Fig. 1.1. The daily average for the tropics is about 400 langleys/day.* Seasonal variation depends primarily on rainfall distribution patterns. In areas with even rainfall distribution, such as rainforests or deserts, there is little seasonality in solar radiation. In areas with distinct rainy and dry seasons, on the other hand, cloudiness causes considerable seasonality. For example, at Los Baños, Philippines, the average solar radiation for the dry season is 417 langleys/day as compared with 341 langleys/day for the rainy season. These differences have a tremendous impact on crop yields and fertilizer response.

In the temperate region, the daily solar radiation average is half that of the tropics (200 langleys/day) but with great seasonal variability between summer and winter (close to 500 in the summer and 150 in the winter, according to Landsberg, 1961). Table 1.3 shows the monthly averages of

Table 1.3   Mean Monthly Solar Radiation Values at Several Stations (langleys/day)

Month	Yurimaguas, Peru 2087 mm rainfall (no dry season)	Los Baños, Philippines 1847 mm rainfall (4 month dry season)	Lambayeque, Peru 19 mm rainfall (desert)	Ithaca, N.Y. 766 mm rainfall (temperate)
January	308	295	487	136
February	309	361[a]	498	214
March	237	379[a]	482	273
April	287	492[a]	456	359
May	249	439[a]	405	470
June	263	377	355	515
July	342	383	321	492
August	324	405	378	412
September	345	333	435	348
October	379	355	481	242
November	326	317	484	107
December	309	263	503	106
Total	111,923	142,593	160,679	111,923
Daily mean	306	366	440	306

[a] Dry season in Los Baños.

* 1langley = 1 gram-calorie per square centimeter.

Fig. 1.1 Generalized isolines of solar radiation reaching the earth's surface (in 100 langleys/year). *Source:* Landsberg (1961).

three tropical locations with different rainfall patterns compared with the values for a temperate location.

The highest solar radiation values are received at the fringes of the tropics in arid environments. A location in the Sudan has recorded 600 langleys/day, while locations in the Sahara, the Arabian peninsula, northern India, northern Venezuela, and the Kalahari Desert have recorded more than 500, according to Landsberg (1961). The lowest solar radiation figures are those for part of the Amazon and Congo rainforests.

On the basis of solar radiation and growing season length, DeWitt (1967) estimated the potential food crop yields by latitudinal belts. Calculations from his data indicate that tropical areas have approximately twice the yield-producing potential per hectare per year of temperate areas, assuming no additional limiting factors. The average annual yield potential for the tropical latitudes so calculated was 60 tons/ha of total dry matter. Approximately half of that amount is considered the economic yield. Such yield levels have already been obtained or approximated in the tropics. Vicente-Chandler et al. (1964) have reported annual production of 60 tons/ha of dry forage in Puerto Rico. At the International Rice Research Institute in the Philippines, 24 tons/ha of rice grain have been produced in one field during one year (IRRI, 1969). These figures surpass by far annual production records of the temperate region and underline the importance of year-round solar radiation as the chief asset of tropical agriculture. Instead of aiming at maximum yields per crop, tropical agronomists should aim at maximum yields per year. The challenge to agricultural scientists is the elimination of the many limiting factors preventing the full use of year-round solar radiation.

Photoperiod

Days in the tropics are generally shorter throughout the year than days during the growing season in the temperate region. Day length changes during the year, ranging from zero at the Equator to 2 hours and 50 minutes at 23.5° latitude, as shown in Table 1.4. The variation in day length in the temperate region is wider. Tropical plants are considered "short day" plants, but many are very sensitive to photoperiod. Some rice varieties, for example, are so photoperiod sensitive that a 10 minute change in day length prevents flowering.

In the tropics, unlike the temperate region, day length and solar radiation are not well correlated. Therefore a wider range of illumination-radiation regimes occurs in the tropics.

RAINFALL

Rainfall is the most important climatic parameter for tropical agriculture, in terms of both excesses and deficits. Given the relative uniformity in temperature, rainfall distribution is the main criterion used to classify tropical climates. The seasons in the tropics are rainy or dry, not cold or hot. As a carryover of temperate influence, the term "summer" is synonymous to "dry season" and the term "winter" is "rainy season" in many tropical countries.

The extent of arable land in the world where moisture limits crop growth was calculated by the President's Science Advisory Committee (1967). A summarized version of the committee's data for the tropics appears in Table 1.5. On the average no moisture limitations throughout the year occur in 28 percent of the tropics. Moisture limits growth from 4 to 6 months in 42 percent of the area and from 8 to 12 months in the remaining 30 percent. Therefore, moisture is a limiting factor in about three-fourths of the arable land in the tropics.

Annual rainfall varies from zero to 10,000 mm in the tropics. In general, rainfall decreases with increasing latitude, but local relief and other conditions severely limit such a relationship. The length of the dry season increases also latitudinally, beginning at zero close to the Equator. The periods of heaviest rainfall occur when the sun is directly overhead.

An idealized diagram appears in Fig. 1.2. Although such generalization is of limited specific value, it correlates fairly well with the timing and length of dry seasons, including the occurrence of two dry seasons and two rainy seasons in some areas within 5° of the Equator.

Table 1.4   Minimum and Maximum Day Lengths (Sunrise to Sunset) at Different Latitudes (hours:minutes)

Latitude	Maximum	Minimum
0°	12:10	12:10
5°	12:30	11:50
10°	12:40	11:30
15°	13:00	11:10
23.5°	13:30	10:40
40°	15:00	9:20
50°	16:20	8:00
65°	22:00	3:30

Source: Blumenstock (1958).

Table 1.5 *Distribution of Potentially Arable Land in the Tropics with No Temperature Limitations According to the Extent of Soil Moisture Limitation (million ha)*

Moisture Limiting Crop Growth (months)	Tropical America	Tropical Africa	Tropical Asia and Pacific	Total	Percent
0	315	109	81	505	28
4	82	0	56	138	8
6	260	223	130	613	34
8	32	206	89	327	18
10	16	114	72	202	11
12	1	16	12	29	1
Total	706	668	440	1814	100

Source: Calculated from President's Science Advisory Committee (1967).

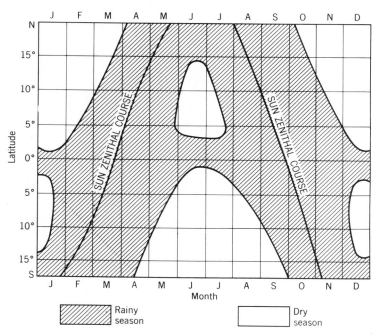

Fig. 1.2 Diagram of march of seasons in the intertropical regions. *Source:* De Martonne (1958).

The distribution, rather than the total amount, is the most important rainfall parameter. Five major rainfall patterns are recognized in the tropics; they are based on the length of the dry season (President's Science Advisory Committee, 1967). A dry month is arbitrarily defined as one with less than 100 mm of rainfall. The distribution of such rainfall patterns is summarized in Table 1.6. These patterns are essentially climatic classifications. Fig. 1.3 shows their geographical distribution.

The Rainy Climates

The rainy climates occupy roughly one-fourth of the tropics, mostly near the Equator. The larger areas are the upper Amazon Basin, the Congo Basin, most of Indonesia, Malaysia, and part of the Philippines. Smaller areas include the Atlantic coast of Central America, the Pacific coast of Colombia, coastal West Africa, and many Pacific Islands. The climax vegetation consists of evergreen tropical rainforests, although such forests have been replaced by cropland in many areas, particularly in Asia. Three examples of this rainfall regime are shown in Fig. 1.4, one for each major tropical region. Rainfall exceeds potential evapotranspiration in all or most months. Rice, cassava, and yams are the predominant food crops. Cacao, bananas, rubber, coconuts, and other plantation crops are grown for export.

The Seasonal Climates

The seasonal climates cover about one-half of the tropics. They include large areas of the Cerrado and Mato Grosso in Brazil, the Llanos of

Table 1.6   Distribution of Major Climatic Regions in the Tropics, Based on the Landsberg-Troll Classification (million ha)

Climate	Humid Months	Predominant Vegetation	Tropical America	Tropical Africa	Tropical Asia	Total	Percent
Rainy	9.5–12	Rainforest and forest	646	197	348	1191	24
Seasonal	4.5–9.5	Savanna or deciduous forest	802	1144	484	2430	49
Dry	2–4.5	Thorny shrubs and trees	84	486	201	771	16
Desert	0–2	Desert and semidesert scrub	25	304	229	558	11
Total			1557	2131	1262	4950	100

Source: Adapted from the President's Science Advisory Committee (1967).

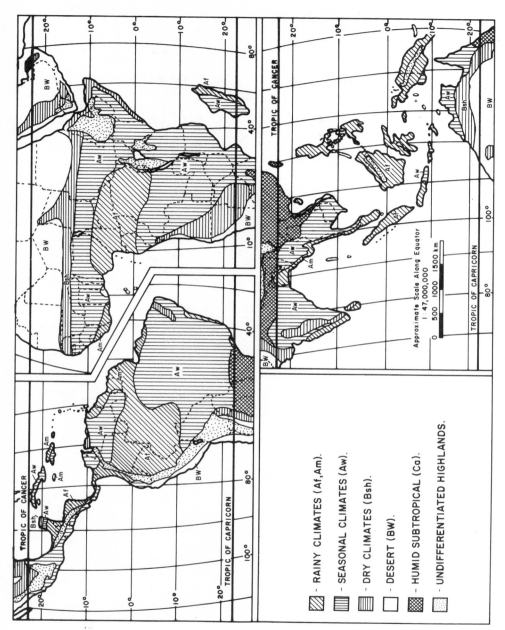

Fig. 1.3 Tropical climates. *Source:* Adapted from Landsberg et al. (1963).

11

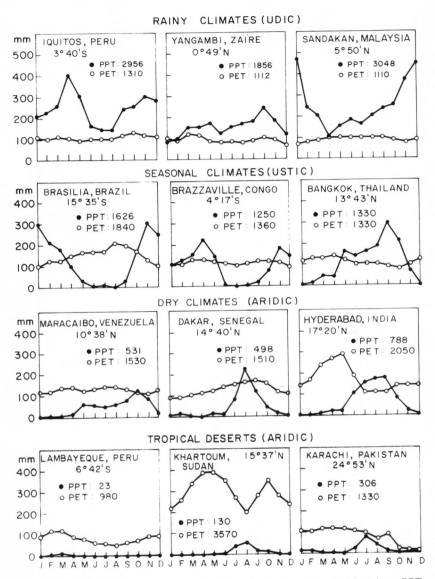

Fig. 1.4 Monthly rainfall–evapotranspiration balances at selected tropical locations. PPT = precipitation, PET = potential evapotransporation. Annual totals in numbers.

Colombia and Venezuela, the Pacific coast of Central America and Mexico, Veracruz, the Yucatan peninsula, and Cuba in tropical America. In Africa they include most of the continent between the Sahara and Kalahari deserts except for part of the Congo Basin. In Asia the seasonal climates cover most of India, inland Indochina, and a belt in northern Australia. The climax vegetation is either semideciduous or deciduous forest or savanna. The wet and dry seasons are well defined. High temperature and solar radiation characterize the dry season, while lower temperatures and lower solar radiation are indicative of the wet. The monsoonic climates of Asia are included in this classification. In spite of seasonal variations, the annual precipitation is equal to or lower than the annual potential evapotranspiration. Most tropical crops are grown in this region, although the root crops and perennials that predominate in the rainy climates are less important in the seasonal climates. Examples of three locations with this rainfall regime appear in Fig. 1.4.

The Dry Climates

The dry climates cover about 16 percent of the tropics. The largest areas are the Sahel, located between the savanna belt and the Sahara Desert in equatorial Africa, the Kalahari Desert in southern Africa, a large proportion of Australia, parts of central India, northeast Brazil, northern Venezuela, and northern Mexico. The climax vegetation consists of sparse, thorny shrubs and trees. The short wet season has high monthly rainfall during which one crop of unirrigated corn, sorghum, millet, or rice can be grown. Potential evapotranspiration exceeds precipitation in most months and in annual amounts as well. Three examples are shown in Fig. 1.4.

The Tropical Deserts

The tropical deserts, defined as those areas having 2 rainy months or less, cover about 11 percent of the tropics. The Sahara, Arabian, Somali, and Australian deserts compose the greatest part of this area. Narrow coastal desert strips are found in Peru, Chile, and Southwest Africa. Usually only nomadic grazing is possible without irrigation. When irrigated, many of the better soils of these deserts are extremely productive. High cotton yields are obtained in the Sudan Gezira, and very high rice and sugarcane yields on the coast of Peru. In the tropical deserts, unlike other tropical climates, potential evapotranspiration may vary substantially from month to month. Three examples of this climate appear in Fig. 1.4.

The Tropical Highlands

The tropical highlands, defined as those areas above 900 m (3000 ft) of elevation, cover approximately 23 percent of the tropics. As mentioned before, the temperatures are low but constant. A wide variety of rainfall patterns is found, including great changes in short distances due to "rain shadow" effects. These highland areas are important in Mexico, Guatemala, Costa Rica, Colombia, Ecuador, Peru, Bolivia, Kenya, and Ethiopia, where a large part of the agriculture and population is located. The highest rainfall areas in the tropics are found where moist air moves upslope and discharges most of the precipitation between 300 and 1000 m. Some mountain areas are dry and desertic, such as the western flank of the Andes in Peru and northern Chile. For a more detailed discussion of the tropical highland climate, the reader is referred to the articles by Blumenstock (1958) and Budowski (1966).

Rainfall Variability

This discussion so far has been limited to average values. Rainfall averages are useful in describing the broad picture but are practically worthless in predicting moisture adequacy for a specific crop. Year-to-year variability in rainfall is extremely large and important in the tropics. In most places that I have visited in the past 10 years, rainfall records have been broken. Mohr et al. (1972) report that total rainfall in the Dutch Antilles varies from 0.2 to 2.5 times the annual means. The monthly variation is much greater.

The information presented in Tables 1.5 and 1.6 and in Fig. 1.4 might imply that no crop moisture stress is likely to occur in the "rainy" climates or during the rainy months in the other climatic types. Nothing could be further from the truth. Severe drought periods, sometimes lasting for a few weeks, frequently occur during the rainy seasons and affect crop yeilds. In some areas certain wind patterns make such droughts more or less predictable, but in most areas they are not. Conversely, heavy rains occur during the dry seasons. In some years the rainy or dry seasons simply do not take place.

The best predictive tool for the tropical agronomist is an estimate of weekly rainfall probabilities. This can be fairly well developed with 15 to 20 years of rainfall data by rather simple statistical procedures. An example of such a study was conducted in the Philippines by Yñiguez and Sandoval (1966). The results can be graphed in a simple form, like that of Fig. 1.5 for Muguga, Kenya.

Reliance on average values has led to disastrous results. The most noto-

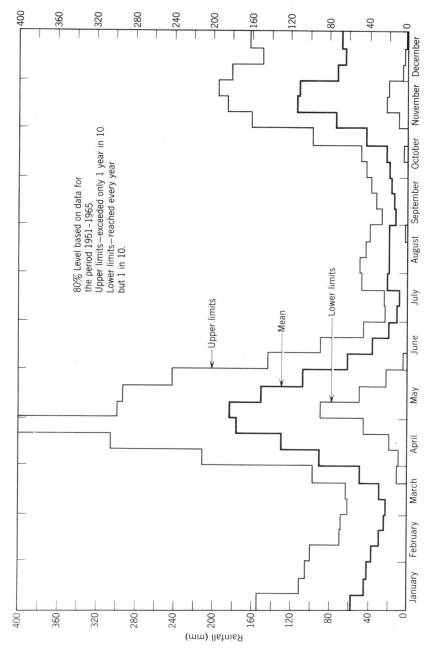

80% Level based on data for
the period 1951-1965.
Upper limits—exceeded only 1 year in 10.
Lower limits—reached every year
but 1 in 10.

Upper limits

Mean

Lower limits

Rainfall (mm)

January February March April May June July August September October November December

Fig. 1.5 Confidence limits of expected rainfall at Muguga, Kenya. *Source:* Lawes (1969).

rious is probably the British Groundnut Scheme in what is now Tanzania during the late 1940s (Wood, 1950). The area had an average rainfall of about 700 mm during the rainy season, which was sufficient in quantity and length to grow peanuts. On this basis several thousand hectares were planted to peanuts. They received 250 mm of rain during the first year and 300 mm the second, resulting in massive crop failures costing about 100 million dollars.

Rainfall intensity is also variable. Advective rains caused by air masses moving inland from the ocean or up a mountain range cause drizzle, fog, and high humidity for several days. Convective rains often develop in localized thunderstorms which cause high-intensity rainfall for short periods. Such intense rains may result in direct crop damage and runoff even in flat areas. It has been estimated in Surinam that 80 percent of a 10 mm/hour rain is retained by the soil, whereas only 32 percent is retained during a 50 mm/hour rain (Mohr et al. 1972).

The contribution of fog and dew to soil moisture status has not been well studied in tropical areas. A review of this subject by Blumenstock (1958) suggests that these sources may contribute larger quantities of moisture than in the temperate region.

Rainfall–Soil Moisture Relations

Many attempts have been made to classify rainfall quantitatively in tropical climates. The reader is referred to the work of Lee (1957), Landsberg et al. (1963), and Williams and Joseph (1970) as examples of such efforts. Since all are based on rather arbitrary assumptions and since our interest in rainfall lies in its role as a source of soil moisture to crops, the accepted soil moisture terminology of the U.S. Soil Taxonomy System is used in this text. Four soil moisture regimes are common in the tropics:

1. Udic: The control section of the soil is dry for no more than 90 cumulative days during the year.
2. Ustic: The control section of the soil is dry for more than 90 cumulative days but less than 180 cumulative days or 90 consecutive days during the year.
3. Aridic: The control section of the soil is dry for more than 180 cumulative days or moist for less than 90 consecutive days per year.
4. Aquic: The soil is saturated with water long enough to cause reduced soil conditions.

The control section is defined as that part of the soil between the following two depths: below that reached by 2.5 cm of water in 24 hours and above that reached by 7.5 cm of water in 48 hours (Soil Survey Staff, 1970). Roughly this corresponds to depths of 10 to 30 cm for clayey soils, 20 to 60 cm for loamy soils, and 30 to 90 cm for sandy soils. The term "dry" refers to a soil moisture tension of 15 bars or more, that is, at or above the wilting coefficient.

These rather complicated criteria were developed because of the relevance of moisture supply to common crops. The udic soil moisture regime implies that during most of the year water stress will be absent (on the average). It is roughly equivalent to the rainy climates for most soils. The ustic regime implies a strong dry season of 3 to 6 months and is well correlated with the seasonal climates. The aridic regime implies a longer dry season, and it is well correlated with the dry and desert climates.

Variation in soil properties and topography permit the existence of different soil moisture regimes under the same rainfall regime. A deep sandy soil may be ustic in a rainy climate because of rapid drainage. The aquic regime is typical of poorly drained sites and occurs even in deserts.

A rough estimate of the extent of these soil moisture regimes in the lowland tropics was calculated from the tentative map of Aubert and Tavernier (1972). Excluding the undifferentiated highland areas, the distribution of tropical areas as follows: udic, 29 percent; ustic, 34 percent; aridic, 29 percent; and aquic, 8 percent. These percentages correlate fairly well with the distribution of climates: rainy, 24 percent; seasonal, 49 percent; and dry and desert, 27 percent. Such estimates are extremely rough, however, because only the dominant soil suborders have been considered.

VEGETATION

Natural vegetation in the tropics is closely correlated with climate. In fact, the two main classification systems of tropical climates presently in use, those of Köppen and Geiger (1936) and of Holdridge (1967), employ vegetation names for the different climatic regions. Tropical vegetation can be grouped into the five general categories shown in Fig. 1.6: savannas and other grasslands, which cover 43 percent of the area; broad-leaved evergreen "rainforests," which cover 30 percent; semideciduous and deciduous forests, which cover 15 percent; desert shrubs and scattered grasses, which cover 7 percent; and no vegetation, which is found in 5 percent of the tropics. A glance at Fig. 1.6 shows that most of the tropics are not covered by tropical rainforests, as commonly conceived. For instance, most of tropical Africa consists of grassland savannas.

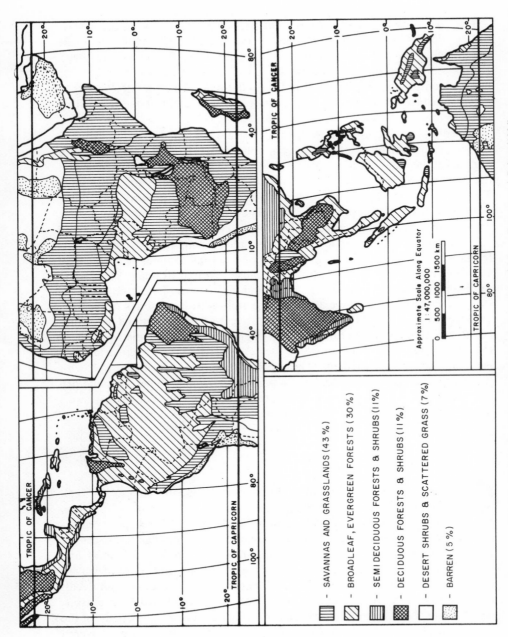

Fig. 1.6 The tropics: Natural vegetation. *Source:* Adapted from Köppen and Geiger (1936).

- SAVANNAS AND GRASSLANDS (43 %)

- BROADLEAF, EVERGREEN FORESTS (30 %)

- SEMI DECIDUOUS FORESTS & SHRUBS (11 %)

- DECIDUOUS FORESTS & SHRUBS (11 %)

- DESERT SHRUBS & SCATTERED GRASS (7 %)

- BARREN (5 %)

Approximate Scale Along Equator
1 : 47,000,000

0   500 1000 1500 km

TROPIC OF CANCER

TROPIC OF CAPRICORN

TROPIC OF CANCER

TROPIC OF CAPRICORN

Tropical Savannas

The term "savanna" probably originated in Central America, where the Caribs used it to refer to any area not covered by forests. According to Eyre (1962), it is now used throughout the tropics to designate plant communities in which grasses and sedges are important. These range from pure grass stands to tree savannas where two canopies (trees and grasses) are almost continuous.

The savannas comprise approximately 28 percent of the American tropics. The largest expanse is the Cerrado of Brazil with over 200 million ha, followed by the Llanos of Colombia and Venezuela, most of Cuba, and parts of the Pacific coasts of Mexico and Central America. High grassland areas above the timberline occupy much of the Andean highlands, locally known as "punas" and "páramos." About 57 percent of tropical Africa consists of savanna vegetation. This includes most of subsahara Africa except the Congo Basin, the Kalahari Desert, and a forested portion of south central Africa. The size of trees decreases toward the deserts. About 34 percent of tropical Asia and the Pacific is covered by savannas, mostly in Australia, with small but significant areas in the Asian mainland and Oceania.

Three types of savannas are generally recognized. They have been sum-

Plate 1   "Cerrado" vegetation near Brasilia, Brazil, a form of short-grass savanna.

marized by Eyre (1962). The "tall grass–low tree" type occurs extensively in Africa. The main grasses belong to the genera *Pennisetum, Andropogon, Imperata,* and *Hyparrhenia,* and they form a dense canopy 2 to 4 m high during the rainy season. Relatively short deciduous trees (10 to 15 m high) are interspersed among the grass. These savannas flank the rainforests of Africa and are referred to as "elephant grass" or "guinea savanna" in certain localities. The second type is the "Acacia–tall grass" savannas, where tussock grasses form a continuous cover reaching up to 1 to 2 m during the rainy season and are dotted with tall deciduous or evergreen trees. In Africa this type is found in areas of lower rainfall than the first one and is called "Sudan savanna." In South America this pattern is typical of the Llanos and Cerrado (Eiten, 1972). The third type is the "Acacia–desert grass" savanna, which occupies areas on the fringes of the African deserts and in northern Australia.

Ecologists have tried to correlate climate, fire, and soil conditions with the genesis of savannas. Strong arguments are still being advanced on behalf of each of these factors. Most savanna areas have either ustic or aridic soil moisture regimes, but they occur in high-rainfall areas after the destruction of rainforests. Annual burning practices by most cropping and grazing enterprises help to maintain the species balance. Savannas also occur in the seasonally flooded plains of the northern Llanos of Colombia, where trees apparently do not tolerate half a year of waterlogging and half a year of almost complete dryness. Low soil fertility and laterite formation have been considered factors favoring savanna vegetation. The origin of the Cerrado vegetation of central Brazil has recently been attributed to the high levels of exchangeable aluminum present in the soil (Goodland, 1971), and low nutrient availability (Lopes, 1975). For additional material on this controversy, the reader is referred to publications by Budowski (1956), Bartlett (1957), Arens (1963), Eyre (1962), and Bouliere and Hardy (1970).

In spite of conflicting theories, there is ample evidence that man-made savannas are rapidly replacing forests when the shifting cultivation system breaks down because of demographic pressure. The unfortunate aspect of this widespread practice is that coarse, unpalatable grasses of the genus *Imperata* dominate the grasslands, rendering them practically worthless for grazing.

Rainforests

Tropical rainforests are the type of natural vegetation found in areas characterized by udic environments with high annual rainfall. The term is essentially synonymous to "equatorial forests," "broadleaf evergreen forests," and "moist tropical forests." The primary characteristic of a rainforest is the low proportion of deciduous trees. As mentioned earlier, this type of vegetation covers about 30 percent of the tropics. Rainforests cover 52 percent of tropical America, primarily the Amazon Basin, as well as the Pacific coast of Colombia, and parts of Central America. Only 12 percent of Africa consists of rainforests, mainly in the lower Congo Basin and parts of the West African coast. A larger area along the West African coast was previously under rainforests that have been destroyed by man in the recent past. About 38 percent of tropical Asia and the Pacific consists of rainforests, mainly in the Malayan Peninsula, Sumatra, Borneo, New Guinea, parts of the Philippines, and many smaller islands. In spite of the wide distribution of rainforests, the structure is essentially the same everywhere. A three-layer canopy is usually present, consisting of tall trees about 30 m high overlying two shorter layers of about 22 and 14 m, respectively. The undergrowth is almost exclusively tree saplings, with no grass. Rainforests are extremely diverse; almost 2500 tree species are found in Malaysia and the Amazon Jungle, as compared with about 12 species in temperate forests. This is one reason why timber production has been limited in these regions. In many areas there are fewer than five marketable trees per hectare, making logging unprofitable.

The tropical rainforests have several unique structures adapted to this

Plate 2 Virgin tropical rainforest near Yurimaguas, Peru.

particular environment. Many leaves have "drip tips" that accelerate drying and reduce mold infestations. Lianas and strangler vines climb the trees and bind many together. Epiphytes or rootless plants develop in upper branches, catch leaves in their aerial roots, and produce a "soil" from which they draw nutrients.

Although tropical rainforests may appear impenetrable, they are easy to walk through with the aid of a machete because of the almost total absence of tangled vegetation at ground level. About 15 percent of the solar radiation reaches the soil surface. Air temperatures are substantially lower, but the humidity is higher than in surrounding cleared areas. Detailed studies of rainforests have been made by Richards (1952) and Odum (1972).

The nutrient cycle between the forest and the soil is essentially closed. Constant litter fall and decomposition throughout the year and the virtual absence of leaching permit the development of luxuriant forest with no nutrient deficiency symptoms in soils of low native fertility. The only marked vegetation differences outside of swamps that can be correlated with soils is the reduced size of rainforests growing on very sandy Spodosols.

Deciduous and Semideciduous Forests

This general classification comprises the forests of ustic environments where the dry season is sufficiently strong to exclude the rainforests but grasses are absent. Approximately 15 percent of the tropics is covered by semideciduous, deciduous, and thorn forests. The semideciduous forests are transitional between the rainforests and the completely deciduous ones. At the wetter end the three-story structure can be observed, but few specialized organs such as epiphytes exist. The nutrient cycle, however, is markedly different from that of the rainforest. With substantial leaf fall during the dry season, the solar radiation reaching the soil surface increases drastically, and the litter layer does not decompose during the dry season. The forest has an appearance similar to that of the mixed temperate forests during the winter. This vegetation type is also called "semievergreen seasonal forest," "monsoon forest," and "dry evergreen forest." In parts of Asia, teak, a valuable timber species, may make up 10 percent of the trees. Large tracts of these forests have been converted by shifting cultivation into savannas.

The deciduous seasonal forests occur in areas with stronger dry seasons and are particularly widespread in southern Africa, India, Indochina, and throughout tropical America. A two-story canopy is still present, except where soils have a root-impeding layer, in which case no understory is present. The forest is bare during the long dry season. In Africa this type is called "miombo forest" or "tree steppe." As the tree canopy becomes discontinuous and grasses begin to be conspicuously present, the forest grades into savanna. The boundary between these two is difficult to establish because it is affected by man-made fires.

The thorn forests occupy the driest range of the ustic environment and part of the aridic. They are widespread throughout tropical America, particularly in the northeast part of Brazil, where they are known as "caatingas." They also occur in northern Colombia and Venezuela, large parts of the Caribbean, the Yucatan Peninsula, and western Mexico. In Asia they cover large parts of India, Burma, Thailand, and Queensland. In Africa they are interspersed with the drier savannas. Although there is no closed canopy, grasses are generally absent. Bottle-shaped trees are often present. These forests also occur in shallow or extremely permeable soils where drought is common in spite of high precipitation.

Desert and Semidesert Scrub

Desert and semidesert scrub cover about 7 percent of the tropics in the extreme aridic regions. Steppes with cacti and other succulent species occur in wide areas of the Sahara and at the fringes of tropical deserts. Thorny vegetation capable of tapping soil moisture at great depths can be seen in most tropical deserts. Much of the vegetation is dormant but grows luxuriantly during sporadic rains.

Areas essentially devoid of vegetation account for about 5 percent of the tropics.

Productivity of Natural Vegetation

The recent emphasis on quantitative ecology has produced several estimates of the annual dry matter production of various tropical vegetation types. These figures are of interest when compared with annual crop production and the productivity of temperate vegetation. Golley and Leith (1972) have developed a "net primary productivity map" of the world, which gives estimates of annual biomass increase plus litter production. A summary of their work appears in Table 1.7. These figures show that tropical forests grow substantially faster than their temperate counterparts, whereas no great differences exist between the tropical savannas, temperate prairies, and croplands.

Table 1.7   Annual Increases in Dry Matter Production of Selected Types of Vegetation (tons/ha)

Type of Vegetation	Mean	Range
Tropical rainforests	20	10–35
Tropical deciduous forests	15	6–35
Temperate deciduous forests	10	4–25
Temperate mixed forests	10	6–25
Tropical savannas	7	2–20
Temperate prairies	5	1–18
Croplands	6.5	1–40

Source: Golley and Leith (1972).

Classification of Vegetation

In addition to the presently used terminology and many similar ones, there is a classification system that defines ecological zones in terms of dominant vegetation and climatic parameters: Holdridge's "life zones" (Holdridge, 1947, 1967). His scheme is illustrated in Fig. 1.7 where mean annual temperature, precipitation, and evapotranspiration are plotted in a triangular fashion that provides the limits for each "life zone." This system has been widely used in the American tropics and has been applied to other regions of the world. The main disadvantage is that it ignores rainfall distribution, which is probably the paramount climatic parameter. For example, the savannas of the Llanos Orientales of Colombia appear as "moist tropical forest" because of their high annual rainfall and thus in the same category as the Amazon Jungle rainforests.

GEOLOGY

The tropical lands are extremely complex geologically and geomorphologically. Approximately 77 percent can be classified as "lowlands" with elevations below 900 m (Fig. 1.8). In 20 percent of the tropics altitudes range from 900 to 1800 m. This category comprises about half of the Andean highlands of Central and South America, parts of Venezuela and Brazil, and the mountain regions of the Caribbean. Most of East Africa, the Cameroons, and certain desert areas of the Sahara and Arabia fall into this category. In Asia the Deccan Plateau of India plus the central backbone of the islands off the Southeast Asian mainland is classified as highland.

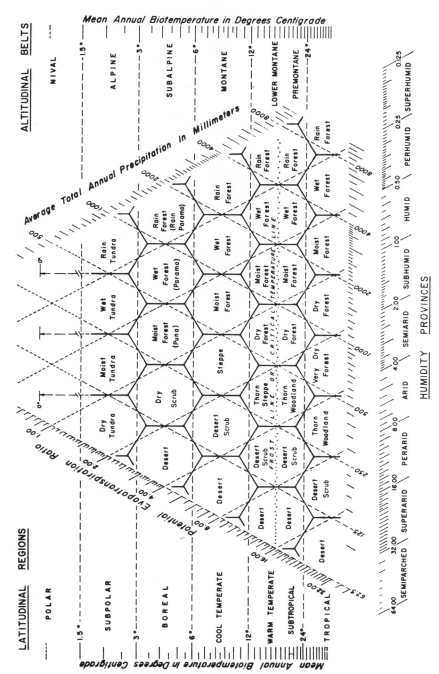

Fig. 1.7 Holdridge's classification of world life zones or plant formations. *Source:* Holdridge (1967).

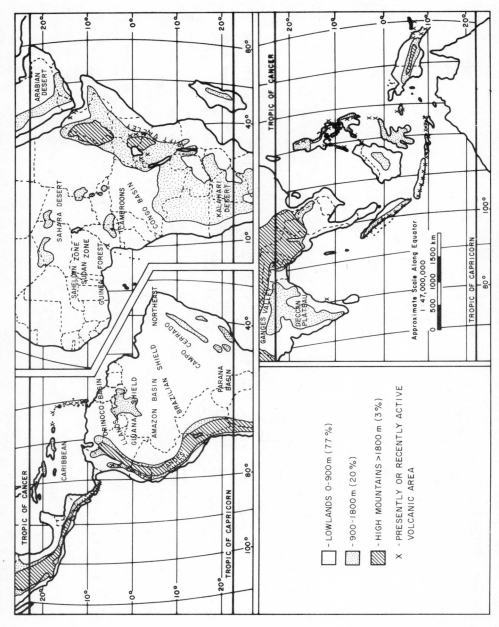

Fig. 1.8 The tropics: physical setting.

In about 3 percent of the tropics altitudes exceed 1800 m. These regions are the High Andes of Central and South America, the Ethiopian and Kenya highlands in Africa, northern Burma, and parts of Indochina, and central New Guinea. The Himalayas are located outside the tropics.

The 20 percent of the tropics located between 900 and 1800 m of elevation has one of the most pleasant climates on earth, with mild but constant temperatures throughout the year. This fact is seldom recognized by persons in the temperate region. A Kenyan visiting England in July used to amaze his British friends when he told them that he wanted to go back to Africa to get cool.

The following is a brief summary of the geology and geomorphology of the three tropical continents. The reader is referred to the book by Thomas (1974) for detailed information on tropical geomorphology.

Tropical America

The main surface features of tropical America are the Andean Cordillera and its continuations in Central America and Mexico; the Guyana and Brazilian shields, the Orinoco, Amazon, and Paraná basins, and the Caribbean Islands.

The Andes and their counterparts in Central America and Mexico extend as a unit throughout the western edge of tropical America. The Andes rose from the sea during the Tertiary period with great volcanic activity. Presently, there are active or recently active volcanos in Mexico, throughout Central America, Colombia, Ecuador, Peru, and Bolivia. Some extensive portions of the Andes such as central Peru, however, have not been affected by recent volcanic activity. Glaciers occur at the highest elevations with the snow line at about 4800 m. Although relatively narrow, the Andes have important and relatively flat intermountain valleys where the population density is high. Examples of these are Mexico City Valley, the Guatemalan Altiplano, the Meseta Central of Costa Rica, the "Sabana" of Bogotá, the Cauca Valley of Colombia, and the Mantaro and Cajamarca valleys in Peru. High-altitude plateaus dominate southern Peru and northern Bolivia.

The Guyana and Brazilian shields are much older land surfaces but with low elevations. They originated during the Archean and Paleozoic periods and consist mainly of granite and gneiss partially covered by stratified rocks, mainly sandstone. Large areas of the central plateau of Brazil have gently undulating topography where two or three erosion surfaces can be seen. The Guyana and Brazilian shields are the oldest land surfaces in tropical America.

The Amazon Basin is limited by the Andes and the two shields. The upper portion west of Manaus is quite wide; the basin narrows considerably east of the city. Many of the present sediments originate from erosion of the Andean uplifts, but some are derived from the shield areas. The Orinoco Basin is separated from the Amazon by the Guyana highlands. Both areas are covered by Tertiary deposits with bands of recent alluvium. The Paraná Basin to the south is rich in basalt deposits.

The Caribbean Islands formed primarily from uplifted limestone or volcanic deposits. The present volcanic activity is limited to the lesser Antilles.

Tropical Africa*

Compared with other continents, Africa is geologically and topographically the most uniform. There are only two intensely folded areas: The Atlas Mountains in the north and the Cape Province ranges in the extreme south, both outside the tropics. The rest of the continent is one vast rigid block of ancient rock, two-thirds of which is covered by sediments. The old crystalline rock is mainly granite but also contains much metamorphic schist and gneiss. Because of the lack of folding and the continuous weathering, most of the continent appears as a worn-down plateau. The plateau is modified by volcanic eruptions, ranging from Precambrian to Recent, and related fault lines such as the Rift Valley, which splits the plateau from the Dead Sea in the north to the Zambezi River in the south. It is now partly occupied by some extensive lakes.

The oldest rocks can be found in the east and west, parallel to the coast. This material dates from the Precambrian period. Some of these were folded before or during the Mesozic period. The granites, gneisses, and schists of the continental plateau are mostly of Archean age. During the Cambrian, Ordovician, and Silurian periods, much of the continent was covered by sea, and some of the sandstones and dolomites in South Africa are considered to be from this period. After the elevation of the continent during the late Carboniferous times, during which Africa and India were united as Gondwanaland, the interior of Africa was occupied by large lakes, resulting in the formation of enormous layers of sandstone and marl. This is referred to as the "Karroo system" of South Africa, but the same sitution is found in the Congo basin and areas in East Africa. During the following Jurassic period, great volcanic activity caused the basalt flows of West Africa. It was during this time that Gondwanaland broke up and the continents drifted apart. Afterwards certain areas were covered by the sea again, such as the Sahara and the west coast. These areas are the only ones

---

* Adapted from a summary prepared by M. Drosdoff, Cornell University.

with marine deposits of Postcambrian age. Continental sediments, on the other hand, are very extensive and range from Cambrian to Pleistocene age, accumulated in the basins of the Middle Niger, in Chad, the Congo, the Sudan, and the Kahalari Desert.

Tropical Asia

The Indian subcontinent, which collided with the Asian mainland during the Eocene period, is a very old land mass and has not been under water since the Carboniferous period. The Deccan Plateau, with elevations lower than 1800 m, remains in the highest area of the peninsula. During the Cretaceous period, volcanoes were very active and covered central India with basalt.

Mainland Southeast Asia is a combination of mountain ranges and large rice-growing valleys such as the Ganges, Brahmaputra, Irrawaddy, Chao Phya, and Mekong. These are the "rice bowls" of tropical Asia and are subject to periodic flooding. Most of the present volcanic activity is limited to the island regions, particularly Indonesia and the Philippines, as well as many Pacific islands, including Hawaii.

D'Hoore (1956) provides an interesting comparison between the geology of tropical America and that of Africa. He points out that most of tropical Africa is a giant crystalline plateau, two-thirds of which is covered by sediments. Continental sediments filled in the big depressions or basins. In tropical America, however, the equivalent crystalline plateaus are limited to the Guyana and Brazilian shields. They are also partially covered by sediments of more recent origin caused by the Andean uplift. Mountain uplifts and volcanic activity are less extensive in Africa than in America. The basaltic deposits in Paraná are contemporary to those in Basutoland in Africa and the Deccan Plateau in India. In conclusion, D'Hoore considers the geologic history of tropical America to be more favorable than that of tropical Africa because the Andes provide a constant source of new materials. This situation is similar to that of the Ganges plains in northern India, which receive the sediments from the Himalayas. However, as a whole, tropical America has a larger proportion of acid soils than tropical Africa. This is apparently due to the aeolian deposits from the Sahara, which affect a large part of tropical Africa.

AGRICULTURE

The economies of most tropical countries are based on agriculture. Crop yields and food production can be drastically increased by the application of presently known technology. The need for such increases is compounded by extremely high population growth rates. This section briefly outlines the human and agricultural aspects of the tropical environment. Figure 1.9 shows the political divisions of the tropics, including the principal written languages.

The World Food Problem: Production versus Population

The ability of the world to feed its human population has been of major concern during the past 200 years. In 1798 Malthus predicted that the world's population would reach its maximum at 1.5 billion people and that worldwide famine would follow. This prediction has approached reality only in specific areas. As recently as 1967 the Paddock brothers predicted worldwide starvation by 1975 (Paddock and Paddock, 1967).

Happily these prophesies have not been fulfilled, nor are they likely to be in the near future. However, a glance at the world population increases depicted in Fig. 1.10 should be sufficient to frighten even the most optimistic reader. World population growth rates were steady at about 1 percent per year until 1930, when a population of two billion was reached. The third billion was reached in 1960, the fourth billion was reached in 1975, and the sixth billion is expected before 2000, assuming moderate gains in birth control programs (President's Science Advisory Committee, 1967). More than twice the number of people will have to be fed in the year 2000 as were fed in 1960.

The bulk of this population explosion is taking place in the tropics, primarily in Asia. This can be appreciated by glancing at Fig. 1.10, where the world population increase is shown from 1920 to 1970, and the best projections for the remainder of this century. According to Cochrane (1969), the "population explosion" that began in the 1950s is due primarily to decreasing death rates through improved health services and reduction of widespread wars. Birth rates have remained nearly constant in the developing countries but have decreased in the developed areas, particularly the United States, Europe, and Japan. Consequently, the world population growth rates increased from 1.2 percent in the 1940s to 2.2 percent in the 1950s and 2.4 percent in the 1960s. It is estimated that present birth control programs are not going to have a worldwide impact in the developing countries until the 1980s. There is little doubt that this planet will have six billion inhabitants before the year 2000.

Although world food production has generally kept pace with increasing population, the quantuum jump in population experienced since World War II has precipitated a closely contested race between population and food production. This can best be estimated by the ratio between the two, which

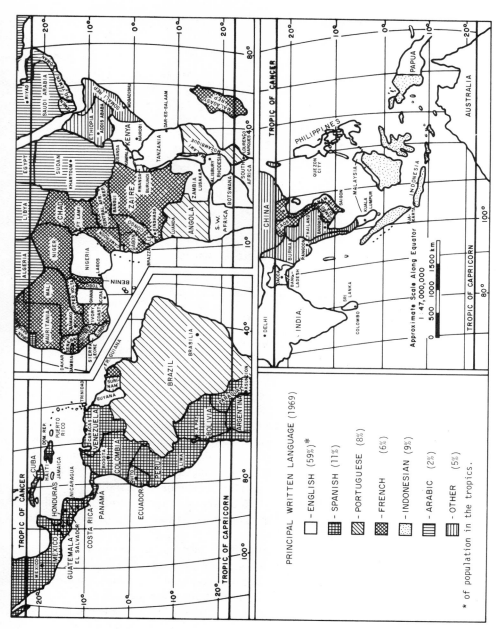

Fig. 1.9 The tropics: political subdivisions.

PRINCIPAL WRITTEN LANGUAGE (1969)

- ENGLISH (59%)*
- SPANISH (11%)
- PORTUGUESE (8%)
- FRENCH (6%)
- INDONESIAN (9%)
- ARABIC (2%)
- OTHER (5%)

* of population in the tropics.

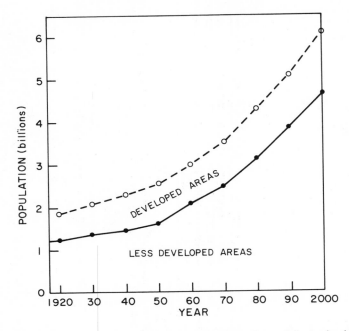

Fig. 1.10 Pattern of world population growth (dotted line) and growth in the developing countries, including projections for the rest of the century. *Source:* Based on FAO (1972) data.

is shown in Fig. 1.11. A careful analysis of the historical trends led Cochrane to conclude that the race was essentially even during the 1950s. Since then, steady and sometimes dramatic progress has taken place in the developed countries, where the application of technology caused production to easily outstrip population growth. In many developing areas, however, population began to outstrip food supplies, particularly in India, where two consecutive droughts in 1965 and 1966 created a crisis situation. This led to widespread concern and to the Paddocks' prediction of worldwide catastrophe by 1975.

Since 1967 the situation has improved, particularly in Asia, with the widespread adaptation of high-yielding varieties of rice and wheat. In 1971 about 33 percent of South Asia's wheat acreage and 13 percent of tropical Asia's rice acreage, totaling about 20 million ha, were planted with the new varieties (Barker, 1972). Similar rates of adoption have also taken place in parts of Latin America. The "green revolution" resulted in a drastic reversal of the decreasing trend in per capita food production shown in Fig. 1.11, so that the decade of the 1960s ended with food production slightly

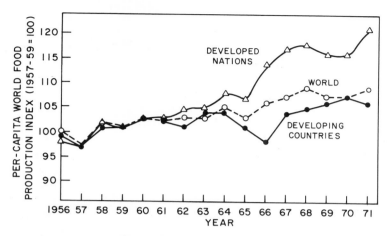

Fig. 1.11 Changes in per capita world food production for the period from 1956 to 1971. Calculated from FAO (1972) data.

ahead of population growth in the developing countries. This is shown in Table 1.8.*

The estimated increase in the use of improved technology, including fertilizers, during the balance of this decade, indicated that food production would keep slightly ahead of population. The worldwide energy crisis of 1973, however, shattered this prediction by creating serious fuel and fertilizer shortages. Again, world food supplies have become critical. What will happen in the 1980s is anybody's guess.

Table 1.8   Actual Stages of the Food–Population Race from 1961 to 1970: Average Annual Rates (%)

Region	Population Growth	Food Production	Per Capita Food Production
Latin America	2.8	3.1	0.4
Asia[a]	2.6	2.7	0.3
Africa	2.4	2.3	0.0
All developing countries[a]	2.4	2.7	0.4
World	2.0	2.5	0.4

Source: Calculated for the FAO 1971 *Production Yearbook.*
[a] Includes the Middle East.

* Examination of more recent FAO data available in 1975 showed no significant changes in trends from the data reported in Tables 1.8 to 1.15.

Unless major successful efforts in agricultural development are attained, population is bound to outstrip food production. Although variety–cultural practice packages triggered the "green revolution," one should not forget that these have to be supplemented with credit, prices, inputs, and infrastructure at the farm level to be successful.

There is no question that similar yield breakthroughs and follow-ups are needed in food crops other than rice and wheat. Tropical soil scientists will play a major role in developing the management practices necessary to utilize fully the yield potential of new varieties or species.

This discussion so far has been limited to the broad quantitative picture. Localized food shortages leading to starvation levels do occur, however, particularly during wars, as happened in eastern Nigeria and Bangladesh, and in the Sahelian countries because of drought. Although, on the average, the daily caloric standard* is achieved by countries covering 97 percent of the world, it is estimated that about 20 percent of the world's population, mostly in Asia, is deficient in calories. Malnutrition, particularly protein deficiency, is widespread throughout the tropics, affecting probably two-thirds of the population. The bloated bellies and the reddish streaks in the hair of many children are clear symptoms of protein deficiency that one finds almost anywhere in rural areas.

The demand for more and better food is rapidly increasing in the developing countries, as part of the development process. This results in a new demand dimension along with population growth. Meeting such demands is both nutritionally and politically sound. The fact that total demand for food is definitely increasing more rapidly than food production (FAO, 1972) could lead to food shortages, higher prices, and serious social unrest. Although the spectre of mass starvation is over for the present, the liklihood of serious food shortages is very real in the developing countries. The President's Science Advisory Committee (1967) estimates that, when both population and increased demand per capita are taken into account, a compound annual growth rate of food production of 4 percent is needed. This is far beyond the average of about 2.7 percent shown in Table 1.8. The same report also indicates that the total demand for food in developing countries could double between 1965 and 1985.

Land Use

Although agriculture is the main economic activity in the tropical regions, the proportion of cultivated land is virtually the same as in the temperate

---

* This standard is 2440 cal for the "average" person of normal activity and health, taking account of environmental temperature, body weight, sex, and age (Cochrane, 1969).

region: about 10 percent. Pastures and meadows account for an additional 20 percent, but many of these areas are not used to any measurable extent. Tables 1.9 and 1.10 show the degree of land use by regions or countries. The President's Science Advisory Committee (1967) has estimated that there are about 1.7 billion ha of potentially arable land in the tropics, of which only about 500 million are presently in use. It also estimates a potential grazing area of about 1.6 billion ha, of which about 1 billion are presently grazed. The criteria used to determine what is potentially arable or grazing land refer to the average level of agricultural technology in the United States. All arable soils, of course, can be used for grazing. These figures indicate that there is a tremendous potential for expanding agricultural output in the tropics by bringing new lands into production with a reasonable degree of expected success.

An analysis of this subject by Kellogg and Orvedal (1969) indicates that most of the potentially arable but presently unused soils of the world are in the tropics. With the exception of Asia, most of the tropical population is concentrated in coastal areas or high-base-status alluvial plains or inter-mountain valleys, with reasonably adequate means of transportation. In spite of the wide differences in population density, the cultivated area per capita in tropical America is not different from that in tropical Asia, as Table 1.9 indicates. The differences in population density are reflected in the percentage of the total area under production, which ranges from 27 percent in Asia to 8 percent in Africa and 5 percent in Latin America.

Table 1.9   Land Use (million ha) and Population in the Tropical Areas

Total	Total Land Area	Culti- vated Area[a]	Pastures and Meadows	Forests	Area Culti- vated (%)	Popu- lation (1969) (millions)	Cultivated Area per Capita (ha)
Tropical America	1,683	83	282	914	5	239	0.35
Tropical Africa	2,212	166	652	571	8	275	0.60
Tropical Asia and Pacific[b]	931	256	21	412	27	956	0.26
Tropics	4,826	503	955	1897	10	1470	0.34
World	13,392	1424	3001	4091	11	3647	0.41
Percent Tropics	36	35	32	46	—	40	—

Source: Calculated from data in the FAO 1970 *Production Yearbook.*
[a] Annual and perennial crops.
[b] Except Australia.

Table 1.10   Land Use (million ha) and Population Patterns in Countries That Are Wholly or Mostly in the Tropics, Arranged by Decreasing Size

Country or Region	Total Land Area	Presently Culti-vated	Pastures and Meadows	Area Culti-vated (%)	Popu-lation (1969) (millions)	Cultivated Area per Capita (ha)
Tropical America						
Brazil	851	30	107	1	90.8	0.3
Mexico	197	24	79	12	48.9	0.5
Peru	128	3	27	2	13.1	0.2
Colombia	114	5	15	4	20.4	0.2
Bolivia	110	3	11	3	4.8	0.6
Venezuela	91	5	14	6	10.6	0.5
Central America	52	5	8	10	16.1	0.3
Guyanas	47	0.2	3	0.5	1.2	0.2
Paraguay	41	1	10	2	2.3	0.4
Ecuador	28	3	2	9	5.9	0.5
Caribbean	24	4	6	18	24.9	0.2
Total	1683	83	282	5	239	0.3
Tropical Asia and Pacific						
India	327	163	14	50	537	0.3
Indonesia	190	13	—	7	117	0.1
Pakistan[a]	95	28	—	30	128	0.2
Indochina	75	6	5	8	48	0.1
Burma	68	16	—	24	27	0.6
Pacific islands	55	2	1	4	4	0.5
Thailand	51	11	—	22	35	0.3
Malaysia	33	4	—	10	11	0.4
Philippines	30	9	1	29	37	0.2
Sri Lanka	6	2	—	30	12	0.1
Total	931	256	21	27	956	0.3

Kellogg and Orvedal recognized that one of the factors presently limiting the utilization of tropical areas suited for crop production is inadequate knowledge of how to manage the highly weathered Oxisols and Ultisols presently under rainforest or savanna vegetation. This subject will cover a major portion of this book, but it is sufficient to state at this point that very high yields have been obtained in such areas when the proper technology is applied.

Another approach for increasing food production is to intensify the annual yields per hectare of the 10 percent of the tropics actually under

Table 1.10 (*Continued*)

Country or Region	Total Land Area	Presently Culti-vated	Pastures and Meadows	Area Culti-vated (%)	Popu-lation (1969) (millions)	Cultivated Area per Capita (ha)
Tropical Africa						
Sudan	250	7	24	3	15	0.4
Zaïre	234	7	66	3	17	0.4
Chad	128	7	45	6	3	2.3
Niger	127	11	3	9	4	2.7
Angola	125	1	29	1	5	0.2
Ethiopia	122	13	69	10	25	0.5
Mauritania	103	0.3	39	0.3	11	0.0
Tanzania	94	12	45	13	13	0.9
Nigeria	92	22	26	24	66	0.3
Mozambique	78	3	44	3	7	0.4
Zambia	75	5	34	6	4	1.2
Madagascar	59	3	34	5	7	0.4
Kenya	58	2	4	3	11	0.1
Cameroon	47	4	8	9	6	0.6
Rhodesia	38	2	5	5	5	0.4
Ivory Coast	32	9	8	27	5	1.8
Upper Volta	27	10	14	35	5	1.9
Ghana	24	3	—	12	9	0.3
Uganda	23	5	5	21	8	0.6
Senegal	19	6	6	29	4	1.5
Others	237	34	85	10	29	0.8
Total	1992	166	593	8	257	0.6

Source: Calculated from the FAO 1970 *Production Yearbook.*
[a] Includes Bangladesh.

cultivation. Green revolution-type programs have shown that yields can be doubled or tripled when new variety–cultural practice packages suited to small farms are rapidly adopted.

Farming Systems

Tropical agriculture is far from homogeneous. Even within short distances one can observe large sugar cane plantations being harvested by huge machines, minute and carefully hand-tended paddy rice fields, and a wide array of crops growing together on steep hillsides cleared by slash and burn.

Just as it is relevant to point out the major differences in climate and vegetation, the relative importance and distribution of the main farming systems should be emphasized even at the expense of broad generalizations. Six main tropical farming systems can be recognized at the broadest level of generalization:

1. Shifting cultivation, covering 45 percent of the tropical area.
2. Settled subsistence farming, covering 17 percent of the tropical area.
3. Nomadic herding, covering 14 percent of the tropical area.
4. Livestock ranching, covering 11 percent of the tropical area.
5. Plantation systems, covering 4 percent of the tropical area.

The distribution of these farming systems is shown in Fig. 1.12. Commercial family-farm-type operations analogous to those common in United States agriculture are present only in very small proportions.

Shifting cultivation occurs in almost half of the tropical world in both forested and savanna areas. These are usually the least developed areas, where farmers cut and burn a small area, plant several crops in the same field, and abandon the fields when native soil fertility, weeds, or other factors decrease yields. Although very little of the production flows through established markets, millions of people in the tropics depend on shifting cultivation as their means for subsistence. Shifting cultivation is most widespread in West and Central Africa under udic and ustic regimes. It is also very common in the Amazon Basin, Central America, and the hill country of Southeast Asia and the Pacific.

Nomadic herding covers approximately 14 percent of the tropics, mostly in the aridic areas of Africa and the Arabian Peninsula. It is essentially absent in tropical America and tropical Asia.

The traditional settled systems cover the remaining areas. About 11 percent is predominantly livestock ranching; this is essentially limited to tropical America and Australia. Settled, small-scale cropping systems can be subdivided into those based on paddy rice, which cover about 5 percent of the tropics (mostly in Asia), and those based on crops other than rice (corn, beans, sorghum, etc.). The latter cover about 12 percent of the tropics throughout the Andes, most of India, the hill country of Southeast Asia, and scattered areas in Africa.

The most sophisticated management systems, in terms of the use of machinery and fertilizer, are the modern plantations growing export crops such as sugarcane, coffee, cacao, rubber, bananas, and pineapple. They

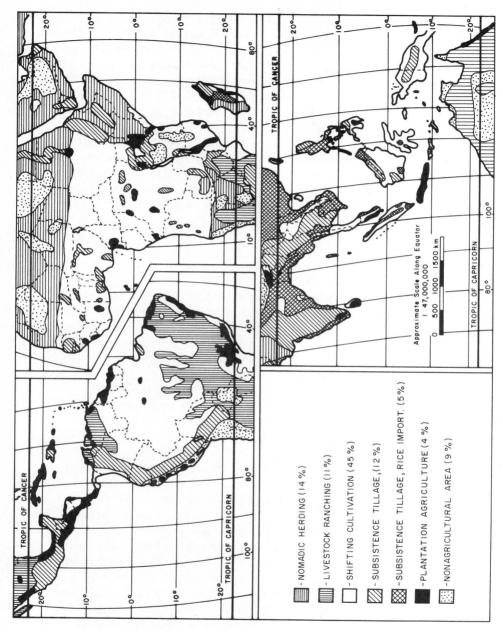

Legend:

- NOMADIC HERDING (14 %)
- LIVESTOCK RANCHING (11 %)
- SHIFTING CULTIVATION (45 %)
- SUBSISTENCE TILLAGE, (12 %)
- SUBSISTENCE TILLAGE, RICE IMPORT. (5 %)
- PLANTATION AGRICULTURE (4 %)
- NONAGRICULTURAL AREA (9 %)

Approximate Scale Along Equator
1 : 47,000,000
0   500 1000 1500 k m

Fig. 1.12 The tropics: management systems adapted from Whittesley (1936).

occur in small areas throughout the tropics but are most prevalent in Central America, the Caribbean, southern Brazil, Malaysia, and Indonesia.

The relative importance and extent of these farming systems must be recognized by temperate and tropical agronomists alike. The American family-farm enterprise is almost absent in the tropics except in northern Australia and southern Brazil. Its closest tropical equivalent is the paddy rice system of Southeast Asia, but there one family usually operates only 1 to 3 ha. Sociologists tell us that it is very difficult to change these traditional systems even with political revolutions. One must recognize them and make attempts to improve them rather than trying to transplant United States or European practices without significant adaptive research. On the other hand, in the relatively unpopulated areas of South America, such as parts of the Cerrado of Brazil, anything is possible since there are no traditional farming systems.

Tropical Food Crops

When the term "tropical crop" is mentioned, most people probably think of mangoes, coffee, papayas, sugarcane, or exotic tropical fruits. Because of the rapid development of plantation agriculture for export purposes, a substantial amount of knowledge exists about crops such as sugarcane, pineapples, bananas, coffee, cacao, and tea. Unfortunately these crops contribute little to the nutrition of tropical people, because even the ones with nutritional value are largely exported to the temperate region.

Since our interest lies in food production, the relative importance of the main tropical food crops should be examined. Tables 1.11 and 1.12 show the production, area planted, and average yield of the 12 most important food crops in the tropics as a whole and by regions. While examining these figures, one should not forget the billion hectares devoted to pastures and meadows for animal production. The total area covered by these 12 crops is approximately 300 million ha. The rest of the total cultivated area (500 million ha) is devoted to plantation crops or minor crops. Because of the frequency of intercropping, it is difficult to be more specific.

Rice (*Oryza sativa*) is the food crop produced in largest quantities in the world and occupies the largest area in the tropics. Rice is not restricted to the tropics; in fact, production in the temperate region, particularly in China and Japan, accounts for 45 percent of the world's total. About 90 percent of the world's rice is produced in Asia under a variety of management systems. Nevertheless, rice is an important food crop in tropical America and Africa, occupying the third and sixth place, respectively, in those areas. Most of the rice grown in these two regions is produced under

Table 1.11   Production, Area, and Yields of the Main Food Crops Grown in the Tropics

Food Crop	Production (million metric tons)	Area (million ha)	Yields		World Production (%)
			ton/ha	bu/acre	
1. Rice	169.4	94.0	1.80	36	55
2. Cassava	90.5	9.6	9.42		99
3. Corn	63.8	53.0	1.20	19	24
4. Sweet potatoes and yams	35.4	4.7	7.53		26
5. Wheat	34.6	28.9	1.20	18	11
6. Sorghum	20.3	29.2	0.70		46
7. Millets	15.4	29.9	0.52		81
8. Peanuts	13.8	16.1	0.86		76
9. Potatoes	13.7	1.7	8.06	120	5
10. Beans	10.3	16.5	0.62	9	88
11. Other grain legumes	7.3	11.2	0.65	10	33
12. Soybeans	2.7	2.6	1.03	15	6

Source: Computed from the FAO 1970 *Production Yearbook.*

upland (nonirrigated) conditions. Publications by the International Rice Research Institute provide excellent information about this crop.

Cassava (*Manihot esculenta*) is the food crop produced in next largest quantity in the tropics. This root crop is essentially limited to the tropics, where it is also known as "yuca", "tapioca", "manioc", and "mandioca". It occupies the first place in total production in tropical America and Africa and the third place in Asia, after rice and wheat. In spite of its importance as a major source of energy (and because of the large quantities consumed, a major source of protein), cassava has received little research attention and improvement. Fortunately major efforts are presently under way in Colombia. For additional information, the reader is referred to articles by Cours (1951), Jennings (1970), and Hendershott et al. (1972).

Corn (*Zea mays*) ranks third in total production in the tropics, after rice and cassava, but it is second in total area planted. Most of the tropical corn is used as food, either directly or as flour. In the more developed areas corn serves as cattle and poultry feed, but the bulk is still consumed directly as food. Unlike the situation with cassava, corn improvement has received much attention in the tropics. A report by CIMMYT (1971b) summarizes much of the present knowledge.

Two starchy root crops, sweet potatoes (*Ipomoea batatas*) and yams (*Dioscorea* spp.) together occupy the fourth place in importance. Yams

Table 1.12   Relative Importance of the 12 Major Tropical Food Crops by Continents

Food Crop	Production (million metric tons)	Area (million ha)	Yield (ton/ha)
Tropical America			
1. Cassava	34.5	2.5	13.80
2. Corn	30.9	23.8	1.30
3. Rice	11.4	6.2	1.84
4. Potatoes	6.6	0.8	8.25
5. Wheat	4.3	3.0	1.44
6. Beans	4.1	6.8	0.60
7. Sweet potatoes and yams	3.7	0.4	8.50
8. Sorghum	2.8	1.3	2.17
9. Soybeans	1.9	1.4	1.35
10. Peanuts	1.1	0.8	1.49
11. Other grain legumes	0.5	0.6	0.83
12. Millets	0.2	0.2	0.81
Tropical Africa			
1. Cassava	36.0	4.9	7.39
2. Sweet potatoes and yams	21.6	3.0	7.09
3. Corn	16.6	15.1	1.10
4. Sorghum	7.1	8.5	0.83
5. Millets	5.4	8.1	0.66
6. Rice	4.8	3.4	1.43
7. Peanuts	4.7	5.8	0.81
8. Wheat	2.7	2.8	0.94
9. Potatoes	1.8	0.2	7.39
10. Beans	0.9	1.4	0.65
11. Other grain legumes	0.6	0.8	0.77
12. Soybeans	0.03	0.04	0.67
Tropical Asia			
1. Rice	153.2	84.4	1.81
2. Wheat	27.6	23.1	1.19
3. Cassava	20.0	2.2	8.82
4. Corn	16.3	14.1	1.16
5. Sorghum	10.4	19.4	0.54
6. Sweet potatoes and yams	10.1	1.3	7.93
7. Millets	9.8	21.6	0.45
8. Peanuts	8.0	9.5	0.84
9. Chickpeas	6.2	8.8	0.71
10. Potatoes	5.3	0.7	8.14
11. Beans and other legumes	3.6	9.3	0.39
12. Soybeans	0.8	1.2	0.67

Source: Computed from the FAO 1970 *Production Yearbook.*

contribute a major portion of the diet in tropical Africa but are also important in the rest of the tropics. Sweet potatoes are more widespread in tropical America, Southeast Asia, and the Pacific islands. Several other root crops such as *Xanthosoma mafaffa* (malanga, yautia, tanier) and *Colocasia esculenta* (dasheen, taro, cocoyam) are usually important where sweet potatoes and yams are grown. With the possible exception of sweet potatoes, these crops have received very limited attention by research workers. Tropical root crop symposia at the University of West Indies (Tai et al. 1967), and in Hawaii (Plucknett, 1970) summarize much of the present knowledge.

Wheat (*Triticum aestivum*) is restricted to the tropical highlands or the edges of the tropics where temperatures are sufficiently low to permit the growth of this basically temperate crop. It is particularly important in the northern parts of the Indian subcontinent, northern Mexico, the Andes, and southern Brazil. Research efforts, originating in Mexico and now spread worldwide, have resulted in more dramatic yield increases in this crop than in any other. A report by CIMMYT (1971a) summarizes the present status of wheat improvement.

Sorghum (*Sorghum* spp.) and millets (*Pennisetum* spp.) occupy the sixth and seventh place, respectively, in importance. They are presently consumed as food by millions of Africans and Asians, principally in the ustic and aridic environments. In areas with progressive livestock enterprises they are beginning to be used as feed.

Peanuts or groundnuts (*Arachis hypogaea*) are also important in the ustic and aridic areas of Africa and Asia, providing a major portion of the protein requirements. They are less important in the humid tropics. Irvine's (1969) book on West African crops describes sorghums, millets, and peanuts in detail.

Potatoes (*Solanum tuberosum*) are a major food source in the tropical highlands, particularly in the Andes, where they originated. Although they are grown to only a limited extent in the lowlands, there seems to be some potential for the lowland tropics. Research in the tropics has increased recently, particularly in Peru. A compilation by the International Potato Center (CIP, 1972) summarizes the present knowledge.

Dry beans (*Phaseolus vulgaris*) are an extremely important protein source in the tropics, especially in Latin America, where they are grown on almost every farm, usually intercropped with corn. The main research effort has been concentrated in Central America but with limited success in increasing yields. Other pulses such as cowpeas (*Vigna sinensis*), lima beans (*Phaseolus limensis*), lentils (*Lens esculentax*), broad beans (*Vicia faba*), peas (*Pisum sativum*), chickpeas (*Cicer arietum*), and pigeon peas (*Cajanus cajan*) are locally important in specific areas. Soybeans (*Glycine max*) have been a traditional source of food in Asia and are presently being introduced into

Latin America for industrial uses. The worldwide demand for this crop and its versatility undoubtedly will elevate it to a more important role in the tropics. A symposium at CIAT (1973) is a good reference for grain legumes for Latin America.

Not included in Tables 1.10 and 1.11 because of lack of statistics, but extremely important, are bananas, both the fruit species (*Musa sapientum*) and the cooking bananas or plantains (*Musa paradisiaca*). A total of 16.3 million tons was produced in tropical America in 1970, but it is not possible to ascertain how much of its was exported. Qualitatively at least, plantains and bananas are probably as important as sweet potatoes and yams as sources of carbohydrates in the tropics. Simmonds' (1967) book is the most up-to-date source on the subject.

The nutritive values of the main tropical food crops appear in Table 1.13. Of special interest is the wide difference in carbohydrate content between the starchy cereals, root crops, and fruits due to moisture content. The protein content of the starchy foods is of great importance since these crops provide the bulk of the protein intake in tropical diets. New varieties and cultural practices can markedly affect this picture. For example, a combination of a high-protein rice variety and delayed timing of nitrogen applications increased the protein content of rice from 7 to 9 percent (DeDatta et al. 1972). Although cassava tubers are very low in protein, a study at CIAT (1969) showed a range from 0.5 to 7.2 percent protein in their germplasm collection. Both breeding and managing for increased protein content have bright futures, but the data in Table 1.13 show the actual situation.

These crops can further be compared in terms of their energy, carbohydrate, and protein production per hectare. Table 1.14 shows such calculations, based on 1970 average yields in the tropics. It may be noted that the root crops, particularly cassava, and the plantains far outstrip the cereals in energy produced per hectare. Plantains, bananas, and cassava produce more than 2.5 tons/ha of carbohydrates per crop. In terms of protein production, soybeans are far ahead of the other crops in the list. Potatoes are the second highest protein producer, far outstripping the other grain legumes and doubling or tripling the protein-producing capacities of cassava, bananas, rice, and corn.

A further refinement of this analysis involves the calculation of production per hectare per day to take into account the wide differences in average growth duration between these crops. Table 1.15 shows the average energy, carbohydrate, and protein production capacity on a daily basis. Sweet potatoes and soybeans produce more than 45 kcal/ha a day, followed by rice, potatoes, plantains, corn, wheat, and cassava. Soybeans, potatoes, field beans, and wheat are the top protein producers per day.

Table 1.13 Nutritive Value of Main Food Crops Grown in the Tropics

Crop	Food	Energy (cal/100 g)	Carbohydrate (%)	Protein (%)	H₂O (%)	Fe (mg/100 g)	A (mg)	B₁ (mg/100 g)	B₂ (mg/100 g)	Niacin (mg/100 g)	C (mg/100 g)
Cereals:											
Rice, milled	Grain	364	79	7.2	12	1.3	0	0.08	0.03	1.6	0
Corn, whole	Grain	361	74	9.4	11	2.5	70	0.43	0.10	1.9	Tr.
Wheat, whole	Grain	330	69	14.0	13	3.1	0	0.57	0.12	4.3	0
Sorghum, whole	Grain	342	76	8.8	10	3.7	10	0.41	0.12	3.2	0
Root Crops											
Cassava, peeled	Tubers	132	32	1.0	65	1.4	Tr.	0.05	0.04	0.6	19
Sweet potato, peeled	Tubers	116	28	1.3	69	1.0	1815[a]	0.11	0.04	0.8	31
Yams, peeled	Tubers	100	24	2.0	73	1.3	Tr.	0.13	0.02	0.4	3
Potatoes, unpeeled	Tubers	79	18	2.8	78	1.0	Tr.	0.11	0.04	1.5	20
Starchy Fruits											
Bananas, peeled	Fruit	110	29	1.2	69	0.5	65	0.04	0.04	0.7	15
Plantains, peeled	Fruit	122	32	1.0	66	0.8	175	0.06	0.04	0.6	20
Grain Legumes											
Beans, field	Grain	337	60	22.0	12	7.6	5	0.54	0.19	2.1	3
Peanuts, with skin	Grain	543	21	25.5	7	3.0	10	0.91	0.21	17.6	1
Soybeans	Grain	398	35	33.4	9	11.5	Tr.	0.88	0.27	0.22	0
Pigeon peas	Grain	337	63	19.2	12	5.0	20	0.72	0.17	2.6	0
Cowpeas	Grain	341	60	24.1	11	7.2	10	0.87	0.23	1.9	3
Chickpeas	Grain	364	61	18.2	12	7.3	15	0.46	0.16	1.7	1

Source: Based on data from Wu Leung and Flores (1961).
[a] Yellow varieties; white varieties average 30.

Table 1.14 Comparative Carbohydrate- and Protein-Producing Capacities of Main Tropical Crops at Average Yield Levels

Crop	Average Crop Yield (tons/ha)	Edible Portion (%)	Energy per Crop (Mcal/ha)	Carbohydrate Production per Crop (tons/ha)	Protein Production per Crop (kg/ha)
Rice	1.80	70	4.6	1.00	91
Corn	1.20	100	4.3	0.89	113
Wheat	1.20	100	3.9	0.83	168
Sorghum	0.70	100	2.4	0.53	62
Cassava	9.42	83	10.3	2.56	78
Sweet potato	6.50	88	6.6	1.63	74
Yams	8.00	85	6.8	1.65	136
Potatoes	8.06	100	6.4	1.47	226
Bananas	15.20	65	10.9	2.86	118
Plantains	15.20	69	12.8	3.38	104
Beans, field	0.62	100	2.1	0.38	136
Peanuts	0.86	71	3.3	0.13	156
Soybeans	1.03	100	4.1	0.36	344
Cowpeas	0.36	100	1.2	0.22	87

Table 1.15 Comparative Carbohydrate and Protein Production per Day of Main Tropical Crops at Average Yield Levels

Crop	Average Growth Duration (days)	Energy (kcal/ha)	Carbo-hydrates (kg/ha)	Protein (kg/ha)
Rice	120	38.2	8.3	0.76
Corn	125	34.6	7.0	0.90
Wheat	120	33.0	6.9	1.40
Sorghum	135	17.6	3.9	0.46
Cassava	330	31.2	7.8	0.23
Sweet potatoes	135	49.1	12.4	0.55
Yams	280	24.3	5.8	0.49
Potatoes	180	35.5	9.7	1.50
Bananas	365	29.8	9.4	0.32
Plantains	365	35.0	9.3	0.28
Beans, field	90	23.3	4.2	1.51
Peanuts	120	27.5	1.0	1.30
Soybeans	90	45.5	4.0	3.82
Cowpeas	90	13.3	2.4	0.96

Although not all these crops can be grown in the same area, such an analysis points out the nutritional advantages and limitations when there is a choice. Multiple cropping systems or crop rotation schemes can be designed to produce the desired amount of energy or protein. Similar calculations can be made for iron and vitamins, based on the data in Table 1.13. Improved varieties and cultural practices often double or triple the average yields on which these calculations are based.

SUMMARY AND CONCLUSIONS

1. The tropics are not uniform. Great variability in temperature and rainfall regimes exists. The only unifying property of tropical climates is their relatively uniform temperature regimes throughout a given year. Tropical climates are not always uncomfortable to man. Temperatures in rainy lowland areas are often high, but seldom as high as those encountered in temperate areas during the summer. It is hard to find a more comfortable climate than the tropical highlands of, for example, San Jose, Costa Rica, or Nairobi, Kenya.

2. High constant temperatures throughout the year characterize the tropical lowlands, which account for 87 percent of the area. The lack of temperature limitations gives the tropics a much higher agricultural potential for year-round production than exists in the temperate region.

3. Rainfall distribution is the principal parameter used to differentiate tropical climates for agricultural purposes. Approximately half of the tropics has pronounced wet and dry seasons (ustic soil moisture regime), one-fourth has high rainfall distributed throughout the year (udic soil moisture regime), and one-fourth has semiarid or desert climate (aridic soil moisture regime).

4. The wet and dry seasons exert on plant growth an influence similar to that of the winter and summer seasons in the temperate region. Much plant growth and animal production stops during the dry season. The beginning of the rains produces an outburst of activity not unlike that marking the arrival of spring in the temperate region.

5. Most of the tropics are not covered by jungles. Savannas are the most extensive type of vegetation, accounting for 43 percent of the area. The tropical rainforests cover 30 percent; semideciduous or deciduous forests cover 15 percent, with desert shrubs and grasses or no vegetation on the rest.

6. Land surfaces in the tropics range from the oldest on earth (ancient crystalline rocks in South America and Africa) to the most recent volcanic deposits.

7. The race between the world's population growth and food supplies remains dangerously close during the decade of the 1970s. Its outcome will be decided in the tropical regions. The population growth rates are twice as high in the tropics as in the temperate region. Food production has generally kept up with population increases.

8. The demand for more and better food in the tropics, in addition to the recent fuel and fertilizer shortage, has disrupted this balance to a danger point. The situation is bleaker when diet adequacy is considered. Although only 20 percent of the world's population is undernourished (in terms of daily calorie intake), more than two-thirds is malnourished (protein and other deficiencies). The bulk of the undernourished and malnourished population is in the tropics.

9. Only one-tenth of the potentially arable land of the tropics is cultivated. Most of the potentially arable but presently unused land in the world lies in the tropics. The long-term solution to the world's food problem probably consists of putting such land into production, lowering the rates of population growth, and intensifying production in the land already under cultivation.

10. Tropical agriculture is far from homogeneous, except in terms of low average yields. The most important agricultural systems in terms of percentage of the area are as follows: shifting cultivation (45 percent), settled subsistence farming systems (17 percent), nomadic herding (14 percent), livestock ranching (11 percent), and plantation crops primarily for export (4 percent).

11. The most important food crops produced in the tropics, in their probable order of importance are as follows: rice, cassava, corn, sweet potatoes and yams, wheat, sorghum and millet, bananas and plantains, peanuts, potatoes, and grain legumes. The nutritive values of these crops vary considerably. At present average yield levels, the root crops and plantains far outstrip the cereals in total energy production per crop. Soybeans and potatoes are the highest protein producers per crop. On a daily basis the most productive crops in terms of both energy and protein production are sweet potatoes and soybeans. Other differences in terms of vitamins, minerals, and protein quality are also important. Recent yield breakthroughs have taken place with rice and wheat with resulting dramatic increases in production. Similar breakthroughs are expected for other tropical crops, particularly the root crops, within the next decade.

REFERENCES

Arens, K. 1963. The dwarfed plants of the "cerrado" fields as flora adapted to mineral deficiencies in the soil. Pp. 285–303. In M. G. Ferri (coord.), *Simpósio sôbre o Cerrado*. Edgar Blücher, São Paulo.

Aubert, G. and R. Tavernier. 1972. Soil survey. Pp. 17–44. In: *Soils of the Humid Tropics.* U.S. National Academy of Sciences, Washington.

Aubréville, A. 1965. Princípes de úne systematique des formations vegetales tropicaux. *Adansonia* 5:153–196.

Barker, R. 1972. The economic consequences of the green revolution in Asia. Pp. 115–126. In *Rice, Science and Man.* International Rice Research Institute, Los Baños, Philippines.

Bartlett, H. H. 1957. Fire, primitive agriculture and grazing in the tropics. Pp. 692–720. In W. L. Thomas (ed.), *Man's Role in Changing the Face of the Earth.* University of Chicago Press, Chicago.

Blumenstock, D. I. 1958. Distribution and characteristics of tropical climates. *Proc. Ninth Pacific Sci. Congr. (Bangkok)* 20:3–23.

Bouliere, F. and M. Hardy. 1970. The ecology of tropical savannas. *Ann. Rev. Ecol. Systematics* 1:125–152.

Budowski, G. 1956. Tropical savannas, a sequence of forest feelings and repeated burnings. *Turrialba* 6 (1–2):23–33.

Budowski, G. 1966a. Some ecological characteristics of higher tropical mountains. *Turrialba* 16 (2):159–163.

Budowski, G. 1966b. Los bosques de los trópicos húmedos de America. *Turrialba* 16:278–285.

CIAT. 1969. Cassava production systems. Annual Report. Pp. 42. Centro Internacional de Agricultura Tropical, Cali, Colombia.

CIAT. 1973. *Potential of Grain Legumes in Latin America.* Centro Internacional de Agricultura Tropical, Cali, Colombia. 388 pp.

CIMMYT. 1971a. *Proceedings of the First Wheat Workshop.* Centro Internacional de Mejoramiento de Maiz y Trigo, El Batán, Mexico. 88 pp.

CIMMYT. 1971b. *Proceedings of the First Maize Workshop.* Centro Internacional de Mejoramiento de Maiz y Trigo, El Batán, Mexico. 107 pp.

CIP. 1972. *Prospects of the Potato in the Developing World.* Centro Internacional de la Papa, Lima, Peru.

Cochrane, W. W. 1969. *The World Food Problem—A Guardedly Optimistic View.* Crowell, New York.

Cours, G. 1951. Le Manioc a Madagascar. *Mem. Inst. Sci. Madagascar,* Ser B 3(2):203–399.

DeDatta, S. K., W. N. Obcemea, and R. K. Jana. 1972. Protein content of rice grain as affected by nitrogen fertilization and some triazines and substituted ureas. *Agron. J.* 64:785–788.

D'Hoore, J. D. 1956. Pedological comparisons between tropical South America and tropical Africa. *Afr. Soils* 4:5–18.

DeMartonne, E. 1957. *Traité de Géographie Physique.* Armand Colin, Paris.

DeVries, C. A., J. A. Ferwerda, and M. Flach. 1967. Choice of food crops in relation to annual and potential production in the tropics. *Netherl. J. Agr. Sci.* 15:241–248.

DeWitt, C. T. 1967. Photosynthesis: Its relationship to overpopulation. Pp. 315–320. In F. A. San Pietro et al. (eds), *Harvesting the Sun.* Academic Press, New York.

Eiten, G. 1972. The cerrado vegetation of Brazil. *Bot. Rev.* 38(2):201–341.

Eyre, S. R. 1962. *Vegetation and Soils–A World Picture.* Aldine, Chicago.

FAO. 1970, 1971. *Production Yearbook.* Food and Agricultural Organization of the United Nations, Rome.

FAO. 1972. *The State of Food and Agriculture.* Food and Agricultural Organization of the United Nations, Rome.

Francis, C. A. 1972. Natural daylength for photoperiod-sensitive plants. *CIAT Tech. Bull.* 2.

Golley, F. B. and H. Leith. 1972. Bases of organic production in the tropics. Pp. 1–26. In *Tropical Ecology.* University of Georgia, Athens.

Goodland, R. 1971. Oligotrofismo e alumínio no Cerrado. Pp. 45–60. In M. G. Ferri (coord.), *III Simpósio sôbre o Cerrado.* University of São Paulo, Brazil.

Harris, D. R. 1972. The origins of agriculture in the tropics. *Amer. Scientist* **60**:180–193.

Hendershott, Ch. H. et al. 1972. *A Literature Review and Research Recommendations on Cassava.* University of Georgia, Athens. 326 pp.

Holdridge, L. R. 1947. Determination of world plant formations from simple climatic data. *Science* **105**:367–368.

Holdridge, L. R. 1967. *Life Zone Ecology,* rev. ed. Tropical Science Center, San José, Costa Rica.

IRRI. 1969. Annual Report. Pp. 109–111. Agronomy Section, International Rice Research Institute, Manila, Philippines.

IRRI. 1971 *Rice Breeding.* International Rice Research Institute, Los Baños, Philippines. 798 pp.

Irvine, F. R. 1969. *West African Crops.* Oxford, London.

Jennings, D. L. 1970. Cassava in Africa. *Field Crops Abs.* **23**(3):267, 271–278.

Kellogg, C. E. and A. C. Orvedal. 1969. Potentially arable soils of the world and critical measures for their use. *Adv. Agron.* **21**:109–170.

Köppen, W. and R. Geiger. 1936. *Hanbuch der Klimatologie,* Vol. 1. Borntraeger, Berlin.

Landsberg, H. E. 1961. Solar radiation at the earth's surface. *Solar Energy* **5**:95–98.

Landsberg, H. E., H. Lippman,K. H. Patten and C. Troll. 1963. In E. Rodenwalt and A. Justaz (eds.), *Die Jahreszeitenklimate der Erde.* Heidelberg Akademie der Wissenshaften, Springler Verlag, Heidelberg.

Lawes, E. F. 1969. Confidence limits for expected rainfall at Muguga, Kenya. *East Afri. Meteorol. Dept. Tech. Mem.* 15 pp.

Lee, D. H. K. 1957. *Climate and Economic Development in the Tropics.* Harper, New York.

Lopes, A. A. 1975. A survey of the fertility status of soils under "Cerrado" vegetation in Brazil. M.S. Thesis, North Carolina State University, Raleigh. 138 pp.

Mohr, E. C. J., F. H. van Baren, and J. Von Schuylerborgh. 1972. *Tropical Soils: A Comprehensive Study of Their Genesis,* 3rd rev., enlarged ed. Mouton-Ichtlar Baru-Van Hoeve, The Hague.

Myers, W. M. 1970. World food supplies and population growth. Pp. 3–30. In D. G. Aldrich (ed.), "Research for the World Food Crisis." *Amer. Assoc. Adv. Sci. Publ.* 92.

Odum, H. T. (ed.). 1972. *A Tropical Rainforest.* U.S. Atomic Energy Commission, Washington.

Paddock, P. and W. Paddock. 1967. *Famine, 1975: America's Decision—Who will Survive?* Little Brown, Boston.

Plucknett, D. L. (ed.). 1970. *Tropical Root and Tuber Crops Tomorrow,* Vol. 1. College of Tropical Agriculture, University of Hawaii.

President's Science Advisory Committee. 1967. *The World Food Problem,* Vol. 2. The White House, Washington.

Richards, P. W. 1952. *The Tropical Rainforest.* Cambridge University Press, London.

Simmonds, N. W. 1967. *Bananas.* Longmans, London.

Smith, G., D. F. Newhall, L. H. Robinson, and D. Swanson. 1964. Soil temperature regimes—

their characteristics and predictability. *U.S. Dept. Agr. Soil Conserv. Service Tech. Paper* 144. 14 pp.

Soil Survey Staff. 1970. *Soil Taxonomy* (draft). Soil Conservation Service, U.S. Department of Agriculture.

Tai, E. A. et al. (eds.). 1967. *Proceedings of the International Symposium on Tropical Root Crops*, Vols. 1 and 2. University of West Indies, Trinidad.

Thomas, M. F. 1974. *Tropical Geomorphology*. Wiley, New York. 332 pp.

Vicente-Chandler, J., R. Caro-Costas, R. W. Pearson, et al. 1964. The intensive management of tropical forages in Puerto Rico. Univ. Puerto Rico Bull. 202.

Whittlesey, D. 1936. Major agricultural regions of the earth. *Ann. Assoc. Amer. Geogr.* **26**:199–242.

Williams, C. N. and K. T. Joseph. 1970. *Climate, Soil and Crop Production in the Humid Tropics*. Oxford University Press, Singapore.

Wood, A. 1950. *The Groundnut Affair*. Bodley Head, London. 264 pp.

Wu Leung, W. T. and M. Flores. 1961. *Tablas de Composición de Elementos para Uso en América Latina*. Instituto de Nutrición de Centro América y Panamá, Guatemala.

Yñiguez, A. D. and A. R. Sandoval. 1966. Rainfall probabilities at the U.P. College of Agriculture. *Philipp. Agr.* **49**(8):681–695.

2

SOILS OF THE TROPICS

One of the main obstacles to understanding tropical soil science is the extremely confusing and inaccurate terminology existing throughout the literature. The terms "tropical soils," "Latosols," "laterite," and "lateritic soils" mean very different things to different people. The widespread use of at least five major soil classification systems increases the confusion, and in essence prevents meaningful comparisons and extrapolation of data from one place to another. It is appropriate at this point to discuss the events leading to this confusion and to suggest means for clarification.

Soil science developed first in the temperate region, particularly in the Soviet Union, Europe, and the United States. Travel of nineteenth century soil scientists to the tropics resulted in a misguided view about the uniformity of tropical soils and an exaggeration of the presence and importance of hardened layers rich in iron oxides called "laterites." When these scientists returned home, they tended to emphasize in their teachings and their publications what they saw as unique: the occurrence of laterites. As a result of these reports on unique soils, vast areas of the tropics with soils similar to those found in the temperate region were essentially ignored. Thus the term "tropical soil" developed in the literature to mean soils high in iron that harden irreversibly upon exposure. "Latosols" and "lateritic soils" became soils in the process of developing into laterite. "Laterization" became an accepted process of soil formation. Such reports are well summarized in reviews by Alexander and Cady (1962) and Sivarajasingham et al. (1962).

As early as 1933, Frederick Hardy published an article emphasizing that laterites have very limited areal extent in the tropics. Feuer (1956), in an exploratory study of the Federal District of Brazil, where laterites are commonly mentioned, found that they may cover only 5 percent of this area. In West Africa a generalized map by Segalen (1970) indicates that the total area with laterite outcrops or laterite layers close to the surface is about 15 percent. Even in South India, where the original studies of laterites were made, soils previously known as "laterite soils" are now classified as Alfisols, Inceptisols, or Ultisols (Gowaikar, 1973). In a review of Latin American soils Buol (1973) emphasized that the present definition of "laterite" or "plinthite" refers to a material and not to a kind of soil. He also emphasized the limited areal extent of laterites and the fact that they occur in definite geomorphic positions, not only in the tropics but also in the United States.

Unfortunately, little attention has been paid to Hardy's article and subsequent publications. This general belief has led many popular articles and scientific publications like those of McNeil (1964) and Kamarck (1972), to conclude that most tropical soils, when cleared of vegetation, will become worthless brick pavement in a few years. Although this situation undoubtedly exists, laterites are found in less than 7 percent of the tropics, more often in the subsoil (Sanchez and Buol, 1975). Nevertheless, laterites constitute a serious management problem in such localities.

From the information presented in Chapter 1, it is obvious that tropical soils cannot be uniform because of the wide variety in climate, vegetation, parent material, geomorphology, and age. In fact, the only property common to all tropical soils is lack of seasonal soil temperature variation (Buol, 1973). All other generalizations are essentially incorrect. To state that all tropical soils are highly leached and infertile is equivalent to stating that all temperate soils are young and fertile. The vague and varied meanings of the terms "Latosols," "laterite," and "lateritic soils" deserve careful scrutiny by modern workers. The only valid exception to this rule is the well-quantified definition of Latosols in the Brazilian soil classification system (Bennema and Camargo, 1964; Beinroth, 1975).

In spite of the voluminous literature, comprehensive knowledge of the characteristics of soils in the tropics is quite limited. In addition to the exaggerations about the importance of laterites, there are other reasons for the lack of integrated knowledge. Full-time tropical soil scientists who work only in a specific area or country develop strong local biases. When they meet with colleagues at international meetings, the lack of common soil classification terminology creates a Tower-of-Babel situation. Most of them give up trying to relate their results to other areas and return home to work on their own problems. The lack of common language has thus impeded the transfer of many important management findings from one area to another. This need has been recognized by soil fertility specialists such as Crowther (1949) and Mukherjee (1963). Crowther went as far as to state that soil fertility problems in the tropics are essentially problems of soil classification.

The purpose of this chapter is to attempt to clarify this communication problem by describing the different terminology used and the geographical distribution of soils in the tropics according to the most recent information. Just as it is important to describe the tropical environment, it is relevant to know the principal characteristics of tropical soils. The genesis of tropical soils will not be discussed in depth because it is outside of this book's objectives. Readers interested in genesis are referred to books by Mohr et al. (1972) and Buol et al. (1973).

SOIL CLASSIFICATION SYSTEMS USED IN THE TROPICS

The strong genetic bias of the early Russian and American pedologists led to the "zonality" concept. Zonal soils are those that have properties according to what genesis theories dictate soils should have, in terms of climate, vegetation, topography, parent material, and age. The simplistic concept of a hot, humid climate, lush vegetation, old parent materials, and old landscapes resulted in a zonal tropical soil concept analogous to that of lateritic soils, which persists in some modern systems. The present classification system used in the Soviet Union has only three categories for the tropics: "Tropical humid savanna and forest soils," "Tropical dry forest and savanna soils," and "Tropical desert soils," (Ivanova, 1956). Some subdivisions of these categories have been proposed (Gerasimov, 1973).

Three major classification systems with strong genetic bias have been and are still in use in tropical regions: the 1938 USDA System, primarily in tropical America and Asia, and the French (ORSTOM) and the Belgian (INEAC) systems, primarily in Africa.

The USDA System: 1938 to 1960

The 1938 U. S. Department of Agriculture System developed by Baldwin et al. (1938) and modified by Thorp and Smith (1949) considers only one suborder for the tropics within the order "Zonal soils": "Lateritic soils of forested warm-temperate and tropical regions." Within it, five great soil groups are recognized: Lateritic, Reddish Brown Lateritic, Yellowish Brown Lateritic, Red Podzolic, and Yellow Podzolic soils. Significantly, this suborder brings together soils from the tropics and southeastern United States. They were later separated in the 1949 modification. Within the "Intrazonal" order, the USDA System recognizes the suborders "Rendzina" and "Ground-Water Laterites" as present in the tropics. Azonal soils, such as the Alluvial soils, Lithosols, and Regosols, are known to exist in all parts of the world. Thorp and Smith expressed their desire to fit more tropical data into the system but were unable to do so.

A second modification of this system was made by Kellogg (1949, 1950) and by Cline et al. (1955), the latter in the report of the soil survey of Hawaii. The suborder "Lateritic soils" was replaced by "Latosols." Latosols were defined as the predominant soils of the humid and subhumid tropics, occuring at elevations from 0 to 2000 m with annual rainfall from 250 to 10,000 mm and under a great variety of vegetation. Laterization was still considered the main process of soil formation, perhaps under a previous climate. These soils are red or reddish brown in color without a strong textural illuvial horizon. Although clays, their high degree of aggregation makes them feel coarser. They have low silica–sesquioxide ratios (1:5 to 1:15) because of either silica depletion or the basic composition of the Hawaiian volcanic parent materials.

Four subgroups were recognized by Cline et al. Low Humic Latosols are found in ustic areas with moderate annual rainfall. Their main characteristic is high base saturation. Humic Latosols occur in forested areas with udic soil moisture regimes. They have higher organic matter contents and lower base saturation than the Low Humic Latosols. Also, clay increases with depth in the Humic Latosols. Neither of these two subgroups hardens irreversibly upon drying. Hydrol Humic Latosols are continuously wet soils with amorphous clay minerals, which harden irreversibly upon drying. These soils are not extensive. Humic Ferruginous Latosols are the oldest and supposedly represent the end product of laterization, with a concentration of iron and aluminum in the subsoil. In places, massive crusts are found. A fifth subgroup, Aluminous Ferruginous Latosols, was later added by Hawaiian scientists to categorize soils high in both aluminum and iron.

Cline and his coworkers recognized the presence of the following intrazonal soils in the tropics: Latosolic Brown Forest, Grey Hydromorphic, and Dark Magnesium Clays (Vertisols).

This nomenclature is of particular significance because of its use in other tropical areas and the excellent fertility work conducted in Hawaii when this system was in use. The Low Humic Latosols now belong to the Ustox and Torrox suborders of Oxisols. Most of the Humic Latosols are Humults. The Hydrol Humic Latosols are Hydrandepts; the Humic Ferruginous

Table 2.1   Orders of the Soil Taxonomy System in Relation to Great Soil Groups of the Previous USDA Schemes and Other Classification Systems

Order	Former Great Soil Groups Included
Entisols	Azonal soils, some Low Humic Gley, Lithosols, Regosols
Vertisols	Grumusols, Tropical Dark Clays, Regur, Black Cotton Soils, Dark Magnesium Clays
Inceptisols	Andosols, Hydrol Humic Latosols, Sol Brun Acide, some Brown Forest, Low Humic Gley, Humic Gley
Aridisols	Desert, Reddish Desert, Sirozem, Solonchak, some Brown and Reddish Brown soils, associated Solonetz
Mollisols	Chestnut, Chernozem, Brunizem, Rendzina, some Brown Forest, Brown, associated Humic Gley and Solonetz
Spodosols	Podzols, Brown Podzolic, Ground-Water Podzols
Alfisols	Gray-Brown Podzolic, Gray Wooded, Noncalcic Brown, Degraded Chernozem, associated Planosols and Half Bog, some Terra Roxa Estruturada and eutric Red-Yellow Podzolics, some Latosols and Lateritic soils.
Ultisols	Red Yellow Podzolic, Reddish Brown Lateritic, Humic Latosols, associated Planosols, and some Half Bogs, Latosols, Lateritic soils, Terra Roxa, and Ground-Water Laterites
Oxisols	Low Humic Latosols, Humic Ferruginous Latosols, Aluminous Ferruginous Latosols, some Latosols, Lateritic soils, Terra Roxa Legítima, Ground-Water Laterites
Histosols	Bog soils, Organic soils, Peat, Muck

Source: Adapted from Soil Survey Staff (1960), Thorp and Smith (1949), and Cline et al. (1955).

Latosols and Aluminum Ferruginous Latosols are Humoxes. Although the Hawaiian modification was not intended to be used elsewhere, it received widespread attention (Kellogg, 1950). Its limitations lie in the narrow range of parent materials found in these islands.

Table 2.1 shows the equivalent terminology between the USDA System and the Soil Taxonomy System orders.

The French System (ORSTOM)

The system developed by the Office de la Recherche Scientifique et Technique d'Outre-Mer is widely used in French-speaking Africa (Aubert, 1968). An understanding of it is essential for evaluating the voluminous literature published in French, which unfortunately is commonly ignored by English-speaking workers. This system also has a strong genetic bias very similar to that underlying the zonality concept of the American and Russian workers. Soils are separated by climatic breaks and vaguely defined criteria such as "slightly weathered" (Sols peu évolués).

A much wider range of soils is recognized in the tropics under this system than in the USDA one. Aubert and Tavernier (1972) list 7 of their 11 higher categories as occurring in the tropics. Classes I (Sols Minéraux Bruts) and II (Sols Peu Évolués) include most of what are now called "Entisols." Class IV (Andosols) and Class VI (Sols Brunifiés des Pays Tropicaux) are readily translated into other systems.

Class IX (Sols Ferrugineux Tropicaux: Ferruginous tropical soils) and Class X (Sols Ferralitiques: Ferralitic soils) deserve special attention.. The Ferruginous soils are high-base-status reddish soils and probably belong to the orders Inceptisols and Alfisol in the Soil Taxonomy System. The Ferralitic soils are those with similar color but lower base saturation. They include most of the Ulsitols and Oxisols, but many Alfisols as well. The distinction of base status at the highest categorical level is a very practical one.

Class XI (Sols Hydromorphes) consists of all the poorly drained soils at the highest categorical level.

Aubert and Tavernier (1972) have developed a correlation table for the French and the U. S. Soil Taxonomy systems (Table 2.2). These relationships are approximate, however, because the French system does not use quantitative criteria for its limits.

The Belgian System (INEAC)

The Institut National pour l'Etude Agronomique du Congo developed its own classification system in the course of its work in Africa (Sys et al. 1961). Although a strong genetic bias is again present, there is a higher degree of quantifiable criteria than in the previous classification systems. Seven orders are recognized: Recent Tropical soils, Brown Tropical soils, Black Tropical soils, Recent Textural soils, Podzols, and Kaolisols. The first five are easily correlated with other systems. The Kaolisols are soils having kaolinitic clay mineralogy without an illuvial horizon. They are subdivided into three great soil groups: Ferrisols, those having indications of argillic horizons and the presence of weatherable minerals in sand fractions; Ferralsols, those without argillic horizons and with no more than traces of weatherable minerals; and Arenoferrals, those with less than 20 percent clay. An approximate correlation of the INEAC System with the French and Soil Taxonomy systems is shown in Table 2.3. Correlation between these systems is difficult because of conflicting criteria related to the argillic horizon concept.

*Table 2.2   Approximate Correlation between the French Soil Classification
System and the Soil Taxonomy of the United States, with Special
Reference to the Tropics*

French Classification	U.S. Taxonomy (Orders, Suborders, or Great Groups)
I. Sols Minéraux Bruts (as far as recognized as soil)	Orthents, Psamments, Fluvents
II. Sols Peu Evolués	
Humifères	Orthents, Humitropepts
A allophanes	Andepts, Eutrandepts, Vitrandepts
Non-climatiques	Orthents, Fluvents, Psamments, Tropepts
IV. Andosols	Andepts
Saturés	Eutrandepts
Désaturés	Hydrandepts, Dystrandepts
VII. Sols Brunifiés des pays Tropicaux	Eutropepts, Tropudalfs
IX. Sols Ferrugineux Tropicaux	
Peu Lessivés	Ustropepts
Lessivés	Haplustalfs, Paleustalfs, Plinthustalfs
Appauvris à pseudogley	Tropaqualfs
X. Sols Ferralitiques	
Fiablement désaturés	
Typiques	Eutrorthox, Eutrustox
Appauvris, remaniés	Alfic Eutrustox
Rajeunis	Ustropepts, Eutropepts
Moyennement désaturés	
Typiques	Haplorthox, Haplustox
Humifères	Haplohumox, Sombrihumox

The Brazilian System

Brazilian pedologists have further subdivided the well-drained Latosols of
the U. S. Department of Agriculture System into more quantitative groups
and have conserved the other U. S. Department of Agriculture units found
in the American tropics (Bennema and Camargo, 1964; Costa de Lemos,
1968). Their concepts of a latosolic B horizon and the textural B horizon
are practically identical to the oxic and argillic horizons of the Soil
Taxonomy System. The Brazilian Latosols, therefore, correspond to the
United States Oxisols. The 10 divisions at the highest categorical order are
closely associated with the U. S. Soil Taxonomy terminology. At lower
categorical levels the Brazilians place special emphasis on color, base satu-

Table 2.2 (*Continued*)

French Classification	U.S. Taxonomy (Orders, Suborders, or Great Groups)
Appauvris	Ultic and Alfic Haplorthox
Remaniés	Oxic subgroups of Udults, Haplorthox, Haplustox
Rajeunis	Typic Dystropepts and Oxic Dystropepts
Fortement désaturés	
Typiques	Haplorthox, Acrorthox, Oxic Psammentic Dystropepts
Humifères	Haplohumox, Acrohumox, Sombrihumox
Appauvris	Ultic subgroups of Haplorthox
Remaniés	Haplorthox, Acrorthox
Rajeunis	Oxic Dystropepts
Lessivés	Paleudults, Oxic Tropudults, Oxic Rhodudults

XI. Sols Hydromorphes (with the exception of the Sols Hydromorphes Organiques et Moyennement Organiques)

Minéraux ou peu humifères	
À gley	Tropaquents, Tropaquepts
Lessivés	Tropaqualfs, Tropaquults
À pseudogley	Aquic Subgroups of Tropudalfs and Tropudults
À accumulation de fer en carapace ou cuirasse	Petroferric subgroups of Aquox, Aquults and Aquepts

Source: Aubert and Tavernier (1972).

ration, and vegetation. An approximate correlation between the Brazilian system and others is shown in Table 2.4.

The U. S. Soil Taxonomy

The mounting evidence against the usefulness of genetic theories in practical classification gradually eroded the confidence of United States scientists in the 1938 system and its modifications. Soils considered zonal for the northern United States, such as Podzols and Chernozems, were found in the tropics, where theories dictated that they should not exist. Likewise, lateritic soils were found in the United States. Many other,

Table 2.3   Correlation of the Belgian Soil Classification System Used in the Congo with the French and U.S. Soil Taxonomy Systems

Belgian (INEAC)	French (ORSTOM)	U.S. (Soil Taxonomy)
Recent Tropical Soils	*Sols Peu Evoulés*	Entisols
Nonhydromorphic	D'aport modal	Fluvents
Hydromorphic	D'aport hydromorphique	Aquents
Brown Tropical Soils	*Sols Bruns Eutrophes Tropicaux*	Eutropepts
Black Tropical Clays	*Vertisols*	Vertisols
Recent Textural Soils	*Sols Halomorphes*	Natrustalfs
Solonetz	A alcali lessivés	
Podzols	*Podzols à Alios*	Aquods
Kaolisols	*Sols Ferralitiques and*	Oxisols, Ultisols,
Ferrisols	*Sols Ferrugineaux*	Alfisols, Entisols
Hydroferrisols	Sols hydromorphes mineraux à gley de surface	Aquepts, Aquults
Hygroferrisols	Sols fiablement ferralitique	Dystropepts, Tropudults
Intergrading to Brown and Recent Tropical Soils		
Typic and intergrading to hygroferralsols	Sols ferralitiques typiques	Orthox; oxic subgroups of Udults
Hygro-xero ferrisols		
Intergrading to Brown and Recent Tropical Soils	Sols fiablement fer-ralitiques	Ustropepts, Tropustults
Typic and intergrading to ferralsols	Sols ferralitiques typiques rouges et jauns	Ustox; oxic subgroups of Ustults
Humiferous fer-risols	Sols ferralitiques humiféres d'altitude	Humox, Humults
Xeroferrisols	Sols ferrugineaux tropi-caux lessivés	Ustalfs
Ferralsols		
Hygroferralsols	Sols ferralitiques lessivés en argille	Orthox, Tropudults
Typic, with plinthite		
Hygroxeroferralsols	Sols ferralitiques lessivés modal	Ustox
-with plinthite	En argille	Tropustults
Arenoferralsols	Sols ferraltique lessivés podzoliques	Oxic quartzi-psamments

Source: **Adapted from Jurion and Henry (1969), Sys et al. (1961), Aubert (1968).**

similar examples forced the placement of similar soils into different orders (Zonal vs. Intrazonal). This led to the development of a completely new soil classification system based on morphological properties that can be quantified by accepted techniques. This new system ended the grouping of soils according to *what they should be* and concentrated on *what they are*.

After about 10 years of development, the first published version, called the Seventh Approximation (Soil Survey Staff, 1960), was presented at the International Soil Science Congress in Madison, Wisconsin. Since then it has been supplemented by several revisions, but extensive distribution has not been made except in the United States. In each revision more attention has been paid to soils of tropical regions. Many tropical scientists have cooperated in this venture. A good description of this fairly complex but more objective system appears in Buol et al. (1973).

Although in use for less than 15 years, the U.S. Soil Toxonomy has made a significant contribution to the understanding of tropical soils by essentially eliminating the genetic bias. Tropical and temperate soils are separated at the third or the fifth categorical level, mainly on the basis of temperature regimes. For example, many soils of the Amazon Basin and North Carolina are grouped in the same categories down to the family level, where they are separated by their temperature regimes.

All 10 orders are found in the tropics. Thirty-nine suborders and 136 great soil groups can occur in the tropics, according to calculations by Cline (1972), based on the temperature and moisture definitions of these categories. Table 2.5 provides a simplified definition of the main suborders and great groups found in the tropics. This table is an interpretation for the purposes of this book and does not include all the classification criteria. The complete definitions are given in the U.S. Soil Taxonomy (Soil Survey Staff, 1970).

The Soil Taxonomy terminology will be used in this book because its quantitative criteria makes it the most relevant classification system for management interpretations.

The FAO Legend

A soil map of the world is in the process of being published by a joint effort of two United Nations agencies, FAO and UNESCO. A legend was developed by Dudal (1968, 1970) to correlate all units of the various soil maps in the world and to obtain a worldwide inventory of soil resources with a common legend. The soil units have two levels, roughly equivalent to the suborder and great soil group levels of the Soil Taxonomy. The defini-

Table 2.4 Approximate Correlation of the Brazilian Soil Classification System with the U.S. Soil Taxonomy, the French System, and the FAO Legend

Brazilian System	U.S. Soil Taxonomy	French System	FAO Legend
Latosols (soils with latosolic B horizon with 6.5 meq/100 g of CEC of clay)	Oxisols	Sols ferralitiques fortement desatures, typiques ou humifères	Ferralsols
Latosol Vermelho Escuro (Dark Red Latosol)	Ustox or Orthox	Sols ferralitiques fortement desatures typiques ou humifères	Orthic or Acric Ferralsols
Latosol Vermelho Amarelo (Red-Yellow Latosol)	Ustox or Orthox	Sols ferraltiques fortement desatures typiques ou humifères	Orthic or Acric Ferralsols
Latosol Amarelo (Yellow Latosol)	Ustox or Orthox	Sols ferralitiques fortement desatures typiques ou humifères	Xanthic Ferralsols
Latosol Roxo or Terra Roxa Legítima (Dusky Red Latosol)	Eutrustox or Eutrorthox	Sols ferralitiques fortement desatures typiques ou humifères derivés de basalte	Rhodic Ferralsols

Podzólico Vermelho Amarelo (Red-Yellow Podzolic)	Ultisols	Sols ferralitiques moyennement désaturés eluvies	Acrisols Dystric Nitosols
Podzólico Vermelho Amarelo equivalente eutrófico (Eutrophic Red-Yellow Podzolic)	Alfisols	Sols ferrugineux, tropicaux lessivés	Luvisols Eutric Nitosols
Terra Roxa Estruturada	Alfisol	Sols ferrugineux tropicaux lessivés	Luvisols Eutric Nitosols
Red and Yellow Sands	Psamments	Sols ferralitiques moyenement on fortement desaturés de texture sableuse	Ferralic Arenosols
Podzols	Spodosols	Podzols	Podzols
Grumusols	Vertisols	Vertisols	Vertisols
Soils with incipient B horizon	Inceptisols	(Several)	Cambisols
Soils with natric B horizon	Aridisols	Sols halomorphes	Solonchaks
Regosols	Entisols	Regosols	Regosols
Soils with hardpan	Various	Planosols	Planosols
Other hydromorphic soils	Various	Sols hydromorphes	Gleysols

Source: Adapted from Van Wambeke (1971), Beinroth (1975), Costa de Lemos (1968), Aubert (1968), and M. N. Camargo (personal communication).

Table 2.5 Simplified Definitions of the Soil Taxonomy Orders, Suborders, and Great Groups Found in the Tropics for Management Purposes
For actual classification, the complete definitions as they appear in the U.S. Soil Taxonomy are needed

Order	Suborder	Great Group

Oxisols: Soils with oxic horizons (< 16 meq/100 g clay), consisting of mixtures of kaolinite, iron oxides, and quartz; low in weatherable minerals. Usually deep, well-drained red or yellow soils, excellent granular structure, very low fertility, uniform properties with depth.

	Orthox:	Oxisols with udic soil moisture regimes
		Haplorthox: simple
		Eutrorthox: high base saturation
		Acrorthox: very low base saturation
		Gibbsiorthox: gibbsite dominant
	Ustox:	Oxisols with ustic soil moisture regimes
		Haplustox: simple
		Eutrustox: high base saturation
		Acrustox: very low base saturation
	Torrox:	Oxisols with aridic soil moisture regimes
	Humox:	Oxisols with high organic matter (usually in highlands)
		Haplohumox: simple
		Acrohumox: very low base saturation
		Gibbsihumox: gibbsite dominant
	Aquox:	Oxisols with aquic soil moisture regimes
		Ochraquox: light-colored A horizon
		Umbraquox: dark-colored A horizon
		Plinthaquox: with plinthite
		Gibbsiaquox: gibbsite dominant

Ultisols: Soils with an argillic horizon (20% increase in clay content in the control section) with less than 35% base saturation in the control section. Usually deep, well-drained red or yellow soils, higher in weatherable minerals than Oxisols, with less desirable physical properties, and relatively low native fertility. Ultisols may have oxic horizons above or below the argillic.

	Udults:	Ultisols with udic soil moisture regimes
		Tropudults: simple
		Paleudults: very deep argillic horizon
		Rhodudults: dusky red, high in oxides
		Plinthudults: with plinthite
		Fragiudults: with fragipans

Table 2.5 (*Continued*)

Order	Suborder	Great Group
	Ustults:	Ultisols with ustic soil moisture regimes Haplustults: simple Paleustults: deep argillic horizon Rhodustults: dusky red, high in oxides Plinthustults: with plinthite
	Humults:	Ultisols high in organic matter (usually in highlands) Tropohumults: simple Palehumults: deep argillic horizon Plinthohumults: with plinthite
	Aquults:	Ultisols with aquic soil moisture regimes Tropaquults: simple Paleaquults: deep argillic horizon Pinthaquults: with plinthite Fragaquults: with fragipan Albaquults: light-colored A horizon

Alfisols: Soils with an argillic horizon with more than 35% base saturation. Similar to Ultisols except for considerably higher native fertility.

	Udalfs:	Alfisols with udic soil moisture regimes Tropudalfs: simple
	Ustalfs:	Alfisols with ustic soil moisture regimes Haplustalfs: simple Paleustalfs: deep argillic horizon Rhodustalfs: dusky red, high in oxides Plinthustalfs: with plinthite Natrustalfs: sodic horizon Durustalfs: with duripan
	Aqualfs:	Alfisols with aquic soil moisture regimes Tropaqualfs: simple Plinthaqualfs: with plinthite Natraqualfs: sodic horizon Duraqualfs: with duripan

Aridisols: Soils of aridic moisture regimes, with horizon diferentiation.

	Argids:	Aridisols with argillic horizons Haplargids: simple Paleargids: deep argillic horizon Natrargids: sodic horizon Durargids: with duripan

Table 2.5   (*Continued*)

Order	Suborder	Great Group
	Orthids:	Aridisols without an argillic horizon Salorthids: salic horizon Paleorthids: deep argillic horizon Calciorthids: calcareous horizon Gypsiorthids: gypsic horizon Camborthid: cambic horizon Durorthids: with duripan

Inceptisols: Young soils with cambic horizon but no other diagnostic horizons.

	Andepts:	Inceptisols derived from volcanic materials Cryandepts: cold andepts Vitrandepts: high in volcanic glass Dystrandepts: low base saturation Eutrandepts: high base saturation Hydrandepts: well drained with high water content Durandepts: with duripan
	Aquepts:	Inceptisols with aquic soil moisture regimes Tropaquepts: simple Andaquepts: affected by volcanic ash Halaquepts: saline Sulfaquepts: acid sulfate soils or cat clays Plinthaquepts: with plinthite
	Tropepts:	Other tropical Inceptisols Dystropepts: low base status Eutropepts: high base status Ustropepts: with ustic soil moisture regimes Humitropepts: high in organic matter (highlands)

Entisols: Soils of such slight and recent development that only an ochric (yellowish) epipedon or simple man-made horizons have formed.

| | *Aquents:* | Entisols with aquic moisture regimes
Tropaquents: simple
Fluvaquents: alluvial
Hydraquents: high moisture content, thixotropic
Sulfaquents: acid sulfate soils or cat clays |
| | *Fluvents:* | Entisols of recent alluvial origin
Tropofluvents: simple
Ustifluvents: with ustic moisture regimes
Torrifluvents: with aridic moisture regimes |

Table 2.5 (*Continued*)

Order	Suborder	Great Group
	Psamments:	Sandy Entisols Tropopsamments: simple Ustipsamment: with ustic soil moisture regimes Torripsamments: with aridic soil moisture regimes Quartzipsamments: with mainly quartz sand
	Orthents:	Other tropical Entisols Troporthents: simple Ustorthents: with ustic soil moisture regimes Torriorthents: with aridic soil moisture regimes

Vertisols: Heavy, cracking clayey soils with more than 35% clay and >50% of 2:1 minerals in clay fractions. Usually shrink and swell with changes in moisture contents, have gilgai microreleif and slickensides on peds.

Order	Suborder	Great Group
	Usterts:	Vertisols with ustic soil moisture regimes Chromusterts: light colored Pellusterts: dark colored
	Uderts:	Vertisols with udic soil moisture regimes
	Torrerts:	Vertisols with aridic soil moisture regimes

Mollisols: Soils with a mollic epipedon (i.e., high in organic matter, soft when dry and >50% base saturation).

Order	Suborder	Great Group
	Rendolls:	Mollisols over limestone, formerly called Rendzinas
	Ustolls:	Mollisols with ustic moisture regimes Haplustolls: typical Argiustolls: argillic horizon Paleustolls: deep argillic horizon Calciustolls: calcitic horizon Durustolls: with duripan Natrustolls: sodic horizon
	Aquolls:	Mollisols with aquic moisture regimes Haplaquolls: typical Argiaquolls: argillic horizon Calciaquolls: calcic horizon Duraquolls: with duripan Natraquolls: sodic horizon

Table 2.5   (*Continued*)

Order . Suborder	Great Group

Spodosols: Soils with a spodic horizon (of iron and organic matter accumulation), usually developed on sandy materials. Equivalent to Podzols in all other classification systems.

Aquods:	Spodosols with aquic moisture regimes
	Tropaquods: typical
	Duraquods: with hardpan
Humods:	Spodosols high in organic matter
	Tropohumods: typical
Orthods:	Other tropical Spodosols
	Troporthods: typical

Histosols: Organic soils with a histic epipedon (>20% organic matter.) All have a tropo-great group.

Fibrists:	Over two-thirds fiber
Folists:	Not saturated with water (foggy tops)
Hemists:	Between one-third and two-thirds fiber
Saprists:	Less than one-half fiber

tions are based on diagnostic horizons and quantifiable criteria similar to those of the U.S. System, but the nomenclature has been drawn from a number of national systems in a successful exercise of international diplomacy. An approximate correlation between the FAO legend, the Soil Taxonomy, and the French System appears in Table 2.6. The FAO staff consider this not a classification system, but a two-level legend of map units. The FAO legend correlates fairly well with the Soil Taxonomy nomenclature at the great group level (Buol, 1973; Buol et al. 1973). When the entire FAO-UNESCO world soil map is published, it is likely that its terminology will become the international standard for generalized comparisons of soil properties. Since it contains only 104 subunits, more detailed comparisons of soil properties needed for management purposes can be made with the Soil Taxonomy.

GEOGRAPHICAL DISTRIBUTION

Since 1967 the Soil Geography Unit of the U.S. Soil Conservation Service has correlated the soils of the world in a very generalized way. An estimate

Table 2.6   Approximate Correlation between the FAO, U.S. Soil Taxonomy, and the French Soil Classification Systems with special reference for the Tropics

FAO Legend	U.S. Soil Taxonomy	French Classification
Fluvisols	Fluvents	Sols minéraux bruts et sols peu évolués d'apport alluvial et colluvial
Regosols	Psamments	Sols minéraux bruts et sols peu évolués d'apport éolien
Arenosols		
Ferralic	Oxic Quartzipsamments	Sols ferralitiques moyennement or fortement désaturés, à texture sableuse
Gleysols		
Eutric and Dystric	Tropaquepts	Sols hydromorphes humifères à gley
Humic	Humaquepts	Sols humiques à gley
Plinthic	Plinthaquepts	Sols hydromorphes à accumulation de fer en carapace or cuirasse
Andosols	Andepts	Andosols
Planosols	Paleudalfs and Paleustalfs	Sols ferrugineux tropicaux lessivés (pro parte)
Cambisols		
Dystric	Dystropepts	Sols ferralitiques fortement et moyennement désaturés, rajeunis (pro parte)
Eutric	Eutropepts	Sols ferrugineux tropicaux (non lessivés)
		Sols ferralitiques faiblement désaturés, rajeunis
Humic	Humitropepts	Sols ferralitiques fortement et moyennement désaturés, humifères, rajeunis
Luvisols	Alfisols	Sols ferrugineux tropicaux lessivés
Acrisols	Ultisols	Sols ferralitiques fortement désaturés
Ferralsols	Oxisols	Sols ferralitiques
Lithosols	Lithic subgroups	Lithosols et sols lithiques

Source: Aubert and Tavernier (1972).

of the distribution of soils in the tropics, using the 1938 USDA nomenclature, was published by the President's Science Advisory Committee in 1967. Table 2.7 shows this distribution by tropical climatic regions. Highly weathered, leached soils, previously called "Latosols," occupy 51 percent of the tropics, while sandy, shallow, high-base-status, alluvial and moderately leached soils cover the rest.

Table 2.7 Distribution of Soils in the Tropics by Climatic Regions (million ha)

Soil Groups (Soil Taxonomy) equivalents)	Rainy $(9.5-12)^a$	Seasonal $(4.5-9.5)^a$	Dry and Desert $(0-4.5)^a$	Total	Percent of Tropics
1. Highly weathered, leached soils (Oxisols, Ultisols, Alfisols)	920	1540	51	2511	51
2. Dry sands and shallow soils, (Psamments and lithic groups)	80	272	482	834	17
3. Light-colored, base-rich soils (Aridisols and aridic groups)	0	103	582	685	14
4. Alluvial soils (Aquepts, Fluvents, and others)	146	192	28	366	8
5. Dark-colored, base-rich soils (Vertisols, Mollisols)	24	174	93	291	6
6. Moderately weathered and leached soils (Andepts, Tropepts, and others)	5	122	70	207	4
Total area	1175	2403	1316	4896	100
Percent of tropics	24	49	27	100	

Source: Adapted from the President's Science Advisory Committee (1967).
^{a} Numbers in parentheses refer to number of months with an average rainfall greater than 100 mm.

A tentative map of tropical soils published by Aubert and Tavernier in 1972 shows their distribution at the suborder level of the Soil Taxonomy (Fig. 2.1). Although this is a very generalized map and subject to substantial modifications, it represents the first attempt to quantify the distribution of tropical soils. Calculations based on this map (Table 2.8) show that Oxisols occupy 22 percent of the tropics, mainly in the Amazon and the Cerrado in South America and in Central Africa. No Oxisol areas are noted for tropical Asia at this level of generalization.

After Oxisols, Aridisols (desert soils) are the second most abundant soil order, covering 18 percent of the tropics. Aridisols are found primarily in the African deserts with scattered areas in tropical America and Asia.

Alfisols are the third most common order, covering 16 percent of the tropics. Most of the tropical Alfisols occur in areas previously mapped as Latosols, such as most of West Africa, India, and Sri Lanka, where most of the literature on laterites originates. The presence of plinthite (a quantitative version of "laterite") in many West African subsoils places many of these soils in the Plinthustalf great group (i.e., Alfisols with ustic soil moisture regimes and plinthite at a certain depth). The management properties are completely different from those of other Alfisols because the erosion of the fairly coarse topsoils may expose plinthite or plinthite-gravel layers. Extensive areas of Alfisols with high native fertility and no major management limitations are found in other parts of Africa, as well as in Asia and tropical America.

Ultisols are the fourth most common order, covering 11 percent of the tropics. They are abundant in many areas of tropical America and Africa and also seem to be the dominant upland soils of Southeast Asia. Most of these soils were formerly mapped as Latosols.

Inceptisols and Entisols cover each about 8 percent of the tropics. Their areal extensiveness is undoubtedly greater because they appear in the younger geomorphic surfaces of most other broad mapping units. Many soils previously considered Latosols are now mapped as Tropepts, particularly in Cuba, Colombia, Brazil, and India. Large areas of sandy soils (Psamments) are found in many tropical areas. When red or yellowish in color, these soils were formerly called "sandy Latosols."

Vertisols and Mollisols cover 2 and 1 percent, respectively, of the tropics on this map. Large areas of Vertisols are found in the Sudan, Ethiopia, India, and Java. Mollisols are extensive in Mexico, Paraguay, and northern India. Both orders are also found within the other mapping units.

The last two remaining orders, Spodosols and Histosols, do not occupy large enough areas in the tropics to appear as mapping units at a scale of 1:50 million. They are present and are locally important, however, in

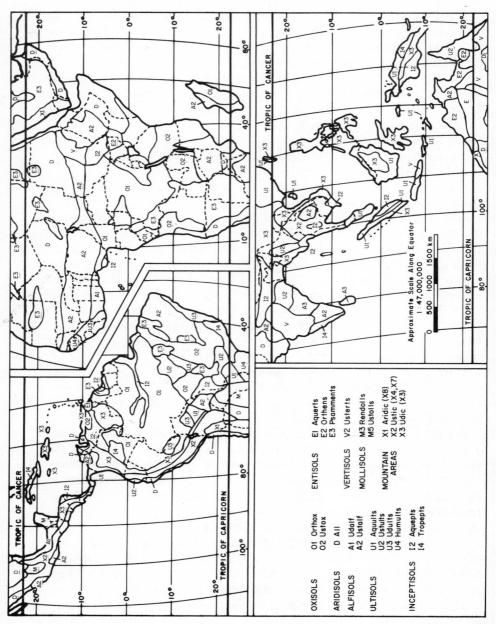

Fig. 2.1 Soils of the tropics. *Source: After Aubert and Tavernier (1972).*

OXISOLS   O1 Orthox   E1 Aquerts
   O2 Ustox   E2 Orthens
   E3 Psamments

ARIDISOLS   D Aii

ALFISOLS   A1 Udalf   VERTISOLS   V2 Usterts
   A2 Ustalf

ULTISOLS   U1 Aquults   MOLLISOLS   M3 Rendolls
   U2 Ustults   M5 Ustolls
   U3 Udults
   U4 Humults   MOUNTAIN   X1 Aridic (X8)
   AREAS   X2 Ustic (X4,X7)
INCEPTISOLS   I2 Aquepts   X3 Udic (X3)
   I4 Tropepts

Approximate Scale Along Equator
1:47,000,000

0   500   1000   1500 km

Table 2.8   Approximate Extent of Major Soil Suborders in the Tropics (million ha)

Order	Suborder	Africa	America	Asia	Total Area	Percent
Oxisols	Orthox	370	380	0	750	15.0
	Ustox	180	170	0	350	7.5
		550	550	0	1100	22.5
Aridisols	All	840	50	10	900	18.4
Alfisols	Ustalfs	525	135	100	760	15.4
	Udalfs	25	15	0	40	0.8
		550	150	100	800	16.2
Ultisols	Aquults	0	40	0	40	1.0
	Ustults	15	35	50	100	2.2
	Udults	85	125	200	410	8.2
		100	200	250	550	11.2
Inceptisols	Aquepts	70	145	70	285	6.0
	Tropepts	0	75	40	115	2.3
		70	225	110	400	8.3
Entisols	Psamments	300	90	0	390	8.0
	Aquents	0	10	0	10	0.2
		300	100	0	400	8.2
Vertisols	Usterts	40	0	60	100	2.0
Mollisols	All	0	50	0	50	1.0
"Mountain areas"		0	350	250	600	12.2
Total		2450	1670	780	4900	100.0

Source: Calculated by M. Drosdoff, Cornell University, on the basis of Aubert and Tavernier's (1972) map.

smaller areas. Several references cited at the end of this chapter describe the properties of these soils.

The "soils of the tropics" map does not differentiate soils in mountain areas, thus excluding 12 percent of the tropics in the above generalization. Important groups such as Andepts (volcanic ash soils) are excluded in this manner.

In spite of its limitations, this map clearly shows that, even at a high level of generalization, tropical soils are not uniform. It also shows that only a fifth of the tropics is covered by Oxisols. Soils previously called "Latosols" are now mapped as Oxisols, Alfisols, Ultisols, and, to a lesser extent, Inceptisols and Entisols.

A further refinement of these relationships can now be calculated from the data presented in the first publication of the FAO–UNESCO soil map of the world for South America (1971). Table 2.9 shows the distribution of soil suborders for the tropical portion of this continent as calculated from this publication. For tropical South America, Oxisols (Ferralsols) cover about 45 percent of the area, followed by Ultisols with 19 percent, Alfisols with 12 percent, Entisols and Inceptisols with 8 percent, Mollisols with 4

Table 2.9   *Approximate Distribution of Soils in Tropical South America*

Order	Suborder	Area (million ha)	Percent
Oxisols	All	636	45.3
Ultisols	Aquults	48	
	Udults, Ustults, and Humults	220	
		268	19.1
Alfisols	Aqualfs	7	
	Udalfs and Ustalfs	164	
		171	12.2
Entisols	Fluvents	19	
	Quartzipsamments	79	
	Orthents and others	21	
		121	8.6
Inceptisols	Aquepts	1	
	Andepts	32	
	Tropets	81	
		115	8.2
Mollisols	All	54	3.8
Aridisols	All	26	1.9
Vertisols	All	9	0.7
Histosols	All	1	0.1
Total		1404	100.0

Source: Calculated from the FAO–UNESCO *Soil Map of the World: South America* and converted to Soil Taxonomy equivalents.

percent, Aridisols with 2 percent, and the rest with less than 1 percent. The dominant soils in the Amazon region are classified as yellow Oxisols (Xanthic Ferralsols) on this map. Recent information suggests, however, that Ultisols rather than Oxisols are the dominant soil order for Amazon Basin areas outside of the influence of the Guyana and Brazilian shields (Benavides, 1973; Sanchez and Buol, 1974). This implies that the percentage of Oxisols will be lower and that of Ultisols higher.

SOIL ASSOCIATIONS IN THE LANDSCAPE

The foregoing statements have been made at the highest level of generalization. It should be obvious that substantial variation occurs within such mapping units. The purpose of this section is to define some of the most frequently observed soil associations in certain typical landscapes. Although many others occur, the ones chosen are considered representative of large areas. The relationship between geomorphological surfaces and soils is of great value in predicting where similar soils will occur. Soil–landscape associations have been recognized and studied for a long time in the tropics. For example, the concept of a drainage catena was developed in East Africa by Milne (1935).

Soil Associations in Udic Environments

Soil-forming processes proceed at faster rates in rainy climates than in other climates because of the almost constant downward water movement, the large amounts of biomass added to the soil, and the constantly high temperatures. The predominant soils of udic regimes are Oxisols, Ultisols, Alfisols, and Inceptisols.

Oxisols are associated with very old, stable land surfaces. Examples of these surfaces occur in both udic and ustic environments. The Saint John Peneplain in central Puerto Rico is an excellent example of a place where Oxisols are found (Beinroth, 1972). Other soils are dominant in younger land surfaces. The well-known Nipe series (Typic Acrorthox), considered by many to represent the end product of soil formation, is found only in the remnants of these peneplains in Puerto Rico and Cuba (Bennet and Allison, 1928). Other soils such as Ultisols and Inceptisols occupy the younger land surfaces.

Oxisols, Ultisols, and Inceptisols are frequently intermixed in udic environments. Recent studies by Lepsch and Buol (1974) in São Paulo, Brazil, by Daniels et al. (1973) in Barranquitas, Puerto Rico, and by Beinroth et al.

(1974) in Kauai, Hawaii, indicate that these orders occur in predictable positions in the landscape. Oxisols occupy the older land surfaces, which may be remnants of a previous peneplain or part of steep backslopes in others. Ultisols occupy the slopes below where Oxisols are found. Their argillic horizons apparently formed after the original peneplain was truncated by erosion. Inceptisols occupy the steeper slopes developing on recently exposed rock. A diagram of this appears in Fig. 2.2. Daniels et al. also observed Oxisols in footslopes located in positions that could receive A-horizon material eroded from Oxisols higher up. Consequently, Oxisols can be recent and alluvial in origin if the material was preweathered before deposition.

In areas where the sediments are probably too young for Oxisols to form, as in the Upper Amazon Basin of Peru and Colombia outside of the influence of the older shields, Ultisols are found to be associated with Alfi-

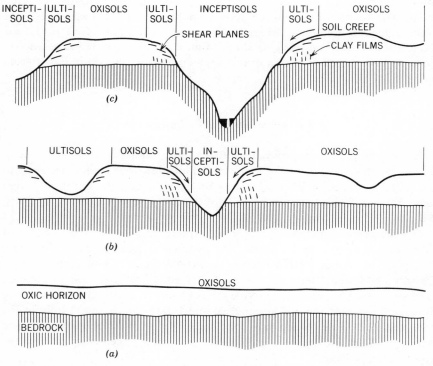

Fig. 2.2 Illustration of the formation of an Oxisol–Ultisol–Inceptisol landscape from an old land surface through dissection in Hawaii. *a* = undissected surface, *b* = moderately dissected, and *c* = strongly dissected *Source:* Beinroth et al. (1974).

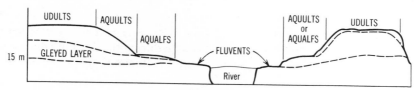

Fig. 2.3 Ultisol–Alfisol–Entisol toposequence found in the Upper Amazon Jungle. Highly weathered acid Udults appear in the flat surfaces, followed by increasing wetness related to the presence of an impermeable gleyed layer. *Source:* Adapted from Tyler (1975).

sols, Inceptisols, and Entisols. Studies by Sanchez and Buol (1974) and Tyler (1975) suggest the following landscape relationships (Fig. 2.3). At the highest topographic positions, the predominant well-drained soils are Udults, which are very acid but have an argillic horizon. In the oldest flat surfaces, the A horizons are sandy, forming Typic Paleudults. Down the slope one finds primarily a drainage catena with increasing wetness and a gleyed mottled horizon that is a mixture of kaolinite and montmorillonite. Tropaquults are found in intermediate positions, and Tropaqualfs at the lower ones. The first terrace and floodplains along the rivers consist of Entisols.

The mottled gray and reddish layer was called "Ground-Water Laterite" by previous workers. Sanchez and Buol (1974) have shown that its mineralogy is primarily 2:1 and 1:1 clay minerals. Although having the color pattern of plinthite, this material does not indurate upon exposure and is obviously not plinthite. Some reports indicate that "Ground-Water Laterites" of similar composition may be quite extensive in the tropics (Marbut and Manifold, 1926; Sombroek, 1966; Benavides, 1973). The presence of these features tends to arouse fear that cultivation will turn the soil into brick. It should be pointed out that cultivation now is, and has been for the past 200 years, carried out on exactly the same type of soils (Ultisols) in the southeastern United States.

Soil Associations in Ustic Environments

The organization of this section into udic and ustic landscapes is merely for the purpose of depicting well-known examples found in each environment. It is not meant to imply that these landscape relations are restricted to such moisture regimes. It is probable that many of the ustic African areas had a udic environment in a previous geological era. In many old landscapes paleoclimate may have had an influence on present soils (D'Hoore, 1956; Ollier, 1959; Mulcahy, 1960).

Ustic environments are characterized by moisture fluctuations in the subsoil during the year. It is reasonable to assume that many weathering processes occur at a substantially slower rate during the dry than during the rainy season. Thus some consequences of these fluctuations can be observed in many landscapes.

Oxisols Landscapes. Parts of the Brazilian and African shields have multicycle landscapes with two or more erosion surfaces (Ruhe, 1954; Feuer, 1956). Figure 2.4 shows a generalized diagram of part of the Cerrado of Brazil close to the capital of Brasilia. This landscape consists of three erosion surfaces, described by Feuer (1956) and Cline and Buol (1973). A first erosion surface believed to be the product of an ancient peneplain marks the skyline at elevations of about 1000 to 1200 m. Locally known as "chapadas," these surfaces have slopes of 0 to 3 percent. The soils are predominantly deep Ustoxes characterized by ultrafine granular structure, high permeability, and extremely low fertility. This surface terminates abruptly at an escarpment of 100 to 150 m depth, followed by a second, gently sloping erosion surface. The escarpments have little soil development in the weathered rock material, and some have plinthite outcrops. The second erosion surface is often several kilometers long with 2 to 8 percent slopes. Ustoxes are also the predominant soils, with inclusions of Histosols and Aquoxes, in poorly drained spots. The third erosion surface consists of 8 to 20 percent valley slopes parallel to narrow floodplains. These soils have higher base status and are probably Ustalfs. The first two surfaces are covered by scrub savanna vegetation indicative of low soil fertility. The third surface consists of semideciduous forests reflecting a high base status.

Very similar landscapes have been described by Ruhe (1954) in eastern

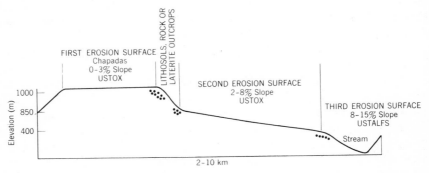

Fig. 2.4 Soil–geomorphology relationships in the Central Plateau of Brazil near Brasilia. Dots represent rock or plinthite outcrops. *Source:* Adapted from Feuer (1956), Cline and Buol (1973), and personal observations.

Zaïre and western Uganda. The formation of plinthite caps at the edges of the first erosion surface is also mentioned in Ruhe's report as well as in others. Sivarajasingham et al. (1962) considered that these caps are formed as follows. During the rainy season iron is reduced and moves laterally along the landscape; ferrous ions are oxidized and then precipitated as ferric oxides and hydroxides at the edges of the erosion surface when in contact with the air. Erosion may lower the soil level enough to form a crust or cap. As the peneplain further erodes, laterite gravels and fragments from the crust roll down into the new erosion surface that is being formed. Stone lines or a thick stratified deposit forms. This layer may be buried by further sediments, thus causing the common stone lines observed in Africa. On low-level plains, soft plinthite may be forming above a water table.

Alfisol Landscapes. Large parts of West Africa consist of Ustalfs with plinthite layers above or below argillic horizons. Classic drainage catenas have been described where the amount of plinthite increases and becomes shallower as drainage becomes progressively worse. Nye (1954) cites one example near Ibadan, Nigeria, on a 5 percent slope landscape underlain by granite–gneiss materials. The soils shown in Fig. 2.5 have a sandy loam A horizon, followed by a clayey argillic horizon that has few iron nodules in the upper parts of the slope. Iron concretions increase in abundance as drainage becomes progressively impeded. In the wetter sites a partly cemented mass of nodules is found at depths of less than 1 m. The International Institute for Tropical Agriculture is located in this general type of soil association. The erosion hazard of such soils is considerable, even in gentle slopes, because of the textural change. Erosion of the sandy A horizon will render the soil practically worthless, as it is very difficult to work the gravelly, clayey B horizon.

Alfisol–Vertisol Landscapes. A very common occurrence in the ustic tropics are the red and black catenas. Red soils (mainly Ustalfs) occupy the better drained sites, and dark, cracking clays (Vertisols) the lower topographic positions. A gradual transition from red to black occurs at intermediate positions. The red soils are predominantly kaolinitic, and the black soils montmorillonitic. Such relationships are common in East Africa and the Sudan (Milne, 1935; Greene, 1947; Radwanski and Ollier, 1959). Since the parent material is probably uniform, movement of soluble silica and bases down the slope and subsequent synthesis of montmorillonite at the lower levels are believed to constitute the main soil-forming process. The fact that montmorillonite can be synthesized in the laboratory in a similar way lends credit to this hypothesis. An example of this relationship is shown in Fig. 2.6.

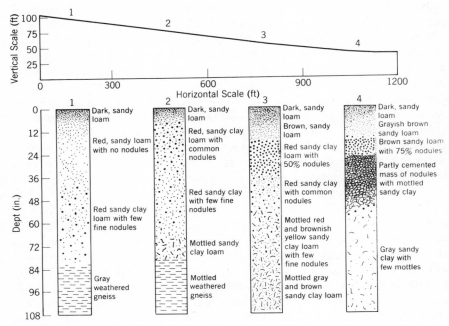

Fig. 2.5 Alfisol catena in Ibadan, Nigeria, with increasing plinthite concretions in the subsoil as drainage gets worse. The well-drained members are Ustalfs; the poorly drained ones, Plinthaqualfs. *Source:* Nye (1954).

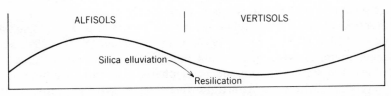

Fig. 2.6 Typical Alfisols (red) and Vertisols (black) catena in ustic areas of Kenya. Arrow indicates the direction of silica movement. In the Sudan, similar Oxisol–Vertisol catenas exist.

80

Soil Associations in the Tropical Highlands

In highland areas affected by volcanic ash, the distribution of soils in the
landscape depends primarily on the age of the ash deposits and the climate.
In humid regions, volcanic ash weathers quickly into allophane, an amorphous aluminum–silicate mixture that forms complexes with organic matter. Such soils form a unique group called Andepts (Andosols). With
further weathering, allophane is converted into kaolinite or halloysite under
well-drained conditions and into montmorillonite in poorly drained ones.
These processes are discussed in great detail by Mohr et al. (1972).
Although Andepts may be located on any topographic position when the
ash is young, a common occurrence in volcanic regions of Southeast Asia is
the association of black soils (Andepts) at higher elevations grading into red
soils (probably Ultisols) with decreasing elevations. This has been observed
in Indonesia across mountain ranges where the volcanic ash is believed to
be of the same age and where enough time has elapsed for weathering to
occur. Dudal and Soepraptohardjo (1960) described the situation depicted
in Fig. 2.7. At elevations above 700 m, Andepts dominate because the high,
constant rainfall regime (udic) and the low temperature regime (isomesic)
favor organic matter accumulation and impede the crystallization of allo-

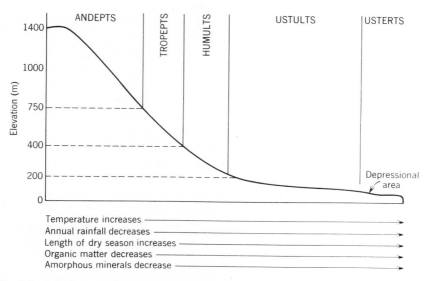

Fig. 2.7 Soil associations along the slope and base of a volcano in Indonesia. Soils developed
from volcanic ash of uniform age. *Source:* Adapted from Dudal and Soepraptohardjo (1960).

phane. At elevations between 350 and 700 m, the predominant soils are called "Latosolic Brown Forests" (probably Tropepts). The change in color from black to brown is associated with lower organic matter, higher clay content, and the presence of kaolinite. This indicates a crystallization of allophane into kaolinite through many intermediate forms. At lower elevations (100 to 350 m) the soils become red with corresponding increases in clay content and decreases in cation exchange capacity. These have been classified as Humic Latosols and are probably Humults. In the lowland areas the moisture regime changes to ustic, and the predominant soils become Ustults. In depressed areas Vertisols may form either by direct transformation of allophane into montmorillonite or by the resilication process mentioned in regard to the East African catena. Similar broad toposequences have been observed in the Philippines.

It is obvious that the above example is limited to areas where elevation and climate dominate the soil-forming processes. Even in areas with relatively uniform parent materials the relationships are more complicated. Tamura et al. (1953) illustrated the distribution of soils in Hawaii according to topography, rainfall, parent material, and age. Figure 2.8, adapted from their report, illustrates a relationship similar to the one described above, with Andepts grading into Humults on the windward side of the island, which has a udic soil moisture regime. On the leeward side the soil moisture regime is ustic or aridic. Differences in the age of parent materials and rainfall produce soils ranging from Oxisols to Aridisols within short distances.

Several typical soil associations in nonvolcanic highlands have been developed by Zamora (1972) for part of the Peruvian Andes. A west-to-east sequence is shown in Fig. 2.9. The westernmost ranges have aridic soil moisture regimes with shallow soils or none at all. In the major highland plateaus such as the Mantaro Valley, a system of valley terraces on calcareous parent material produces the following soil association. At the edges of the valleys, soils shallow to bedrock predominate, particularly the Rendolls or Rendzinas when limestone is the parent material. In the higher terraces, Mollisols (Ustolls and Aquolls) are most common, grading into Fluvents in the valley floodplains. At altitudes above 4000 m, the relatively flat "altiplano" or "punas" are dominated by Cryandepts (cold Andepts) in the well-drained sites and by Histosols in the poorly drained areas. On the eastern flank of the Andes with udic soil moisture regimes, shallow Entisols are found near the high peaks. Below, dystric or eutric Tropepts occur, according to parent materials. At lower elevations, Ultisols occur together with Vertisols in depressed areas with ustic soil moisture regimes, and with Fluvents near the rivers.

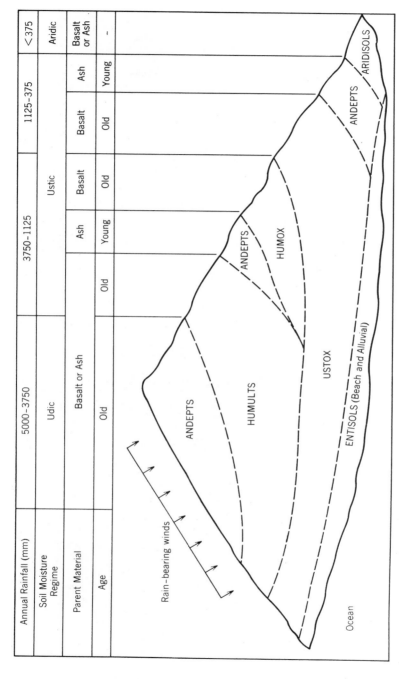

Annual Rainfall (mm)	5000–3750	3750–1125				1125–375			<375
Soil Moisture Regime	Udic	Ustic				Ustic			Aridic
Parent Material	Basalt or Ash	Ash		Basalt		Basalt		Ash	Basalt or Ash
Age	Old	Young		Old		Old		Young	–

Fig. 2.8 Soil associations in a volcanic island, as affected by rainfall, topography, parent material, and age. *Source*: Adapted from Tamura et al. (1953).

83

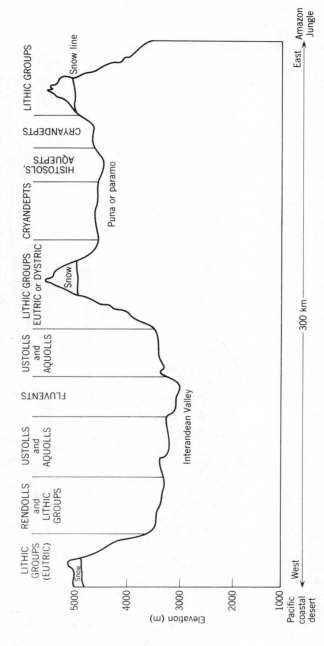

Fig. 2.9 Major soil associations in a transect of the Peruvian Andes. *Source:* Adapted from Zamora (1972) and his unpublished data.

Soil Associations in Tropical Deserts

Less is known about soils of tropical deserts. On the coast of Peru, the following relationships can be found. The bare mountains are essentially soilless or have lithic subgroups of Entisols or Inceptisols. In the valleys, moving sand dunes (Psamments) are found in association with several Aridisols, including Salorthids and Camborthids. In the river floodplains, Fluvents occur, forming the basis for extremely productive irrigated agriculture. The lower parts of these irrigated areas progressively receive salt deposits from higher parts, converting some Fluvents into Salorthids and poorly drained saline soils. Alfisols tend to surround the Aridisols as rainfall increases.

Soils of Tropical Alluvial Plains and Deltas

The geomorphic relationships of river valleys are quite similar throughout the world. The systems of floodplains, levees, terraces, and adjacent uplands in coastal valleys and deltas are basically no different in the tropics than in the rest of the world. The principal difference is in soil properties. In glaciated temperate areas alluvium is generally rich in bases and weatherable minerals, with illite as the predominant clay mineral. In the tropics such rich alluvial deposits are found in watersheds with freshly exposed sediments such as those originating in the Andes and the Himalayas, or in areas where fresh volcanic ash deposits exist. Rivers originating in areas of old landscapes that have undergone several erosion cycles will form alluvial deposits rich in quartz, kaolinite, and iron oxides. Many of them are quite infertile (Edelman and Van der Voorde, 1963). Consequently, high fertility in recent alluvial soils cannot be taken for granted in the tropics.

In many areas the older alluvial terraces have highly weathered soils. Oxisols, Ultisols, and other highly weathered soils occur within valley regions.

The greater abundance of pyrite-rich marine deposits in the tropical plains and coastal areas results in a greater extension of acid sulfate soils (Sulfaquepts) there than in the temperate region. Sulfaquepts, or cat clays, cover large portions of the Mekong Delta, the Bangkok Plain, the Gambia, and parts of the coastal areas of the Guayanas in South America. In Vietnam alone, they occupy over 2 million ha. These soils have such extreme chemical properties that they deserve special attention.

According to Moormann (1963), acid sulfate soils are found where sulfate ions, carried by inundating seawater, are reduced to H_2S under anaerobic conditions in sediments high in organic matter. Hydrogen sulfide then reacts with iron compounds present in the soil, forming pyrite (FeS_2). When these deposits are exposed to air and the soil is low in calcium carbonate, FeS_2 is oxidized to ferric sulfate and free sulfuric acid, producing pH values on the order of 2 or 3. The ferric sulfates are further hydrolized into straw-colored jarosite (a basic ferric sulfate material) and accumulate in the soil, giving it characteristic bright yellow mottles.

Acid sulfate soils are extremely infertile. The sulfate concentration and the depth at which it occurs determine whether crops can be grown at all. The free sulfuric acid dissolves clay minerals and produces large amounts of exchangeable aluminum in quantities toxic to most crops. Iron and manganese toxicities and phosphorus deficiency are common; physical properties are very poor. Flooded rice is often grown, since under constantly reduced conditions the pH increases sufficiently to eliminate aluminum toxicity. Few aerobic crops tolerate cat clays, but some can grow if enough leaching permits a decrease in aluminum toxicity. Pineapples are grown in such soils in the Mekong Delta. Reclamation of these soils is possible in many areas through liming, fertilization, and drainage practices. A symposium on acid sulfate soils (Dost, 1973) has resulted in the most comprehensive treatise on this subject.

SUMMARY AND CONCLUSIONS

1. The strong genetic bias of early tropical pedologists resulted in serious misconceptions, implying that tropical soils are uniform, highly weathered, and subject to being transformed into bricklike laterite when cleared for cultivation. Recent studies demonstrate that tropical soils exhibit as broad a range of properties as soils of the temperate region. The only property common to all tropical soils is their uniform temperature regime.

2. The use of at least six major soil classification systems in the tropics has impeded meaningful correlation and data extrapolation from one part to another. The strong genetic bias of the early classification systems resulted in grouping soils according to stereotyped concepts of a uniform tropical environment. The advent of the quantitative U.S. Soil Taxonomy system and the FAO-UNESCO mapping legend now permits the grouping of tropical soils according to their properties rather than according to what genesis theories dictate. The older terminology ("lateritic soils," "Latosols," "tropical soils") can now be discarded, and the quantitatively defined terms used in a more meaningful way.

3. At a scale of 1 to 50 million, Oxisols are the most abundant soil order found in the lowland tropics. They cover about 22 percent of the area. The areal distributions of the other orders are as follows: Aridisols (18 percent), Alfisols (16 percent), Ultisols (11 percent), Inceptisols (8 percent), Entisols (8 percent), Vertisols (2 percent), and Mollisols (1 percent). Spodosols and Histosols are also found.

4. At the landscape level, several well-defined soil relationships are found in tropical areas. In udic environments Oxisol–Utisol–Inceptisol sequences are common in areas with very old parent materials. On more recent parent materials Ultisol–Alfisol-Inceptisol sequences are found. In ustic environments with very old parent materials Oxisol landscapes occur with rock or plinthite outcrops at the breaks of the slopes separating erosion surfaces. On younger materials Alfisol catenas with plinthite and Alfisol–Vertisol landscapes are common.

5. In the tropical highlands the influence of volcanic materials commonly results in Andept–Tropept–Ultisol–Vertisol landscape sequences, depending on the age of the volcanic ash deposits, rainfall, and elevation. In nonvolcanic highlands a wide diversity of landscape–soil relationships is found. The volcanic ash soils or Andepts are more extensive in the tropics than in the temperate region and are of major economic importance in the highlands of tropical Asia, Africa, and Latin America.

6. Although the geomorphology of alluvial plains and deltas is no different in the tropics than in the temperate region, the high fertility properties of alluvium are not always found in the tropics. In watersheds originating from young materials such as the Andes and the Himalayas, alluvium is rich. In watersheds originating from highly weathered surfaces, alluvial materials are usually infertile.

7. Many tropical coastal and deltaic areas have extensive regions of acid sulfate soils. When drained, these soils are extremely acid with a pH below 3 and are difficult to put into production.

REFERENCES

Abrol, I. P. and D. R. Bhumbla. 1973. Saline and alkali soils in India, their occurrence and management. *FAO World Soil Resources Rept.* 41, pp. 42–51.

Adams, M. E. 1970. The Rewa peat bog and related clay horizons. *Fiji Agr.* 32:3–8.

Aguilera, N. 1963. Los recursos naturales del sudeste de México y su explotación: Suelos. *Chapingo Rev. Esc. Nac. Agr.* 3:1–54.

Ahmad, N., R. L. Jones, and A. H. Beavers. 1962. Some mineralogical and chemical properties of the principal inorganic coastal soils of British Guiana. *Soil Sci.* **96**:162–174.

Ahmad, N. and R. L. Jones. 1969. A plinthaquult of the Aripo savannas, north Trinidad. *Soil Sci. Soc. Amer. Proc.* **33** (5):762–765.

Alexander, L. T. and J. G. Cady. 1962. Genesis and hardening of laterites in soils. *U.S. Dept. Agr. Tech. Bull.* 1282.

Alvarado, A. and C. Lopez. 1971. Química y física de algunos Ultisoles de Costa Rica, America Central. *Turrialba* **21** (3):304–311.

Aubert, G. 1963. Soil with ferruginous and ferralitic crusts of tropical regions. *Soil Sci.* **95** (4):235–242.

Aubert, G. 1968. Classification des sols utilesee par les pedologues francais. *FAO World Soil Resources Rept.* 32, pp. 78–94.

Aubert, G. and R. Tavernier. 1972. Soil survey. Pp. 17–44. In *Soils of the Humid Tropics*. U. S. National Academy of Sciences, Washington.

Augustinus, P. G. E. F. and S. Slager. 1971. Soil formation in swamp soils of the coastal fringe of Surinam. *Geoderma* **6**:203–211.

Baldwin, M., C. E. Kellogg, and J. Thorp. 1938. Soil classification. Pp. 979–1001. In *Soils and Men, the Yearbook of Agriculture*. U.S. Government Printing Office, Washington.

Beaudou, A. G. 1971. Rouge et beige sols: Etude d'une sequence sur quartzite dans la zone forestière de la Republique Centrafricaine. *Cah. ORSTOM, Sèr. Pedol.* **9** (2):147–187.

Beinroth, F. H. 1972. The general pattern of the soils of Puerto Rico. *Trans. Fifth Caribbean Geol. Conf. Geol. Bull.* 5, pp. 225–229.

Beinroth, F. H. 1975. Relationships between the U.S. Soil Taxonomy, the Brazilian Soil Classification System and the FAO/UNESCO soil units. Pp. 92–108. In E. Bornemisza and A. Alvarado (eds.), *Soil Management in Tropical America*. North Carolina State University, Raleigh.

Beinroth, F. H., G. Uehara, and H. Ikawa. 1974. Geomorphic relations of Oxisols and Ultisols of Kauai, Hawaii. *Soil Sci. Soc. Amer. Proc.* **38**:128–131.

Benavides, S. T. 1973. Mineralogical and chemical characteristics of some soils of the Amazonía of Colombia. Ph.D. Thesis, North Carolina State University, Raleigh. 216 pp.

Bennema, J. 1963. The red and yellow soils of the tropical and subtropical uplands. *Soil Sci.* **95** (4):250–257.

Bennema, J., M. N. Camargo, and A. C. S. Wright. 1962. Regional contrast in South American soil formation in relation to soil fertility. *Trans. Comm. IV and V, Int. Soc. Soil Sci.* (*New Zealand*): 453–506.

Bennema, J., and M. N. Camargo. 1964. *Segundo Esboço Parcial de Classificação de Solos Brasileiros*. Ministério da Agricultura, Río de Janeiro.

Bennet, H. H. and R. V. Allison. 1928. *The Soils of Cuba* Tropical Plant Research Foundation, Washington.

Bloomfield, C. 1972. The oxidation of iron sulphides in soils in relation to the formation of acid sulfate soils and of ochre deposits in field drains. *J. Soil Sci.* **23**:1–6.

Boer, M. W. H. de. 1972. Land forms and soils in eastern Surinam (South America). *PUDOC* 13. Wageningen, Netherlands. 169 pp.

Bonnet, J. A. 1960. *Edafología de los Suelos Salinos y Sódicos*. University of Puerto Rico, Rio Piedras. 337 pp.

Bornemisza, E. and J. C. Morales. 1969. Soil chemical characteristics of recent volcanic ash. *Soil Sci. Soc. Amer. Proc.* **33**:528–531.

Brammer, H. 1971. Coatings in seasonally flooded soils. *Geoderma* **6**:5–16.

Buol, S. W. 1973. Soil genesis, morphology, and classification. Pp. 1–37. In P. A. Sanchez (ed.), *A Review of Soils Research on Tropical Latin America.* North Carolina Agr. Exp. Sta. Tech. Bull. 219.

Buol, S. W., F. D. Hole, and R. J. McCracken. 1973. *Soil Genesis and Classification.* Iowa State University Press, Ames. 360 pp.

Calhoun, F. G., V. W. Carlisle, and C. Luna. 1972. Properties and genesis of selected Colombian Andosols. *Soil Sci. Soc. Amer. Proc.* **36**:480–484.

Camargo, M. N. and I. C. Falesi. 1975. Soils of the Planalto and Transamazonic Highway of Brazil. Pp. 25–44. In E. Bornemisza and A. Alvarado (eds.): *Soil Management in Tropical America.* North Carolina State University, Raleigh.

Chatelin, Y. 1969. Contribution à l'etude de la sequence sols ferralitiques et ferrugineaux tropicaux beiges: Examen de profils Centraficains. Cah. *ORSTOM, Sér. Pedol.* **7**:449–453.

Cline, M. G. 1972. *Some Potential Soil Taxa of the Tropics.* Mimeographed. Department of Agronomy, Cornell University, Ithaca, N.Y. 6 pp.

Cline, M. G. 1975. Origin of the term Latosol. *Soil Sci. Soc. Amer. Proc.* **39**:162.

Cline, M. G. et al. 1955. Soil survey of the Territory of Hawaii. *Soil Survey Ser. 1939,* No. 25. U.S. Department of Agriculture, Washington.

Cline, M. G. and S. W. Buol. 1973. Soils of the Central Plateau of Brazil. *Agron. Mimeo* 73-17. Cornell University, Ithaca, N.Y.

Colmet-Daage, F. et al. 1968. Características de algunos suelos de aluvión de la zona oriental de la provincia de Guayas. *Banano, Ecuador* **1**:9–12.

Colmet-Daage, F. et al. 1969. Characteristiques de quelques sols d'Equateur dérivés de cendres volcaniques. *Cah. ORSTOM, Sér. Pédol.* **7**:494–560.

Colmet-Daage, F., C. de Kimpe, M. Daleune, et al. 1970. Characteristique de quelques sols derivés de cendres volcaniques du Nicaragua. *Cah. ORSTOM, Sér. Pédol.* **8**:113–172.

Colmet-Daage, F. and J. Gaotheyrou. 1974. Soil association on volcanic materials in tropical America with special reference to Martinique and Guadaloupe. *Trop. Agr. (Trinidad)* **51**:121–128.

Comerma, J. A. 1971. Los suelos de Venezuela y la séptima aproximación. *Agron. Tropical (Venezuela)* **21**:365–377.

Cortés, A. and D. P. Franzmeier. 1972. Climosequence of ash-derived soils in the Central Cordillera of Colombia. *Soil Sci. Soc. Amer. Proc.* **36**:653–659.

Costa de Lemos, R. 1968. The main tropical soils of Brazil. *FAO World Soil Sources Rept* 32, pp. 95–106.

Coutinet, S. and J. H. Durant. 1966. Les sols de mangrove de la côte Nord-Ouest de Madagascar, region d'Ambanja. *Agron. Tropicale (France)* **21**:345–360.

Crécy, J. de, 1970. Les vertisols sur calcaire aux Antilles: problémes d'utilization agricole. Pp. 251–265. In *Proceedings of the Seventh Annual Meeting.* C. F. C. S., Martinique, Guadaloupe.

Crowther, E. M. 1949. Soil fertility problems in tropical agriculture. *Commonwealth Bur. Soil Sci. Tech. Commun.* **46**:134–192.

Daniels, R. B., F. H. Beinroth, L. H. Rivera, and R. B. Grossman. 1973. Landscape and soils in an area of East Central Puerto Rico. *Soil Survey Investigations* (in press).

Desphande, S. B., J. B. Fehrenbacher, et al. 1971. Mollisols of the Tarai Region of Uttar Pradesh, Northern India. 1. Morphology and mineralogy. 2. Genesis and classification. *Geoderma* **6**:179–201.

D'Hoore, J. L. 1956. Pedological comparisons between tropical South America and tropical Africa. *Afr. Soils* **4**:5–18.

D'Hoore, J. L. 1965. Soils map of Africa—1:5,000,000 with explanatory monograph. *Comm. Tech. Cooperation in Africa, Joint Project* 11. Lagos, Nigeria. 205 pp.

D'Hoore, J. L. 1968. The classification of tropical soils. Pp. 7–28. In R. P. Moss (ed.), *The Soil Resources of Tropical Africa*. Cambridge University Press, London.

Dijkerman, J. C. 1969. Soil resources of Sierra Leone, West Africa. *Afr. Soils* **14**:185–205.

Donahue, R. L. 1972. Ethiopia: Taxonomy, cartography, and ecology of soils. *Afr. Studies Center Monogr.* 1. Michigan State University, East Lansing.

Dost, H. (ed.). 1973. Acid sulphate soils, Vols. 1 and 2. *Int. Inst. Land Reclamation and Improvement Publ.* 18. Wageningen, Netherlands.

Dudal, R. 1963. Dark clay soils of tropical and subtropical regions. *Soil Sci.* **95**:264–271.

Dudal, R. 1965. Dark clay soils of tropical and subtropical regions. *FAO Agr. Dev. Paper* 83. 161 pp.

Dudal, R. 1968. Definitions of soil units for the soil map of the world. *FAO World Soil Resources Rept.* 33.

Dudal, R. 1970. *Key to Soil Units for the Soil Maps of the World*. Food and Agricultural Organization of the United Nations, Rome. 16 pp.

Dudal, R. and M. Soepraptohardjo. 1960. Some considerations of the genetic relationship between Latosols and Andosols in Java, Indonesia. *Trans. Seventh Int. Congr. Soil Sci. (Madison)* **4**:229–237.

Dudal, R. and F. R. Moormann. 1964. Major soils of Southeast Asia, their characteristics, distribution, use and agricultural potential. *J. Trop. Geogr.* **18** (2):54–80.

Edelman, C. E. and P. K. J. van der Voorde. 1963. Important characteristics of alluvial soils in the tropics. *Soil Sci.* **95** (4):258–263.

Fadl, A. E. 1971. A mineralogical characterization of some Vertisols in the Gezira and Kenana clay plains of the Sudan. *J. Soil Sci.* **22**:129–135.

FAO. 1964. Meeting on the classification and correlation of soils from volcanic ash (Tokyo). *FAO World Soil Resources Rept.* 14. 169 pp.

FAO. 1971. *Production Yearbook*. Food and Agricultural Organization of the United Nations, Rome.

FAO-UNESCO. 1971. *Soil Map of the World*, Vol. IV: *South America*. United Nations Educational, Scientific, and Cultural Organization, Paris.

Feuer, R. 1956. An exploratory investigation of the soils and agricultural potential of the soils of the future Federal District in the Central Plateau of Brazil. Ph.D. Thesis, Cornell University, Ithaca, N.Y. 432 pp. *University Microfilms Publ.* 16,254.

Gerasimov, I. P. 1973. A genetic approach to the subdivision of tropical soils, regolith and their products of redeposition. *Soviet Geogr.: Rev. and Translation* **14**:165–177.

Goedert, W. J. and M. T. Beatty. 1971. Caracterização de grummussolos no sudoeste do Rio Grande de Sul. II. Mineralogía e génese. III. Morfología e classificação. *Pesq. Agropec. Bras.* **6**:183–193, 243–251.

Gouvela, D. G. 1968. Vertisolos do norte de Moçambique. *Agron. Moçambicana* 2:139–147.

Gowaikar, A. S. 1973. Influence of moisture regime on the genesis of laterite soils in South India. III. Soil classification. *J. Indian Soc. Soil Sci.* 21:343–347.

Greene, H. 1945. Classification and use of tropical soils. *Soil Sci. Soc. Amer. Proc.* 10:392–396.

Greene, H. 1947. Soil formation and water movement in the tropics. *Soils and Fert.* 10:253–256.

Guerrero, R. 1964. *Suelos de Colombia y su Relación con la Séptima Aproximación.* Instituto Geográfico "Agustin Codazzi," Bogotá, Colombia.

Guerrero, R. 1975. Soils of the Eastern Region of Colombia. Pp. 61–90. In E. Bornemisza and A. Alvarado (eds.), *Soil Management in Tropical America.* North Carolina State University, Raleigh.

Habibullah, A. K. M., D. J. Greenland, and H. Brammer. 1971. Clay mineralogy of some seasonally flooded soils of East Pakistan. *J. Soil Sci.* 22:179–190.

Hardy, F. 1933. Cultivation properties of tropical red soils. *Emp. J. Exptal. Agr.* 1:103–112.

Harpstead, M. I. 1973. The classification of some Nigerian soils. *Soil Sci.* 116:437–443.

Hart, M. G. R. 1959. Sulfur oxidation in tidal mangrove soils of Sierra Leone. *Plant and Soil* 11:215–235.

Herath, J. W. and R. W. Grimshaw. 1971. A general evaluation of the frequency distribution of clay and associated minerals in the alluvial soils of Ceylon. *Geoderma* 5:119–130.

IICA. 1969. *Panel on Soils Derived from Volcanic Aşh in Latin America* (English and Spanish editions). Inter-American Institute of Agricultural Sciences, Turrialba, Costa Rica.

Ivanova, E. N. 1956. An attempt at a general classification of soils (translated from the Russian). *Pochvovedeniye* 6:82–102.

Jacomine, P. K. T. 1969. Descrição das caracteristicas morfológicas, físicas, quimicas e mineralógicas de algunos perfiles de solos sob vegetação de cerrado. *Equipe de Pedología e Fertilidade do Solo Bol. Tec.* 11. Ministério da Agricultura, Rio de Janeiro, Brazil. 126 pp.

Jahn, R. E. 1970. Los suelos orgánicos o turbas del area depresional, Buena Vista, Estado Sucre. *Agron. Tropical (Venezuela)* 20:299–309.

Jurion, F. and J. Henry. 1969. Can primitive farming be modernized? *INEAC, HORS Ser.* 1969. pp. 427–457. Institut National pour L'Etude Agronomique du Congo, Brussels.

Kamarck, A. M. 1972. Climate and economic development. *EDI Seminar Paper* 2. Economic Development Institute, International Bank for Reconstruction and Development, Washington.

Kanapathy, K. 1971. Reclamation of acid swamps soils. *Malaysian Agr. J.* 48:33–46.

Kawaguchi, K. and K. Kyuma. 1969. *Lowland Rice Soils in Thailand.* Center for Southeast Asian Studies, Kyoto University, Japan. 270 pp.

Keivie, W. van der. 1971. Acid sulphate soils in Central Thailand. *FAO World Soil Resources Rept.* 41, pp. 32–41.

Kellogg, C. E. 1949. Preliminary suggestions for the classification and nomenclature of great soil groups in tropical and equatorial regions. *Commonwealth Bur. Soils Tech. Commun.* 46:76–85.

Kellogg, C. E. 1950. Tropical soils. *Trans. Fourth Int. Congr. Soil Sci. (Amsterdam)* 1:266–276.

Kellogg, C. E. and F. D. Davol. 1949. An exploratory study of soil groups in the Belgian Congo. *INEAC, Sér. Sci.* 46. Institut National pour L'Etude Agronomique du Congo, Brussels. 76 pp.

Kesseba, A., J. R. Pitblado, and A. P. Uriyo. 1972. Trends in soil classification in Tanzania. I. The experimental use of the Seventh Approximation. *J. Soil Sci.* **23**:235–247.

Klinge, H. 1965. Podzol soils in the Amazon Basin. *J. Soil Sci.* **16**:95–103.

Klinge, H. 1967. Podzol soils: A source of blackwater rivers in Amazonia. *Atas do Simpósio sobre a Biota Amázonica* **3**:117–125.

Ku Wing Leung. 1971. A synthesis study of the genesis of reddish brown latosols, yellowish brown latosols and red-yellow podzolic soils of Taiwan. *J. Agr. Assoc. China* **76**:63–77.

Kyuma, K. and K. Kawaguchi. 1966. Major soils of Southeast Asia and the classification of soils under rice cultivation. *Kyoto Univ. Southeast Asian Studies* **4** (2):290–312.

Laganathan, P. and L. D. Swindale. 1969. The properties and genesis of four middle altitude Dystrandept volcanic ash soils from Mauna Kea, Hawaii. *Pacific Sci.* **23**:161–171.

Lepsch, I. F. and S. W. Buol. 1974. Investigations in an Oxisol–Ultisol toposequence in São Paulo State, Brazil. *Soil Sci. Soc. Amer. Proc.* **38**:491–497.

Marbut, C. F. and C. B. Manifold. 1926. The soils of the Amazon Basin in relation to agricultural possibilities. *Geogr. Rev.* **15**:239–244.

Martini, J. A. and L. Mosquera. 1972. Properties of five Tropepts in a toposequence of the humid tropics in Costa Rica. *Soil Sci. Soc. Amer. Proc.* **36**:473–477.

Martini, J. A. and M. Macias. 1974. A study of six "Latosols" from Costa Rica to elucidate the problems of classification, productivity and management of tropical soils. *Soil Sci. Soc. Amer. Proc.* **38**:644–652.

McNeil, Mary. 1964. Lateritic soils. *Sci. Amer.* **211**(5):96–102.

Meigs, P. 1966. Geography of coastal deserts. *UNESCO Arid Zone Res. Publ.* 28. 140 pp.

Milne, G. 1935a. Composite units for the mapping of complex soil associations. *Trans. Third Int. Congr. Soil Sci.* **1**:345–347.

Milne, G. 1935b. Some suggested units of classification and mapping, particularly for East African soils. *Soil Res.* **4**:183–198.

Mohr, E. D. J., F. A. Van Baren, and J. Von Schuylerborgh. 1972. *Tropical Soils: A Comprehensive Study of Their Genesis,* 3rd rev., enlarged ed. Mouton-Ichtiar Baru-Van Hoeve, The Hague.

Moormann, F. R. 1963. Acid sulfate soils (cat-clays) of the tropics. *Soil Sci.* **95**(4):271–275.

Moss, R. P. 1968. Soils, slopes and surfaces in Tropical Africa. In R. P. Moss (ed.), *The Soil Resources of Tropical Africa.* Pp. 29–60. Cambridge University Press, London.

Moura-Filho, W. and S. W. Buol. 1972. Studies of a Latosol Roxo (Eutrustox) in Brazil. *Experientiae* **13** (7):201–247.

Mukherjee, H. N. 1963. Determination of nutrient needs in tropical soils. *Soil Sci.* **95**:276–280.

Mulcahy, M. J. 1960. Laterite and lateritic soils in S. W. Australia. *J. Soil Sci.* **11**:206–225.

Murthy, R. S. 1971. Acid sulphate soils in India. *FAO World Soil Resources Rept.* 41, pp. 24–29.

Natohadiprawiro, T. 1972. Problems and perspectives of agriculture on alluvial soils of Indonesia. *Ilmo. Pertanian (Agr. Sci.)* **1**:247–257.

Nye, P. H. 1954. Some soil forming processes in the humid tropics. I. A field study of a catena near Ibadan. *J. Soil Sci.* **5**:7–21.

Nye, P. H. 1955. Some soil forming processes in the humid tropics. II. The development of the upper member of the catena. III. Laboratory studies of a typical catena over granite gneiss. *J. Soil Sci.* **6**:51–62, 63–72.

Ollier, C. D. 1959. A two-cycle theory of tropical pedology. *J. Soil Sci.* **10**:137–148.

Pagel, H. and I. Insa. 1970. Physikalish-chemische und chemische Eigenschaften einiger Mangrovenboden Guineas. *Beitr. Trop. Subtrop. Landwitsch.* **8**:93–106.

Palencia, J. A. and J. A. Martini. 1970. Características morfológicas, químicas, y físicas de algunos suelos derivados de ceniza volcánica en Centroamérica. *Turrialba* **20**:325–332.

Pantastico, E. B. et al. 1968. A fertility record of the Taal volcano ejecta after the eruption of 1965. *Philipp. Agr.* **51**:17–23.

Porrenga, D. H. 1967. Clay mineralogy and geochemistry of recent marine sediments in tropical areas as exemplified by the Niger delta, the Orinoco shelf, and the shelf off Sarawak. *Publ. Fysisch-Geogr. Bodemk. Lab. Univ. Amsterdam* **9**:1–145.

Post, J. L. and R. L. Sloane. 1971. The nature of clay soils from the Mekong Delta, An Giang Province, South Vietnam. *Clays and Clay Minerals* **19**:21–29.

President's Science Advisory Committee. 1967. *The World Food Problem,* Vol. 2. The White House, Washington.

Quantin, P. 1972a. Les andosols: Revue bibliographique des connaissances actuelles. *Cah. ORSTOM, Sér. Pédol.* **10**:273–301.

Quantin, P. 1972b. Nature and fertility of the volcanic ash soils from recent eruptions in the New Hebrides Archipelago. *Cah. ORSTOM, Sér. Pédol.* **10**:123–134.

Quiñones, H. and R. Allende. 1974. Formation of the lithifield carapace of calcareous nature which covers most of the Yucatan Peninsula and its relation to the soils and geomorphology of the region. *Trop. Agr. (Trinidad)* **51**:94–101.

Radwanski, S. A. 1971. East African catenas in relation to land use and farm planning. *World Crops* **25**:265–273.

Radwanski, S. A. and C. D. Ollier. 1959. A study of an East African catena. *J. Soil Sci.* **10**:149–168.

Raychaudhury, S. P. and M. M. Patel. 1969. Comparative study of some typical saline and alkali soils occurring in certain climatic regions in India. *J. Indian Soc. Soil Sci.* **17**:291–299.

Raychaudhury, S. P. and S. V. Govinda Rajan. 1971. Soils of India. *Indian Council Agr. Res. Tech. Bull.* 25. 45 pp.

Richards, P. W. 1941. Lowland tropical podzols and their vegetation. *Nature* **148** (3774):129–131.

Roldan, J. 1959. Fertility status of organic soils of Puerto Rico. *J. Agr. Univ. Puerto Rico* **43**:255–267.

Ruhe, R. V. 1954. Erosion surfaces of the Central African interior high plateaus. *INEAC, Ser. Sci.* 59. Institut National pour l'Etude Agronomique du Congo, Brussels.

Ruhe, R. V. 1956. Landscape evolution in the high Ituri, Belgian Congo. *INEAC, Ser. Sci.* 66, Institut National pour l'Etude Agronomique du Congo, Brussels.

Ruhe, R. V. 1959. Stone lines in soils. *Soil Sci.* **87**:223–231.

Sanchez, P. A. and S. W. Buol. 1974. Properties of some soils of the Amazon Basin of Peru. *Soil Sci. Soc. Amer. Proc.* **38**:117–121.

Sanchez, P. A. and S. W. Buol. 1975. Soils of the tropics and the world food crisis. *Science* **188**:598–603.

Sehgal, J. L. and C. Sys. 1970. The soils of Punjab (India). I. Geographical conditions. II. Application of the Seventh Approximation to the classification of the soils of Punjab: Some problems, considerations and criteria. *Pédologie* **20**:178–203, 244–267.

Segalen, P. 1970. *Pédologie et Developpement: Techniques Rurales en Afrique* 10. Secretariat d'Etat aux Affaires Extrangeres, Paris.

Sherman, G. D., Z. C. Foster, and C. K. Fujimoto. 1949. Some of the properties of the Fer-
 ♣ ruginous Humid Latosols of the Hawaiian Islands. *Soil Sci. Soc. Amer. Proc.* **13**:471–456.

Sherman, G. D. and L. T. Alexander. 1959. Characteristics and genesis of Low Humic Latosols. *Soil Sci. Soc. Amer. Proc.* **23** (2):168–170.

Sivarajasingham, S., L. T. Alexander, J. G. Cady, and M. G. Cline. 1962. Laterite. *Adv. Agron.* **14**:1–60.

Slager, S. and J. Van Schuylenborgh. 1970. *Morphology and Geochemistry of Three Clay Soils of a Tropical Coastal Plain in Surinam.* Centre for Agricultural Publications and Documentation, Wageningen, Netherlands.

Soil Survey Staff. 1960. *Soil Classification—A Comprehensive System: Seventh Approximation.* U.S. Department of Agriculture, Washington.

Soil Survey Staff. 1970. *Soil Taxonomy* (draft). Soil Conservation Service, U.S. Department of Agriculture, Washington.

Sombroek, W. G. 1966. *Amazon Soils.* Centre for Agricultural Publications and Documentation, Wageningen, Netherlands. 292 pp.

Sombroek, W. G. 1971. Ancient levels of plinthisation in Northwest Nigeria. Pp. 329–337. In D. H. Yaalon (ed.): *Paleopedology: Origin, Nature, and Dating of Paleosols.* Israel University Press.

Srinivasan, T. R. et al. 1969. Placement of black soils of India in the comprehensive soil classification system—Seventh Approximation. *J. Indian Soc. Soil Sci.* **17**:323–331.

Sys, C. 1972. Caractérisation morphologique et physico-chimique de profils types de l'Afrique centrale. *INEAC, HORS Ser.* 1972. Institut National pour l'Etude Agronomique du Congo, Brussels. 497 pp.

Sys, C., A. Van Wambeke, R. Frankart, et al. 1961. La cartographie des sols au Congo, ses principes et ses methodes. *INEAC, Ser. Tech.* 66. Institut National pour l'Etude Agronomique du Congo, Brussels.

Tamura, T., M. L. Jackson, and G. D. Sherman. 1953. Mineral content of Low Humid, Humic, and Hydrol Humic Latosols of Hawaii. *Soil Sci. Soc. Amer. Proc.* **17**:343.

Tay, T. H. 1972. Crop production on West Malaysian peat. *Proc. Int. Peat Congr.* (*Helsinki*) **4**:75–88.

Thorp, J. and G. D. Smith. 1949. Higher categories of soil classification: Order, suborder, and great soil groups. *Soil Sci.* **67**:117–126.

Turenne, J. F. 1974. Molecular weights of humid acids in podzols and ferralitic soils of the savannas of French Guyana and their evolution related to moisture. *Trop. Agr.* (*Trinidad*) **51**:133–144.

Tyler, E. J. 1975. Genesis of the soils with a detailed soil survey in the Upper Amazon Basin, Yurimaguas, Peru. Ph.D. Thesis, Soil Science Department, North Carolina State University, Raleigh. 171 pp.

UNESCO. 1969. Soils and tropical weathering. *UNESCO Natural Resources Res.* 11.

Van der Voorder, P. K. J. 1956. Podzolen in Suriname. *Surinamse Landbouw* **4**:45–51.

Van Wambeke, A. 1967. Recent developments in the classification of soils of the tropics. *Soil Sci.* **104**:309–313.

Van Wambeke, A. 1971. Recherches sur la mise en valeur agricole des sols acides des savanes arborees du Brazil. *Pédologie* **21**:211–255.

Veen, A. W. L., S. Slager, and A. G. Jongmans. 1971. A micromorphological study of four pleistocene alluvial soils of Surinam. *Geoderma* **6**:81–100.

Velázquez, A. and E. Ortega. 1968. Características físicas y químicas de algunos suelos del valle del Rio Fuerte. *Agr. Técnica (Mexico)* **2**:400–406.

Vine, H. 1966. Tropical soils. Pp. 28–67. In C. C. Webster and P. N. Wilson (eds); *Agriculture in the Tropics*. Longmans, London.

Vine, H. 1974. Toposequences in soils of central Trinidad. *Trop. Agr. (Trinidad)* **51**:109–120.

Watson, J. P. 1964. A soil catena on granite in southern Rhodesia. I. Field observations. II. Analytical data. *J. Soil Sci.* **15**:238–257.

Watson, J. P. 1965a. A soil catena on granite in southern Rhodesia. III. Clay minerals. IV. Heavy minerals. V. Soil Evolution. *J. Soil Sci.* **16** (1):158–169.

Watson, J. P. 1965b. Soil catenas. *Soils and Fert.* **28** (4):307–310.

Zamora, C. 1972a. *Esquema de los Podzoles de la Región Selvática del Perú*. Oficina Nacional de Evaluación de Recursos Naturales, Lima.

Zamora, C. 1972b. *Regiones Edáficas del Perú*. Oficina Nacional de Evaluación de Recursos Naturales, Lima. 15 pp.

Zamora, C. 1974. *Los Suelos, Uso y Problemas de las Tierras Áridas del Perú*. Oficina Nacional de Evaluación de Recursos Naturales, Lima. 18 pp.

Zamora, C. 1975. Soils of the lowlands of Peru. Pp. 46–60. In E. Bornemisza and A. Alvarado (eds.), *Soil Management in Tropical America*. North Carolina State University, Raleigh.

Zein, A. and G. H. Robinson. 1971. A study of cracking in some vertisols of the Sudan. *Geoderma* **5**:229–241.

3

SOIL PHYSICAL PROPERTIES

Soil physical properties are those responsible for the transport of air, heat, water, and solutes through the soil. They are widely variable in tropical soils, including some that are unknown in the temperate region. Oxisols and Andepts are generally considered to have excellent physical properties in their natural state. Many Ultisols and Alfisols are susceptible to erosion because of sharp textural changes. Several physical properties can and do change with management. In the tropics severe dessication and high temperatures at the soil surface may be followed by abrupt changes due to high-intensity rainstorms. Many soil physical properties deteriorate with cultivation, rendering the soil less permeable and more susceptible to runoff and erosion losses. The ability of the soil to retain water and supply it to plants is one of the main limiting factors in tropical agriculture. The purpose of this chapter is to describe the differences in physical properties among tropical soils and the management of these parameters.

DEPTH OF ROOTING

The most obvious physical limitation—insufficient soil depth for adequate root development is often ignored. Physical barriers to root development are probably less common in the tropics than in the temperate region because of the generally deeper subsoils in highly weathered areas, particularly in Oxisols, Ultisols, Alfisols, and some Inceptisols. For example, in tropical South America about 10 percent of the soils is classified by the FAO as having lithic or paralithic contact at 50 cm or less. These Lithosols, however, cover 140 million ha, as shown in Table 2.9. Studies at the International Rice Research Institute (1964) showed that rice yields increased linearly as the effective depth of rooting increased from 10 to 40 cm. Droughty subsoils or aluminum-toxic subsoils are often more common barriers to root development than shallow rocks or hardpans. A good illustration of aluminum-toxic subsoil is given in Table 7.10. Corn growing in Brazilian Oxisols strongly responded to deep liming applications. Yields were correlated with the increased rooting depth associated with liming.

Although little can be done to increase rooting depth in Lithosols, the removal of some chemical barriers to root development can be accomplished by proper management.

SOIL STRUCTURE

Soil structure is a poorly defined and quantified concept. This writer prefers Brewer and Sleeman's (1960) definition: "the size, shape, and arrangement of primary particles to form compound particles and the size, shape, and arrangement of compound particles." What is considered good structure depends on the desired speed with which the air and water move through the soil. Good structure for growing flooded rice is that attained by puddling (i.e., the destruction of aggregates) to eliminate downward water movement. For other tropical crops good structure is that which maintains aggregate stability upon abrupt changes of moisture and intense rainfall (Pereira, 1956).

The Problem of Measuring Soil Structure

There is no single parameter that adequately measures soil structure. Although aggregate stability is often used as a measure of soil structure, it is at best an empirical one. Pereira (1955, 1956), working in East Africa, compared several dry-sieving and wet-sieving techniques to measure aggregate stability with porosity determinations, infiltration rates, and a rainfall acceptance test in an attempt to separate "good" and "bad" soil structure as visually observed in the field. All parameters but the last-named were unsatisfactory. The influence of root or insect channels was the primary cause for the failure in percolation rates. Pereira's rainfall acceptance test consisted of subjecting a core sample to heavy simulated rainfall and measuring both percolation and runoff. This technique, however, has not been used elsewhere. The French-speaking workers in Africa also have their own system, based on silt-plus-clay content divided by the mean of three aggregate stability determinations. The application of this technique is discussed in Baver's review (1972). It is used only in French-speaking Africa.

Field tests using double-ring infiltrometers, when adequately replicated, are perhaps the most pragmatic means of evaluating soil structure. Field infiltration measurements integrate many of the physical variables and provide a soil volume large enough to reduce the excessive variability caused by insect and root channels. Lugo-Lopez et al. (1968) applied this technique to 740 soils throughout Puerto Rico and found that infiltration rates during the first 3 hours depended on the initial soil moisture content. After the fourth hour infiltration rates became constant and approximated the hydraulic conductivity of the saturated soil. A summary of their results by soil order appears in Table 3.1. The high infiltration rates for Oxisols, Ultisols, and Mollisols reflect their good structure; the low range for the Vertisols indicates their high content of expanding 2:1 clays.

Structure of Major Soil Groups

In general, Oxisols and oxidic families of Ultisols, Alfisols, and Inceptisols possess excellent structure in terms of Pereira's concept. Certain Oxisols

Table 3.1   Ranges in Infiltration Rates of Puerto Rican Soils, Grouped by Soil Orders (Ranges of 57 Soil Types and 740 Tests)

	Infiltration Rate (cm/hour)	
Soil Order	Minimum	Maximum
Oxisols	8.4	15.4
Ultisols	7.4	23.6
Mollisols	8.2	19.5
Alfisols	2.7	11.5
Inceptisols	2.7	13.2
Entisols	2.3	27.5
Vertisols	0.1	9.5

Source: Lugo-Lopez et al. (1968).

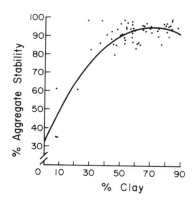

Fig. 3.1 Relationship between aggregate stability and clay content of "red" soils of Luzon. *Source:* Briones and Veracion (1965).

can be plowed with heavy machinery a day after a heavy rain with little aggregate disruption. Oxisols with over 80 percent clay feel like loams or silt loams when estimating field texture. If one continues to work the wet soil with his fingers, however, it gradually feels more clayey as the aggregates are progressively destroyed.

The excellent structure of these soils is caused by primary particles being aggregated in very stable sand-sized granules. Their high stability is associated with their high clay content and cementing or coatings of amorphous iron and aluminum oxides. Briones and Veracion (1965), working with Philippine "red" soils, found that the aggregate stability of these soils increased linearly with clay content up to about 50 percent clay. This is shown in Fig. 3.1. Although quite obvious, the fact that high clay contents are needed for these desirable structural properties is often forgotten. Many sandy Oxisols and Ultisols suffer readily from compaction and erosion.

Organic matter content is also correlated with aggregate stability, as it plays a cementing role between primary mineral particles (Lugo-Lopez and Juarez, 1959; Briones and Veracion, 1965). Other soils high in clay and organic matter lack such excellent physical properties. When Oxisols are subjected to particle size analysis without pretreatment with sodium dithionate to remove free iron oxides, the sand and silt contents are considerably higher than when iron oxides are removed. In an example given by Moura and Buol (1972), the clay content of an Eutrustox from Brazil increased from 40 to 83 percent when iron oxides were removed. Even in the subsoil of this Eutrustox, which has about 0.6 percent organic carbon, the clay content increased from 49 to 65 percent. In general the actual clay content of such a soil can be estimated by multiplying the 15 bar water content by a factor of 2.5 (Soil Survey Staff, 1970).

The above generalizations do not hold throughout the Oxisol order.

Plate 3 The excellent granular structure of Oxisols such as this one from Brasilia, Brazil, permits tillage operations within a short interval after a heavy rain.

Uehara et al. (1962) observed different structural characteristics in Hawaiian Oxisols of similar clay and organic matter contents and mineralogical composition (primarily kaolinite and amorphous iron and oxides). Differences in aggregate stability could not be correlated with free Fe_2O_3 contents beyond the 5 percent level. Cagauan and Uehara (1965) then determined that these differences were associated with the degree of anisotropy (clay orientation) in the aggregates. This relationship is illustrated in Fig. 3.2. A more oriented system of kaolinite particles

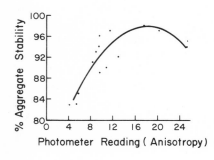

Fig. 3.2 Dependence of aggregate stability on soil anisotropy in Hawaiian Oxisols. *Source:* Cagauan and Uehara (1965).

cemented or coated with iron oxides provides stronger aggregates than a less oriented arrangement. The greatest degree of anisotropy has been found in ustic soil moisture regimes with high annual rainfall; udic regimes exhibit less clay orientation. Apparently Oxisols in constantly humid climates do not retain as strong a structure as those with marked seasonal moisture fluctuations. Many Oxisols in udic environments belong to the tropeptic subgroups because the structure tends toward the blocky type. Consequently, clay contents, organic matter, free iron oxides, and particle orientation within the aggregates are the main factors associated with the excellent structure of Oxisols, but the degree of manifestation of this structure varies.

Ultisols and Alfisols in general have less desirable structural properties because of lower clay contents in the A horizon and an absence of clay orientation within their blocky peds. The clay skins that coat these peds, however, are anisotropic and provide increased aggregate stability, although this is seldom as strong as in the Oxisols. The coarser textured topsoils of Ultisols and Alfisols present serious erosion and compaction hazards.

Andepts also have extremely stable structure. The high organic matter content of these soils is intimately associated with allophane, an amorphous

Plate 4 Surface of a sandy Alfisol in Ile-Ife, Nigeria, compacted by tillage and exposure, which caused sheet erosion in a flat area.

mixture of silica and aluminum oxides. The aggregates in Andepts are somewhat larger than those in Oxisols, but their stability is very similar. A study of 18 volcanic ash soils from Pasto, Colombia, showed that 81 percent of the stable aggregates were greater than 2 mm in diameter (Escobar et al., 1972).

Andepts usually have very low bulk density values (0.4 to 0.8 g/cc) because of their high organic matter content and porosity. The absence or near absence of layer silicate minerals in many of these soils makes particle size analysis practically meaningless. Furthermore, these soils are very difficult to disperse because of this intimate association between allophane and organic matter. Attempts to estimate clay content by multiplying the 15 bar water content by 2.5 failed in these soils (Alvarado and Buol, 1975).

A few volcanic ash soils essentially devoid of crystalline materials, such as the Hydrandepts of Hawaii, have thixotropic properties. Upon a certain level of compaction, the soil suddenly becomes fluid (Swindale, 1964). In most Andepts, however, this situation does not occur.

The structural properties of other tropical soils dominated by layer silicate mineralogy, such as Vertisols, Mollisols, and Aridisols, are no different from those observed in the temperate region. In these systems increased

Plate 5 The poor physical properties of Vertisols, such as this one in Kavado, Kenya, make tillage operations difficult.

anisotropy means lower aggregate stability, as little or no iron coats are present to cement these particles. The poor aggregate stability of Vertisols and associated soils is associated with low organic matter contents and high levels of soluble sodium (Lugo-Lopez and Juarez, 1959; Lugo-Lopez and Perez, 1969). Vertisols can be tilled adequately at a much narrower moisture range than other soils. When they are dry, shrinkage and cracking breaks them into huge massive clods of such strength as to make most tillage practices unfeasible. When wet, they are soft and sticky.

Changes of Soil Structure with Cultivation

It is generally believed that soil structure deteriorates under cultivation in the tropics. Aside from the localized importance of plinthite formation, a fair amount of evidence indicates that the degree of structural change varies with soil properties and management practices.

The effects of intense cultivation on the size distribution of water-stable aggregates have been studied by Grohmann (1960) in an Oxisol and an Ultisol from Brazil. The results illustrated in Table 3.2 indicate that cultivation reduced the percentage of aggregates larger than 2 mm by about half in both soils. This was reflected in a uniform increase of other aggregate sizes in the Oxisol, and in a drastic increase of the aggregates smaller than 0.21 mm in the Ultisol. These smaller aggregates can clog the large pores between the larger aggregates and decrease infiltration.

Moura and Buol (1972) compared the effects of 15 years of annual crop-

Table 3.2   Effect of Intense Cultivation in a Brazilian Oxisol and Ultisol on Wet-Sieved Aggregate Size Distribution without Pretreatment (% distribution of water-stable aggregate)

Aggregate Size (mm)	Terra Roxa Legitima (Oxisol)		Massapé Soil (Ultisol)	
	Forest	Under Cultivation	Pasture	Under Cultivation
>2	84.2	48.2	80.8	36.0
2–1	1.1	13.2	7.2	11.1
1–0.5	0.5	13.0	3.9	6.6
0.5–0.21	0.5	15.1	4.2	12.5
<0.21	13.7	10.5	3.9	33.8

Source: Grohmann (1960).

ping in a Brazilian Eutrustrox and observed that infiltration rates decreased from 82 to 12 cm/hour with intensive cropping (Table 3.3). The decrease in infiltration was associated with a sharp decrease in macropores greater than 0.05 mm in diameter in both A and B horizons, whereas the micropores remained essentially unchanged. Compaction by machinery was considered the cause of the decreased macroporosity. Moura and Buol also observed that water-dispersible clay contents decreased in the A horizon with cultivation and increased in parts of the B horizon. Apparently, some clay translocation may have also reduced porosity. A drop in infiltration rate from 82 to 15 cm/hour could be considered beneficial because of reduced percolation and leaching losses.

The effects of clearing a forest soil were evaluated by Cunningham (1963) on a sandy loam of Ghana, probably an Alfisol. Uncultivated plots were monitored for 3 years under total, partial, and no exposure through the use of artificial shade. The results shown in Table 3.4 indicate a sharp decrease in noncapillary porosity from 15 to 10 percent and a small but significant decrease in total porosity in the bare soil. The percentage of water-stable aggregates greater than 3 mm also decreased. The detrimental effects of exposure were also associated with a decrease in silt and clay content of 3 percent in the A horizon and an increase of 6 percent in the B horizon. Cunningham's studies suggest that clay movement reduced porosity by clogging large pores and attribute this to a sharp drop in organic carbon content with exposure. This type of evidence does not include the effects of tillage or crop growing after exposure but suggests, along with other studies, that changes in porosity can occur quite quickly in these sandy topsoils.

Consequently the changes in structural properties after clearing depend on the intrinsic properties of the soil.

Table 3.3   Effects of Cultivation on the Physical Properties of an Eutrustox from Minas Gerais, Brazil

Soil Property	Recently Cleared	Annual Cropping for 15 Years
Infiltration rate (cm/hour)	82	12
Pores > 0.05 mm, A horizon (%)	25	11
Pores > 0.05 mm, B horizon (%)	34	13
Pores > 0.05 mm, A horizon (%)	33	32
Pores > 0.05 mm, B horizon (%)	30	33
H_2O—dispersible clay, A horizon (%)	13	7
H_2O—dispersible clay, B horizon (%)	1	7

Source: Adapted from Moura Filho and Buol (1972).

Table 3.4   Physical Changes Observed in the Top 7.5 cm after Clearing an Alfisol in the Forest Zone of Ghana

Treatment	Capillary Porosity (%)	Noncapillary Porosity (%)	Water-stable Aggregates, > 3 mm (%)
Shade	37	14.7	55.3
Half exposure	35	16.4	50.2
Full exposure	32	10.1	48.7

Source: Cunningham (1963).

Control and Correction of Structural Deterioration

The magnitude of soil compaction problems after clearing Alfisols and Ultisols with coarse topsoil textures is a major management problem that has received considerable attention in West Africa. Intensive rainfall accompanied by a rapid decrease in organic matter content is considered the reason for this phenomenon.

Perhaps the best management practice is to prevent compaction by protecting the soil surface with mulches or plant canopies which decrease the energy of raindrops and the rate of organic matter decomposition as well. Lal (1975) has shown that mulching also prevented excessively high soil temperatures, increased soil water storage, decreased weed infestations, and prevented runoff and erosion in sandy-textured Alfisols of Nigeria. Evidence accumulating from other tropical areas with similar coarse-textured topsoils supports Lal's findings. In traditional shifting cultivation systems the soil is never tilled but is protected with an ash mulch and later with a constant crop canopy in intercropped patterns. Mulching (along with minimum or no tillage) can prevent these damaging compaction effects.

In areas where mulching and intercropping are not practiced because of the large scale of farm operations, attempts have been made to correct soil compaction by deep plowing. In Senegal, with shallow hoe cultivation, the bulk density of the first few centimeters decreased from 1.6 to 1.4 g/cc. With tractor plowing the same values were reached to a depth of 10 or 30 cm (Charreau and Nicou, 1971; Charreau, 1972). Significant yield increases from uncultivated plots have been attained by Charreau and his coworkers with both shallow hand cultivation and deep tractor tillage. The results are shown in Table 3.5. In these soils decreases of 0.1 g/cc in bulk density have a beneficial effect on root development and yields of crops such as sorghum and peanuts, as indicated in Fig. 3.3.

Root density in the top 20 or 30 cm was closely correlated with grain

Table 3.5   Effects of Superficial Tillage and Deep Plowing on Yields of Several Crops Grown on Sandy Alfisols of West Africa (tons/ha)

Crops	No Tillage	Manual Tillage (< 5 cm)	No Tillage	Mechanical Deep Plowing (15–25 cm)
Millet	1.40	1.74	1.31	1.60
Sorghum	1.93	2.42	1.52	1.88
Corn	2.59	3.49	1.86	3.21
Rice	1.16	2.36	1.62	2.80
Cotton	1.34	1.67	1.30	1.80
Peanuts	1.45	1.77	1.62	1.76

Source: Nicou (1972).

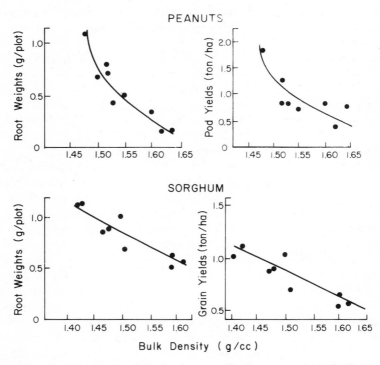

Fig. 3.3   Relationships between bulk density, root development, and yields of peanuts and sorghum in sandy Alfisols from Bambey, Senegal. *Source:* Charreau and Nicou (1971).

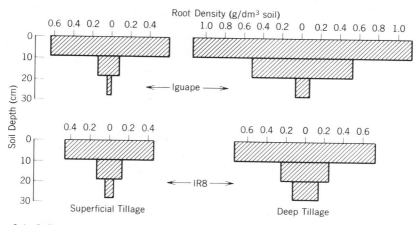

Fig. 3.4 Influence of plowing depth on the root density of two rice varieties grown in upland conditions in South Senegal. *Source:* Adapted from Nicou et al. (1970).

yields in these soils (Nicou, 1972). A schematic representation of this effect in two rice varieties grown under upland conditions appears in Fig. 3.4.

Studies in completely different soils support Nicou's observations that relatively small changes in bulk density have a marked effect on root development. In a study of the effect of tractor-caused compaction on root development of sugarcane in Hawaii, Trouse and Humbert (1961) showed that small changes in bulk density caused roots to become flattened, while substantial compaction caused actual root restriction (Table 3.6). Another interesting aspect of Table 3.6 is the widely different bulk density values at

Table 3.6 Bulk Densities of Various Hawaiian Soils and the Development of Sugarcane Roots

Soil	Bulk Density (g/cc) at Which Rootlets:		
	Grew Normally	Became Flattened	Were Restricted
Low Humic Latosol (Ustox)	1.02	1.12	1.52
Hydrandept (Andept)	0.58	0.70	1.08
Gray Hydromorphic (Aquept)	1.17	1.32	1.76

Source: Trouse and Humbert (1961).

which sugarcane roots are affected; they range from 0.7 g/cc in an Andept to 1.3 in an Aquept.

In addition to soil properties, the problem of soil structure degeneration is also a function of management.

SOIL WATER RETENTION

Moisture Retention Patterns

The differences in structure between different broad types of tropical soils produce dramatic differences in their water retention properties. Figure 3.5 shows the classic soil moisture retention curves of conventional layer silicate sandy and clayey soils, a clayey Oxisol, and an Andept. The sandy soil empties its pores of gravitational water at tensions close to 0.1 bar, while the layer silicate clay does this at 0.5 bar. The moisture retention pattern of a clayey, well-aggregated Oxisol is a hybrid of these two. It holds as much water as a layer silicate clay up to approximately 0.1 bar, but it drains its

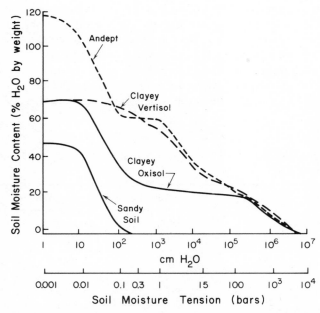

Fig. 3.5 Moisture retention curves of a sandy soil, a clayey Oxisol, a clayey Vertisol, and an Andept. *Source:* Compiled from unpublished data of Uehara and from Yamanaka (1964).

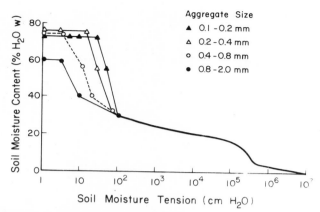

Fig. 3.6 Moisture retention curves of different aggregate fractions of an Oxisol from Hawaii.
Source: Sharma and Uehara (1965).

macropores at about that tension, thus reaching "field capacity" like the
sands. At higher tensions the well-aggregated clayey Oxisol holds a larger
amount of water than sands but less than layer silicate clays of similar clay
content. These well-aggregated oxidic families act like sands in terms of
water movement at low tensions but hold water like clays at higher tensions.
Therefore they have a narrower available water range than other clays.
Many of these well-aggregated soils have drought problems completely out
of proportion to their clay and water contents. The aggregates may be close
to saturation, while the pores between aggregates are totally depleted of
moisture available to crops. These unique properties are due to the
aggregate size distribution. Figure 3.6 shows that the rapid drainage at soil
moisture tensions between 0.01 and 0.1 bar is due to the predominance of
sand-sized water-stable aggregates in these soils. At moisture tensions
greater than 0.2 bar, aggregate size is irrelevant because it is the micropores
within the aggregates that are being drained.

Andepts also possess unique water retention characteristics. They hold
considerably more water at lower tensions than other soils because of their
high porosity and the size of water-stable aggregates. If water content is calculated on a weight basis, the figures may range from 100 to 300 percent
H_2O. If calculated on a volume basis, their moisture contents at 0.3 bar
may run as high as 84 percent and at 15 bar to 45 percent because of their
low bulk density. Nevertheless, Andepts hold more water at a given tension
than most other soils, as shown in Fig. 3.5. Although the quick achievement
of "field capacity" is due to the drainage of large pores between large and
very stable aggregates, the high macroporosity within the aggregates makes

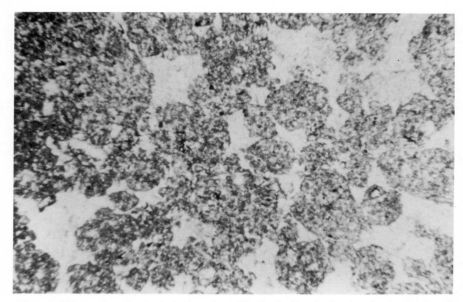

Plate 6 Thin section of an oxic horizon from São Paulo, Brazil, showing the aggregation of clay particles into strong sand-size granules. Courtesy of Dr. Stanley Buol, North Carolina State University.

them hold more water (Forsythe et al., 1969; Swindale, 1969). Andepts generally have more water available to plants than Oxisols because organic matter–allophane aggregates are more porous than kaolinite–iron oxide aggregates.

Ultisols and Alfisols with partial sequioxide coatings seem to have better moisture retention properties than Oxisols in general. The large areas of tropical soils with conventional layer silicate mineralogy have moisture retention curves quite similar to those presented in standard textbooks.

Available Water

The definition of available water as the diference between water held at field capacity (0.3 bar) and the permanent wilting percentage (15 bar) is still widely used by soil scientists everywhere. The original work of Briggs and Shantz in 1914, using one plant (sunflower) and one loamy soil, has become a general principle. There are, however, many important limitations to this principle. Available water is not uniformly available in soils, roots do not extract it uniformly, and the replenishment of water in the depleted areas may be slow enough to cause wilting in the presence of ample water. For example, sometimes one can observe rice with wilted leaves growing in a flooded soil. There are also rather large differences between crops and varieties in the ability to extract water at different soil moisture tensions.

A study with several Oxisols, Vertisols, Inceptisols, and Entisols of Zambia by Maclean and Yager (1972) showed that soil moisture tensions between 0.1 and 0.2 bar approximate actual field capacity better than the classic 0.3 bar. When available water was calculated as the difference between 0.3 and 15 bars, the result was an underestimation of the available water range by 35% compared with actual field measurements in these soils.

The original sunflower test that established 15 bars as the permanent wilting point is exceptionally accurate. Sharma and Uehara (1968) in Hawaii and Wolf (1975) in Puerto Rico have also shown very little difference in water content between 1 and 15 bars. It may be possible that the available water range is shifted for 0.1 to 1 bar in these highly aggregated soils of the tropics. A series of *in situ* studies on Oxisols and Ultisols of Puerto Rico and Brazil led Wolf (1975) to recommend that the available water range of these soils be estimated as the differences between 0.1 and 15 bars. The 0.1 bar water content has to be estimated in undisturbed samples or in the field, whereas the 15 bar water content can be determined by conventional pressure membrane techniques in the laboratory. Wolf noted that the water held between 1 and 15 bars is less available to crops than that held between 0.1 and 1 bar. This amount was approximately equal in three Oxisols of widely different texture.

Table 3.7 shows some representative available water ranges. Regardless of actual texture the available water range of Oxisols is usually about 10 percent. Ultisols generally have higher available water capacities, Andepts have much more, and other soils approximate the amounts found in the temperate region as a function of texture.

SOIL WATER MOVEMENT

Very little field information exists on the patterns of soil water movement in the tropics. A study by Wolf and Drosdoff (1974) on Puerto Rican Oxisols and Ultisols, however, has provided valuable information. Initial water infiltration rates depend on soil moisture contents. For example, when the top 30 cm of an Ultisol contained 40 percent H_2O (0.08 bar), the infiltration rate was 38 cm/hour. The rate dropped by half when the moisture content increased to 44 percent (0.04 bar) and down to 3 cm/hour when the soil was essentially saturated with 50 percent H_2O (0.006 bar). In general, initial infiltration rates averaged about 20 cm/hour in clayey Ultisols, sandy

Table 3.7   *Available Water Ranges of Some Tropical Soils*

Soil	Horizon (cm)	Clay (%)	0.3 bar	15 bars	Difference
			% H_2O (Weight)		
Haplustox	0–23	13	10.4	4.6	5.8
(São Paulo,	23–62	14	10.0	4.7	5.3
Brazil)	62–120	15	10.8	4.9	5.9
Haplustox	0–14	47	27.6	17.7	9.9
(São Paulo,	14–41	54	26.4	19.2	7.2
Brazil)	41–80	56	28.4	20.0	8.4
	80–120	54	29.0	20.8	8.2
Haplorthox	0–15	77	41.8	31.9	9.9
(Puerto Rico)	15–33	77	40.3	33.6	6.7
	33–51	84	43.7	35.7	8.0
	51–86	83	42.1	34.2	7.9
	86–117	70	35.1	30.0	5.1
Eutrustox	5–15	71	27.7	23.5	4.2
(Minas Gerais,	20–30	72	27.9	24.8	3.1
Brazil)	40–50	68	26.3	24.7	1.6
	60–70	70	26.6	24.7	1.9
	70–80	60	28.1	25.0	3.1
Tropohumult	0–10	64	47.5	32.1	15.4
(Puerto Rico)	10–23	69	39.8	31.4	8.4
	23–38	63	39.0	31.4	7.6
	38–64	48	36.7	30.1	6.6
	64–81	39	40.7	28.6	12.1
	81–114	33	39.6	28.0	11.6
Eutrandept	0–5	—	47.1	29.9	17.2
(Hawaii)	5–12	—	41.5	29.1	12.4
	12–20	—	46.1	32.8	13.3
	20–90	—	61.4	48.3	13.1
	90–100	—	77.4	59.2	18.2
Aquept	0–20	61	48.5	29.9	18.3
(Philippines)	20–37	57	58.9	34.2	24.6
	37–58	50	47.0	36.3	20.7
Ustert	0–18	52	32.4	20.4	12.0
(Puerto Rico)	18–36	55	40.6	23.5	17.1
	36–58	62	40.8	27.1	13.7
	58–86	63	38.6	26.2	12.4

Oxisols, and a clayey Oxisol. It is fascinating to realize that soils with such a range in texture had similar water movement properties. By comparison a clayey montmorillonitic soil averaged 10 cm/hour. The higher rates in the Ultisols were attributed to lateral water movement, primarily through cracks present because of mixed 1:1 and 2:1 mineralogy and the heavier argillic horizon below. The importance of such cracks in water movement is great. Later in the rainy season many of these cracks seal and reduce lateral water movement, which is probably less important in Oxisols, although it may occur in deeper layers.

ADAPTING CROPS TO VARIABLE WATER SUPPLY

Rainless periods of up to 2 weeks occur sporadically during most rainy seasons, causing severe water stress in many crops. These drought periods can cause severe yield reductions when they occur at critical growth stages. Often only the A horizon is affected; shallow-rooted crops suffer greatly, while the subsoil remains relatively well supplied with water. Root development in the subsoil may be restricted because high levels of exchangeable aluminum or low calcium levels prevent roots from reaching the place where available water is plentiful.

Results published in Wolf's thesis (1975) on Oxisols of Brasilia, Brazil, show that, in over half of the 30 years for which rainfall data were available, severe drought periods of 1 week or more occurred during the rainy season. Because of the low available water range of these soils and their aluminum-toxic subsoil, corn starts wilting after 6 rainless days. An example of the effect of short-term droughts on soil moisture storage is shown in Fig. 3.7.

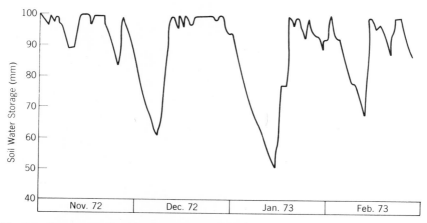

Fig. 3.7 Computed soil moisture storage during the first 4 months of the rainy season in a clayey Oxisol of Brazil. *Source:* North Carolina State University (1973).

Management practices designed to promote deeper rooting, such as deeper lime incorporation in acid soils, drastically decrease water stress.

Time of planting and crop growth duration should be manipulated to match the crop requirements with the probable moisture supply expected for the rainy season. In ustic environments the beginning of the rainy season is marked by sporadic and intense rain showers. Farmers usually wait for the rains to become firmly established and the soil adequately moistened before they start their planting operations. A study by Kowal and Andrews (1973) in an Alfisol of northern Nigeria illustrates a fairly precise matching of cultural practices for a locally grown sorghum variety, which consumes about 700 mm of water during its 6 month growing season, and a total rainfall of about 1100 mm during a 5 month rainy season, followed by almost no rain during the rest of the year.

Four distinct periods are recognized by Kowal and Andrews and are illustrated in Table 3.8. Period I comprises the first 3 weeks of rains, which are too sporadic for actual planting, but sufficient to moisten the topsoil for adequate land preparation. A surplus of 35 mm of soil water accumulates during this period. Period II comprises the first 2 months after sorghum is seeded. Heavy rainfall (350 mm) recharges the profile since the crop evapotranspiration is only 214 mm. Period III of 70 days' duration further recharges the profile with an accumulated surplus of 309 mm. The crop begins its reproductive phase during the middle of this period with little risk of water stress. The valley water table also rises during this period. Excess moisture is lost as runoff, and erosion occurs. Period IV consists of the last

Table 3.8   Average Water Supply and Use by a Local Sorghum Variety under Rainfed Conditions at Samaru, Nigeria (mm H$_2$O)

Period	Rainfall	Evapotranspiration	Soil Water Surplus or Deficit	
			Period	Cumulative
I. (May 1–20)	76	41[a]	+ 35	+ 35
II. (May 21–July 21)	350	214	+136	+171
III. (July 21–Sept. 30)	607	298	+309	+480
IV. (Oct. 1–Nov. 30)	36	155	−119	+361
V. (Dec. 1–April 30)	38	823[b]	−785	−424
Total	1107	1531	−424	

Source: Adapted from Kowal and Andrews (1973).
[a] Moistening topsoil for cultivation.
[b] Evaporation only.

60 days of crop growth from flowering to harvest. The dry season has definitely started, but the crop does not suffer from water stress because the ample stored moisture is used. The high solar radiation during this period is also helpful in increasing yields and facilitating ripening. The 6 month growing season in this case is beautifully synchronized with the 4 month rainy season. A fifth period consists of the bulk of dry season, when moisture surpluses are exhausted by evaporation.

One may expect high-yielding, earlier maturing varieties to have problems in adapting to this rainfall regime. If planted at the same time, they mature during the peak of the rainy season, making harvest difficult and producing moldy, poor-quality grain. If they are planted late to coincide with the harvest of the local varieties, the soil is too wet for good seedbed preparation.

The local 6 month variety has a yield potential of about 2.2 tons/ha of grain with optimum fertility and 0.7 ton/ha without fertilization. The 100 day American hybrids can yield from 3 to 5 tons/ha in Samaru when fertilized. If such hybrids can be adapted to the moisture regime, production may increase substantially. Is this possible? The hybrids can be planted at the same time (May 20) and harvested in the middle of August at the peak of the rainy season. They will probably yield about 2 to 3 tons/ha of poor-quality grain not suitable for human consumption but good for feed. By cutting the stalks by hand and giving an extra application of nitrogen, a ratoon crop can be grown without land preparation. This ratoon crop will probably mature in about 60 days and have the full benefit of the high solar radiation and ample soil moisture supply of Period IV. A total of 8 or 9 tons/ha of grain sorghum per year could be collected, or 10 times the average yield, with change in variety and fertilization, as long as the timing of operations was adapted to the moisture regime. This example shows how a rational adaptation of new varieties, coupled with environmental realities, can be achieved.

Rainfall variability in udic soil moisture regimes is also considerable and affects yields in rain-fed areas. Monthly plantings of traditional and improved upland rice varieties in Yurimaguas, Peru, over a period of 18 months demonstrated that high yields were correlated with monthly rainfall values of 150 mm or more during the 4 month growing season (Fig. 3.8). When rainfall distribution was of this order, the traditional variety yielded 2 or 3 tons/ha, while the improved variety doubled that amount. Under poor rainfall distribution the traditional variety averaged 0.5 to 0.8 ton/ha; the improved variety, 1.0 to 1.2 tons/ha. This example shows the close association between soil moisture supply and yields in udic regimes, when other factors are held constant, and the consistent benefits of planting a high-yielding short-statured variety adapted to local conditions. Several studies

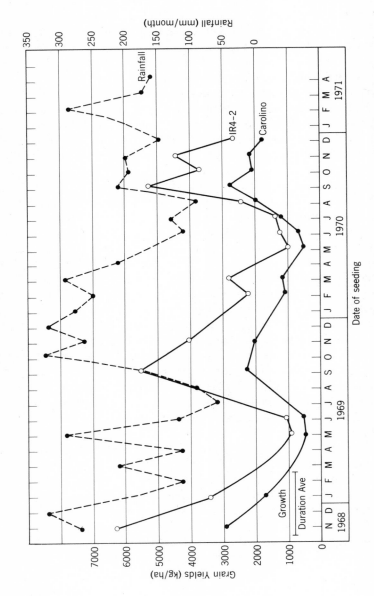

Fig. 3.8 Performances of IR4-2, a short-statured rice variety, and Carolino, the local variety, as a function of date of planting under upland conditions and available rainfall on a Tropaqualf from the Amazon of Peru. *Source:* Sanchez (1972).

cited in the references provide additional examples of the effect of planting dates.

LEACHING LOSSES

Rainfall in excess of the soil's storage capacity causes a series of soil and nutrient losses through leaching, runoff, and erosion. Estimates of leaching losses available from a few tropical locations are summarized in Table 3.9. The amounts vary considerably with annual leaching volume, soil properties, and crops grown. The three West African examples were obtained with *in situ* lysimeters (Charreau, 1972). They show progressively higher nutrient losses with increasing annual rainfall. The high leaching losses of nitrogen and potassium in the Ivory Coast are probably due to the fertilization program on banana plantations, as compared with the rubber plantation, which was not fertilized.

The data from the Philippines were obtained on a much younger soil grown to flooded rice (Reyes et al., 1961). The low nitrogen values reflect the absence of easily leached nitrates under reduced conditions. The differences between the two controlled leaching rates suggests a similar composition of the leachate.

The Colombian study was carried out in highly porous Andepts under heavy annual rainfall, which produced extremely high leaching losses for nitrates, calcium, and magnesium (Suarez de Castro and Rodriguez, 1955,

Table 3.9   Average Annual Leaching Losses at Several Tropical and Temperate Locations

Location	Soil	Cover	Annual Rainfall (mm)	Annual Leaching (mm)	Nutrients Leached (kg/ha)					
					N	P	K	Ca	Mg	S
Senegal	Ustalf	Cropped	660	118	6	0.13	8	31	12	7
Ivory Coast	Ultisol	Rubber	1569	845	79	2.90	63	31	40	—
Ivory Coast	Ultisol	Bananas	2040	828	235	1.20	24	256	113	—
Philippines	Alfisol	Flooded rice	2000	312	2	0.03	16	78	91	8
Philippines	Alfisol	Flooded rice	2000	1248	11	0.13	60	391	91	89
Colombia	Andept	Bare	2530	1771	249	0.23	202	776	232	—
Colombia	Andept	Pasture	2530	1450	204	0.21	163	878	251	—
New York	Ochrept	Bare	810	623	62	Tr.	65	358	57	48
New York	Ochrept	Cropped	810	467	7	Tr.	51	207	40	39

Sources: Adapted from Charreau (1972), Reyes et al. (1961), Suarez de Castro and Rodriguez (1958), and Buckman and Brady (1969).

1958). For comparison, temperate data are included when large calcium losses are reported in a young soil. When bare and cropped soils are compared, leaching losses generally decrease because of crop uptake. In general, calcium, magnesium, potassium and nitrogen are the main nutrients lost through leaching. Phosphorus losses are negligible except in extremely sandy soils. Micronutrient leaching losses were measured only in the Philippine study, which gave the following ranges: 19 to 144 kg Fe/ha a year and 4 to 30 kg Mn/ha a year. Reduced soil conditions probably favored iron and manganese movement in this flooded soil.

The high leaching losses reported for calcium in the well-aggregated Ultisols from the Ivory Coast and in the Colombian Andept suggest that calcium applied as lime to the soil surface may move downward and increase the base status of the subsoil. Morelli et al. (1971) studied this phenomenon in an Andept from Costa Rica and showed that when lime was incorporated in the topsoil, the calcium content of the subsoil increased when measured 3½ years later. Relatively low rates of 2.8 tons/ha of lime increased the pH from 4 to over 5.5 in the 20 to 60 cm layer with a corresponding increase in exchangeable calcium. This downward calcium movement can decrease aluminum saturation and consequently increase the effective rooting depth and water uptake of some soils. These effects have practical importance and should be given careful attention in areas of acid subsoils and well-aggregated soils.

RUNOFF

When the soil becomes saturated quickly, high-intensity rains may cause considerable runoff even on gentle slopes. The values will depend on the porosity of the soil, its moisture content, rainfall intensity, and soil cover. Table 3.10 shows the average values for runoff in four West African locations under forest, cultivation, and bare soil conditions. The data show that soil cover is the predominant factor affecting erosion, in spite of differences in slope and rainfall. The lower runoff values in cultivated soils than in bare ones underscore the need to keep a constant soil cover in these sandy Alfisols or Ultisols with fairly poor structure.

In a well-aggregated Andept from Chinchiná, Colombia, considerable runoff amounting to 62 percent of the annual rainfall was measured by Suarez de Castro and Rodriguez (1955, 1958) on bare soils with 22 percent slopes. Table 3.11 shows that large quantities of calcium and magnesium were washed away in the process. When pastures or coffee plantations were established, runoff, erosion, and leaching losses decreased drastically.

Runoff losses from artificial high-intensity rainstorms were evaluated by

Table 3.10   *Magnitudes of Soil Runoff in Four Locations of West Africa*

Locality	Slope (%)	Annual Rainfall (mm)	Runoff (% annual rainfall) Forest Land	Cultivated Land	Bare Soil
Ouagadougou, Upper Volta	0.5	850	2	2–32	40–60
Sefa, Senegal	1.2	1300	1	21	40
Bouaké, Senegal	4.0	1200	0.3	0.1–26	15–30
Abidjan, Ivory Coast	7.0	2100	0.1	0.5–20	38

Source: Charreau (1972).

Barnett et al. (1972) on steep Ultisols and Inceptisols of Puerto Rico. Substantially lower runoff and nutrient removal were observed in the better aggregated Ultisols. Runoff losses are often deposited in the same fields, resulting in no net farm losses although important local effects are created.

EROSION

In spite of the generally higher aggregate stability of many tropical soils, the extent of erosion in the tropics is reaching alarming proportions. For example, the average value of annual soil erosion in Africa and South

Table 3.11   *Effect of Cropping Systems and Conservation Practices on Annual Soil Erosion and Nutrients in the Runoff on a Dystrandept in Chinchiná, Colombia: 6 Year Average; 2775 mm Rainfall*

Treatment	Slope (%)	Soil Eroded (tons/ha)	Runoff (mm)	Nutrient (kg/ha) N	P	K	Ca	Mg
Bare soil, tilled monthly	22	225.4	1730	25	0.98	24	238	152
Pasture	22	7.1	513	7	0.15	6	25	26
Young coffee plantation	45	1.8	190	8	0.04	2	6	7
Young coffee plantation with terraces	45	0.2	410	4	0.14	4	8	9
Old coffee plantation without soil conservation practices	55	0.6	59	1	0.08	1	2	2

Source: Suarez de Castro and Rodriguez (1955), 1958).

America is about 7 tons/ha, whereas in Europe it is 0.8 ton/ha (Dudal, unpublished data). The principal causes of erosion are deforestation, overgrazing of pasture lands, and poor use of shifting cultivation practices. Deforestation is associated primarily with logging operations and to a lesser extent with agricultural practices. Erosion resulting from soil exposure in shifting cultivation systems occurs only when population pressures reduce the length of the fallow period or when displaced urban dwellers unfamiliar with the system are forced to practice shifting cultivation. Traditional shifting cultivation systems cause minimal soil erosion even on steep slopes and in high-rainfall areas because of the short time that the soil is actually exposed. Overgrazed pastures result in bare spots and compacted paths in which gullies eventually form. This may be the primary form of soil erosion in savanna areas.

In many highland areas where arable land is scarce, extremely sophisticated soil conservation systems have been developed and practiced for centuries. The terraced areas in the Andes of Peru, central Luzon, and other places are examples of excellent soil conservation practiced in traditional systems. It may be safe to generalize that traditional management systems are well geared toward soil conservation. The problems arise when new practices or new people move into an area or when the old systems break down because of overpopulation.

General estimates of the magnitude of erosion losses in West Africa are shown in Table 3.12. The importance of a continuous soil cover is also underlined by this information. The extent of erosion in Africa is probably the greatest of all tropical regions because of changes from shifting to permanent cultivation on easily erodible soils. That research workers are aware of this problem is shown by the fact that 78 papers on erosion were presented at the two Inter-African Soils Conferences held in Leopoldville in 1954 and Dalaba in 1959. The deep plowing practices recommended for sandy soils of Senegal, illustrated in Table 3.5, resulted in a reduction of soil erosion from an average of 10 tons/ha a year with shallow plowing to

Table 3.12   Magnitudes of Annual Soil Erosion Losses in Four Localities of West Africa (tons/ha)

Locality	Forest Land	Cultivated Land	Bare Soil
Ouagadougou, Upper Volta	0.1	0.6–8.0	10–20
Sefa, Senegal	0.2	7.3	21
Bouaké, Senegal	0.1	0.1–26	18–30
Abidjân, Ivory Coast	0.03	0.1–90	108–170

Source: Charreau (1972).

3.7 tons/ha a year with deep plowing because of an increase in soil porosity and better crop growth (Charreau, 1972).

In Oxisols of São Paulo State, Brazil, Marques and Bertoni (1961) determined that annual erosion losses increased from 8 to 16 tons/ha with deeper plowing in cotton fields. The apparent contradiction between these two reports is due to the drastically different physical properties of the soils: highly aggregated Oxisols in the latter case and poorly aggregated sandy Alfisols in the former one.

The principles of soil conservation in the tropics are no different from those developed in the temperate region. The high intensity of some tropical rainstorms and the lower availability of capital, however, require some special adaptations. Planting along the contours, instead of up and down the slope, is usually a feasible and sound practice, but a difficult one to introduce to farmers used to plowing up and down the slopes. Keeping the soil covered with crop residues or under cultivation during periods of expected high rainfall is another sound principle. Intensifying cattle grazing and shifting cultivation beyond its limits are probably the greatest problems. Planting tree crops or pastures on steep slopes, however, seldom requires additional conservation practices, such as terracing, in well-aggregated soils. Compaction by heavy tractor traffic, particularly on sugarcane plantations, has caused severe problems, and adequate alternatives must be developed. Excessive use of machinery for land preparation is common in areas where tractors have recently been introduced. Studies by Marques et al. (1961) and Primavesi (1964) in Brazil, Roose (1967, 1970) and Pereira et al. (1967) in Africa, Smith and Abruña (1955) in Puerto Rico, and Alles (1958) in Sri Lanka provide good evidence for the principles stated above.

Minimum or no tillage is an ancient practice in the tropics. In many shifting cultivation systems the planting stick is the only tool inserted into the soil. Fertilization practices, however, generally require some degree of tillage. Vicente-Chandler et al. (1966) compared tilled and untilled practices with several crops in three steep mountain soils of Puerto Rico. Weeds were controlled by herbicides, and both fertilizers and lime were applied to the surfaces of these porous soils. The results shown in Table 3.13 indicate no differences in yields between tillage and no tillage for corn, plantains, several root crops, tobacco, and sugarcane. The harvesting of root crops, however, requires significant soil disturbance.

SOIL TEMPERATURE

Soil temperature is seldom considered a serious limiting factor in the tropics. As mentioned in Chapter 1, soil temperatures approximate air temperatures at about 50 cm depth; these are usually adequate for most

Table 3.13   Effects of Tillage on Crop Yields in Steep Soils of Central Puerto Rico (tons/ha)

Crop	Catalina Clay (Orthox)		Mucara Clay (Tropept)	
	Tilled	Untilled	Tilled	Untilled
Sugarcane	111	109	66[a]	80[a]
Corn	3.92	4.14	6.72	8.16
Plantains	19.6	17.0	19.2	18.9
Sweet potatoes	10.1	9.5	—	—
Yams	15.9	15.1	14.9	13.5
Taniers	7.5	8.3	13.3	15.2
Tobacco, cured leaves	1.74[a]	1.55[a]	1.92	1.85

Source: Adapted from Vicente Chandler et al. (1966).
[a] Statistically significant differences.

tropical crops. There are two instances, however, in which soil temperatures can be limiting: excessively high temperature in certain sandy topsoils, and cool temperatures in the tropical highlands.

Controlling Excessively High Soil Temperatures

Large areas of West Africa are covered with Alfisols having a high sand content and gravel in the topsoil. The thermal diffusivity of these horizons is much lower than that of loamy or clayey topsoils. Consequently, they can retain large quantities of heat, particularly when dry. Studies in Ibadan, Nigeria, have shown that soil temperatures can reach 42°C at 5 cm depth and 38°C at 10 cm depth in such soils when bare or recently planted (IITA, 1972; Lal, 1974; Lal et al., 1975). These temperatures were found to inhibit the emergence of yam sprouts and to virtually stop the growth of corn and soybean seedlings when topsoil temperatures were about 35 to 38°C. Furthermore, Lal and his coworkers observed that nutrient uptake, nutrient translocation, and water uptake were severely affected by such high temperatures. Topsoil temperatures decrease as crops grow and a canopy is established. The solution to this problem consists of mulching with straw or stover from a previous crop. Mulching rapidly decreased soil temperatures down to 20 cm depth, as shown in Fig. 3.9. The unmulched plots had supraoptimal high temperatures down to 20 cm depth during early stages of growth. As crops develop a canopy, soil temperatures decrease; but if a severe drought occurs later in the season, the mulch can prevent excessively high temperatures (Lal, 1974). Mulching also increases soil moisture storage throughout the rainy season, as shown in Fig. 3.9. These increases in soil moisture were equivalent to about 8 to 12 percent of the rainfall received. In this example the overall effect of mulching was to increase corn yields from 3.6 to 5.4 tons/ha and to decrease weed growth to one-third of that observed in the unmulched plots (Lal, 1974).

There is evidence which suggests that these mulching principles should be applied to a wider range of soils. When a forest is cleared, topsoil tempera-

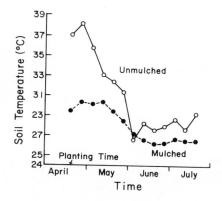

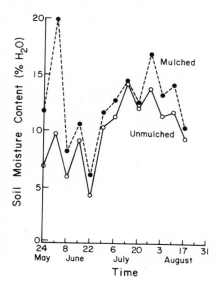

Fig. 3.9 Effect of mulching on soil temperatures at 3 P.M. at 5 cm depth (above) and on soil moisture storage (below) in the top 10 cm of a sandy Alfisol of Ibadan, Nigeria. *Source:* IITA (1972).

Plates 7 and 8 Beneficial effect of mulching on the early growth of corn in a sandy Ultisol of Yurimaguas, Peru. Plate 7 shows a bare field, and Plate 8 the effect of mulching in an adjacent plot.

124

tures increase by 7 to 11°C (Sanchez, 1973) because of the higher solar radiation reaching the soil surface upon exposure. These higher temperatures undoubtedly accelerate organic matter decomposition rates and in some cases adversely affect soil structure (Cunningham, 1963). Nothing is gained and much is to be lost by maintaining a bare soil surface. Consequently, the control of soil temperature should be an important component of soil management systems. This can be accomplished by keeping a crop canopy the year round and by mulching during exposed periods. The advantages of crop residue mulches are many: lower soil temperatures, higher soil moisture storage, fewer weed problems, lower structural deterioration because of rainfall impact, lower organic matter decomposition rates, and even increases in nutrients released while the mulch decomposes.

Controlling Cool Soil Temperatures

In the tropical highlands with isothermic or isomesic soil temperature regimes, many adapted crops suffer from low-temperature limitations, particularly during wet periods. A wet soil also has low thermal diffusivity. When certain crops are grown close to their minimum temperatures, soil temperature becomes a limiting factor. Such is the case with certain pineapple plantations on Oxisols in the highlands of Hawaii. Low soil temperatures exert a negative influence on nutrient uptake and consequently result in longer growth duration and lower yields (Ravoof et al., 1973).

The solution to this problem is also mulching, but this time with clear plastic. Unlike the type discussed above, a clear plastic mulch is a vapor barrier that substantially increases soil temperatures through a greenhouse effect. Although expensive, certain techniques developed in Hawaii enable the mulch to last for a sufficiently long time to make its use profitable. Transparent polyethethylene mulches have also increased mean soil temperatures and decreased irrigation requirements of potatoes in northern India, where low temperatures in the late fall season limit potato growth. Grewal and Singh (1974) observed that potato yields increased from 14 to 20 tons/ha when such a mulch was used. This increase was positively correlated with the rise in minimum soil temperatures caused by the polyethylene mulch. Lower yield responses and soil temperature increases were obtained with a straw mulch, which can be used, however, if clear plastic is not available.

SUMMARY AND CONCLUSIONS

1. The management of soil physical properties in the tropics is highly soil-specific. Attention should be diverted from the exaggerated concern regard-

ing plinthite hardening at the soil surface to the specific physical peculiarities of three groups of tropical soils.

2. The highly aggregated Oxisols, Andepts, and oxidic families of other orders present tremendous advantages and serious disadvantages. The advantages lie in the very stable aggregates coated with oxides and organic matter, which extend the time period of tillage operations because they drain gravitational water much like sands. Strong aggregation also diminishes compaction and erosion problems. The disadvantages lie in the low range of available moisture of these soils in spite of their capability to hold large quantities of water at high tensions inside the aggregates. Many clayey Oxisols are actually droughty soils, and leaching in these soils is considerable. Their unique properties require that the available moisture range be redefined as 0.1 bar for "field capacity" to about 1 bar as the upper limit.

3. The large areas of Ultisols and Alfisols with sandy topsoil textures and similar soils present a completely different set of management problems. Exposure and cultivation can easily lead to serious soil compaction, runoff, and erosion. The moisture range suitable for tillage operations is narrower than in the case of the first group. On the positive side, many of these soils hold more available water in their profiles than do clayey Oxisols, particularly when they have a clayey argillic horizon. Protecting the soil surface with mulches and/or a continuous crop canopy is the best management alternative. In steep areas minimum or no tillage is almost essential. In certain exposed and compacted areas deeper plowing is required in order to increase porosity, promote root development, and increase crop yields.

4. The third large group of tropical soils for the purposes of this chapter consists of loamy to clayey soils with layer silicate mineralogy and other properties similar to those of temperate-region soils. These include other Alfisols, Ultisols, and Inceptisols, as well as the other orders. Among them, Vertisols present special physical limitations due to severe shrinking and swelling and the narrower moisture range suitable for tillage operations. Management practices developed in the temperate region for such soils should apply with a minimum of adjustment to tropical soils with layer silicate mineralogy.

5. Water stress during the rainy season or throughout the year in udic environment is common because of short-term droughts. When acid subsoils prevent deep root development, the attenuation of subsoil acidity may permit roots to tap the stored subsoil moisture.

6. Many traditional cropping systems show excellent synchronization between crop moisture requirements and available soil moisture supply.

When attempting to increase yields through the introduction of high-yielding varieties of different growth duration, success will depend on the degree to which the new cultural practices mesh with the moisture regime.

7. Soil temperature, either too high or too low, is a limiting factor in certain areas of the tropics. The solution is usually to mulch with various materials, depending on whether temperatures should be decreased or increased. In soils recently cleared for cultivation, management of soil temperature via straw mulches or by keeping a crop canopy as long as possible is extremely important to neutralize the increases in topsoil temperatures with exposure and the correspondingly faster rate of organic matter decomposition, which can reduce infiltration rates in soils low in iron and aluminum oxide coatings. In addition, mulching decreases water consumption and the need for weed control and often increases yields. This practice should receive serious consideration in most tropical soil management systems.

8. The management of soil physical properties is usually of lower priority than the management of chemical properties in traditional agricultural systems. Such farmers always prefer soils of higher native fertility because physical problems are of little concern when only a planting stick and the harvesting of root crops disturb the soil. As management systems become more intensive and mechanized and fertility problems are solved economically through fertilization and liming, soil physical properties that limit the efficient use of machinery become a critical management concern.

REFERENCES

Abruña, F. and J. Lozano. 1974. Effect of season of the year on yields of 13 varieties of rice growing in the humid region of Puerto Rico. *J. Agr. Univ. Puerto Rico* **58**:11–18.

Akehurst, B. C. and A. Sreedharan. 1965. Time of planting—a brief review of experimental work in Tanganyika, 1956–1962. *East Afr. Agr. For. J.* **30**:189–201.

Alles, W. S. 1958. Some studies on runoff and infiltration. *Trop. Agriculturalist (Ceylon)* **114**:197–206.

Alvarado, A. and S. W. Buol. 1975. Toposequence relationships of Dystrandepts in Costa Rica. *Soil Sci. Soc. Amer. Proc.* **39**:932–937.

Ballico, P. 1971. Considerazioni sull andamento delle temperature del terreno in regioni temperate e tropicale. *Rev. Agr. Subtrop. e Trop. (Italy)* **65**:315–329.

Barnett, A. P., J. R. Carrecker, F. Abruña, et al. 1972. Soil and nutrient losses in runoff with selected cropping treatments in tropical soils. *Agron. J.* **64**:391–395.

Baver, L. D. 1972. Physical properties of soils. Pp. 50–62. In *Soils of the Humid Tropics*. National Academy of Sciences, Washington.

Bertoni, J. and F. I. Pastana. 1964. Relação chuvas perdas por eroçào em diferentes tipos de solo. *Bragantia* 23:3–11.

Bertrand, R. 1971. Response de l'enraicinement du riz de plateau aux caractères physique et chemique du sol. *Agron. Tropicale (France)* 26:376–386.

Bhushan, C. S. and B. P. Ghildyal. 1971. Influence of shape of implements on soil structure. *Indian J. Agr. Sci.* 41(9):744–751.

Biswas, T. D., B. L. Jain, and S. S. Bains. 1971. Response of wheat and potato to physical properties of the soils obtained under different crop rotations. *Proc. Int. Symp. Soil Fert. Eval. (New Delhi)* 1:475–485.

Brams, E. and R. Weikel. 1971. Maize as a second crop for the uplands of Sierra Leone, West Africa. *Prairie View A & M Coll. Bull.* 1. Prairie View, Tex.

Brewer, R. and J. R. Sleeman. 1960. Soil structure and fabric: Their definition and description. *J. Soil Sci.* 11:172–185.

Briggs, L. J. and H. L. Shantz. 1914. Relative water requirements of plants. *J. Agr. Res.* 3:1–63.

Briones, A. A. and B. G. Cagauan, Jr. 1964. Moisture retention of some paddy soils in Laguna. *Philipp. Agr.* 48:61–81.

Briones, A. A. and J. G. Veracion. 1965. Aggregate stability of some red soils of Luzon. *Philipp. Agr.* 49:153–167.

Buckman, H. O. and N. C. Brady. 1969. *The Nature and Properties of Soils.* Macmillan, New York.

Cagauan, B. and G. Uehara. 1965. Soil anistropy in relation to aggregate stability. *Soil Sci. Soc. Amer. Proc.* 29:198–200.

Chan, P. Y. and E. Z. Arlidge. 1971. Moisture relations in some Mauritius soils. Pp. 74–78. Annual Report, 1970. Mauritius Sugar Industry Research Institute.

Charreau, C. 1972. Problemes poses par l'utilisation agricole des sols tropicaux par des cultures annuelles. *Agron. Tropicale (France)* 27:905–929.

Charreau, C. and R. Nicou. 1971. L'amélioration du profil cultural dans les sols sableux et sablo-argileux de la zone tropicale seche Ouest-Africane et sis incidence agronomiques. *Agron. Tropicale (France)* 26:209–255, 903–978, 1183–1247.

Craufurd, R. Q. 1964. The relationship between sowing date, latitude, yield and duration of rice. *Trop. Agr. (Trinidad)* 41:213–224.

Cunningham, R. K. 1963. The effect of clearing a tropical forest soil. *J. Soil Sci.* 14:334–345.

DeDatta, S. K. and P. M. Zarate. 1970. Environmental conditions affecting growth characteristics, nitrogen response, and grain yield of tropical rice. *Biometeorology* 4:71–89.

Deshpande, T. L., D. J. Greenland, and J. P. Quick. 1968. Changes in soil properties associated with the removal of iron and aluminum oxides. *J. Soil Sci.* 19:108–122.

Dowkar, B. D. 1964. A note on the reduction of yield of Taboran maize by late planting. *East Afr. Agr. For. J.* 30:33–34.

Drinkow, R. 1970. Some net-radiation and soil thermal diffusivity measurements for the Port Moresby area. *Papua New Guinea Agr. J.* 21:112–117.

El-Swafy, S. A. 1973. Structural changes in tropical soils due to anions in irrigation water. *Soil Sci.* 115:64–72.

Escobar, G., R. Jurado, and R. Guerrero. 1972. Propiedades físicas de algunos suelos derivados de ceniza volcánica del Altiplano de Pasto, Nariño, Colombia. *Turrialba* 22:338–346.

Farbrother, H. G. 1972. Field behavior of Gezira clay under irrigation. *Cotton Growing Rev.* **49**:1–27.

Fernandes, B. E. and J. D. Sykes. 1968. Capacidade de campo e retenção de agua en tres solos de Minas Gerais. *Ceres* **15**:1–39.

Forsythe, W. and R. Diaz-Romeu. 1969. La densidad aparente del suelo y la interpretación del análisis de laboratorio para el campo. *Turrialba* **19**:128–131.

Forsythe, W. M., S. A. Gavande, and M. Gonzalez. 1969. Physical properties of soils derived from volcanic ash. Pp. B3.1–B3.7. In *Panel on Soils Derived from Volcanic Ash in Latin America*. Inter-American Institute of Agricultural Sciences, Turrialba, Costa Rica.

Forsythe, W. M. and O. Vazquez. 1973. Effect of air drying on the water retention curves of disturbed samples of three Costa Rican Soils derived from volcanic ash. *Turrialba* **23**:200–207.

Gavande, S. A. 1968. Water retention characteristics of some Costa Rican soils. *Turrialba* **18**:34–38.

Glover, J. 1967. The relationship between total seasonal rainfall and maize yields in the Kenya highlands. *J. Agr. Sci.* **49**:285.

Godefroy, J., M. Muller, and E. Roose. 1970. Estimation des partes par lixiviation des elements fertilisants daris un sol bananerie de basse Cote D'Ivoire. *Fruits* **25**(6):403–420.

Godoy, O. P. 1961. Influencia de epoca de semeadura na produção das variedades de arroz. *Rev. Agr. (Brazil)* **36**:171–177.

Goedert, W. and M. T. Beatty. 1971. Caracterização de grummusolos no sudoeste do Rio Grande do Sul. I. Propiedades fisicas adversas ao uso. *Pesq. Agropec. Bras.* **6**:91–102.

Grohmann, F. 1960. Distribução de tamanho de poros en tres tipos de solos do Estado de São Paulo. *Bragantia* **19**:319–328.

Goldson, J. R. 1963. The effect of time of planting on maize yields. *East Afr. Agr. For. J.* **29**:160–163.

Gray, R. W. 1970. The effect of yield of the time of planting of maize in southwest Kenya. *East Afr. Agr. For. J.* **35**:291–298.

Green, R. E., R. L. Fox, and D. D. F. Williams. 1965. Soil properties determine water availability to crops. *Hawaii Farm Sci.* **14**:6–9.

Grewal, S. S. and N. T. Singh. 1974. Effect of organic mulches on the hydrothermal regime of soil and growth of potato crop in northern India. *Plant and Soil* **40**:33–47.

Hanna, L. W. 1971. The effects of water availability on tea yields in Uganda. *J. Appl. Ecol.* **8**:791–813.

Hardy, F. 1933. Cultivation properties of tropical red soils. *Emp. J. Exptal. Agr.* **1**:103–112.

Hardy, F. 1936. Soil crumb. *Trop. Agr. (Trinidad)* **13**:143–145.

Hardy, F. and L. F. Derraugh. 1947. The water and air relations of some Trinidad sugarcane soils. *Trop. Agr. (Trinidad)* **24**:76–87, 111–121.

Haridasan, M. and R. K. Chibber. 1971. Effect of physical and chemical properties on the erodibility of some soils of the Malwa Plateau. *J. Indian Soc. Soil Sci.* **19**:293–298.

Hazra, C. R., S. B. Ray, and R. D. Biswas. 1973. Effect of stored soil moisture reserves through conservation practices and supplemental irrigation on wheat yields under dryland farming conditions. *J. Indian Soc. Soil Sci.* **21**:9–22.

Holford, I. C. R. 1971. Effect of rainfall on the yield of groundnuts in Fiji. *Trop. Agr. (Trinidad)* **48**:171–175.

IITA. 1972. Farming systems program. Annual Report. International Institute for Tropical Agriculture, Ibadan, Nigeria.

IRRI. 1964. Annual Report. P. 190. International Rice Research Institute, Los Baños, Philippines.

Jaquot, M. 1972. Quelques observations sur l'influence du milieu dans des cultures de riz pluvial. *Agron. Tropicale (France)* 27:1007- 1002.

Joshi, S. N. and M. M. Kabaria. 1972. Effect of rainfall distribution on the yield of groundnuts. *Indian J. Agr. Sci.* 42:681–685.

Kalma, J. D. 1971. The annual course of air temperatures and near surface soil temperature in a tropical savanna environment. *Agr. Meteorol.* 8:293–303.

Kolarkar, A. S. and N. Singh. 1971. Moisture storage capacities in relation to textural composition of western Rajahstan soil. *Ann. Arid Zone* 10:29–32.

Kowal, J. 1968–1969. Some physical properties of soils at Samaru, Zaria, Nigeria. I. Physical status of the soil. II. Storage of water and crop use. *Nigerian Agr. J.* 5:13–20, 6:18–29.

Kowal, J. 1970. The hydrology of a small catchement basin at Samaru, Nigeria. III. Assessment of surface runoff under varied land management and vegetation cover. IV. Assessment of soil erosion under varied land management and vegetation cover. *Nigerian Agr. J.* 7:120–147.

Kowal, J. 1972. Effect of an exceptional storm on soil conservation at Samaru. *Samaru Res. Bull.* 141.

Kowal, J. and D. F. Andrews. 1973. Pattern of water availability and water requirement for grain sorghum production at Samaru, Nigeria. *Trop. Agr. (Trinidad)*, 50:89–100.

Krishman, A. and R. S. Kushwaha. 1972. Analyses of soil temperature in the arid zone of India. *Agr. Meteorol.* 10:55–64.

Lal, R. 1973. Effects of seedbed preparation and time of planting maize in Western Nigeria. *Exptal. Agr.* 9:303–314.

Lal, R. 1974. Soil temperature, soil moisture, and maize yields from mulched and unmulched tropical soils. *Plant and Soil* 40:129–143.

Lal, R. 1975. Role of mulching techniques in tropical soil and water management. *IITA Tech. Bull.* 1.

Lal, R., B. T. Kang, F. R. Moorman, A. S. R. Juo, and J. C. Moomaw. 1975. Soil management problems and possible solution in western Nigeria. Pp 372–408. In E. Bornemisza and A. Alvarado (eds.); *Soil Management in Tropical America*. North Carolina State University, Raleigh.

Lugo-Lopez, M. A. 1951. Functional relationships between moisture at several equilibrium points and the clay contents of tropical soils. *J. Agr. Univ. Puerto Rico* 35:66–70.

Lugo-Lopez, M. A. 1953. Moisture relationships of Puerto Rican soils. *Univ. Puerto Rico Agr. Exp. Sta. Tech. Paper* 9.

Lugo-Lopez, M. A. 1950. Pore size and bulk density as mechanical soil factors impeding root development. *J. Agr. Univ. Puerto Rico* 44:40–44.

Lugo-Lopez, M. A. and G. Acevedo. 1956. Effects of tractor traffic compaction on the physical properties of an irrigated soil in southwestern Puerto Rico. *J. Agr. Univ. Puerto Rico* 40:235–244.

Lugo-Lopez, M. A. and J. Juarez, Jr. 1959. Evaluation of the effects of organic matter and other soil characteristics upon the aggregate stability of some tropical soils. *J. Agr. Univ. Puerto Rico* 43:268–272.

Lugo-Lopez, M. A., J. Juarez, Jr., and J. A. Bonnet. 1968. Relative infiltration rates of Puerto Rican soils. *J. Agr. Univ. Puerto Rico* **52**:233–240.

Lugo-Lopez, M. A. and R. Perez-Escolar. 1968. Functional relationships between the contents of particles smaller than 0.5 mm and 0.002 mm in size and the plasticity index of soils. *J. Agr. Univ. Puerto Rico* **52**:343–350.

Lugo-Lopez, M. A. and R. Perez-Escolar. 1969. A mathematical approach to evaluating the influence of various factors on the stability of aggregates in Vertisols. *J. Agr. Univ. Puerto Rico* **53**:57–60.

Lugo-Lopez, M. A. and M. Capiel. 1972. Seasonal changes in soil and air temperatures at three locations in Puerto Rico. *J. Agr. Univ. Puerto Rico* **56**:307–317.

Maclean, A. H. and T. V. Yager. 1972. Available water capacities in Zambian soils in relation to pressure plate measurements and particle size and analysis. *Soil Sci.* **133**:23–29.

Marques, J. Q. A. and J. Bertoni. 1961. Sistemas de preparo do solo em relação a produção e à erosão. *Bragantia* **20**:403–459.

Marques, J. Q. A. et al. 1961. Perdas por erosão no Estado de São Paulo. *Bragantia* **20**:1144–1181.

McCown, R. L. 1971. Available water storage in a range of soils in northeastern Queensland. *Austr. J. Exptal. Agr. Anim. Husb.* **11**:343–348.

McCown, R. L. 1973. An evaluation of the influence of available water storage capacity on growth season length and yield of tropical pastures using simple water balance models. *Agr. Meteorol.* **11**:53–64.

Mohr, E. C. J., F. A. Van Baren, and J. Van Schuylerborgh. 1972. *Tropical Soils*, 3rd ed. Pp. 5–13. Van Hoeve, The Hague.

Morelli, M., K. Igue, and R. Fuentes. 1971. Effect of liming on the exchange complex and on the movement of calcium and magnesium. *Turrialba* **21**:317–322.

Moura Filho, W. and S. W. Buol. 1972. Studies of a Latosol Roxo (Eutrustox) in Brazil. *Experientiae* **13**:201–234.

Mukhtar, O. M. A., A. R. Swoboda, and C. L. Godfrey. 1974. The effects of sodium and calcium chlorides on the structural stability of two Vertisols: Gezira clay from Sudan, Africa, and Houston Black Clay from Texas, USA. *Soil Sci.* **118**:109–120.

Nicou, R. 1972. *Sythese des Estudes de Physique du Sol Realisés par l'IRAT en Afrique Tropicale Seche.* Seminar on Tropical Soils Research. International Institute for Tropical Agriculture, Ibadan, Nigeria. 19 pp.

Nicou, R., L. Seguy, and G. Hadad. 1970. Comparison d l'enracinemente de cuatre varietes de riz pluvial en presence on absence de travail du sol. *Agron. Tropicale (France)* **25**:639–659.

Nyandat, N. N. 1972. Gypsum as improved of the permeability of grumosol (Typic Pellustert) in the Kano Plains of Kenya. *East Afr. Agr. For. J.* **38**:1–7.

Pereira, H. C. 1955. The assessment of structure in tropical soils. *J. Agr. Sci.* **45**:401–410.

Pereira, H. C. 1956. A rainfall test for structure of the tropical soils. *J. Soil Sci.* **7**:68–74.

Pereira, H. C., E. M. Chenery, and W. R. Mills. 1954. The transient effects of grasses on the structure of tropical soils. *Emp. J. Exptal. Agr.* **22**:148–160.

Pereira, H. C., P. H. Hosegood, and M. Dagg. 1967. Effects of tied ridges, terraces and grass leys on a lateritic soil in Kenya. *Exptal. Agr.* **3**:89–98.

Perez-Escolar, R. and M. A. Lugo-Lopez. 1968. The nature of aggregation in tropical soils of Puerto Rico. *J. Agr. Univ. Puerto Rico* **52**:227–233.

Perez-Escolar, R. and M. A. Lugo-Lopez. 1969. Availability of moisture in aggregates of various sizes in a typical Ultisol and a typical Oxisol of Puerto Rico. *J. Agr. Univ. Puerto Rico* **53**:113–117.

Pidgeon, J. D. 1972. The measurement and prediction of available water capacity of ferralitic soils in Uganda. *J. Soil Sci.* **23**:431–441.

Plá, I. and G. Campero. 1971. Algunas propiedades de la estructura de suelos de los Llanos Occidentales de Venezuela y su relacion con características del suelo. *Agron. Trop.* (*Venezuela*) **21**:433–437.

Prihar, S. S., M. R. Chowdhary, and T. M. Varghese. 1971. Effect of post-planting loosening of unstable soil on the anatomy of corn roots. *Plant and Soil* **35**:57–63.

Primavesi, A. 1964. Factors responsible for low yields of sugarcane in old cultivated "Terra Roxa Estruturada" soil. *Soil Sci. Soc. Amer. Proc.* **28**:579–580.

Ravoof, A. A., W. G. Sanford, H. Y. Young, and J. A. Silva. 1973. Effects of root temperature and nitrogen carriers on total nitrogen uptake by pineapple. *Agron. Abs.* **1973**:194.

Reyes, E. D., N. C. Galvez, and N. B. Nazareno. 1961. Lysimeter studies of Lipa clay loam grown to paddy rice. I. Leaching losses of some soil constituents. *Philipp. Agr.* **45**:246–257.

Roberts, R. C. 1933. Structural relationships in a lateritic profile. *Amer. Soil Survey Assoc. Bull.* 14, pp. 88–90.

Roose, E. 1967a. Dix anees de mesure de l'erosion et du ruissellment au Senegal. *Agron. Tropicale* (*France*) **21**:123–152.

Roose, E. 1967b. Quelques examples des effects de l'erosion hydrique sur le cultures. Pp. 1385–1404. In *Colloque sur la Fertilite des Sols Tropicaux (Tanarive).*

Roose, E. J. 1970. The relative importance of erosion, latent and vertical drainage on present pedogenesis of ferralitic soil of central Ivory Coast, *Cah. ORSTOM, Sér. Pedol.* **8**:469–482.

Roose, E. and J. Godefroy. 1968. Lessivage des elements fertilisants en bananeire. *Fruits* **23**:580–584.

Roose, E. J. and R. Bertrand. 1971. Contribution a l'etude de la methode des bandes d'arrêt pour lutter contre l'erósion hidrique en Afrique de L'Ouest. *Agron. Tropicale* (*France*) **26**:1270–1283.

Samuels, G. 1967. The influence of time of planting on food production in Puerto Rico. *Proc. Caribbean Food Crops Soc.* **5**:128–133.

Samuels, G. 1972. Influence of water excess or deficiency on leaf nutrient content and plant growth of sugarcane and other crops. *J. Agr. Univ. Puerto Rico* **56**:81–84.

Sanchez, P. A. 1972. Técnicas agronómicas para optimizar el potencial productivo de las nuevas variedades de arroz en América Latina. Pp. 27–43. In *Políticas Arroceras en América Latina.* Centro Internacional de Agricultura Tropical, Cali, Colombia.

Sanchez, P. A. 1973. Soil management under shifting cultivation. Pp. 46–48. In P. A. Sanchez (ed.), "A Review of Soils Research in Tropical Latin America." *North Carolina Agr. Exp. Sta. Tech. Bull.* 219.

Semb, G. and P. K. Garberg. 1969. Some effects of planting date and nitrogen fertilizer in maize. *East Afr. Agr. For. J.* **34**:371–379.

Sharma, M. L. and G. Uehara. 1968. Influence of soil structure on water relations in Low Humic Latosols. I. Water Retention. II. Water Movement. *Soil Sci. Soc. Amer. Proc.* **32**:765–774.

Sharma, R. B., G. P. Verma, and M. B. Russell. 1971. Effect of varying bulk density of surface and subsurface soil on wheat growth under field conditions. *Int. Symp. Soil Fert. Eval. Proc. (New Delhi)* **1**:519–527.

Silva, S., J. Vicente-Chandler, F. Abruña, and J. A. Rodriguez. 1972. Effect of season and yields of intensively managed soybeans under tropical conditions. *J. Agr. Univ. Puerto Rico* **56**:365–370.

Singh, N. T. and G. S. Dhaliwal. 1972. Effect of soil temperature on seedling emergence of different crops. *Plant and Soil* **37**:441–444.

Singh, Y. P. and R. N. Gupta. 1971. Fertilizer response to physical effects of soil compaction. *J. Indian Soc. Soil Sci.* **19**:345–352.

Smith, G. D., F. Newhall, L. H. Robinson, and D. Swanson. 1964. Soil temperature regimes—Their characteristics and predictability. *U.S. Dept. Agr. Soil Conserv. Service Tech. Paper* 144. 14 pp.

Smith, R. M. and C. F. Cernuda. 1952. Some characteristics of the macrostructure of tropical soils in Puerto Rico. *Soil Sci.* **73**:183–192.

Smith, R. M. and F. Abruña. 1955. Soil and water conservation research in Puerto Rico, 1938–1947. *Univ. Puerto Rico Agr. Exp. Sta. Bull.* 124.

Soil Survey Staff. 1970. *Soil Taxonomy* (Draft). U.S. Department of Agriculture, Washington, D. C.

Suarez de Castro, F. and A. Rodríguez. 1955. Perdidas por erosión de elementos nutritivos bajo diferentes dubiertas vegetales y con varias prácticas de conservación de suelos. *Federación Nacional de Cafeteros de Colombia Bol. Tec.* 14.

Suarez de Castro, F. and A. Rodríguez. 1958. Movimiento del agua en el suelo. Estudios con lisímetros monolíticos. *Federación Nacional de Cafeteros de Colombia Bol. Tec.* 2(19), 1–19. 1–19.

Suri, J. B. and H. Singh. 1970. Effect of dates of sowing and nitrogen levels on the growth and yield of wheat variety Sonora 64. *Indian J. Agron.* **15**:106–111.

Swindale, L. D. 1964. The properties of soil derived from volcanic ash. *FAO World Soil Resources Rept.* 14; pp. 82–87.

Swindale, L. D. 1969. Properties of soils derived from volcanic ash. Pp. B10.1–B10.8. In *Panel on Soils Derived from Volcanic Ash in Latin America*. Inter-American Institute of Agricultural Sciences, Turrialba, Costa Rica.

Tomar, V. S. and B. P. Ghildyal. 1973. Short note on the wilting phenomenon in crop plants. *Agron. J.* **65**:514–515.

Trouse, A. C., Jr., and R. P. Humbert. 1961. Some effects of soil compaction on the development of sugarcane roots. *Soil Sci.* **91**:208–217.

Turner, D. J. 1966. An investigation into the causes of low yields in late planted maize. *East Afr. Agr. For. J.* **31**:249–260.

Uehara, G., K. W. Flach, and G. D. Sherman. 1962. Genesis and micromorphology of certain structural soil types in Hawaiian latosols and their significance to agricultural practices. Pp. 264–270. In *Trans. Comm. IV and V, Int. Soc. Soil Sci. (New Zealand)*.

Van Wambeke, A. 1970. *Soil Studies in Tropical Latin America*. III. *Soil Physics*. National Research Council, Committee on Tropical Soils, Washington. 65 pp.

Vicente-Chandler, J. et al. 1966. High crop yields produced with or without tillage on three typical soils of the humid mountains of Puerto Rico. *J. Agr. Univ. Puerto Rico* **50**:146–150.

Vizier, J. F. 1971. Changes in the apparent specific volume of hydromorphic soils of Chad. *Cah. ORSTOM, Série Pédol.* **9**:133–145.

Wang, C. D., K. Y. Lee, C. C. Yang, et al. 1969. The effect of asphalt barriers on the moisture and nutrient retention in rice and sugarcane fields in sand soils. *Taiwan Sugar Exp. Sta. Res. Rept.* 5.

Wild, A. 1972. Nitrate leaching under bare fallow at a site in northern Nigeria. *J. Soil Sci.* **23**:315–324.

Wilkinson, G. E., 1970. The infiltration of water into Samaru soils. *Samaru Agr. Newsletter* **12**:81–83.

Winkler, E. I. G. and W. J. Goedert. 1972. Características hídricas de dois solos de Pelotas, Río Grande do Sul. *Pesq. Agropec. Bras.* **7**:1–14.

Wolf, J. M. 1975a. Soil–water relations in Oxisols of Puerto Rico and Brazil. Pp 145–153. In E. Bornemisza and A. Alvarado (eds.); *Soil Management in Tropical Latin America.* North Carolina State University, Raleigh.

Wolf, J. M. 1975b. Water constraints to soil productivity in Central Brazil. Ph.D. Thesis, Cornell University, Ithaca, N.Y. 199 pp.

Wolf, J. M. and M. Drosdoff. 1974. Soil–water studies on Oxisols and Ultisols of Puerto Rico. *Agron. Mimeo* 74–22. Cornell University, Ithaca, N.Y. 58 pp.

Wood, H. B. 1971. Land use effects on the hydrologic characteristics of some Hawaiian soils. *J. Soil Water Conserv.* **26**:158–160.

Yamanaka, K. 1964. Physical properties. Pp. 69–91. In *Volcanic Ash Soils in Japan.* Ministry of Agriculture and Forestry, Tokyo.

Zein, A. and G. H. Robinson. 1971. A study in cracking in some Vertisols of the Sudan. *Geoderma* **5**:229–241.

4

CLAY MINERALOGY AND ION EXCHANGE PROCESSES

The nature and properties of clay minerals are more varied in the tropics than in the temperate region. They are also less well defined and understood than in temperate glaciated areas, where most of the ion exchange concepts have developed. Mineralogy and ion exchange studies have been largely limited to soils high in layer silicates, like kaolinite, montmorillonite, and illite; they are directly applicable to tropical soils high in these minerals.

Up to 1940, soil clays were considered amorphous (Mattson and Wicklander, 1940). With the advent of X-ray spectrometry this notion was quickly discarded, and mineralogists concentrated on studying the various crystalline forms. For this purpose, soil clays had to be treated with a series of reagents to eliminate inorganic material that was not crystalline. These materials were literally thrown away in laboratory sinks and ignored. Gradually, however, soil scientists working with soils high in sesquioxides became aware that these amorphous materials play a very significant role in the ion exchange properties of soils now classified as Oxisols, Ultisols, Alfisols, and Andepts (Tanada, 1952; Sherman et al., 1964).

In the late 1950s, Coleman and his coworkers demonstrated the presence of amorphous or semicrystalline minerals between layer silicate lattices in Ultisols of the southeastern United States. This finding led to fundamental changes in our understanding of the chemistry of acid soils (Coleman and Thomas, 1967). Also, outside of the tropics, New Zealand and Japanese scientists conducted intensive studies on the nature of allophane, an amorphous aluminosilicate mineral dominant in Andepts. Hawaiian scientists using high-resolution electron microscopes have demonstrated the existence of amorphous oxide coatings over many layer silicate clays (Jones and Uehara, 1973). An almost completely new set of soil chemical relationships has been applied to soils exhibiting such coatings (Uehara et al., 1972; Uehara and Keng, 1975).

The purpose of this chapter is to summarize the predominant ion exchange reactions taking place in tropical soils, relate them to specific soil groups, and describe some management practices that improve their cation exchange capacity.

The principal clay minerals occurring in the tropics can be divided into two major groups: those with mainly constant or permanent charge (cation exchange capacity) and those with mainly variable charge.

The permanent charge minerals are the 2:1 layer silicates (montmorillonite, vermiculite, illite), the 2:2 layer silicates (chlorites), and to a lesser extent the 1:1 layer silicates (kaolinite, halloysite). A review by Zelazny and Calhoun (1971) summarizes the present knowledge. The variable charge minerals are the intergrade minerals (mixtures of 2:1 minerals with iron and aluminum hydroxides), several species of iron and aluminum oxides and hydroxides, both crystalline and amorphous, and allophane. The oxide species found extensively in the tropics are shown in Table 4.1.

The charge characteristics of pure soil minerals are illustrated by examples in Table 4.2. Total cation exchange capacity (CEC) refers to the value determined by $BaCl_2$-triethanolamine extraction, buffered at pH 8.2. Permanent CEC is the sum of cations extracted by unbuffered salts such as normal KCl at the actual pH of the soil. This is normally referred to as "effective CEC." Effective CEC is not an accurate measure of permanent charge, but perhaps it is a realistic one because the soils are at their field pH. The difference between CEC estimated at pH 8.2 and effective CEC is considered the variable or pH-dependent charge. Table 4.2 shows that most of the CEC of 2:1 layer silicate minerals is "permanent." The 1:1 minerals show both permanent charge due to isomorphous substitution and variable charge due to edge effects and broken lattices. In sharp contrast the iron and aluminum hydroxides, allophane, and peat have predominantly variable charge.

Most minerals also have some degree of anion exchange capacity (AEC). In the layer silicates this arises out of net positive charges due to broken edges or a replacement of OH^- by other anions. In oxides it is due to the

Table 4.1 *Ionic, Amorphous, and Crystalline Species of Oxide Minerals in Humid Tropical Soils*

| Ionic | Amorphous | | Crystalline | |
	Gel	Criptocrys-taline	Primary	Aged or Dehydrated
Al^{3+} $\xrightarrow{OH^-}$	$Al(OH)_3$	$\longrightarrow$?	$\longrightarrow Al(OH)_3$ gibbsite	$\longrightarrow AlOOH$ boehmite
Fe^{3+} $\xrightarrow{OH^-}$	$Fe(OH)_3$	$\longrightarrow Fe(OH)_3$	$\longrightarrow FeOOH$ goethite	$\longrightarrow Fe_2O_3$ hematite
Fe^{2+} $\xrightarrow{OH^-}$	$Fe(OH)_2$	$\longrightarrow Fe(OH)_2$	$\longrightarrow FeOOH$ lepidocrocite	$\longrightarrow Fe_2O_3$ maghemite
Si^{4+} $\xrightarrow{OH^-}$	SiO_2 silica gel	$\longrightarrow$ Opal silcrete	$\longrightarrow SiO_2$ quartz	

Source: Sherman et al. (1964).

Table 4.2   Charge Characteristics of Some Clay Minerals Separated from Soils of Kenya (meq/100 g clay)

Material	Cation Exchange Capacity			Anion Exchange Capacity
	Permanent	Variable	Total	
Montmorillonite	112	6	118	1
Vermiculite	85	0	85	0
Illite	11	8	19	3
Halloysite	6	12	18	15
Kaolinite	1	3	4	2
Gibbsite	0	5	5	5
Goethite	0	4	4	4
Allophanic colloid	10	41	51	17
Peat	38	98	136	6

Source: Mehlich and Theisen (unpublished).

latter mechanism. It is relevant to note that some oxides have as much anion exchange as cation exchange capacity.

Clay minerals do not occur alone in soils. They interact with each other and with organic matter. The Soil Taxonomy System has established several mineralogy classes at the family level. A simplified definition of these mineralogy classes appears in Table 4.3.

Vertisols, Mollisols, Aridisols, and some Alfisols, Inceptisols, and Entisols have montmorillonitic, vermiculitic, chloritic, illitic, carbonatic, siliceous, or mixed mineralogical families. Such families have primarily permanent charge with a small degree of variable charge contributed by their organic matter contents. Their chemical properties are similar to those of most soils of glaciated temperate areas, and thus the ion exchange concepts developed in these areas are entirely applicable to these families.

Oxisols, Ultisols, and some Alfisols and Inceptisols have kaolinitic, halloysitic, oxidic, gibbsitic, ferritic, siliceous, and mixed mineralogy families. These "red" soils have both permanent and variable charge because of the presence of pH-dependent minerals such as iron and aluminum hydroxides as well as organic matter. Andepts belong to the ashy and cindery families if they are coarse textured. Unfortunately no allophanic family is defined in the Soil Taxonomy System. This is a serious oversight since Andepts present a particularly high amount of pH-dependent charge because of allophane and high organic matter contents. Another major omission is the lack of "interlayer" family to separate soils high in an intimate mixture of 2:1 layer silicates and oxides. Many Ultisols, Alfisols, and Oxisols have such properties.

Table 4.3   Simplified Definitions of Mineralogy Classes in the U.S. Soil Taxonomy That Are Extensive in the Tropics

Class	Simplified Definition	Orders Most Commonly Found
Halloysitic (1:1)	>50% nontabular halloysite in clay fraction by weight	Oxisol, Ultisol, Inceptisol
Montmorillonitic (2:1)	>50% montmorillonite or nontronite in clay fraction by weight, or a mixture with more montmorillonite than any other single clay mineral	Vertisol, Mollisol, Alfisol, Aridisol, Inceptisol, Entisol
Illitic (2:1)	>50% illite (hydrous mica) in clay by weight or >4% K_2O	Mollisol, Alfisol, Aridisol, Inceptisol, Entisol
Vermiculitic (2:1)	>50% vermiculite in clay by weight, or more vermiculite than any other single clay mineral	Mollisol, Ultisol, Alfisol, Aridisol, Inceptisol, Entisol
Chloritic (2:2)	>50% chlorite in clay by weight, or more chlorite than any other mineral	Mollisol, Alfisol, Aridisol, Inceptisol, Entisol
Kaolinitic (1:1)	>50% clay fraction kaolinite or tabular halloysite by weight	Oxisol, Ultisol, Alfisol, Inceptisol
Oxidic	>20% free Fe_2O_3 + Al_2O_3 in clay fraction	Oxisol, Ultisol, Alfisol
Gibbsitic	>40% gibbsite and boehmite of soil by weight	Oxisol, Ultisol
Ferritic	40% free Fe_2O_3 of soil by weight	Oxisol, Ultisol
Ashy	>60% volcanic ash, cinders, and pumice in soil	Andept
Cindery	>35% cinders larger than 2 mm by volume	Andept
Siliceous	>90% by weight of quartz, chalcedony, or opal of 0.02–2 mm fraction by weight	Ultisol, Alfisol, Spodosol, Inceptisol, Entisol
Carbonatic	>40% $CaCO_3$ by weight	Mollisol, Aridisol
Mixed	Soils with less than 40% of any mineral in the 0.02–2 mm fraction and less than 50% single layer silicate mineral in the clay fraction	Ultisol, Alfisol, Aridisol, Spodosol, Inceptisol, Entisol

Source: Soil Survey Staff (1970).

For the purpose of this book, soils of the tropics can be grouped into three ion exchange systems: the layer silicate systems, the oxide systems, and the oxide-coated layer silicate systems.

LAYER SILICATE SYSTEMS

These systems occur in Entisols, Vertisols, Aridisols, Mollisols, and other soils that have little or no iron and aluminum oxides or allophane. The classic concepts developed in the northern United States are entirely applicable to these soils. These clay systems have only negative charge because of isomorphous substitution of silica or aluminum ions for other ions of lower valence. These soils exhibit a slight amount of variable charge that is due to organic matter contents and broken edges of kaolinite.

The pH-dependent properties of organic matter are due to the reactions that take place when organic matter–aluminum complexes are subjected to an increase in pH. The following reaction, adapted from Coleman and Thomas (1967), illustrates the increases in net negative charges of carboxyl radicals when the complexed aluminum ions are precipitated:

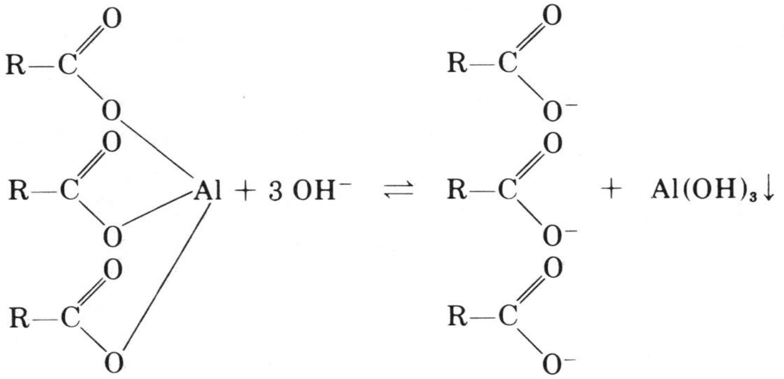

The extent of this reaction is limited. Coleman and Thomas (1967) reported that the effective CEC of 2:1 layer silicate acid soils of North Carolina accounts for 80 percent of the total CEC at pH 8.2.

Layer silicate systems such as these are extensive in the tropics. They occupy large areas covered by Aridisols, Vertisols, Mollisols, and soils of alluvial plains and deltas.

OXIDE SYSTEMS

Oxide systems are those in which the entire clay particles consist of iron and aluminum oxides or allophane, or those in which the layer silicates are covered by thick, stable coats of such oxides. They occur in oxidic, gibbsitic, and ferritic families of Oxisols, Ultisols, and Alfisols and also in "allophanic" families of Andepts. The occurrence of this ion exchange system is well correlated with the excellent granular structure found in well-aggregated tropical soils. Although many of the exchange reactions characteristic of this system were developed by Mattson several decades ago (Mattson, 1926; Mattson and Pugh, 1934; Mattson and Wicklander, 1940), they have been revived in the light of more recent concepts by Uehara and his coworkers (Mekaru and Uehara, 1972; Jones and Uehara, 1973; Keng and Uehara, 1974; Uehara and Keng, 1975). The following discussion is largely based on these publications.

In oxide systems the charge is entirely pH dependent. Oxide systems in soils may exhibit net negative charge (CEC), net positive charge (AEC), or no net charge at all. The surface chemistry of these oxides is similar. An iron hydroxide is used in the following example, but an aluminum species could serve as well. According to Keng and Uehara (1974), the following changes occur at the surface of an uncharged oxide when its pH changes by the addition of H^+ or OH^- ions:

Net positive charge   Net zero charge   Net negative charge

Thus, when the pH decreases, positive charges are created. When the pH increases by the additions of OH^-, net negative charges form.

ΔpH and the Zero Point of Charge

The charge status of an oxide system can be easily determined by measuring its pH in water and in a neutral salt such as normal KCl. Mekaru and

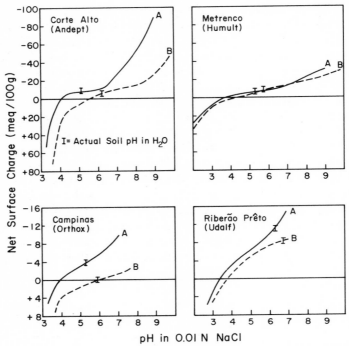

Fig. 4.1 Changes in net surface charge with pH in A and B horizons of four soils high in allophane or oxides. *Source:* Adapted from Carrasco (1972) and Van Raij and Peech (1972).

Uehara (1972) defined ΔpH as the difference between pH in KCl and pH in water. If the ΔpH is positive, there is net positive charge (anion exchange capacity). If the ΔpH is negative, there is net negative charge (cation exchange capacity). Thus the sign of ΔpH corresponds to the sign of the colloidal charge.

In layer silicate systems ΔpH is always negative. The following simplified reactions illustrate the reason why:

$$\exists\ H^+ + H_2O \rightleftharpoons\ \exists\ H^+ + H_2O$$
$$\exists\ H^+ + KCl \rightleftharpoons\ \exists\ K^+ + Cl^- + H^+$$

Therefore the pH is higher in water than in KCl in this system.

In oxide systems ΔpH can be positive or negative, depending on the actual pH of the soil, as illustrated previously. If the oxide system has net positive charge, the following reactions occur when the pH is measured:

$$\overset{+}{\exists}\ OH^- + H_2O \rightleftharpoons\ \overset{+}{\exists}\ OH^- + H_2O$$
$$\overset{+}{\exists}\ OH^- + KCl \rightleftharpoons\ \overset{+}{\exists}\ Cl^- + K^+ + OH^-$$

These relationships can be graphically illustrated when the total charge of the soil is determined over a wide pH range. Figure 4.1 shows an example of these charge–pH relationships for an Andept and an Ultisol from Chile, and an Oxisol and an Alfisol from Brazil. All soils exhibit marked degrees of variable charge. The actual field pH of each soil at sampling time is marked as "I." All topsoils and subsoils exhibit net CEC at their field pHs except the Oxisol subsoil, which has net AEC. This indicates the predominance of cation exchange capacity in the four topsoils and the other three subsoils.

The vast majority of ΔpH data reviewed by this writer indicates no net positive changes in the A horizon of tropical soils except for Hydrandept from Hawaii, and few instances of net positive changes in subsoils of soils high in oxides or allophane. Table 4.4 shows some pH values of representative soils. Even in the largest Oxisol area of the world, the Central Plateau of Brazil, the vast majority of the soils have negative ΔpHs (Camargo and Falesi, 1975).

Having a negative ΔpH does not imply a total absence of positive charges on the clay surfaces. A small number of positive charges do occur, perhaps in areas isolated from the negative charges. For the soils illustrated in Fig. 4.1, Carrasco (1972) and Van Raij and Peech (1972) measured less than 1 meq/100 g of AEC in the Ultisol, Oxisol, and Alfisol, and 6.8 meq/100 g in the Andept, at the field pHs of the soils.

Factors Affecting pH–Charge Relationships

The pH–charge relationships of oxide systems can be illustrated by the following equation, according to Keng and Uehara (1974):

$$\sigma = \frac{kDRT}{4\pi F} \cdot \frac{pH_0}{pH}$$

where σ = the surface charge (meq/100 g)
k = the reciprocal of the double layer thickness
D = the dialectric constant
R = the gas constant
T = the absolute temperature
F = the Faraday constant
pH = the soil pH
pH_0 = the soil pH at the isoelectric point, that is, the pH at the zero point of charge (ZPC)

The magnitudes of D, R, T, F, and 4π are constant in soil systems; con-

Table 4.4   Net Positive or Negative Charges in Some Tropical Soils

Soil Name and Country	Great Group	Family	Horizon (cm)	pH 1:1 H₂O	pH 1 N KCl	ΔpH
Oxisols						
Molokai (Hawaii)	Haplustox	Halloysitic	0–20	6.2	5.4	–
			41–76	5.8	5.6	–
Latosol (Brazil)	Acrohumox	Kaolinitic	0–20	4.9	4.1	–
			200–230	4.9	4.3	–
Dark Red Latosol (Brazil)	Haplustox	Kaolinitic	0–35	4.8	4.2	–
			35–70	4.9	4.2	–
Red-Yellow Latosol (Brazil)	Acrustox	Kaolinitic	0–20	4.9	4.4	–
			40–60	5.6	4.7	–
Halii (Hawaii)	Gibbsihumox	Gibbsitic	0–32	5.0	4.5	–
			75–110	5.0	5.2	+
Ultisols						
Paaloa (Hawaii)	Tropohumult	Oxidic	0–20	4.3	3.9	–
			76–122	4.8	4.1	–
Humatas (Puerto Rico)	Tropohumult	Mixed	0–10	4.6	3.9	–
			180–240	4.7	3.7	–

Soil (location)	Subgroup	Mineralogy	Depth			
Polovorin (Colombia)	Paleudult	Mixed	0–17	4.6	3.8	—
			157–210	4.7	3.5	—
Yurimaguas (Peru)	Paleudult	Mixed	0–12	4.2	3.8	—
			35–85	4.0	3.8	—
Alfisols						
Terra Roxa Estruturada (Brazil)	Tropudalf	Kaolinitic	0–20	6.3	5.4	—
			37–71	6.7	5.9	—
Egbeda (Nigeria)	Paleustalf	Mixed	0–5	6.5	5.9	—
			45–65	6.4	5.1	—
Inceptisols						
Birrisito (Costa Rica)	Dystrandept	Ashy	A	5.3	5.0	—
			B	5.2	5.3	—
Waimea (Hawaii)	Eutrandept	Ashy	0–18	5.7	5.4	—
			76–102	5.8	5.4	—
Hilo (Hawaii)	Hydrandept	Thixotropic	0–25	4.5	4.7	+
			25–51	5.4	5.4	0
Ekiti (Nigeria)	Ustropept	Mixed	0–16	6.9	6.1	—
			46–66	6.2	4.9	—
Vertisols						
Laulualei (Hawaii)	Chromustert	Montmorillonitic	0–15	7.4	6.6	—
			122–152	7.9	7.1	—

145

sequently they can be combined into one constant, Q. The equation is thus simplified as follows:

$$\sigma = \frac{k \mathrm{pH}_0 Q}{\mathrm{pH}}$$

The three remaining parameters vary with soil properties. The reciprocal of the double layer thickness k depends on the electrolyte concentration in the soil solution. Its value increases with soluble fertilizer applications and soil drying. An example of this is the use of water and N KCl in determining pH. The soil pH can be increased by liming or decreased by the residual effect of certain nitrogen fertilizers. In an oxide system, therefore, changes in pH and salt content will alter the exchange properties.

Alteration of the Zero Point of Charge

The third parameter, the pH at the zero point of charge, can also be altered. When organic anions and certain inorganic anions are absorbed by the oxide surfaces, the pH_0 shifts to lower values. The reaction between organic matter and oxide surfaces responsible for this shift is the following:

The net negative charge produced by this reaction increases the CEC of the soil without changing its pH. Thus the pH–charge curve shifts to the left, and the pH at the zero point of charge is lower. Indirect evidence of this phenomenon can be seen in Fig. 4.1 when the lower pH at the zero point of charge of the topsoil is compared with the pH_0 of the subsoil. All soils in this figure showed no differences in texture or clay mineralogy between the A and B horizons. The main difference between these horizons lies in their organic matter contents. The difference between 2.5 percent organic matter in the A horizon of the Oxisol and 0.7 percent in the B horizon was responsible for the net CEC of the topsoil and the net AEC of the subsoil at approximately the same field pH. In the Alfisol of Fig. 4.1, there were similar differences in organic matter contents, but both horizons were well above the zero point of charge at the field pH of the soil. This soil is probably a combination of oxides and layer silicate minerals, which will be discussed in the next section. The Orthox and Andept of Fig. 4.1 are true oxide systems.

These relationships have some practical implications. The organic matter contents of such soils can be increased by management, thus raising significantly the CEC without altering the pH. Unfortunately, there is little evidence available that such a shift in the zero point of charge can be accomplished through agronomic practices. This is a worthwhile area of research.

In several Andepts, Mekaru and Uehara (1972) and Schalscha et al. (1972, 1974) have shown that additions of 310 ppm P increase cation exchange capacity by about 0.7 meq/100 g. These increases are of little practical importance, however, because Andepts usually have high CECs, and the cost associated with adding 310 ppm P (1410 kg P_2O_5/ha) is probably too high to be economically sound. Similar information is needed for Oxisol topsoils with very low CECs. It is possible that the residual effect of prolonged phosphorus applications in such soils might increase the CECs over several years.

OXIDE-COATED LAYER SILICATE SYSTEMS

An ion exchange system intermediate between the pure layer silicates and the pure oxide systems probably describes more adequately than either of these the majority of tropical soils, which contain measurable quantities of oxides but do not have their surface area completely covered by them. This system probably now represents the bulk of "red" tropical soils, Oxisols, Ultisols, Alfisols, and some Inceptisols that have kaolinitic, halloysitic, or mixed mineralogy.

The layer silicates are partially coated by thin, sometimes monomolecular coats of iron and aluminum oxides. Electron micrographs show that in certain soils the oxides form "islands" or discrete particles attached to the surface of kaolinite (Greenland et al., 1968). In other cases aluminum oxides are sandwiched between two layer silicate lattices (Coleman and Thomas, 1967). High-resolution electron micrographs obtained by Jones and Uehara (1973) suggest that the oxide gels are quite dynamic, responding to changes in moisture and even stretching out and breaking like chewing gum. The dynamic nature of these thin coats implies that at a certain time they might completely cover layer silicate surfaces, whereas in other areas or at a different time these surfaces might be exposed.

The exchange properties of this system are intermediate between those of the other two. Its permanent CEC is considerably lower than that of pure layer silicates because some negative charges arising from isomorphous substitution are balanced by net positive charges of the oxides. The oxide coats or particles also prevent expansion and contraction of 2:1 minerals, thus changing drastically their physical properties.

Coleman and Thomas (1967) proposed the following model, based on voluminous research conducted on Ultisols and Alfisols of North Carolina and Virginia:

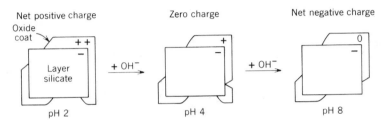

Although the layer silicate component has permanent negative change, the oxide coats have net positive charges up to pH's of 7 or 8, where they reach their zero point of charge. Consequently, at pH's lower than 4 the system may have net positive charge; it may reach its zero point of charge at pH 4 and have net negative charge at higher pH's. In the field it will exhibit a negative ΔpH.

Observation of the behavior of most "red" soils not having oxidic, gibbsitic, or ferritic mineralogy suggests that this mixed system describes their ion exchange more adequately than does the pure layer silicate or the pure oxide system. The latter system is probably restricted to Andepts and the oxidic, gibbsitic, and ferritic families of Oxisols, Ultisols, and Alfisols.

CATION AND ANION EXCHANGE LEVELS

The description of the three ion exchange systems suggests a wide range in cation and anion exchange capacity in tropical soils. Mehlich and Theisen (1973) studied 44 soils of Kenya and related the charge relationships to the content of amorphous or crystalline minerals. Table 4.5, extracted from this study, shows representative values for the main mineralogical groupings. Permanent CEC was determined by using unbuffered neutral salts at the field pH of the soil. Total CEC is the value obtained by $BaCl_2$–triethanolamine extraction at pH 8.2. The variable or pH-dependent charge was calculated as the difference between the two. Anion exchange capacity was determined with 0.05 N NH_4VO_3.

Mehlich and Theisen's data suggest that these parameters, rather than semiquantitative estimates of mineral components, should be used to define mineral family criteria. Montmorillonitic families have more than 50 meq CEC/100 g of clay, with more than 70 percent of the CEC as permanent charge, and little AEC. Mixed families of 2:1 and 1:1 layer silicates have 30 to 50 meq CEC/100 g of clay, of which 30 to 70 percent is permanent charge. Anion exchange capacity is less than variable charge.

Soils with kaolinitic or halloysitic mineralogy have total CECs of less than 30 meq/100 g clay, of which 30 to 50 percent is permanent. Anion exchange capacity is greater than 50 percent of the total CEC in kaolinitic families and lower in halloysitic ones. Anion exchange in these soils is attributed to broken edges, which produce net positive charges.

Soils with mixed crystalline and amorphous materials have CECs from 30 to 50 meq/100 g of clay, most of which corresponds to variable charge.

Predominantly amorphous volcanic soils, which should be called "allophanic" families, have CECs greater than 50 meq/100 g of clay, but more

Table 4.5 Charge Characteristics of Selected Kenya Soils in Relation to Crystalline and Amorphous Soil Constituents (meq/100 g)

Soil	CEC			AEC	CEC of Clay
	Permanent	Variable	Total		
Predominantly montmorillonitic (Songhor, 60% clay)	44	3	47	3	75
Mixed crystalline minerals (Mogutato, 67% clay)	17	12	29	10	43
Predominantly kaolinitic–halloysitic (Ishiara, 64% clay)	7	10	17	4	26
Mixed crystalline and amorphous (Eldoret, 56% clay)	4	17	21	11	39
Predominantly amorphous (Chinga, 62% clay)	6	32	38	20	61
Predominantly organic (Gathaithi, 12% clay)	8	30	38	7	100

Source: Mehlich and Theisen (unpublished).

than 70 percent is variable charge. Anion exchange capacity is more than half of the total CEC. Organic soils have more than 70 percent of their CECs as variable change, but AEC seldom reaches 20 percent of the total CEC.

Measurement Problems

Cation exchange capacity is most commonly determined as the quantity of cations absorbed from salt solutions buffered at pH 7 with NH_4OAc or at pH 8.2 with $BaCl_2$-triethanolamine (Ba–TEA). For soils with field pH values of 7 or 8.2, these measurements adequately reflect the CECs. They are also adequate if the soil has little or no variable charge. For acid soils these methods are adequate for some layer silicate systems that have no pH-dependent charge, such as montmorillonitic soils low in organic matter. For other soils with pHs lower than 7 or 8.2, however, these methods grossly overestimate the exchange capacity. This is particularly relevant in the tropics, where most soils exhibit a significant amount of variable charge. Table 4.6 illustrates the magnitude of this discrepancy for several Colombian soils.

The measurement that determines more accurately the total charge at the actual soil pH involves leaching with a neutral, unbuffered salt such as KCl or $CaCl_2$, determined at the pH of the soil. It has been called the "effective CEC" by Coleman and Thomas (1967). Estimates of effective CEC usually show strikingly lower values than are obtained by the other methods, as Table 4.6 indicates. Effective CEC will be used in this book unless otherwise indicated. In general, effective CEC values higher than 4 meq/100 g suggest sufficient cation exchange capacity to prevent serious leaching losses.

The method used seriously affects the interpretation not only of CEC values, but also of percentage base saturation. Base saturation is calculated as the sum of exchangeable bases divided by the CEC of the soil. When Ba–TEA extraction buffered at pH 8.2 is used to estimate the CEC, essentially all the variable charge is considered as extractable acidity. Base saturation calculated in this manner is grossly underestimated, making a soil seem more acid than it actually is. A similar situation occurs in acid soils for acetate extraction buffered at pH 7. With these two extractants, a soil is 100 percent base saturated at pH 7 or at pH 8.2. When the effective CEC is used, base saturation is calculated as the sum of bases divided by the sum of bases plus H^+ and Al^{3+}. This method more accurately reflects what is happening in the soil and the fact that a soil is 100 percent base saturated at pH 5.5 to 6.0 because all the aluminum has been precipitated out.

Table 4.6 compares the base saturation values obtained when CEC was

Table 4.6 Comparison of Different Extraction Methods on the Cation Exchange Capacity and Percent Base Saturation of Several Colombian Soils

Soil	Mineralogy	pH	Organic Matter (%)	Cation Exchange Capacity (meq/100 g)			Base Saturation	
				N KCl (ECEC)	CaOAc pH 7	Ba-TEA pH 8.2	N KCl (ECEC) (%)	Ba-TEA pH 8.2 (%)
Andept	Allophanic	6.0	6	3	19	23	88	16
Andept	Kaolinite-intergrade	5.2	36	8	65	110	4	1
Tropept	Kaolinite-intergrade	4.4	4	5	13	18	6	2
Aquept	Kaolinite-intergrade	4.5	18	22	42	62	81	29
Humult	Kaolinite-intergrade	4.5	15	6	15	21	50	16
Ustox	Kaolinite-intergrade	4.9	2	1	6	7	23	14
Orthox	Goethite	5.1	10	3	17	24	74	10

Source: Adapted from León (1967).

151

determined at pH 8.2 and with an unbuffered salt. The Andept with a pH of 6 is considered 16 percent saturated with bases when Ba–TEA was used and 88 percent base saturated when the effective CEC was used. A soil at that pH has few, if any, acidity problems.

Base saturation methods based on CEC determined at pH 7 or 8.2 are satisfactory for most soils of the midwestern and western United States, Europe, and the Soviet Union. They are still used as classification criteria in the U.S. Soil Taxonomy in spite of objections from southeastern United States and tropical soil scientists. For example, the criterion to separate Ultisols from Alfisols is 35 percent base saturation calculated by the Ba–TEA method at pH 8.2. Correlations between this and base saturation calculated from effective CEC determinations, using 88 soils from the midwestern and southeastern United States and Puerto Rico, indicated that 35 percent base saturation at pH 8.2 is equivalent to 55 percent base saturation obtained with effective CEC (Buol, 1973). This relationship, illustrated in Fig. 4.2, holds very well for loamy and clayey soils low in organic matter. For sandy soils or soils with more than 1 percent organic matter, the relationship is different.

On the basis of similar calculations, the limit for separating Mollisols and Alfisols, 50 percent base saturation at pH 7, corresponds to 90 percent base saturation using effective CEC, as Fig. 4.3 shows.

The interpretation of CEC and base saturation values, therefore, depends heavily on the methods used to obtain them.

Anion Exchange Capacity

It is significant to note that many tropical soils possess some degree of anion exchange capacity. A small number of net positive charges are found even in soils with net CEC because some positive charges are physically too distant from negative charges to be neutralized. Soils having a negative ΔpH, therefore, may exhibit a significant degree of AEC, particularly at low pH levels. Anions such as sulfate, chlorides, and nitrates can be exchanged by these positive charges just as cations are exchanged by negative charges. In oxidic subsoils with positive ΔpH, cations are freely leached whereas anions undergo exchange reactions. Consequently, some Oxisols and Andepts may retain appreciable quantities of nitrates and sulfates in exchangeable form (León and Coleman, 1972; Kinjo and Pratt 1971).

Sulfate, phosphate, and silicate anions undergo stronger adsorption to positively charged oxide surfaces that impede conventional anion exchange. The management implications of these sorption or fixation processes will be discussed in other chapters.

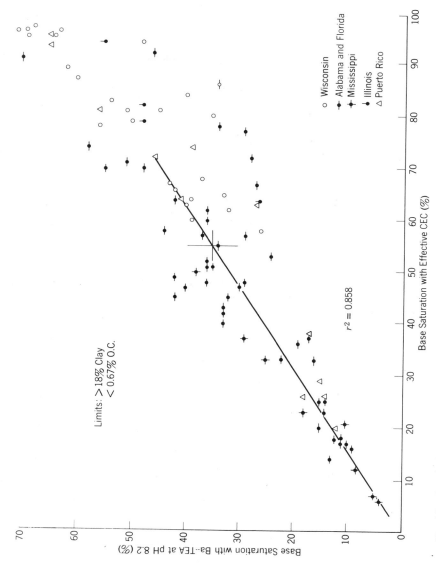

Fig. 4.2 Correlation between base saturation with Ba-TEA at pH 8.2 and 1 *N* KCl extraction. *Source:* Buol (1973).

153

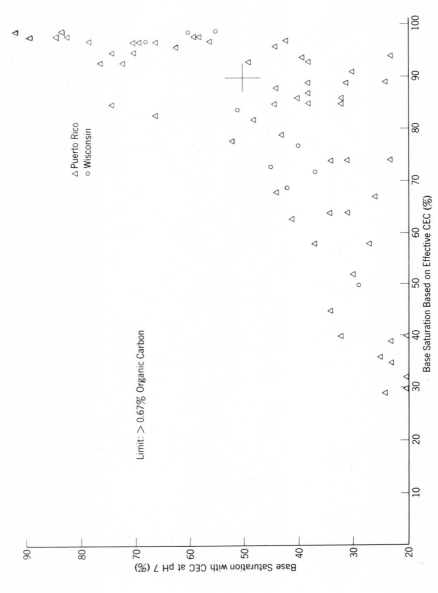

Fig. 4.3 Correlation between base saturations calculated with Na$_4$OAc buffered at pH 7 and with 1 *N* KCl extraction. *Source:* Buol (1973).

154

MANAGEMENT OF CATION EXCHANGE CAPACITY

An effective cation exchange capacity of at least 4 meq/100 g is needed to retain most cations against leaching. Higher CEC values are even better, especially if the exchangeable cations present are basic. Because of highly weathered minerals or sandy textures, many tropical soils commonly have effective CEC values lower than 4. In such soils, increasing CEC is an important management goal. This can be accomplished by two processes: liming acid soils with oxide or oxide-coated layer silicate systems, and increasing soil organic matter content.

The CECs of very acid soils with oxide or oxide-coated systems, particularly Oxisols, Ultisols, and Andepts, can be increased by liming, because of the predominance of pH-dependent charge. Table 4.6 shows how large these increases can be if the soils are limed to pH 7. Unfortunately, liming to pH 7 is not a recommended practice for such soils because it can have other detrimental effects. Liming to pH 5.5 or 6.0 seems to be the most appropriate level, as will be discussed later. Liming soils with pHs on the order of 4.0 or 4.5 to pH 5.5 can sometimes produce a large increase in CEC, as shown in Fig. 4.1. In some cases it has changed the charge relationships from net positive to net negative charge, but the occurrence of positively charged topsoils seems limited to Hydrandepts of Hawaii.

In many soils with oxide or oxide-coated systems, organic matter contributes the bulk of the net negative charges. For example, Martini (1970) showed that organic matter contributed from 45 to 85 percent of the total CEC of volcanic ash soils from Panama, whereas in alluvial soils with layer silicate systems the contribution of organic matter ranged from 10 to 28 percent of the total CEC.

In many highly weathered tropical soils, particularly Oxisols, the maintenance of organic matter is almost synonymous to the maintenance to CEC. Brams (1971) studied the process of organic matter diminution in Oxisols of Sierra Leone after clearing the forest. Organic matter contents decreased 50 percent in a 5 year period, while effective CEC decreased 30 percent. The close correlation between organic matter content and cation exchange capacity in the Sierra Leone Oxisol and a Nigerian Alfisol is illustrated in Fig. 4.4.

SUMMARY AND CONCLUSIONS

1. The kinds and properties of clay minerals are much more varied in the tropics than in glaciated temperate areas, where most of the concepts of ion

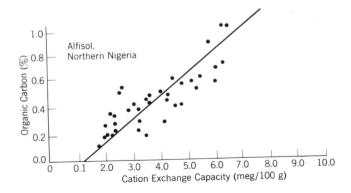

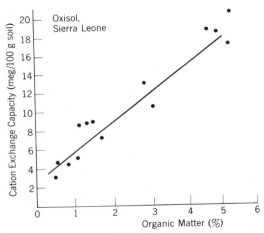

Fig. 4.4 Relationship between cation exchange capacity and organic matter contents in the topsoils of an Alfisol from northern Nigeria and an Oxisol from Sierra Leone. *Source:* Brams (1971) and unpublished data of Brams.

exchange were developed. Mineralogical classes at the family level of the Soil Taxonomy System permit the grouping of tropical soils into three major ion exchange systems: layer silicate systems, oxide systems, and oxide-coated layer silicate systems.

2. The layer silicate systems include Entisols, Vertisols, Aridisols, Mollisols, and other tropical soils with little or no iron and aluminum oxides or allophane. They exhibit permanent negative charges that vary little with soil pH. The concepts developed in glaciated temperate areas are entirely applicable to such soils.

3. The oxide systems include soils in which the clay particles consist of iron and aluminum oxides or allophane, or layer silicate minerals are covered with thick, stable coats of such oxides. They are typical of Andepts and of oxidic, ferritic, and gibbsitic families of Oxisols, Ultisols, and Alfisols. They exhibit a strong relationship between charge and pH. In some cases these soils may exhibit net positive charge at low pH, and net negative charge at high pH, or no charge at all. The actual charge depends on soil pH, electrolyte concentration, and the pH at the zero point of charge. Additions of organic matter and strongly adsorbed phosphate anions can increase net negative charge without altering the pH of the soil. These relationships are completely different from those of layer silicate systems.

4. An intermediate ion exchange system consisting of a mixture of the other two, called the "oxide-coated layer silicate system," is believed to predominate in the bulk of "red" soils of the tropics, that is, Oxisols, Ultisols, Alfisols, and some Inceptisols of kaolinitic, halioysitic, or mixed mineralogy with a significant amount of iron and aluminum oxides. The layer silicates are partially coated with oxide minerals or have discrete oxide particles attached to their surfaces or sandwiched in between two layer silicate crystals. The exchange properties are intermediate between those of the other two systems, with both permanent and pH-dependent charge.

5. Substantial confusion exists in the tropical literature because of the different methods used for estimating cation exchange capacity. The "official" methods recommended by the U.S. Soil Taxonomy involve extracting cations with solutions buffered at pH 7 or 8.2. Such extractions adequately reflect the charge status of a layer silicate system or of any soil at pH 7 or 8.2. Unfortunately they grossly overestimate the CEC of acid, oxide or oxide-coated systems because of the strong pH dependence of their charge. The CEC estimated with an unbuffered salt at or near the actual field pH, called the "effective CEC," is a much more realistic parameter. Base saturation calculations based on the buffered extractants tend to grossly underestimate the base status of acid soils. Some tentative correlations between extractants are presented.

6. Many tropical soils exhibit significant anion exchange capacity, particularly in the subsoil of oxide systems. Leaching losses of nitrates may be decreased in these subsoils.

7. The management of cation exchange capacity is crucial in many tropical soils that have low effective CEC values due to mineralogy or coarse texture. Liming acid soils to pH levels of 5.5 or 6.0 can increase the CEC without causing adverse effects. Organic matter contributes the bulk of the exchange sites in many highly weathered Oxisols, Ultisols, and Alfisols. The

maintenance of organic matter in such soils is the principal way to keep CEC values within reasonable levels.

REFERENCES

Askenasy, P. E., S. E. Dixon, and T. R. McKee. 1973. Spheroidal halloysite in a Guatemalan soil. *Soil Sci. Soc. Amer. Proc.* **38**:799–803.

Barber, R. G. and D. L. Rowell. 1972. Charge distribution and the cation exchange capacity of iron-rich kaolinitic soil. *J. Soil Sci.* **23**:135–146.

Besoaín, E. 1969. Clay mineralogy of volcanic ash soils. Pp. B1.1–B1.17. In *Panel on Soils Derived from Volcanic Ash in Latin America.* Inter-American Institute of Agricultural Sciences, Turrialba, Costa Rica.

Blasco, M. et al. 1969. Mineralogy of the soils of the Cauca Valley, Colombia. *Turrialba* **19**:332–339.

Bleeker, P. 1972. The mineralogy of eight latosolic and related soils from Papua New Guinea. *Geoderma* **8**:191–205.

Bornemisza, E. 1969. Minerales de arcilla en suelos centroamericanos y de Panamá. *Turrialba* **19**:97–102.

Bornemisza, E. and K. Igue. 1967. Oxidos de hierro y aluminio libre en suelos tropicales. *Turrialba* **17**:23–30.

Bornemisza, E., F. A. Laroche, and H. W. Fassbender. 1968. Effects of liming on some chemical characteristics of a Costa Rican Latosol. *Proc. Crop Soil Sci. Soc. Fla.* **27**:219–226.

Brams, E. 1971. Continuous cultivation of West African soils: organic matter diminution and effects of applied lime and phosphorus. *Plant and Soil* **35**:401–414.

Buol, S. W. 1973. Soil laboratory needs in tropical research. *Agron. Abst.* **1973**:111.

Camargo, M. N. and I. C. Falesi. 1975. Soils of the Central Plateau and Transamazonic Highway of Brazil. Pp. 25–44. In E. Bornemisza and A. Alvarado (eds.), *Soil Management in Tropical America.* North Carolina State University, Raleigh.

Carrasco, A. 1972. Distribution of electric charges in Chilean soils derived from volcanic ash. M.S. Thesis, Cornell University, Ithaca, N.Y. 101 pp.

Chatterjee, R. K. and D. S. Gupta. 1970. Clay minerals in some western Uttar Pradesh soils. *J. Indian Soc. Soil Sci.* **18**:391–396.

Coleman, N. T. and G. W. Thomas. 1967. The basic chemistry of soil acidity. *Agron. Monogr.* **12**:1–41.

Davis, C. E., N. Ahmad, and R. L. Jones. 1971. Effects on exchangeable cations on the surface area of clays. *Clay Min.* **9**:258–261.

DeVilliers, J. M. 1971. The problem of quantitative determination of allophane in soils. *Soil Sci.* **112**:2–7.

D'Hoore, J. L., J. J. Fripiat, and M. C. Gastuche. 1954. Tropical clays and their iron oxide coverings. *Proc. Second Inter-Afr. Soil Conf.* **1**:257–260.

Dobrorolski, V. V. 1973. Mineralogical and geochemical properties of Kenya, Uganda, and Tanzania black soils. *Pochvovedenie* **1973**(8):14–25.

Eswaran, H. and C. Sys. 1970. An evaluation of the free iron in tropical basaltic soils. *Pedologie* **20**:62–85.

Eswaran, H. and F. de Coninck. 1971. Clay mineral formations and transformations in basaltic soils in tropical environments. *Pedologie* **21**:181–210.

Fieldes, M., L. D. Swindale, and J. P. Richardson. 1952. Relations of colloidal hydrous oxides to the high cation exchange capacities of some tropical soils of the Cook Islands. *Soil Sci.* **74**:197–205.

Fox, R. L. 1974a. Examples of anion and cation adsorption by soils of tropical America. *Trop. Agr.* (*Trinidad*) **51**:200–210.

Fox, R. L. 1974b. Chemistry and management of soils dominated by amorphous colloids. *Proc. Soil Crop Sci. Soc. Fla.* **33**:112–119.

Gaikawad, S. T. and S. V. Govinda Rajan. 1971. Nature and distribution of silicon, aluminum and iron oxides in the lateritic soils from Durg district, Madhya Pradesh. *Indian J. Agr. Sci.* **14**:1079–1084.

Gebhardt, H. and N. T. Coleman, 1974. Anion adsorption by allophanic tropical soils. I. Chloride adsorption. II. Sulphate adsorption. III. Phosphate adsorption. *Soil Sci. Soc. Amer. Proc.* **38**:255–265.

Greenland, D. J., J. M. Oades, and T. W. Sherwin. 1968. Electron-microscope observations of iron oxides in red soils. *J. Soil Sci.* **19**:123–126.

Harada, Y. and K. Wada. 1973. Release and uptake of protons by allophanic soils in relation to their cation exchange capacity and anion exchange capacity *Soil Sci. Plant Nutr.* (*Tokyo*) **19**:73–82.

Hingston, F. J., A. M. Posher, and J. P. Quirk. 1972. Anion adsorption by goethite and gibbsite. I. The role of the proton in determining adsorption envelopes. *J. Soil Sci.* **23**:177–192.

Jones, R. C. and G. Uehara. 1973. Amorphous coatings on mineral surfaces. *Soil Sci. Soc. Amer. Proc.* **38**:792–798.

Kaloga, B. and C. Thomann. 1971. La physico-chimie du complexe absorbent dans les sols bruns eutrophes. *Cah. ORSTOM, Sèr. Pedol.* **9**:461–507.

Kanehiro, J. and G. D. Sherman. 1956. Effect of dehydration-rehydration on cation exchange capacity of Hawaiian soils. *Soil Sci. Soc. Amer. Proc.* **26**:341–344.

Keng, J. C. W., and G. Uehara. 1974. Chemistry, mineralogy and taxonomy of Oxisols and Ultisols. *Proc. Soil Crop Sci. Soc. Fla.* **33**:119–126.

Kinjo, T. and P. F. Pratt. 1971. Nutrate adsorption. I. In some acid soils of Mexico and South America. II. In competition with chloride, sulfate and phosphate. III. Desorption, movement and distribution in Andepts. *Soil Sci. Soc. Amer. Proc.* **35**:722–732.

Kitagawa, Y. 1971. The unit particle of allophane. *Amer. Mineralogist* **56**:465–478.

León, L. A. 1967. Chemistry of some acid tropical soils from Colombia. Ph.D. Thesis, University of California at Riverside. 191 pp.

León, L. A. and N. T. Coleman. 1972. Adsorción aniónica en suelos ácidos de Colombia. *Fitotecnia Lat.* **8**(3):70–77.

Martini, J. A. 1970. Allocation of cation exchange capacity to soil fractions in seven surface soils of Panama and the application of the cation exchange factor as a weathering index. *Soil Sci.* **109**:324–331.

Matsusaka, Y. and G. D. Sherman. 1950. Titration curves and buffering capacities of Hawaiian soils. *Hawaii Agr. Exp. Sta. Tech. Bull.* 11.

Mattson, S. 1926. The relations between the electrokinetic behavior and the base exchange capacity of soil colloids. *J. Amer. Soc. Agron.* **18**:458–512.

Mattson, S. and A. J. Pugh. 1934. The laws of colloidal behavior. XIV. The electrokinetics of hydrous oxides and their ionic exchange. *Soil Sci.* **38**:229–313.

Mattson, S. and L. Wicklander. 1940. The pH and amphoteric behavior of soils in relation to the Donnan equilibrium. *Ann. Agr. Coll. Sweden* **8**:1–54.

Mehlich, A. and A. A. Theisen. 1973. Charge characteristics in relation to crystalline and amorphous soil constituents. Unpublished manuscript, North Carolina Department of Agriculture, Soil Testing Division, Raleigh.

Mekaru, T. and G. Uehara. 1972. Anion adsorption in ferruginous tropical soils. *Soil Sci. Soc. Amer. Proc.* **36**:296–300.

Miller, E. V. and N. T. Coleman. 1952. Colloidal properties of soils from western Equatorial South America. *Soil Sci. Soc. Amer. Proc.* **16**:239–244.

Mitchell, B. D., V. C. Farmer, and W. J. Machardy. 1964. Amorphous inorganic materials in soils. *Adv. Agron.* **16**:327–383.

Parfitt, R. L. 1972. Amorphous minerals in some Papua and New Guinea soils. *Soil Sci. Soc. Amer. Proc.* **36**:683–686.

Perez-Escolar, R. and M. A. Lugo-Lopez, 1968. Influence of the degree of clay mineral crystallization and free oxides on the cation exchange capacity of Catalina and Cialitos soils. *J. Agr. Univ. Puerto Rico.* **52**:148–154.

Pratt, P. F. and R. Alvahydo. 1966. Cation exchange characteristics of soils of São Paulo, Brazil. *IRI Res. Inst. Bull.* 31.

Pratt, P. F., F. F. Peterson, and C. S. Holzley. 1969. Qualitative mineralogy and chemical properties of a few soils from São Paulo, Brazil. *Turrialba* **19**:491–496.

Reeve, N. G. and M. E. Summer. 1971. Cation exchange capacity and exchangeable aluminum in Natal oxisols. *Soil Sci. Soc. Amer. Proc.* **35**:38–42.

Renau, R. B., Jr., and J. G. A. Fiskell. 1972. Mineralogical properties of clays from Panama soils. *Soil Sci. Soc. Amer. Proc.* **36**:501–505.

Russell, J. R. M. 1973. Cation exchange capacity measurements of some noncalcareous Rhodesian subsoils. *Rhodesian J. Agr. Res.* **11**:77–82.

Sawhney, B. L., C. R. Frink, and D. E. Hill. 1970. Components of pH-dependent cation exchange capacity. *Soil Sci.* **109**:272–278.

Sawhney, B. L. and K. Norrish. 1971. pH-dependent cation exchange capacity: minerals and soils of tropical regions. *Soil Sci.* **112**:213–215.

Schalscha, E. B., C. Gonzalez, I. Vergara, et al. 1965. Effects of drying on volcanic ash soils of Chile. *Soil Sci. Soc. Amer. Proc.* **29**:481–482.

Schalscha, E. B., P. F. Pratt, T. Kinjo, and J. Amarn. 1972. Effect of phosphate salts as saturating solutions in cation exchange capacity determinations. *Soil Sci. Soc. Amer. Proc.* **36**:912–914.

Schalscha, E. B., P. F. Pratt, and D. Soto. 1974. Effect of phosphate adsorption on the cation exchange capacity of volcanic ash soils. *Soil Sci. Soc. Amer. Proc.* **38**:539–540.

Sherman, G. D., Y. Matsusaka, H. Ikawa, and G. Uehara. 1964. The role of the amorphous fraction in the properties of tropical soils. *Agrochimica* **8**:146–163.

Soil Survey Staff. 1970. *Soil Taxonomy* (draft). Soil Conservation Service, U.S. Department of Agriculture, Washington.

Tan, K. H. et al. 1970. The nature and composition of amorphous materials and free oxides in some temperate regions and tropical soils. *Commun. Soil Sci. Plant Anal.* **1**:225–238.

Tanada, T. 1952. Certain properties of the inorganic colloidal fraction of Hawaiian soils. *J. Soil Sci.* **2**:83–96.

Uehara, G., L. D. Swindale, and R. C. Jones. 1972. Mineralogy and behavior of tropical soils. Seminar on Tropical Soils Research, International Institute for Tropical Agriculture, Ibadan, Nigeria (mimeo).

Uehara, G. and J. Keng. 1975. Management implications of soil mineralogy in Latin America. Pp. 351–362. In E. Bornemisza and A. Alvarado (eds.), *Soil Management in Tropical America.* North Carolina State University, Raleigh.

Van Raij, B. 1969. Capacidade de troca de fracciones organicas e minerais do solos. *Bragantia* **28:**85–112.

Van Raij, B. and M. Peech. 1972. Electrochemical properties of some Oxisols and Alfisols of the tropics. *Soil Sci. Soc. Amer. Proc.* **36:**587–593.

Wada, K. 1967. A structural scheme of soil allophane. *Amer. Mineralogist* **52:**670–708.

Wada, K. and Y. Harada. 1971. Effects of temperature on the measured cation exchange capacities of ando soils. *J. Soil Sci.* **22:**109–117.

Wada, K. and Y. Tokashiki. 1972. Selective dissolution and difference infrared spectroscopy in qualitative mineralogical analysis of volcanic ash soil clays. *Goederma* **7:**199–213.

Weaver, R. M. 1972. Chemical and clay mineral properties of highly weathered soils from the Colombian Llanos Orientales. *Agron. Mimeo* 72–19. Cornell University, Ithaca, N.Y.

Weaver, R. M. 1974. Soils of the Central Plateau of Brazil: Chemical and mineralogical properties. *Agron. Mimeo* 74–8. Cornell University, Ithaca, N.Y.

Wright, E. H. M. and M. N. Fawty. 1971. Physical and physicochemical characteristics of some Sierra Leone soils. *Sols Afr.* **16:**5–30.

Yuan, T. L. 1974. Chemistry and mineralogy of Andepts. *Proc. Soil Crop Sci. Soc. Fla.* **33:**101–108.

Zelazny, L. W. and F. G. Calhoun. 1971. Mineralogy and associated properties of tropical and temperate soils in the Western Hemisphere. *Proc. Soil Crop Sci. Soc. Fla.* **31:**179–189.

5

SOIL ORGANIC MATTER

When the subject of organic matter in tropical soils is mentioned, one commonly held view immediately arises: tropical soils have low organic matter contents because of the high temperatures and decomposition rates. Thus, every effort must be made to maintain what little organic matter exists, and therefore organic matter conservation is essential to the productivity of tropical soils. When the latter statement was challenged by British soil scientists at a meeting in Ibadan, Nigeria, in 1972, many workers from Africa protested strongly and requested that the heretical remarks be stricken from the record. Each side, of course, has evidence to support its contentions. This chapter examines the contents and changes of organic matter in tropical soils and attempts to specify their relevance to management practices.

CONTENTS

Contrary to commonly held views, organic matter contents in tropical soils are not very different from those in the temperate region. In 1930 Dean found that the average content of the top 30 cm of 223 Hawaiian soils was 3.75% O.M. The mean contents of several hundred topsoils from udic areas of Puerto Rico were 3.54% O.M. and 0.19% N, and from ustic areas 1.84%

O.M. and 0.20% N (Smith et al., 1951). The mean content of 570 topsoils (0 to 30 cm) collected in East Africa by Birch and Friend (1956) was 3.36% O.M. Approximately half of them contained more than 4% O.M. in the top 15 cm. All these figures compare very favorably with their temperate-region counterparts.

Extensive studies by Jenny in the 1940s and 1950s related organic matter content inversely to mean annual temperature in the United States. These relationships broke down, however, when Jenny included soils from Colombia and Costa Rica. These soils were higher in organic matter than their mean annual temperatures would predict (Jenny et al., 1948).

Organic carbon values of randomly chosen Oxisols, Ultisols, and Alfisols from Brazil and Zaïre are compared with those of Mollisols, Ultisols, and Alfisols from the United States in Table 5.1. This table indicates that the organic carbon content of the top meter of the Oxisols is not significantly different from that of the temperate-region Mollisols and that there is no difference in this respect between tropical and temperate Ultisols and Alfisols. Buol (1973) found a similar lack of differences in organic carbon content between major temperate and tropical soils.

There are several explanations as to why tropical soils are higher in

Table 5.1 Comparison of Average Organic Carbon Contents of Several Soil Orders in the United States, Brazil, and Zaïre: Each Figure is the Average of 16 Randomly Chosen Profiles (%)

Soil Order	United States	Brazil	Zaïre	Means
0–15 cm depth				
Mollisols	2.44	—	—	2.44
Oxisols	—	2.01	2.13	2.07
Ultisols	1.58	1.61	0.98	1.39
Alfisols	1.55	1.06	1.30	1.30
	$LSD_{0.05} = 0.38$			
0–100 cm depth				
Mollisols	1.11	—	—	1.11
Oxisols	—	1.07	1.03	1.05
Ultisols	0.49	0.88	0.45	0.61
Alfisols	0.52	0.53	0.55	0.53
	$LSD_{0.05} = 0.19$			

Source: Wade and Sanchez (unpublished data).

organic matter than generally believed. The most obvious one is the absence of a direct relationship between color and organic carbon content. Table 5.2 shows that many red Oxisols and Ultisols have higher organic carbon contents than black Vertisols. As a group, Vertisols contain the least organic matter. As early as 1930 Vageler suggested that humus may be colorless.

As a group, Andepts have the highest organic matter contents of mineral soils. Allophane reacts with organic radicals to form complexes that remain relatively resistant to mineralization. Thus organic matter tends to accumulate in these soils. Bornemisza and Pineda (1969) showed an inverse relationship between organic matter mineralization and allophane content. Although the reasons for this relationship are not entirely clear, a physical blocking of organic particles by allophane may render the organic materials partly inaccessible to microorganisms. Recently, Munevar and Wollum (1976) proved that the extreme phosphorus deficiency typical of these soils inhibits microbial growth, resulting in a lower mineralization rate. Jenny's samples from Colombia and Costa Rica came from soils high in allophane, according to Bornemisza and Pineda (1969).

The relatively high organic carbon contents of some Oxisols have not been adequately explained. The interaction between oxides and organic materials, on the one hand, and low available nutrient levels, on the other, may be involved. Unpublished results at North Carolina State University indicate that phosphorus and calcium additions increase nitrogen mineralization in an Oxisol from Brazil that is extremely low in these nutrients. As a group, Oxisols rank third in organic carbon after Andepts and Mollisols in Table 5.2.

The organic carbon distribution in the profile due to native vegetation is essentially the same in tropical and temperate regions. Forested areas show a marked organic matter accumulation in the topsoil as a result of litter fall and the superficial nature of forest tree roots. Savannas and prairies generally produce more carbon in the subsoil because of the decomposition of deep grass roots. The effect may be less marked in tropical savannas with aluminum toxicity problems in the subsoil.

ADDITIONS AND DECOMPOSITION

The organic carbon content of a soil in equilibrium with the vegetation is a function of the annual additions and decompositions of organic carbon. The following formulae explains these relationships:

$$C = \frac{bm}{k}$$

Table 5.2   Organic Carbon and Nitrogen Contents of Some Representative Tropical Soil Profiles

Soil	Location and Reference	Horizon (cm)	Organic C (%)	Total N (%)	C:N Ratio
OXISOLS					
Red-Yellow Latosol	Brazil (FAO-UNESCO, 1971)	0–4	7.8	0.69	11
		4–12	2.7	0.20	13
		12–25	1.7	0.15	11
		25–80	1.0	0.08	12
		80–200	0.6	0.06	10
		200–270	0.3	0.06	5
Eutrustox	Brazil (Moura and Buol, 1972)	0–10	2.8	0.25	11
		20–30	1.3	0.15	9
		40–50	1.0	0.11	9
		80–90	0.6	0.07	8
		100–110	0.6	0.05	12
		120–130	0.6	0.05	12
		140–170	0.3	0.04	7
Ferralitic soil (sandy)	Zaïre (D'Hoore, 1964)	0–8	1.3	0.10	13
		8–23	0.8	0.05	16
		23–38	0.3	0.02	15
		38–85	0.1	0.01	10
		85–200	0.1	0.01	10
Haplorthox (Coto clay)	Puerto Rico (Beinroth, 1972)	0–13	2.4	0.25	10
		13–26	1.7	0.20	9
		26–43	1.0	0.14	7
		43–63	0.6	0.12	5
		63–91	0.5	0.11	4

Table 5.2 (*Continued*)

Soil	Location and Reference	Horizon (cm)	Organic C (%)	Total N (%)	C:N Ratio
ULTISOLS					
Tropohumult (Humatas clay)	Puerto Rico (Beinroth, 1972)	0–10	4.9	0.43	11
		10–23	2.0	0.21	9
		23–38	1.2	0.12	10
		38–63	0.6	0.07	9
		63–82	0.3	0.05	7
Red-Yellow Podzol	Brazil (Sombroek, 1967)	0–2	3.0	0.26	12
		2–40	0.9	0.09	10
		40–100	0.5	0.05	10
		100–180	0.3	0.04	9
		180–230	0.2	0.02	9
Orthic Acrisol (Profile 13)	Brazil (FAO-UNESCO, 1971)	0–10	1.3	0.13	10
		10–30	0.7	0.10	7
		30–45	0.5	0.07	7
		45–75	0.5	0.08	6
		75–155	0.2	0.06	3
		155–195	0.1	0.04	2
Dystric Nitosol	Brazil (FAO-UNESCO, 1971)	0–15	3.2	0.28	11
		15–40	1.1	0.12	9
		40–70	0.7	0.10	7
		70–160	0.5	0.07	7
		160–190	0.2	0.04	5

ALFISOLS

Ultustalf
(Ferruginous tropical soil)

Senegal
(D'Hoore, 1964)

Depth			
0–6	1.0	0.07	14
6–13	0.7	0.05	14
13–31	0.4	0.03	13
31–79	0.3	0.03	10
79–117	0.3	0.03	10
117–150	0.2	0.03	7

Eutric Nitosol
(Terra Roxa Estruturada)

Brazil
(FAO-UNESCO, 1971)

Depth			
0–19	1.5	0.18	8
19–80	0.6	0.07	9
80–134	0.4	0.05	8
134–224	0.2	0.03	6
224–250	0.2	0.05	4

MOLLISOLS

Cauca Valley
(ave. 71 samples)

Colombia
(Gomez et al., 1969)

Depth			
0–20	2.5	0.22	11
20–40	1.6	0.16	10
40–70	1.1	0.11	10

Haplic Kastazozem (Haplustoll)

Puno, Peru
(FAO-UNESCO, 1971)

Depth			
0–10	5.1	0.24	21
10–25	2.7	0.15	18
25–60	0.7	0.07	10
60–80	0.1	0.03	3

VERTISOLS

Chromic Vertisol

Bahia, Brazil
(FAO-UNESCO, 1971)

Depth			
0–20	0.4	0.06	7
20–70	0.3	0.05	6
70–132	0.3	0.05	6
132–142	0.2	0.06	3

Table 5.2 (*Continued*)

Soil	Location and Reference	Horizon (cm)	Organic C (%)	Total N (%)	C:N Ratio
Vertisol	Gezira, Sudan (D'Hoore, 1964)	0–2	0.3	0.03	10
		2–90	0.3	0.02	15
		90–130	0.4	0.03	13
		130–165	0.4	0.02	20
		165–185	0.2	0.01	20
		185–220	0.2	0.02	10
Chromusert (Fraternidad clay)	Puerto Rico (Beinroth, 1972)	0–13	2.4	0.25	10
		13–26	1.7	0.20	9
		26–43	1.0	0.14	7
		43–63	0.6	0.13	5
		63–91	0.5	0.11	4
INCEPTISOLS					
Andept	Guatemala (Palencia and Martini, 1970)	0–100	8.2	0.68	12
		100–200	2.4	0.14	17
Eutrandept (Waimea)	Hawaii (IICA, 1969)	0–5	13.0	1.18	11
		5–13	5.0	0.61	8
		13–20	4.8	0.52	9
		20–58	4.3	0.38	11
		58–88	3.2	0.26	12
		88–118	2.9	0.26	11

	Depth (cm)			
Dystrandept				
Antioquia, Colombia	0–10	12.9	0.9	14
(IICA, 1969)	10–30	7.1	0.5	14
	30–60	1.6	0.3	13
	60–80	3.6	0.3	12
	80–120	3.5	0.3	12
	120–125	1.3	0.1	9
	125 +	0.5	0.1	8
Eutropept				
Puerto Rico	0–15	1.7	0.16	11
(Smith et al., 1951)	20–30	1.0	0.11	9
	45–61	0.3	0.04	7
	91–106	0.4	0.03	9
ENTISOLS				
Albic arenosol (Tiwiwid sand)				
Guyana	0–10	2.5	0.04	62
(FAO-UNESCO, 1971)	10–18	1.7	0.02	85
	18–60	0.1	0.01	10
	60–120	0.0	—	—
Ferralic Arenosols				
São Paulo, Brazil	0–15	0.5	0.04	11
(FAO-UNESCO, 1971)	15–49	0.3	0.03	10
	49–112	0.2	0.02	10
	112–148	0.2	0.02	10
	148–600	0.1	0.01	10
Thionic Fluvisol (Mara clay)				
Guyana	0–3	7.2	0.29	25
(FAO-UNESCO, 1971)	3–10	2.1	0.30	7
	10–30	0.8	0.24	3
	30–84	0.5	0.05	10
	84–100	1.4	0.09	15

and

$$a = bm$$

where C = the percentage of soil organic carbon in equilibrium (tons/ha)
$\quad\quad b$ = the annual amount of fresh organic matter added to the soil (tons/ha)
$\quad\quad m$ = the conversion rate of fresh organic matter into soil organic carbon (percent)
$\quad\quad a$ = the annual addition of soil organic carbon (tons/ha)
$\quad\quad k$ = the annual decomposition rate of soil organic carbon (percent)

The magnitudes of these various parameters can be observed in Table 5.3. The annual addition of fresh organic matter as litter, branches, and dead roots (b) is on the order of 5 tons/ha of dry matter in tropical forests and about 1 ton/ha in temperate forests. The actual ranges in the literature are from 3 to 15 tons/ha in tropical forests and 1 to 8 tons/ha in temperate forests. Tropical savannas add from 0.5 to 1.5 tons/ha; temperate prairies, about 1.5 tons/ha.

An opposite relationship, therefore, exists between tropical and temperate vegetation forms. Tropical udic forests furnish about five times as much raw organic matter to the soil as do their temperate counterparts, mostly in the form of litter in both cases. This difference is due to the faster growth rate of tropical forests. Fresh organic matter additions in grasslands are primarily in the form of root decomposition. Temperate prairies develop more abundant root systems, whereas growth in tropical savannas is often limited by low nutrient availability. Annual burning of savannas further reduces the raw organic matter additions.

The conversion rate (m) of fresh organic matter into soil organic carbon (humus) is on the order of 30 to 50 percent per year. The rates are relatively constant in the different environments. The annual additions of soil organic carbon or humus (a), therefore, are about four times higher in tropical than in temperate forests and are fairly similar in tropical and temperate grasslands.

The annual decomposition rates of soil organic carbon (k) vary considerably. They range from 2 to 5 percent in tropical forests with the exception of the Colombian example, which is very low (0.5 percent) because of the presence of allophane in that particular soil. The k values for temperate forests range from 0.4 to 1 percent, probably as a result of temperature limitations. The decomposition rate of tropical savannas averages 1.2 percent or three times that of temperate prairies.

Differences between tropical and temperate regions are primarily a function of temperate fluctuations. It should be remembered, however, that k rates during the temperate summer may be higher than in tropical regions

Table 5.3 Estimates of Annual Additions, Decomposition Rates, and Equilibrium Levels of Topsoil Organic Carbon in Some Tropical and Temperate Locations

Location	b Addition of Undecomposed Organic Matter (tons/ha)	m Decomposition rate of Fresh Organic Matter into Soil Organic C (%)	a Soil Organic C Addition (tons/ha)	k Soil Organic C Decomposition Rate (%)	C Soil Organic C at Equilibrium (tons/ha)	(%)
Tropical forests						
Ghana (Ustic)	5.28	50	2.64	2.5	106	2.4
Zaïre (Udic)	6.05	47	2.86	5.2	55	1.2
Colombia (Udic, Andept)	3.85	51	1.97	0.5	394	9.0
Temperate forests						
California (oak)	0.75	47	0.35	0.4	88	2.0
California (pine)	1.65	52	0.86	1.0	86	1.9
Tropical savannas						
Ghana (1250 mm rain)	1.43	50	0.71	1.3	55	1.2
Ghana (850 mm rain)	0.44	43	0.19	1.2	16	0.4
Temperate prairie						
Minnesota (870 mm rain)	1.42	37	0.53	0.4	134	3.0

Source: Recalculated from data by Greenland and Nye (1959).

because of higher temperatures, but the overall annual decomposition rates are lower because of the winter. The higher k values in udic tropical forests than in ustic ones reflects the effect of the dry season in slowing organic carbon decomposition rates as a result of low moisture. The lower k rates in tropical savannas also reflect the ustic soil moisture regime.

Equilibrium topsoil organic carbon contents (C) can be calculated by the formula given above. The values in Table 5.3 show the interaction between environmental factors and further reflect mineralogy, clay contents, and other soil factors. In general, the higher the clay contents and the higher the proportion of oxides and allophane, the lower will be the k values. Among the three examples from tropical forests, the Colombian soil had high allophane contents, the Ghanian soil was of medium texture, and the soil from Zaïre had a sandy texture.

This similarity between tropical and temperate soils, therefore, can be understood in terms of the temperature and moisture regimes and the empirical rule that for every 10°C increase in temperature the rate of biological activity doubles. In the temperate regions low winter temperatures greatly reduce biological activity. In the 78 percent of the tropics that has an ustic or aridic soil moisture regime, the lack of moisture during this period has a similar effect. Topsoil and air temperatures during the tropical rainy seasons are similar to, but seldom as high as, the corresponding summer temperatures in the temperate regions. For the 22 percent of the tropical areas with udic soil moisture regimes, the explanation is somewhat different. Most of these areas are covered by tropical rainforests. Neither temperature nor moisture limits organic matter accumulation and decomposition at any time. These forests produce about five times as much biomass and soil organic matter per year as comparable temperate forests. The rate of organic matter decomposition, however, is also about five times greater than in temperate forests. Thus the equilibrium contents are similar (Sanchez and Buol, 1975).

EFFECTS OF CULTIVATION

The equilibrium values previously discussed change when a and k change as a consequence of cultivation. Annual k rates increase with higher temperature, greater moisture, aeration, cultivation, uptake, leaching, and denitrification. Annual organic carbon additions are drastically reduced when forests are brought under cultivation; crop residues usually provide only a fraction of the 5 tons/ha or so of dry matter previously supplied by forest litter and root decomposition. Since k increases with temperature and aeration, cultivation accelerates organic carbon decomposition. In the savannas, exposure and plowing results in a fourfold increase in k relative to the equilibrium values. Table 5.4 shows some examples of decomposition rates upon cultivation. A bare soil in Zaïre had an organic carbon decomposition rate of 13 percent in the same site as Table 5.3, where k was 5.7 percent under forest. A corn–cassava rotation of 7 years also doubled the k value of the undisturbed forest in Ghana shown in Table 5.3. Crop rotation with legumes in Trinidad, however, resulted in normal k values according to temperate-region experience. Table 5.4 also shows that k values in savanna areas range from 1 to 4 percent per year, whereas in the temperate region the rates seldom increase to 3 percent. Figure 5.1 shows the general effects of cultivation on profile organic carbon contents in two West African soils.

Shifting cultivation, however, seldom results in substantial soil organic matter depletion. After analyzing 100 traditional shifting cultivation sites in forested areas of Liberia, Reed (1951) found that the carbon contents were maintained at about 75 percent of the equilibrium levels. When overpopulation narrows the crop:fallow ratio, organic carbon contents drop to 50 percent of the original equilibrium values.

Inorganic fertilization can increase a values by a large margin because of its effect in adding more crop residues, including decomposing crop roots, to the soil. In properly managed systems it can also reduce the decomposi-

Table 5.4 *Soil Organic Carbon Decomposition Rates under Several Years of Cultivation in Some Tropical and Temperate Topsoils*

Location	Treatment	Years under Cultivation	Decomposition Rate k (%/year)
1. Zaïre	Clean-weeded bare fallow	3	12.8
2. Ghana	Corn–cassava rotation	7	4.7
3. Trinidad	Crop rotation with legumes	6	2.6
4. Trinidad	Crop rotation with legumes	12	1.8
Tropical savannas			
5. Ghana	Crop rotation	7	4.0
6. Senegal	Continuous peanuts	6	6.6
7. Sudan	Cotton–peanuts rotation	6	2.5
Temperate zone			
8. Missouri	Continuous corn	25	2.8
9. Missouri	Crop rotation	25	0.8
10. France	Crop rotation	14	1.4

Source: Greenland and Nye (1959).

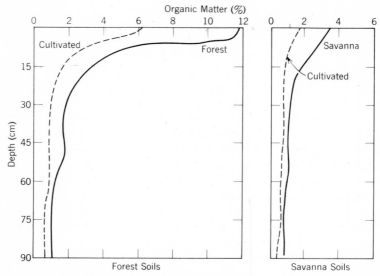

Fig. 5.1   Organic matter distribution in typical West African forest and savanna soil profiles. *Source:* Brams (1972).

tion constant. A long-term study of tea soils in Sri Lanka, illustrated in Fig. 5.2, shows that higher organic nitrogen equilibrium levels were obtained when tea was fertilized annually with a 120-60-60 kg/ha rate of N, P_2O_5, and K_2O. The unfertilized plots kept losing organic nitrogen at an annual rate of 1.5 percent with correspondingly lower total nitrogen contents as equilibrium was approached (Gokhale, 1959). The fertilized plots had a decomposition rate of 1.0 percent. This example underscores the fact that organic matter can be kept at high levels with good management practices in the tropics.

BENEFICIAL EFFECTS

The rapid depletion of topsoil organic carbon at yearly rates of 5 to 10 percent can result in marked detrimental effects to unfertilized crops. Researchers working in areas where this occurs emphasize the need to reduce the decline in soil organic matter. Greenland and Dart (1972) have pointed out the following benefits of soil organic matter for no-fertilizer agriculture:

1. Organic matter supplies most of the nitrogen and sulfur and half of the

phosphorus taken up by unfertilized crops. The slow-release pattern of nitrogen and sulfur mineralization offers a definite advantage over soluble fertilizers.

2. Organic matter supplies most of the cation exchange capacity of acid, highly weathered soils. Rapid decreases in organic matter result in sharp reduction in the CEC.

3. By forming complexes with organic matter, amorphous oxides do not crystallize. Phosphorus fixation by these oxides is decreased by organic radicals blocking the fixation charges.

4. Organic matter contributes to soil aggregation and thus improves physical properties and reduces susceptibility to erosion in sandy soils.

5. Organic matter modifies water retention properties, particularly in sandy soils. In Ghana the soil water-holding capacity decreased from 57 to 37 percent when the soil organic matter decreased from 5 to 3 percent.

6. Organic matter may form complexes with micronutrients which prevent their leaching.

In spite of all these valid reasons, soil organic matter is of minor concern in management schemes where fertilizers are effectively and economically used, as in sandy Ultisols of the southeastern United States. There is no question that organic matter is a good thing if present; but if there is not much of it, the means of increasing it must be weighed against direct fertilizer and mulching practices. Again, good management practices do increase soil organic matter.

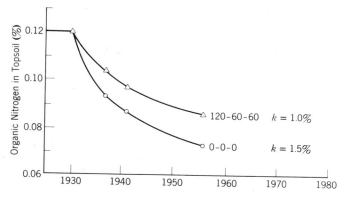

Fig. 5.2 Long-term charges in organic nitrogen contents in fertilized and unfertilized tea plantations in Assam, India. *Source:* Gokhale (1959).

MANURE APPLICATIONS

The traditional way to increase organic matter in cultivated areas is to add undecomposed raw materials in the form of animal manures, composts, or plant materials incorporated as green manure. Manure application is an ancient practice in tropical Asia, related mostly to paddy rice cultivation. Evaluations of such practices versus the use of more concentrated inorganic fertilizers have been carried out principally in India and Africa.

A typical example is shown in Table 5.5, where the effects of continuous manure versus inorganic fertilization for 10 years are shown on paddy rice yields and soil properties. An annual rate of 5.6 tons/ha of animal manures increased yields as much as nitrogen or phosphorus fertilization. Both manure and fertilizer treatments increased soil organic carbon and organic nitrogen slightly.

At the other end of the soil spectrum, Pichot (1971) conducted a similar study on a ferralitic soil (Oxisol) of the Central African Republic, in which corn and upland rice were grown in rotation for 5 years. The results, shown in Table 5.6, indicate slight superiority of an annual application of 60 tons/ha of animal manure over an annual rate of 120-160-160 kg/ha of N, P_2O_5, and K_2O. The manure application increased soil organic carbon, organic nitrogen, and exchangeable calcium, thereby resulting in a significant pH increase. Apparently the manure had considerable quantities of calcium. Pichot was unable to conclude why manuring was slightly superior, but his data strongly suggest that the calcium content of the manure was an important reason.

The foregoing examples are typical of manure research. The overall yield levels were low; the nutrient composition of the manure was not given.

Table 5.5   Effects of 10 Years of Manure and Fertilizer Applications on Lowland Rice Production and the Properties of a Vertisol in Bagwai, Madhya Pradesh, India

Annual Application	Rice Yields (tons/ha)	Organic C (%)	Total N (%)	Available P (Olsen) (ppm)	CEC (meq/100 g)
None	0.88	0.07	0.063	10	31
Manure (5.6 tons/ha)	1.49	1.15	0.066	12	31
N (67 kg/ha)	1.55	1.12	0.066	11	31
P_2O_5 (67 kg/ha)	1.57	1.09	0.066	12	32

Source: Adapted from Shinde and Ghosh (1971).

Table 5.6 *Effects of Fertilizer and Manure Applications on Cereal Production and Soil Properties after 5 Years of Continuous Applications in a Ferralitic Soil (Oxisol) from Boukoko, Central African Republic*

Annual Application	Cummulative Grain Yields in 5 Years (tons/ha)		Soil Organic C (%)	Total N (%)	Soil pH	Exchangeable Ca (meq/100 g)	Exchangeable K
	Corn	Rice					
None	5.82	5.94	0.95	0.097	5.2	1.23	0.11
60-80-80 kg N, P_2O_5, K_2O/ha	9.06	9.07	0.87	0.091	4.8	0.98	0.16
120-160-160 kg N, P_2O_5, K_2O/ha	11.19	11.44	0.93	0.100	4.9	1.44	0.20
Animal manure (60 tons/ha)	13.12	10.47	1.37	0.160	5.7	2.91	0.34

Source: Adapted from Pichot (1971).

177

When manuring was superior to fertilization, no reasons were advanced to explain why. Also, both examples show that soil organic carbon and nitrogen can be increased with manuring but also with fertilization.

Several long-term comparisons in Africa did take into account the nutrient composition of the animal or green manure employed (Djokoto and Stephens, 1961; Stephens, 1969; Heathcote, 1970). They show that crop responses to animal manure can be explained in terms of its nutrient composition, particularly its potassium and phosphorus contents. For example, Stephens (1969) found in 18 long-term experiments in Uganda that manure applications were superior to chemical fertilizers. Manure provided 75 kg K/ha annually, whereas the fertilizer mixture used supplied only 25 kg K/ha to potassium-deficient soils. In such cases the choice is essentially a matter of nutrient composition, availability of chemical fertilizer versus availability of manures, their relative nutrient compositions, fertilizer costs, and transportation costs. In many instances it may be more economical to apply the extra 50 kg K/ha than to apply and incorporate large quantities of manure. The results of Table 5.6 suggest that liming to a pH of 5.5 might do the same job for several years as applying large amounts of manure every year.

Another common reason given for the beneficial effects of manures is the improvement in soil physical properties. Unfortunately, this writer has found no data in which increased crop yields with manure applications are directly related to improvement in soil physical properties. However, there is adequate information that quantifies the beneficial effects of long-term manure applications on the physical properties of several soils. Some results shown in Table 5.7 indicate that long-term manuring increased aggregate stability, pore space, bulk density, and available water range and decreased bulk density in some Indian soils. The results were particularly beneficial in the coarse-textured alluvial soil in terms of decreasing bulk density and in the clayey Vertisol in almost doubling hydraulic conductivity. The physical benefits in the "red" and "lateritic" soils were also considerable.

Incorporation of green manures is often effective only to the next subsequent crop and not to complete rotation schemes. Vine (1953) considered it of doubtful value to spend 3 months growing a green manure crop that provides 25 kg N/ha to the subsequent crop, in view of the possible profits from growing a cash crop during that time and the relatively low cost of such a nitrogen rate when fertilizer is available. In most soils green manure incorporation requires quite a bit of power. It also requires an understanding of the value of a crop that will give no direct income to the farmer.

Perhaps the most valuable approach to maintaining soil organic matter with cropping is to provide a surface mulch that will lower soil temperatures enough to prevent large increases in k decomposition rates and will also provide protection against erosion.

Table 5.7 Effects of Long-Term Manuring Applications on the Physical Properties of Some Indian Soils

Soil and Location	Annual Manure Applied (tons/ha)	pH	Organic C (%)	Water-stable Aggregates (%)	Bulk Density (g/cc)	Pore Space (%)	Hydraulic Conductivity (cm/hour)	Available H_2O (0.1–15 bars) (%)
Alluvial (Saborer)	0	7.8	0.6	26	1.47	46	0.43	28
	74	7.5	2.9	56	1.29	50	0.47	37
Vertisol (Poona)	0	8.2	0.5	22	1.26	53	0.06	29
	45	8.0	0.6	33	1.18	56	0.10	31
"Red" (Bihar)	0	5.8	0.6	29	1.37	49	0.33	21
	9	5.7	0.7	33	1.30	51	0.50	23
"Lateritic" (Orissa)	0	4.8	0.3	22	1.53	42	1.68	13
	9	5.5	0.5	24	1.31	50	2.16	14

Source: Adapted from Biswas and Khosla (1971).

The above recommendations apply to labor-intensive small-scale agricultural systems. When mechanization is feasible and fertilizers are available at reasonable cost, there is no reason to consider the maintenance of organic matter as a major management goal. Experience in low-organic-matter soils of the southeastern United States and the coast of Peru have shown this to be the case. The controversy concerning the importance of organic matter is essentially an issue of economics. With the present high cost of fertilizers, however, this issue is becoming increasing relevant in large areas.

SUMMARY AND CONCLUSIONS

1. Organic matter contents in the tropics are similar to those of the temperate region. Highly weathered Oxisols have higher organic matter contents than their reddish colors would indicate.

2. The factors affecting the organic carbon contents of soils can be analyzed in terms of annual additions of organic carbon and annual decomposition rates. The annual additions of organic carbon received by soils are about five times greater in tropical than in temperate forests. Because of the lack of significant temperature or moisture limitations throughout the year, the rate of organic carbon decomposition is five times greater in udic tropical forests than in temperate forests. Consequently, equilibrium organic matter contents are similar.

3. In ustic environments lack of moisture during the dry season decreases organic carbon decomposition, as low temperatures do during the winter in the temperate region. The annual decomposition rates in ustic tropical environments are about half of those in udic tropical environments. Because of low organic carbon additions caused by annual burning, tropical savannas generally have lower organic carbon contents than tropical forests or temperate prairies.

4. Clearing and cultivation decrease the annual additions and at least double the decomposition rates of organic carbon. Organic matter depletion is quite rapid unless certain management practices are applied. In traditional shifting cultivation systems, organic matter depletion is small; practices aimed at keeping the soil covered and sustained fertilization decrease the organic matter depletion to a level that ceases to be of concern.

5. In unfertilized soils the beneficial effects of organic matter consist of supplying most of the nitrogen and sulfur to plants, maintaining cation exchange capacity, blocking phosphorus fixation sites, improving structure in poorly aggregated soils, and the formation of complexes with micronutrients.

6. The maintenance of organic matter is essential for no-fertilizer agriculture. It is also important in low-CEC soils, in which most of the negative charges are in the organic radicals, and in poorly aggregated sandy soils susceptible to compaction. In soils not exhibiting these problems, economically sound fertilization practices decrease the importance of organic matter conservation. In fact, adequate fertilization practices increase the soil organic matter contents because of increased root decomposition.

7. Animal manure applications in the tropics can be effective. Most experimental data show that the effect of a manure is related to its nutrient composition. The choice between animal manures and inorganic fertilizers is a matter of nutrient content, economics, transportation, and accessibility. Long-term manure applications improve physical soil properties.

8. Green manures are usually effective to the next crop. This practice is limited to mechanized agriculture since the power requirements are usually too high for manual labor and more profitable land use alternatives are available.

9. Mulching conserves organic matter by decreasing soil temperatures. The use of mulches may be applicable to a wider area than are animal or green manures.

REFERENCES

Bartholomew, W. V. 1972. Soil nitrogen and organic matter. Pp. 63–81. In *Soils of the Humid Tropics*. National Academy of Sciences, Washington.

Beinroth, F. H. 1972. The general pattern of the soils of Puerto Rico. *Trans. Fifth Caribbean Geol. Conf. Geol. Bull.* 5:225–229.

Birch, H. F. and M. T. Friend. 1956. The organic matter and nitrogen status of East African soils. *J. Soil Sci.* 7:156–167.

Biswas, T. D. and B. K. Khosla. 1971. Building up of organic matter status of the soil and its relation to the soil physical properties. *Proc. Int. Symp. Soil Fert. Eval. (New Delhi)* 1:831–842.

Blasco, M. 1971. Efecto de la humedad sobre la mineralización del carbono en suelos volcánicos de Costa Rica. *Turrialba* 21:7–12.

Bornemisza, E. and R. Pineda. 1969. The amorphous minerals and the mineralization of nitrogen in volcanic ash soils. Pp. B7.1–B7.7. In *Panel on Soils Derived from Volcanic Ash in Latin America*. Inter-American Institute of Agricultural Sciences, Turrialba, Costa Rica.

Brams, E. A. 1971. Continuous cultivation of West African soils: organic matter diminution and effects of applied lime and phosphorus. *Plant and Soil* 35:401–414.

Brams, E. A. 1972. Cation exchange as related to the management of tropical soils. Mimeographed lecture presented at the Tropical Soils Institute, University of Puerto Rico. Prairie View A and M University, Texas.

Buol, S. W. 1973. Soil genesis, morphology, and classification. Pp. 1–38. In P. A. Sanchez (ed.); "A Review of Soils Research in Tropical Latin America." *North Carolina Agr. Exp. Sta. Tech. Bull.* 219.

Charreau, C. 1972. Problèmes posés par l'utilisation agricoles des sols tropicaux par des cultures annuelles. *Agron. Tropicale (France)* 27:905–929.

Dean, A. L. 1930. Nitrogen and organic matter in Hawaiian pineapple soils. *Soil Sci.* 30:439–442.

D'Hoore, J. D. 1964. Soil map of Africa—explanatory monograph. *Comm. Tech. Cooperation in Africa Publ.* 93. Lagos, Nigeria.

Diaz-Romeu, R., F. Balerdi, and H. W. Fassbender. 1970. Contenido de materia orgánica y nitrógeno en suelos de América Central. *Turrialba* 20:185–192.

Djokoto, R. K. and D. Stephens. 1961. Thirty long term fertilizer experiments under continuous cropping in Ghana. I. Crop yields and responses to fertilizers and manures. II. Soil studies in relation to the effects of fertilizers and manures on crop yields. *Emp. J. Exptal. Agr.* 29:181–196, 245–258.

Endredy, A. S. de. 1954. The organic matter content of Gold Coast soils. *Trans. 5th Int. Congr. Soil Sci.* 2:457–463.

FAO-UNESCO. 1971. *Soil Map of the World, Vol. IV: South America.* United Nations Educational, Scientific, and Cultural Organization, Paris.

Gokhale, N. G. 1959. Soil nitrogen status under continuous cropping and with manuring in the case of unshaded tea. *Soil Sci.* 87:331–333.

Gomez, J. A., D. F. Zorrilla, and C. A. Flor. 1969. Algunas consideraciones sobre la materia orgánica en los suelos cultivados con maíz en el Valle del Rio Cauca. *Rev. ICA* 4:3–10.

Greenland, D. J. and P. H. Nye. 1959. Increases in carbon and nitrogen contents of tropical soils under natural fallows. *J. Soil Sci.* 9:284–299.

Greenland, D. J. and P. J. Dart. 1972. Biological and organic aspect of plant nutrition in relation to needed research in tropical soils. Tropical Soils Research Seminar. International Institute for Tropical Agriculture, Ibadan, Nigeria (mimeo).

Heathcote, R. G. 1970. Soil fertility under continuous cultivation in northern Nigeria. I. The role of organic manures. *Exptal. Agric.* 6:229–237.

IICA. 1969. *Panel on Soils Derived from Volcanic Ash in Latin America.* Inter-American Institute of Agricultural Sciences, Turrialba, Costa Rica.

Jenny, H. 1950. Causes of the high nitrogen and organic matter content of certain tropical forest soils. *Soil Sci.* 69:63–69.

Jenny, H. 1961. Comparison of soil nitrogen and carbon in tropical and temperate regions as observed in India and America. *Missouri Agr. Exp. Sta. Bull.* 765.

Jenny, H., F. T. Bingham, and B. Padilla-Saravia. 1948. Nitrogen and organic matter contents of equatorial soils of Colombia, South America. *Soil Sci.* 66:173–186.

Jenny, H., S. P. Gessel, and F. T. Bingham. 1949. Comparative study of decomposition rates of organic matter in temperate and tropical regions. *Soil Sci.* 68:419–432.

Jones, M. L. 1973. The organic matter content of the savanna soils of West Africa. *J. Soil Sci.* 24:42–53.

Le Mare, P. H. 1972. Long term experiment on soil fertility and cotton yields in Tanzania. *Exptal. Agr.* 8 (4):299–310.

Maurya, P. A. and A. B. Ghosh. 1972. Effect of long-term manuring and rotational cropping on fertility status of alluvial calcareous soils. *J. Indian Soc. Soil Sci.* 20 (1):31–43.

Moura, W. and S. W. Buol. 1972. Studies of a Latosol Roxo (Eutrustox) in Brazil. *Experientiae* 13:201–247.

Munevar, F. and A. G. Wollum. 1976. Effects of the addition of phosphorus and inorganic nitrogen on the carbon and nitrogen mineralization of some Andepts from Colombia. *Soil Sci. Soc. Amer. Proc.* (in press).

Nye, P. H. 1961. Organic matter and nutrient cycles under moist tropical forest. *Plant and Soil* 13:333–345.

Nye, P. H. and D. J. Greendland. 1960. The soil under shifting cultivation. *Commonwealth Agr. Bur. Tech. Commun.* 51, pp. 46–61. Harpenden, England.

Okigdo, N. 1972. Maize experiments in the Nsukka plains, Nigeria. III. The effect of kinds of mulch on the yield of maize in the humid tropics. *Agron. Tropicale (France)* 27:1036–1048.

Palencia, A. J. and J. A. Martini. 1970. Características morfológicas, físicas y quimicas de algunos suelos derivados de cenizas volcánicas en Centroamérica. *Turrialba* 20:325–332.

Pichot, J. 1971. Etude de l'evolution du sol en presénce de fumures organiques ou minérales. Cinq anneés d'experimentation á la Station de Bouakoko (République Centrafricaine). *Agron. Tropicale (France)* 26:736–754.

Raheja, S. K., R. Prasad, and H. C. Jain. 1971. Long term fertilizer studies in crop rotations. *Proc. Int. Symp. Soil Fert. Eval. (New Delhi)* 1:881–903.

Reed, W. E. 1951. Reconnaissance soil survey of Liberia. *U.S. Dept. Agr. Inf. Bull.* 66.

Sanchez, P. A. and S. W. Buol. 1975. Soils of the tropics and the world food crisis. *Science* 188:598–603.

Satyanaraya, K. V. S., K. Swaminathan, and B. Viswa Nath. 1946. Carbon and nitrogen status of Indian soils and their profiles. *Indian J. Agr. Sci.* 16:316–327.

Shinde, D. A. and A. B. Ghosh. 1971. Effect of continuous cropping and manuring on crop yield and characteristics of a medium black soil. *Proc. Int. Symp. Soil Fert. Eval. (New Delhi)* 1:905–916.

Singh, C. and S. S. Verma. 1969. Long range effect of green manuring on soil fertility and wheat yields in black cotton soils under rainfed conditions. *Indian J. Agron.* 14:159–164.

Smith, R. M., G. Samuels, and C. F. Cernuda. 1951. Organic matter and nitrogen build-ups in some Puerto Rican soil profiles. *Soil Sci.* 72:409–427.

Sombroek, W. G. 1967. *Amazon Soils.* Centre for Agricultural Publications and Documentation, Wageningen, Netherlands. 292 pp.

Stephens, D. 1969. The effects of fertilizers, manures and trace elements in continuous cropping rotations in southern and western Uganda. *East Afr. Agr. Exp. For. J.* 34:401–417.

Tan, K. H. and J. Van Schuylenborgh. 1961. On the organic matter in tropical soils. *Netherl. J. Agr. Sci.* 9:174–180.

Turrenne, J. F. 1970. Influence de la saison des pluies sur la dynamique des acides humiques dan des profils ferralitiques et podzoliques sous savanes de Guyane Française. *Cah. ORSTOM, Ser. Pédolog.* 8:419–449.

Vageler, P. 1933. *An Introduction to Tropical Soils* (Translated from the 1930 German edition by H. Greene). Macmillan, London. 240 pp.

Vine, H. 1953. Experiments on the maintenance of soil fertility at Ibadan, Nigeria, 1922–1951. *Emp. J. Exptal. Agr.* 21:65–85.

6
NITROGEN

Nitrogen is the nutrient element that most frequently limits yields in the tropics as well as in the temperate region. With the exception of some recently cleared land, most cultivated soils are deficient in this element. The fact that the nitrogen contents of tropical crops and grasses are in general lower than corresponding values in the temperate region (Webster and Wilson, 1966) contributes to the protein deficit so widespread in the tropics. As seen in Chapter 5, larger quantities of carbon are decomposed in certain tropical areas. This necessitates a correspondingly higher rate of nitrogen mineralization. The purpose of this chapter is to describe the basic soil nitrogen dynamics in the tropics, the fertilizer nitrogen reactions, and the management of both in relation to tropical crops.

SOIL ORGANIC NITROGEN

Additions of nitrogen to soils originate from rain and dust, nonsymbiotic fixation, symbiotic fixation, and animal and human wastes. Losses of nitrogen from the soil are due to volatilization, leaching, denitrification, erosion, and plant uptake. A summary of the relative contribution of each nitrogen source under four tropical conditions is presented in Table 6.1

Table 6.1   Low and High Estimates of the Relative Annual Contributions of Different Sources of Nitrogen in Five Tropical Ecosystems (kg N/ha)

Source	Rain forest	Tall Grass Savanna	Short Grass Savanna	Sugar cane	Rice Paddy
Rain and dust	4–8	4–8	4–8	4–8	4–8
Nonsymbiotic fixation					
In phyllosphere	12–40	0–12	0–4	0–12	0–4
By blue–green algae	0	0–10	0–10	0–10	14–70
In rhizosphere	0–6	0–13	0–6	0–9	0–10
In litter	0–25	0–10	0–6	12–50	0–10
Symbiotic fixation	34–68	0	0–10	0	0
Total	46–147	4–63	4–44	16–89	18–102

Source: Kass and Drosdoff (1970).

Nitrogen in Rain and Dust

These sources contribute an average of 4 to 8 kg N/ha annually. The highest levels, however, are recorded in tropical areas, perhaps because of intense electrical activity during thunderstorms. Dust storms from the Sahara contribute small amounts of nitrogen to the sub-Saharan countries. A question that has not been studied is whether nitrogen volatilized when vegetation is burned is brought back to another spot with rainfall in shifting cultivation areas. Visual observations indicate that smoke is sometimes intercepted by localized thundershowers, but no information is available on this matter. Studies in the temperate region show that most of the nitrogen is volatized as N_2 gas; therefore the possible recycling is limited to the portion of the volatilized nitrogen that is converted into NO, NO_2, and other oxides.

Asymbiotic Fixation

Asymbiotic nitrogen fixation is known to occur in the phyllosphere (leaf canopy), in the litter, in the soil, and in the rhizosphere. Atmospheric nitrogen is "fixed" by large populations of *Azotobacter* and *Beijerinckia* spp. in the leaf blades of many tropical species. The annual contribution of this process ranges from 0 to 8 kg N/ha. In rainforests it may supply up to 40 kg N/ha.

Asymbiotic fixation in the litter layer is on the same order of magnitude, as shown in Table 6.1.

Asymbiotic nitrogen fixation in the soil by blue-green algae is a well-known phenomenon thought to be particularly relevant in flooded rice culture, where the blue-green algae population is large. Estimates shown in Table 6.1 indicate generally low levels for the aerobic ecosystems.

The importance of asymbiotic nitrogen fixation in the rhizosphere has received careful attention from Johanna Döbereiner and her associates in Brazil (Döbereiner, 1968; Döbereiner et al., 1972; Döbereiner and Day, 1975). They identified bacteria of the genera *Azotobacter, Beijerinckia,* and *Derxia* in the rhizosphere of sugarcane, rice, and several tropical grasses, particularly *Paspalum notatum* (Bahia grass). Several species of these bacteria were found to thrive in extremely acid Oxisols, where high aluminum and manganese contents and low phosphorus levels decrease the population of other microbial flora. Theories have been advanced to explain the observed lack of pronounced nitrogen response by some of these grasses in terms of asymbiotic fixation by these bacteria. Findings by Kass et al. (1971) and Döbereiner et al. (1972) show that the annual contribution of asymbiotic nitrogen fixation by these bacteria is on the order of less than 10 kg N/ha. Consequently, this process does not contribute large amounts of nitrogen to crops.

Symbiotic Fixation

As in the temperate region, symbiotic nitrogen fixation is the main mechanism for soil nitrogen additions in the tropics. The magnitude of this phenomenon depends on the amount of legume species, certain nonlegumes such as *Casuarina,* a common coastal pine tree, and some tropical grasses. The large amounts of nitrogen fixed in tropical forests are probably due to the significant proportion of tree legumes present. Symbiotic nitrogen fixation is of little relevance in cropped fields unless a legume is present. Single legume stands contribute from 16 to over 500 kg N/ha a year (Henzell and Norris, 1962). In many cases, however, nitrogen fixation by legumes is substantially lower than would be expected. This may be due to the low phosphorus or high aluminum levels of many soils, which inhibit *Rhizobium* activity, to extremes in soil moisture, and to the lack of a specific inoculum for the crop or variety in question. Very little nitrogen is fixed by field beans (*Phaseolus vulgaris*) in Latin America, partly because of their naturally poor nodulation. Some tropical pasture legumes are well adapted to acid conditions, and when properly or naturally inoculated they produce substantial quantities of nitrogen. This subject is discussed in detail in Chapter 13.

Evidence of symbiotic fixation in tropical grasses is now accumulating. Certain tropical grasses possessing the more efficient C-4 dicarboxylic acid photosynthetic pathway have been shown to enter into loose symbiotic associations with bacteria. Such symbiosis occurs between certain varieties of *Paspalum notatum* and *Azotobacter paspali*. Although nitrogen fixation takes place in the rhizosphere, the association is considered a symbiosis because the nitrogen is taken up directly by the plants and none is secreted into the soil (Döbereiner and Day, 1975). Extrapolations from laboratory and greenhouse experiments suggest that the magnitude of this mechanism may be on the order of 1 kg N/ha per day; however, no field measurements have been made. Also, it is not clear where the energy to fix such large quantities of nitrogen comes from.

Döbereiner and Day (1974) have identified a new symbiotic relationship between certain varieties of Pangola grass (*Digitaria decumbens*) and the bacterium *Spirillum lipoferum*. This microorganism grows only in lactate, malate, or citrate substrates and thus escaped identification in the sugar media commonly used by microbiologists. In tropical grasses with the C-4 dicarboxylic acid pathway, malate is one of the main photosynthetic products. Döbereiner and Day identified this microorganism in a sodium–malate culture. When malate instead of sugar is considered as the energy source for nitrogen fixation, the previous doubts regarding the energy requirements are dispelled. Döbereiner and Day found that *S. lipoferum* is abundant in the rhizosphere of many tropical grasses in Brazil. Nitrogen fixation, however, was observed only inside certain roots of particular varieties of Pangola grass. The authors suggest that the practical application of this recent breakthrough may be the selection of varieties and species that can develop symbiotic relationships with *Spirillum* and perhaps other similar organisms. They also recommend identifying the soil conditions favorable for symbiosis.

These findings have opened an extremely interesting area of research in tropical soil microbiology, and their practical applications may be determined in the near future. Preliminary field trials in Florida by Smith et al. (1975) show a positive dry matter response to *Spirillum* inoculations in two grass species.

In summary, the additions of atmospheric nitrogen into the soil may be as little as 4 but no more than 50 kg N/ha in cropped fields, whereas in tropical rainforests the annual range is between 46 and 147 kg N/ha, as shown in Table 6.1.

Total Nitrogen Additions

The annual additions of soil organic nitrogen in several locations are shown in Table 6.2. Like the annual organic carbon additions, they are greater in tropical forests than in temperate forests or savannas. The reasons are similar to those that account for the differences in soil organic carbon contents. Table 5.2 showed that the range in total nitrogen and in carbon:nitrogen ratio among tropical soils is similar to that found under temperate conditions.

Table 6.2   Estimates of Annual Soil Organic Nitrogen Increments in Several Surface Soils

Location	Annual Soil N increments (kg N/ha)	
	Maximum	Minimum
Tropical lowland forests		
1. Ghana	55	22
2. India	60	24
3. Zaïre	58	23
4. Indonesia	55	22
5. Colombia (Andept)	30	12
Tropical highland forests		
6. Colombia	57	23
7. Indonesia	45	18
8. Madagascar	38	15
Temperate forests		
9. California (oak)	4	2
10. California (pine)	9	4
Tropical savannas		
11. Ghana (1250 mm rain)	15	5
12. Ghana (850 mm rain)	4	2
Temperate prairie		
13. Minnesota (870 mm)	11	5

Source: Greenland and Nye (1959).

Nitrogen Mineralization

The decomposition of soil organic nitrogen into inorganic compounds, called "mineralization," consists of three steps: aminization, the transformation of proteins into amines; ammonification, the transformation of amines into ammonium (NH_4^+); and nitrification, the transformation of ammonium into nitrate (NO_3^-) with a short intermediate stage of nitrite (NO_2^-) formation.

Nitrogen mineralization rates depend on temperature, C:N ratios, soil pH, clay mineralogy, and moisture status. Temperature is seldom limiting in the lowland tropics. The C:N ratios operate somewhat differently in acid soils than in high-base-status soils. The fact that carbon tends to be mineralized at a faster rate than nitrogen at low pH values decreases the C:N ratio and results in an increase in nitrogen mineralization. The total amounts mineralized in high-base-status soils often depend on the total nitrogen contents, except in Andepts, where nitrogen mineralization is inversely proportional to their allophane contents (Bornemisza and Pineda, 1969).

Perhaps the dominant factor affecting nitrogen mineralization rates in the tropics is soil moisture content. Calder (1957) and Semb and Robinson (1969) observed that mineralization may take place at moisture tensions greater than 15 bars. The possible explanation of this phenomenon is the substantial amount of water present at high tensions in well-aggregated soils. Although unavailable to plants, this seems to be available to mineralizing microorgamisms.

Nitrogen mineralization also occurs under flooded conditions, but it stops at the ammonification stage, because only aerobic microorganisms can convert NH_4^+ into NO_3^-. Although the mineralization processes are slower, anaerobic microorganisms apparently can transform organic nitrogen to ammonium at higher C:N ratios than aerobic microorganisms (De Datta and Magnaye, 1969). The net result is a mineralization rate similar to that in aerobic soils.

Between these two extremes in moisture contents, most tropical soils undergo several periods of alternate wetting and drying. Organic carbon and nitrogen mineralization is faster under alternate wetting and drying than under "optimum" moisture conditions. Furthermore, the critical C:N ratios for mineralization also change under these conditions. Birch (1960) found that drying promotes faster carbon than nitrogen mineralization, thus reducing the C:N ratios. Birch also found that the critical C:N ratio above which mineralization stops is higher under alternate wetting and drying. For example, a residue material containing 1.5 percent N was mineralized under alternate wetting and drying but inmobilized under constant moisture. This phenomenon is probably associated with a more active microbial population after drying and rewetting the soil or perhaps an increased accessibility of humus to microorganisms by shrink-swell of clay minerals or thin oxide coats.

SEASONAL FLUCTUATIONS OF INORGANIC NITROGEN

Inorganic nitrogen in most tropical areas shows a marked seasonal fluctuation, as illustrated in Fig. 6.1. The pattern consists of (1) a slow nitrate buildup in the topsoil during the dry season, (2) a large but short-lived increase at the onset of the rainy season, and (3) a rapid decrease during the rest of the rainy season. When short-term droughts occur during the rainy season, they are followed by sharp but smaller increases in inorganic nitrogen and then by gradual decreases. These short-term peaks, called "flushes," were first described by Hardy in 1946. Subsequent work in Africa by Birch and other workers has substantiated their existence in a wide range of soil conditions (Birch, 1958, 1960, 1964). The flushes are sometimes called the "Birch effect" because of the popularity of Birch's articles, although Hardy deserves the original credit.

The following is an analysis of the individual components of the seasonal pattern. Although this pattern is typical of ustic soil moisture regimes, it occurs to a lesser extent in udic regimes also. Apparently, the mechanism is very sensitive to short but pronounced soil moisture changes.

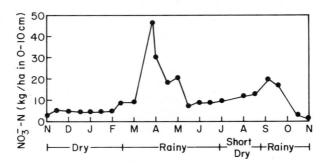

Fig. 6.1   Seasonal pattern of NO_3-N fluctuation in the top 10 cm of a cultivated Alfisol in Ghana. *Source:* Adapted from Greenland (1958).

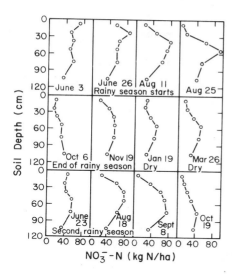

Fig. 6.2 Nitrate fluctuations in a northern Nigerian Alfisol profile. *Source:* Wild (1972).

Nitrate Accumulation during the Dry Season

The accumulation of nitrate in the topsoil during the dry season can be explained by the existence of nitrification at soil moisture tensions of 15 to 80 bars (Semb and Robinson, 1969). Although the topsoil may be drier than those tensions indicate, the subsoil may have enough moisture to support mineralization. Since most of the water movement during the dry season is upward, nitrates previously present or recently mineralized in the subsoil may move up and accumulate in the topsoil. Wetselaar (1961), working with an Alfisol from northern Australia, found evidence of dramatic nitrate buildups in the top 5 cm. He explained that nitrate is mineralized in the subsoil, where adequate moisture existed during the dry season and accumulated just below the soil surface crust, where capillary conductivity is broken. Wild (1972b) followed the nitrate content of a profile in northern Nigeria for 2 years. His results, shown in Fig. 6.2, indicate an upward movement of nitrate during the dry season. Figure 6.2 suggests that this nitrate was leached into the subsoil during the previous rainy season. Hardy's original data (Table 6.3) show actual nitrate levels observed during the dry season in Trinidad.

Table 6.3   Seasonal Levels of Soil Nitrate and Moisture Contents of a Sandy Soil in Trinidad: Mean of 3 Years

Season	Soil Horizon	Cropping System kg N/ha as NO_3^-		
		Fallow	Corn	Pasture
Rainy season	A	18	9	8
(190 mm/month)	B	13	10	7
Dry season	A	35	22	10
(38 mm/month)	B	17	10	9

Source: Adapted from Hardy (1946a).

Nitrogen Flushes at the Onset of the Rainy Season

Within a few days after the first heavy rains, dramatic increases in inorganic nitrogen take place. In the field they may range from 23 to 121 kg N/ha within 10 days (Semb and Robinson, 1969). The sharpness of the peaks is directly proportional to the duration and intensity of the preceding dry period. These sharp increases are accompanied by similarly sharp decreases caused by rapid leaching in the rainy season. Semb and Robinson provided clear evidence of NO_3^- moving into the subsoil after such flushes.

Several reasons were advanced by Birch (1950) to explain these flushes. Active microbial populations build up rapidly when moisture becomes available and easily decomposible substrate is abundant. Intense drying lowers the C:N ratio of humus because carbon decomposes at a faster rate than nitrogen in dry periods. Nitrogen mineralization proceeds faster at lower C:N ratios. Also, the dead microbial population provides additional substrate, which stimulates mineralization further.

Nitrogen Losses during the Rainy Season

As the rainy season progresses, the inorganic nitrogen supply is reduced by plant uptake, leaching, and denitrification. Hardy (1946) showed the effect of crop uptake in decreasing NO_3^- contents in Table 6.3. Leaching depends on a series of soil factors. For fairly well-aggregated Alfisols of northern Nigeria, Wild (1972b) found that peak NO_3^- concentrations in the profile gradually move down as the rainy season progresses. Figure 6.2 shows that at the beginning of the rainy season (June 3) most of the nitrates were concentrated in the top 15 cm. Three weeks later, after the flush, leaching started and the peak concentration moved down to between 15 and 30 cm. By August 11, well within the rainy season, the peak concentration was at 30 to 45 cm, and 2 weeks later at 45 to 60 cm.

In Wild's experiment nitrate moved at the approximate rate of 0.5 mm/mm rain. This is considerably slower than values from temperate-region Ultisols. Terry and McCants (1970), for example, found that nitrate moved 1 to 5 mm/mm rain in sandy soils of North Carolina. The apparent discrepancy is due to soil structure differences. In the well-aggregated clays both downward movement of rainwater through large channels and slow lateral diffusion through the aggregates occur. In sandy soils there is just the former type. Consequently, although many of the well-aggregated clays act like sands in terms of water retention, their leaching losses of mineralized soil nitrogen are bound to be slower than those in sands because the nitrates mineralized inside the granules have to move through the micropores and out to the macropores before they can be susceptible to leaching. If the subsoil has some degree of anion exchange capacity, nitrate leaching may be reduced (Kinjo and Pratt, 1971).

Wild also found that the total nitrate content in the top 120 cm of the soil did not change significantly during the year, except at the end of the rainy season, when larger leaching losses took place because of a drop in the valley water table. In his 2 years of observations, therefore, most of the inorganic nitrogen never left the soil profile. This example should not be misconstrued to mean that no nitrate leaches out of the solum. The data given in Chapter 3 should also be considered.

Denitrification losses are poorly quantified (Greenland, 1958). They are probably important during very active periods of organic matter decomposition and during temporary waterlogging in well-drained soils. Denitrification is probably the main nitrogen loss mechanism in the aquic soil moisture regime.

The magnitude of seasonal fluctuations of inorganic nitrogen varies with the intensity and frequency of rainfall. An even rainfall distribution minimizes fluctuations; consequently, they are more pronounced in ustic than in udic soil moisture regimes. Fluctuations are also minimal if actively growing crops absorb nitrate quickly. Hardy's data in Table 6.3 show that a soil under pasture had a lower nitrate content than one under corn because of year-round uptake.

It is important to realize that the inorganic nitrogen fluctuation pattern just described is quite different from the one commonly found in the temperate region. The growing season in the udic temperate region begins with a cold wet soil, low in inorganic nitrogen because of the slow mineralization rates during fall and winter. As temperatures warm up gradually in the spring, there is a correspondingly gradual increase in organic nitrogen mineralization. Inorganic nitrogen increases slowly, without flushes, because of the high soil moisture status. During the summer, mineralization rates can be very high, and the inorganic nitrogen content will depend on crop uptake, leaching, and denitrification. Most of the nitrogen management practices used in the temperate region have been developed to fit such conditions. Nitrogen management practices in the tropics should also take seasonal patterns into consideration.

NITROGEN FERTILIZER REACTIONS IN SOILS

Nitrogen is the fertilizer nutrient applied in largest quantities in the tropics. Nevertheless, the actual amounts used are much lower than in the temperate region. Calculations from FAO (1971) statistics indicate that only 13 percent of the world's total nitrogen production is consumed in the tropics. In 1971, 2.6 million tons of nitrogen were used in tropical Asia, 1.3 million in tropical America, and 0.2 million in tropical Africa. In certain tropical areas, particularly for irrigated rice, sugarcane, other plantation crops, and some pastures, nitrogen use per unit area rates among the highest in the world. In sharp contrast the bulk of subsistence agriculture is just beginning to use fertilizers, particularly in tropical Asia and Latin America.

The most common fertilizer nitrogen sources used in the tropics are urea and ammonium sulfate. In temperate areas like the United States, on the other hand, ammonium nitrate, anhydrous ammonia, and ammonium phosphates tend to predominate.

The following sections describe the reactions these fertilizers undergo in tropical environments. Included is a discussion of localized nitrogen applications, which are commonly observed in tropical agriculture, particularly in subsistence systems.

Urea Hydrolysis

Urea is the most commonly used inorganic nitrogen source in the tropics. Its popularity is partly due to its high content (46 percent N), low unit cost, and availability in the world market. When applied to a moist soil, urea is hydrolyzed into ammonium carbonate by the enzyme urease in the following way:

$$CO(NH_2)_2 + 2H_2O \xrightarrow{\text{urease}} (NH_4)_2CO_3$$

Ammonium carbonate in the presence of water dissociates into the ammonium and carbonate ions. Before hydrolysis, urea is as mobile as nitrate and may be leached down below the root zone with heavy rainfall if soil structure permits. Tamimi and Kanehiro (1962) showed that urea hydrolysis proceeds at about the same speed in the tropics as in the temperate region and is complete within 1 to 4 days. In flooded soils, Delaune and Patrick (1970) found that the rate of hydrolysis is similar to that in well-drained soils. Consequently, the first reaction of urea is no different in the tropics than in the temperate region.

Volatilization Losses of Ammonia

At soil pH values higher than 7, the NH_4^+ ions can be converted to NH_3 (ammonia gas) and lost to the atmosphere if the soil is dry. Volatilization losses of ammonia were first recognized in the tropics by Jewitt (1942), working with Vertisols in the Sudan. Although ammonia volatilization losses can occur with both urea and ammonium sources, they are particularly important with urea since its hydrolysis increases the pH of the surrounding soil. Broadcast nitrogen applications to the soil surface are very common in the tropics; therefore volatilization losses can be of practical importance in high-pH soils, particularly when high nitrogen rates are applied. Shankaracharya and Mehta (1971), working with a loamy sand of pH 7.1 in Gurajat, India, measured field volatilization losses of 4 percent when 28 kg N/ha were applied as urea to the surface. When the rate was increased to 277 kg N/ha, volatilization losses increased to 44 percent. Such high rates are common in areas where high-yielding rice or wheat varieties are planted.

Urea volatilization losses can be drastically reduced if the material is placed below the soil surface before hydrolysis. This can be accomplished by incorporation, by deep placement, or simply by moving the freshly applied urea down with irrigation water or rainfall. Table 6.4 shows the reduction in volatilization losses when irrigation followed a surface application. The irrigation water simply moved the urea down before it had a chance to be hydrolyzed. In the presence of moisture, volatilization of ammonia does not take place. This table also indicates that urea volatilization losses are essentially eliminated by incorporating the material to about 5 cm depth. The practical implication is that urea should be incorporated into the soil if it is to be applied to a dry calcareous soil.

Table 6.4   Volatilization Losses of Applied Urea as a
Function of Depth and Timing in Relation to Irrigation
in a Calcareous Loamy Sand in Gurajat, India
(N rate: 222 Kg N/ha)

| Placement Depth (cm) | Percent Loss of Applied N | |
	Applied before Irrigation	Applied after Irrigation
Surface	8.1	40.2
1.2	1.2	33.4
2.5	0.6	18.1
5.0	0.05	0.5
7.5	0	0

Source: Adapted from Shankaracharya and Meta (1971).

Nitrification of Broadcast Ammonium Sulfate Applications

Ammonium sulfate when broadcast on the soil surface does not suffer substantial volatilization losses as urea does. The nitrification of NH_4^+ into NO_3^- and the distribution of both ionic species in the profile vary with soil properties and moisture conditions. Wetselaar (1962) followed these changes in Alfisols of northern Australia. Nitrification was very fast in clay loam soils under high rainfall during the rainy season. Most of the applied nitrogen was detected as nitrates in the 60 to 120 cm section of the subsoil. Whereas deep-rooted crops such as cotton were able to utilize this nitrogen, a sorghum crop with an effective rooting depth of 40 cm was not. In a sandy soil Wetselaar found that NH_4^+ ions accumulated in the 15 to 30 cm depth 3 days after application (Table 6.5). Afterwards, a dry period followed. At 21 days after application most of the applied nitrogen was detected in the top 8 cm, probably as a result of nitrification in the subsoil, followed by upward movement during the dry period. When ammonium sources are quickly nitrified, they can move quickly up or down the profile in response to water movement.

Nitrification of Banded Nitrogen Applications

Banded applications of nitrogen fertilizers are being practiced in mechanized agriculture, primarily when complete fertilizer mixtures are used. The usual justification for this practice is the reduction of phosphorus fixation. Also, side-dressed nitrogen applications by hand in traditional systems are localized close to plants. Wetselaar et al. (1972, 1973b) found that banding nitrogen applications affects the rate of nitrification of ammonium sources, and that this practice may result in an increased efficiency of applied nitrogen. Wetselaar and his coworkers found that, when a rate of 80 kg N/ha as ammonium sulfate or urea was incorporated in the top 15 cm of a soil at planting, giving an average concentration of 40 ppm N, the nitrification rate under suitable moisture conditions was above 80 percent within a few days. The nitrate produced may be leached away from the root zone before plants can develop a root system to utilize it. When nitrogen was placed at the same rate in bands spaced at 35 cm, the resulting NH_4^+ concentration could be as high as 1000 ppm N in the band. Wetselaar et al. (1972) observed no nitrification in these bands when the concentration of ammonium sulfate was greater than 80 ppm N or the concentration of urea or aqua ammonium exceeded 400 ppm N. Nitrifying organisms meet a hostile environment because of high osmotic suction and a pH above 8 around the band. The ammonium ions are then stable in a soil with net CEC.

With time the ammonium ion concentration around the band decreases; when the soil pH near the urea band is between 7 and 8, nitrification proceeds to the nitrite stage. Nitrite accumulates in such bands and is toxic to plants. As the pH decreases below 7 because of CO_2 increases, nitrates are formed. Nitrite accumulation does not occur with ammonium sulfate, however, because the pH of the band does not rise above 7 in acid soils. These reactions are illustrated in Table 6.6. Studies with wheat roots by Passioura and Wetselaar (1972) showed that roots were absent from an area 10 cm around bands of either urea or ammonium sulfate during the first 4

Table 6.5   Percent Recovery of Ammonium Sulfate Applied at 80 kg N/ha to the Surface of a Sandy "Lateritic" Red Earth in northern Australia

Depth (cm)	After 3 Days (%)		After 21 Days (%)	
	$NH_4^+ - N$	$NO_3^- - N$	$NH_4^+ - N$	$NO_3^- - N$
0–8	23.7	2.6	26.5	56.3
8–15	15.5	3.1	0.6	5.4
15–30	51.0	5.6	0.4	8.0
30–45	12.1	1.2	0.7	1.7
Total	102.3	12.5	28.2	71.4

Source: Wetselaar (1962).

Table 6.6   Nitrite and Nitrate Formation after Additions of 1000 ppm N as Urea or Ammonium Sulfate in Bands as a Function of Time on Calcareous Soil from northern Australia

Nitrogen Fertilizer		Weeks of Incubation			
		2	4	6	12
Urea	$NO_2^- - N$ (ppm)	170	345	125	0
	$NO_3^- - N$ (ppm)	15	55	330	365
	Soil pH	7.4	7.2	6.0	4.7
$(NH_4)_2SO_4$	$NO_2^- - N$ (ppm)	0	0	0	0
	$NO_3^- - N$ (ppm)	25	85	130	140
	Soil pH	6.2	6.4	5.6	4.8

Source: Wetselaar et al. (1972).

weeks. Afterwards, roots penetrated the ammonium sulfate bands and started taking up nitrogen. Root development around the urea bands was delayed for an additional 4 weeks until the nitrites were converted into nitrate.

These studies suggest that the nitrate supply can be managed by varying the nitrogen concentration in the soil through banding at different spacings. The optimum band spacing will depend on factors such as pH, CEC, and the nitrogen requirements of specific plants. The practice of concentrating ammonium sulfate (not urea) applications in bands or around individual plants by many subsistence farmers appears to have some advantages.

NITROGEN REQUIREMENTS OF TROPICAL CROPS

In spite of large amounts of research on nitrogen fertilization, information about nitrogen uptake patterns of the principal tropical crops is quite limited. Estimates of nutrient removals at harvest, compiled from the available literature, appear in Table 6.7. Whenever possible, nutrient removal at several yield levels is given. Bartholomew (1972b) has compiled estimates of nitrogen uptake by corn, wheat, and rice at several yield levels. His results are reproduced in Fig. 6.3.

Cereals

At the present average yield levels in the tropics of about 1 ton of grain/ha, corn, rice, wheat, and sorghum remove around 30 kg N/ha. This is probably in balance with the nitrogen-supplying capacity of most tropical

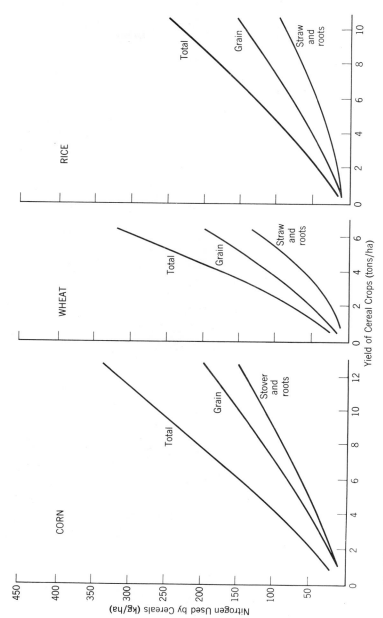

Fig. 6.3 A comparison of nitrogen use by corn, wheat, and rice at a range of yield level. *Source:* Bartholomew (1972b).

199

Table 6.7 Nutrient Removal by Major Tropical Crops

Crop	Part	Yield[a] (tons/ha)	Nutrient (kg/ha)				
			N	P	K	Ca	Mg
Cereals							
Corn	Grain	1.0	25	6	15	3.0	2.0
	Stover	1.5	15	3	18	4.5	3.0
	Total	2.5	40	9	33	7.5	5.0
	Grain	4.0	63	12	30	8.0	6.0
	Stover	4.0	37	6	38	10.0	8.0
	Total	8.0	100	18	68	18.0	14.0
	Grain	7.0	128	20	37	14.0	11.0
	Stover	7.0	72	14	93	17.0	13.0
	Total	14.0	200	34	130	31.0	24.0
Rice	Grain	1.5	35	7	10	1.4	0.3
	Straw	1.5	7	1	18	2.6	2.2
	Total	3.0	42	8	28	4.0	2.5
	Grain	8.0	106	32	20	4.0	1.0
	Straw	8.0	35	5	70	24.0	13.0
	Total	16.0	141	37	90	28.0	14.0
Wheat	Grain	0.6	12	2.4	3	0.3	1.0
	Straw	1.0	3	0.8	14	2.0	2.0
	Total	1.6	15	3.2	17	2.3	3.0
	Grain	5.0	80	22	20	2.5	8.0
	Straw	5.0	38	5	60	10.0	10.0
	Total	10.0	118	27	80	12.5	18.0

Sorghum	Grain	1.0	20	0.9	4	4.0	2.4
	Straw	1.2	6	0.4	2	4.6	3.2
	Total	2.2	26	1.3	6	8.6	5.6
	Grain	8.0	135	10	27	16.0	9.6
	Straw	8.0	65	4	13	18.0	12.8
	Total	16.0	200	14	40	34.0	22.4
Finger millet	Grain	1.1	17	5	59	—	—
Root Crops							
Cassava	Roots	8.0	30	10	50	20	10
	Roots	16.0	64	21	100	41	21
	Roots	30.0	120	40	187	77	40
	Whole plant	59.0	64	19	176	102	26
	Roots	59.0	42	28	291	43	19
Potatoes	Roots	12.0	52	10	80	22	14
	Roots	22.0	120	20	166	40	26
	Roots	40.0	172	34	232	70	48
	Whole plant	62.0	147	19	403	60	31
	Tubers of above	44.0	77	14	224	4	9
Sweet potatoes	Roots	16.5	72	8	88	—	—
Grain Legumes							
Beans	Beans	1.0	31	3.5	6.6	—	—
Soybeans	Beans	1.0	49	7.2	21	—	—
Peanuts	Unhulled nuts	1.0	49	5.2	27	—	—
Grasses	(Annual production, cut every 2 months)						
Guinea	Above ground	10.0	107	27	180	78	49

Table 6.7 (*Continued*)

Crop	Part	Yield[a] (tons/ha)	N	P	K	Ca	Mg
			Nutrient (kg/ha)				
(Panicum	Above ground	23.0	288	44	363	149	99
maximum)	Above ground	35.0	560	77	600	230	133
Pangola	Above ground	10.0	120	22	180	36	28
(Digitaria	Above ground	23.0	299	47	358	109	67
decumbens)	Above ground	31.0	400	53	558	130	87
Elephant	Above ground	10.0	144	24	180	35	30
(Pennisetum	Above ground	25.0	302	64	504	96	63
purpureum)	Above ground	46.0	800	92	900	129	87
Paragrass	Above ground	8.0	80	17	160	28	16
(Panicum	Above ground	24.0	307	43	383	115	79
purpurascens)	Above ground	33.0	600	69	660	135	66
Other Crops							
Sugarcane	Above ground	100	75	20	125	28	10
(2 year crop)	Above ground	200	149	29	316	55	58
	Above ground	300	254	35	499	96	80
Cotton	Seed	0.8	30	4.4	7	—	—
Coffee	Dry beans	1.0	25	1.7	16	1	2
Tea	Dry leaf	0.6	31	2.3	15	2	—
Tobacco	Cured leaf	1.0	116	14	202	—	—
Rubber	Dry latex	3.0	7	1.2	4	4	1
Cacao	Dry beans	0.5	10	2.2	5	1	—
Oil palm	Fruit	15.0	90	8.8	112	28	—

Fruit Crops

Bananas	Bunches	10.0	19	2.0	54	23	30
	Stem and leaves	—	20	1.3	22	1	3
	Total		39	3.3	76	24	33
	Bunches	30.0	56	6.0	161	70	82
	Stem and leaves	—	29	4.0	65	2	8
	Total		85	10.0	226	72	90
Pineapples	Fruits	12.5	9	2.3	29	3	—
Coconuts	Dry copra	1.2	60	7.2	40	—	—

Source: Unpublished compilation by Publio Santiago, Cornell University; Wrigley (1961); Ochse et al. (1961).
[a] Yields of cereals, grain legumes, and grasses on a dry weight basis; root crops and bananas at 15–20% dry matter.

soils. At moderately high yield levels of 4 to 5 tons/ha, nitrogen uptake is on the order of 100 to 150 kg N/ha. This yield level is particularly important, because it can be commonly attained in many tropical areas with new high-yielding varieties and fertilization. At very high yield levels of 8 to 10 tons/ha, total nitrogen uptake exceeds 200 kg N/ha.

The nitrogen uptake pattern of cereals with time has the characteristic sinusoid curve. For rice two periods exists when the nitrogen requirements are highest: at the tillering stage, when secondary shoots appear, and at the panicle primordium initiation stage, which marks the start of the reproductive phase. The number of panicles per unit area is highly correlated with the nitrogen supply at tillering. The number of spikelets per panicle is dependent on the nitrogen supply at panicle initiation (Sanchez, 1972). Rice yields are positively correlated with adequate plant nitrogen levels at these two critical growth stages.

In corn, tillering is an undesirable characteristic and yields are correlated with a maximum number of kernels per unit area. Nitrogen requirements are high during the "grand period of growth" in the vegetative phase and at silking. The nitrogen uptake pattern of wheat is similar to that of rice (Srivatsava, 1969), and the pattern of sorghum to that of corn (Roy and Wright, 1973).

Root Crops

Root crops also remove large quantities of nitrogen. At presently low yield levels of 8 to 10 tons/ha of fresh roots, cassava roots and potato tubers remove about 40 kg N/ha. At the higher yield levels attained with fertilization, these crops can remove over 150 kg N/ha, including the tops. The nitrogen uptake pattern of Peruvian potatoes observed by Ezeta and McCollum (1972) shows that most of the nitrogen was taken up before tuber initiation. During the grand period of growth from 97 to 137 days after planting, nitrogen accumulated at the rate of 2.5 kg N/ha per day.

The nitrogen accumulation pattern in cassava is slow during the first 2 months, then increases linearly, and reaches a maximum at 10 months for varieties that mature in 14 months (Hendershott et al., 1972). Yams show a sharp rise in nitrogen requirements at 3 months after planting, when the leaf area index increases sharply and secondary shoots begin to appear (Chapman, 1965). Cocoyams or taro have a rapid growth rate during the first 6 months, followed by a decline in top growth and a steady increase in corm weights from 3 months until maturity (De la Peña and Plucknett, 1972). Nitrogen fertilization in root crops should take these critical growth periods into consideration.

Grain Legumes

At presently low yield levels of 0.5 to 1.0 ton/ha, beans, soybeans, and peanuts remove about 30 to 50 kg N/ha. At higher yield levels Fassbender (1957) has reported removal rates on the order of 100 to 150 kg N/ha for *Phaseolus vulgaris* beans.

Pastures

Of all tropical crops, tropical grasses are the greatest annual extractors of nitrogen. At low annual yield levels (10 tons dry matter/ha) tropical grasses may remove about 100 kg N/ha a year. At high yield levels of 30 to 50 tons, they may remove from 400 to 600 kg N/ha a year. Since these are perennial species, the nitrogen requirement patterns varies with time and height of cutting or grazing (Vicente-Chandler et al., 1964). When legumes are grown with grasses, the total uptake may be closer to the legume contribution, which can reach 200 kg N/ha in some cases.

Other Crops

The nitrogen requirement of sugarcane is lower than that of tropical grasses. Peak periods coincide with tillering. Nitrogen removal by permanent tree crops is low since most of the plant parts are not removed. Nitrogen uptake of bananas is low, 50–80 kg N/ha, in relation to the amount of dry matter produced.

These figures show some interesting trends. At present average yield levels in the tropics, cereals, root crops, and grain legumes remove from 30 to 50 kg N/ha. Year-long crops, such as pastures and bananas, remove about 60 to 100 kg N/ha annually. The relative constancy of these figures probably reflects the fact that the general nitrogen release of unfertilized soils is on the order of 40 kg N/ha for short-term crops and perhaps twice that amount for long-duration crops. It should be obvious that in order to produce yield increases more nitrogen has to be made available to plants.

MANAGEMENT OF NITROGEN FERTILIZERS

The preceding discussions underline the need for fertilizer nitrogen additions in order to produce high yields in the tropics. Organic sources are unlikely to add enough nitrogen to satisfy crop requirements at higher yield

levels, as shown in Chapter 5. Consequently, this discussion will be limited
to inorganic fertilizers. Nitrogen management practices can be discussed in
terms of rates, sources, timing, placement, and efficiency of utilization.

Determining Application Rates

Since there are no practical soil tests for estimating the available nitrogen
levels in the soil, the determination of optimum nitrogen rates has to be
based on indirect methods, usually field experience. In areas where organic
nitrogen is supposed to be at equilibrium in the soil because of similar
management for many years, three parameters can be used to estimate rates
of applications: (1) the nitrogen uptake required by the crop to produce a
desired yield level, (2) the nitrogen supplied by the soil, and (3) the percent
recovery of added nitrogen.

The nitrogen uptake required for optimum yields can be obtained in the
case of rice, corn, and wheat by referring to Fig. 6.3, by measuring nitrogen
uptake in field experiments, or by calculating the internal nitrogen requirement of the crop in question. The internal nitrogen requirement is the
minimum amount of nitrogen in the above-ground portions of the crop
associated with maximum yields (Stanford, 1966). Only a few estimates are
available: 0.2 percent N for sugarcane, 1.3 to 2.4 percent for some tropical
grasses, 1.4 percent for wheat, 1.2 percent for corn, and 0.8 percent for rice
(Stanford, 1966; Sanchez et al., 1973a). The uptake at the desired yield level
can be obtained by multiplying the internal nitrogen requirement by the
total dry matter produced at the desired yield levels.

The nitrogen supplied by the soil can be estimated from the average yield
without nitrogen or, preferably, by determining the nitrogen uptake of the
check plots in nitrogen fertilizer experiments. In equilibrium conditions this
is a reliable measure, but in newly cleared land or in irrigated soils some
chemical methods are preferred (Bartholomew, 1972b). Measurements of
total soil nitrogen, organic matter, and inorganic nitrogen might be useful
at the research level, but they are too expensive or cumbersome for routine
soil testing.

The efficiency of fertilizer nitrogen utilization can be calculated as the
apparent fertilizer recovery from field experiments with or without the use
of N^{15} radioisotopes. By knowing the nitrogen uptake at the certain rates
and the uptake without added nitrogen, the percent recovery can be calculated as follows:

$$\text{Percent recovery} = \frac{\text{N uptake at applied rate} - \text{N uptake without added N}}{\text{N rate}} \times 100$$

Recovery ranges from 20 to 70 percent. The higher figure is common for crops with extensive root systems such as pastures (Vicente-Chandler et al., 1964; Henzell, 1971). The lowest recoveries are found in areas of alternate wetting and drying.

The optimum nitrogen rate, therefore, can be determined by the following equation:

$$N \text{ rate} = \frac{N \text{ uptake at desired yield level} - N \text{ uptake without added } N}{\text{Percent recovery}}$$

Nitrogen Responses in Cereals

Rates. Corn responses to nitrogen are usually positive, except in newly cleared land, when the profile inorganic nitrogen is very high, or when acidity or serious problems with other nutrients exist. Many of the responses in tropical Africa, although positive, increased yields only from 1.2 to 1.4 tons/ha, according to a summary by Richardson (1966). This clearly indicates that either the rates used were too low or some other factor, such as variety or spacing, was inhibiting yields. In tropical America, where yield responses are generally higher, the recommended rates of application range from 60 to 150 kg N/ha (Sanchez, 1973). In addition to soil factors, the shape of the response curve is affected by variety, plant population, and water regime. For example, the recommended rates of application in Mexico have gradually increased from 40 kg N/ha in the 1940s to 80 kg N/ha in the 1950s and from 80 to 175 kg N/ha in the 1960s (Sanchez, 1973). This is clearly a result of the varietal improvement in that country.

Plant population also affects the shape of the response curve. At low levels of applied nitrogen a population of 30,000 plants/ha is optimum, whereas at higher nitrogen rates populations on the order of 40 to 50,000 plants/ha are best. Figure 6.4 shows the interaction between nitrogen rate and population on corn yields in Guadalajara, Mexico. To achieve yields of over 4 tons/ha, both high populations and high nitrogen rates were needed. Rainfall regimes also exert a marked influence on corn nitrogen responses. Figure 6.5 illustrates the lower responses obtained when either excess moisture or drought occurred in a series of corn experiments in Mexico.

Rice responses to nitrogen are affected more by nonsoil factors than by soil properties. Plant type, solar radiation, spacing, and growth duration essentially determine the shape of the nitrogen response curve (Sanchez, 1972). This topic will be discussed in greater detail in Chapter 11.

Wheat responses to nitrogen are also affected by plant type. Tall, traditional varieties respond positively to relatively low levels of nitrogen. When

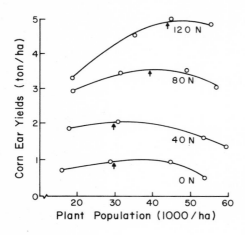

Fig. 6.4 Interaction between nitrogen response and plant population of corn in Guadala;ara, Mexico. The optimum population is indicated by an arrow. *Source:* Laird and Lizárraga (1959).

they receive high rates, they show a yield decrease due to excessive vegetative growth and lodging. On the other hand, the new short, stiff-strawed Mexican varieties respond positively to higher nitrogen rates and produce much higher yields at the optimum application rates. Their stiff straw prevents lodging at higher nitrogen rates. Figure 6.6 illustrates the response

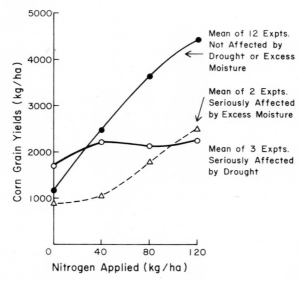

Fig. 6.5 Effects of soil moisture regimes on nitrogen response by unirrigated corn in El Bajío, Mexico. *Source:* Rockefeller Foundation (1963–1964).

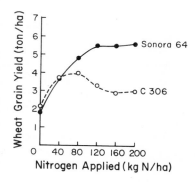

Fig. 6.6 Varietal response to nitrogen by wheat in Pantnagar, U.P., India. *Source:* Sharma et al. (1970).

pattern of a tall variety (C-306) and a short-statured variety (Sonora 64) in Pantnagar, India.

Sources. Much research has been conducted to compare urea, ammonium sulfate, and other nitrogen sources on corn, rice, wheat, and sorghum in the tropics. For all crops the overwhelming evidence indicates no differences between urea and ammonium sulfate or other ammonium sources (DeDatta and Magnaye, 1969; International Atomic Energy Agency, 1970a; Sanchez, 1972, 1973; Khalifa, 1973). In instances where ammonium sulfate was superior to urea, the effect was due to sulfur deficiency or volatilization losses of surface-applied urea. Where urea was superior, differences were due to the acidifying effect of ammonium sulfate in already acid soils or, in the case of rice, to H_2S toxicity in flooded soils very low in iron. The bulk of the evidence indicates few differences in ammoniacal sources when properly applied.

Nitrate sources are usually inferior to ammoniacal sources under conditions favoring leaching or denitrification. Their lower nitrogen content also implies a greater cost per unit of nitrogen.

Organic sources are excellent. However, the quantity required to reach the optimum rates requires a tremendous amount of bulk and may cause problems of incorporation.

Of a series of experimental, slow-release sources of inorganic nitrogen that has been studied (Prasad et al., 1971), the most promising so far seems to be sulfur-coated urea, developed by the Tennessee Valley Authority. Urea granules are coated with a layer of sulfur which decomposes slowly, resulting in a nitrogen release rate of 0.5 to 1 percent per day. Experience to date indicates that this rate is too slow to provide sufficient nitrogen for corn during the grand period of growth (Fox et al., 1974). For rice, preplant applications of sulfur-coated urea are economically superior to preplant or

split applications of conventional urea under conditions of rapid leaching
and/or denitrification (Sanchez et al., 1973b). Sulfur-coated urea is not economically sound, however, under conditions not conducive to these losses
(Diamond and Myers, 1972). The similar nitrogen uptake patterns of corn
and sorghum, and of rice and wheat, suggest that inorganic slow-release
sources may be promising for wheat but less promising for sorghum. In any
case they should be tested under conditions conducive to large losses of
applied nitrogen, where they may have some potential. The cost per unit of
nitrogen of sulfur-coated urea is estimated to be 30 percent higher than that
of conventional urea.

Nitrification inhibitors have received much attention recently. A review
by Prasad et al. (1971) cites several instances of yield responses throughout
the world but no evidence of economical results under field conditions.

Timing and placement.   The most practical way of applying any fertilizer
is to broadcast it and incorporate it into the soil surface before planting.
For nitrogen this procedure is efficient only if the NH_4^+ and NO_3^- ions
released stay in the root zone and are not leached or denitrified to a
considerable extent. Since crop nitrogen requirements are low at early
growth stages, the optimum timing is that which ensures a good nitrogen
supply at the two critical growth stages of the cereals at the lowest possible
cost.

Preplant urea or ammonium sulfate applications incorporated into the
soil were found to be as effective as other timing or placement practices for
high-yielding corn in Samaru, Nigeria, under conditions of little leaching
(Jones, 1973). The same is true of rice under constantly flooded conditions
with essentially no leaching. In most cases, however, delaying the only
nitrogen application until the first critical stage or splitting applications in
two is more efficient. The work of Fox et al. (1974) in Puerto Rico, summarized in Fig. 6.7, shows the substantial benefits of delaying one application for corn in udic regions of Puerto Rico. They also obtained the same
results with sorghum. Similar results were found with rice and wheat
(Sanchez, 1972; Hamid, 1972; Khalifa, 1973). It is seldom profitable,
however, to have more than two split applications.

There is no question that incorporating preplant applications is a good
placement practice. Whether the nitrogen is broadcast or banded depends
on how other nutrients are applied and whether nitrification should be
delayed by locally high rates. Postplant applications are usually broadcast
or can be incorporated lightly in the process of weeding. The International
Atomic Energy Agency (1970) studies with corn and rice emphasize the
importance of incorporation whenever possible.

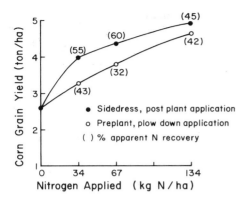

Fig. 6.7 Effect of timing of one nitrogen application on corn yield response in Puerto Rico. Average of several Oxisols and Ultisols that responded to nitrogen. *Source:* Fox et al. (1974).

Efficiency of utilization. The recovery of applied nitrogen is highly variable. Bartholomew (1972) estimates a 50 percent recovery for rice and wheat. These figures are applicable to the temperate region but to only a few places in the tropics. Fox et al. (1974) obtained recoveries of 51 percent with the optimum postplant side-dressed application rate for corn in Puerto Rico but only 33 percent when the same rate was incorporated into the soil surface. Jones (1973) reports a 70 percent recovery for corn under conditions of no leaching, with the nitrogen applied before seeding or side-dressed.

Nitrogen recovery by rice ranges from 30 to 50 percent under constant flooding and from 20 to 30 percent under water management practices conducive to leaching and denitrification. In the latter case use of sulfur-coated urea may increase the recovery rate to 30 or 40 percent (Sanchez et al., 1973b). Nitrogen recovery by wheat may be as high as 50 percent with the best rate, timing, and placement practices (Hamid, 1972).

Nitrogen Response by Root Crops

Nitrogen responses are often negative for cassava in spite of the high nitrogen requirement of this crop. Nitrogen applications increase the top:root ratio and decrease yields in many cases. In others it increases the tuber yield and protein content but not the starch yield (Sanchez, 1973). Long-term experiments in Brazil by Normanha et al. (1968) emphasize the importance of an appropriate N:K ratio. Nitrogen responses were obtained in the presence of ample phosphorus and potassium supplies. Cassava cuttings are very susceptible to salt concentrations when in contact with fertilizers. Nitrogen should be applied to the side of the planting furrow, and the rest side-dressed at later stages.

Potatoes respond to applications of 60 to 120 kg N/ha in southern Brazil and to 160 kg N/ha in the Peruvian Sierra. In Peru, McCollum and Valverde (1968) noted that the magnitude of the response was inversely related to the soil organic matter content. They also observed no differences among nitrogen sources, but with organic manure (guano) the quantity needed for optimum rates was too bulky to be practical. The best timing of applications is half at planting and half at later stages.

Sweet potatoes normally respond to 40 to 60 kg N/ha; excess amounts may increase the top:root ratio and decrease yields. No differences between sources, placement, or timing have been observed. Yams also respond to moderate nitrogen rates but show a rather dramatic effect of timing. Chapman (1965) found in Trinidad that delaying applications until 3 months after planting coincided with sharp increases in leaf area index and produced large yield increases. Applications at planting had no effect, largely because of leaching losses. Taro or cocoyam responds to very high rates, up to 560 kg N/ha, and yields as much as 50 tons/ha under flooded conditions in Hawaii. Under nonflooded conditions maximum yields of 20 tons/ha were obtained with a rate of 280 kg N/ha, according to De la Peña and Plucknett (1972).

Nitrogen Responses by Grain Legumes

Even though legumes supposedly receive their nitrogen from symbiotic fixation, there are two instances in which inorganic nitrogen fertilization is necessary. Throughout Latin America, symbiotic fixation by *Rhizobium* in *Phaseolus vulgaris* is negligible, and nitrogen application rates on the order of 30 to 100 kg N/ha are needed, according to a review by Fassbender (1967). The second case occurs when the soil is severely deficient in nitrogen. In such instances a small application at planting helps to get the plant established. This practice is also helpful with soybeans.

RESIDUAL EFFECTS OF NITROGEN FERTILIZATION

Inorganic Nitrogen Supply

If it is assumed that 30 to 50 percent of the added nitrogen is recovered by plants, the rest either stays in the soil or is lost by leaching and denitrification. The fate of the leftover nitrogen undoubtedly varies with soil and climatic conditions. In some cases large quantities are lost from the profile, and in others a gradual buildup takes place. Fox et al. (1974) report that some Oxisols and Ultisols of Puerto Rico contained over 300 kg N/ha as inorganic nitrogen in their profiles after many years of continuous fertilization. Yields of 5.8 tons/ha of grain sorghum were obtained in these soils without additional nitrogen.

The magnitudes of these nitrogen buildups are not well known or described. They may occur where high rates have been consistently applied over many years, and all the excess nitrogen has not been completely lost by leaching or denitrification.

Changes in Soil Properties

Ammonium sulfate and urea have a net residual acidity because of the following reactions:

$$(NH_4)_2SO_4 + 4O_2 \rightarrow 2NO_3^- + 2H_2O + 4H^+ + SO_4^{2-}$$

$$(NH_2)_2CO + 2H_2O \rightarrow (NH_4)_2CO_3 + 4O_2 \rightarrow 2NO_3^- + 3H_2O + 2H^+ + CO_2$$

Ammonium sulfate generates 2 moles of H^+ per mole of NH_4^+, while urea generates 1 mole of H^+ per mole of NH_4^+. In terms of lime values, 7.1 kg of $CaCO_3$ is required to neutralize 1 kg of nitrogen as ammonium sulfate and half that amount when the source is urea. The nitrification process is the cause of residual acidity. If the NH_4^+ ions are taken directly, there is no acidifying affect. The presence of oxygen is also necessary for residual acidity to occur since denitrification has the opposite effect. Sodium nitrate produces a net increase in pH because of its sodium content.

The residual effects of nitrogen fertilizers on chemical soil properties are most common in heavily fertilized pastures. Villamízar and Lotero (1967) evaluated the effects of annual applications of urea, ammonium sulfate, and sodium nitrates at rates up to 1000 kg N/ha a year for 5 years in an Andept from Colombia. The results shown in Fig. 6.8 indicate that optimum yields of 30 tons/ha a year of Pangola grass were obtained at an annual rate of 500 kg N/ha. At this rate the topsoil pH decreased from 5.8 to 4.2 with ammonium sulfate, remained the same with urea, and increased to about 6.8 with sodium nitrate. In view of the similarity in efficiency between urea and ammonium sulfate, urea is definitely the better source.

Heavy ammonium sulfate applications also affect the base status of the subsoil. Annual rates of 900 and 4000 kg N/ha drastically decreased the pH and base saturation of two subsoils in Puerto Rico (Abruña et al., 1959) as shown in Fig. 6.9. When lime was applied in conjunction with these heavy rates, the base status of acid subsoils increased dramatically. Pearson et al.,

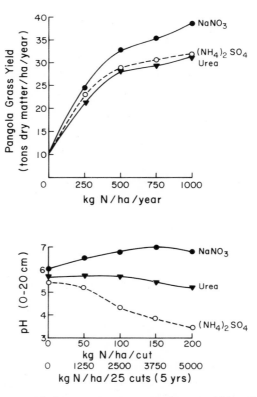

Fig. 6.8 Effects of rates and nitrogen sources on pangola grass yields and changes in soil pH after 5 years in an Andept from Colombia. *Source:* Villamízar and Lotero (1967).

(1962) attributed this effect to the downward movement of calcium and magnesium applied to the surface as lime. In such cases Vicente-Chandler et al. (1964) have recommended the application of a ton of lime per ton of 10-14-10 fertilizer when ammonium sulfate is used in heavily fertilized pastures of Puerto Rico.

SUMMARY AND CONCLUSIONS

1. As in the case of organic carbon, total soil nitrogen contents in the tropics are basically no different from those of the temperate region. The main sources of soil organic nitrogen additions are symbiotic and asymbiotic fixation, rain, and dust. Asymbiotic and loose symbiotic

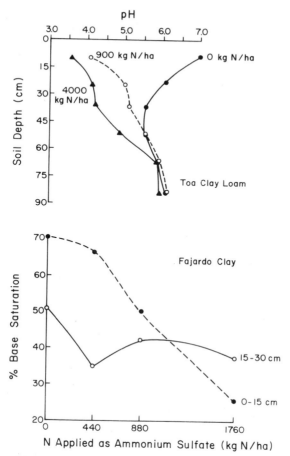

Fig. 6.9 Effects of leaching nitrogen fertilization on pH and base saturation changes in two soils from Puerto Rico. *Source:* Abruña et al. (1968).

nitrogen fixation by several species of bacteria in association with certain tropical grasses provides significant quantities of nitrogen under certain acid soil conditions. The mineralization rates of organic nitrogen seem to be affected to the greatest degree by soil moisture conditions in tropical environments.

2. A cyclical pattern of inorganic nitrogen fluctuations in the soil profile is found in ustic tropical conditions. It includes upward movement of nitrate during the dry season, sharp but short-lived "flushes" of mineralization at the start of the rainy season, and a gradual depletion due to crop uptake

and leaching. The inorganic nitrogen contents in certain profiles are much higher than previously believed.

3. Fertilizer nitrogen behavior is no different in the tropics than in the temperate region. Urea hydrolysis, ammonia volatilization, and broadcast applications of ammonium sulfate support accepted concepts. Localized nitrogen applications around individual plants or through banding result in a temporary decrease in the availability of added nitrogen.

4. The nitrogen requirements of the major tropical crops are variable and depend on the desired yield levels. Some crops show a definite internal nitrogen requirement. Nitrogen application rates can best be determined through field experimentation or through calculations based on the expected nitrogen uptake and the recovery of added nitrogen.

5. Nitrogen fertilizer management practices are a function of crop requirements, soil properties, sources, placement, and timing of applications. There is little difference in the effectiveness of urea and ammonium sulfate in the tropics. Organic manures are too low in nitrogen to supply sufficient quantities for high yields. Inorganic slow-release fertilizers such as sulfur-coated urea seem promising for slow-growing crops under conditions favorable for leaching. Although preplant incorporated nitrogen applications are the easiest to make, they usually result in significant losses due to leaching or denitrification. The timing of nitrogen application should be geared to provide sufficient nitrogen to the plant at critical growth stages. Such stages have been identified for most cereals, certain root crops, and pasture species.

6. Intensive nitrogen fertilization for several years can create marked residual effects. They include high inorganic nitrogen contents in some profiles and, when ammonium sulfate is used, a drastic decrease in pH and base saturation in the profile.

REFERENCES

Abruña, F., R. W. Pearson, and C. B. Elkins. 1958. Quantitative evaluation of soil reaction and base status changes resulting from field applications of residually acid forming nitrogen fertilizers. *Soil Sci. Soc. Amer. Proc.* **22**:539–542.

Acquaye, D. K. and R. K. Cunningham. 1965. Losses of nitrogen by ammonia volatilization on surface-fertilized tropical forest soils. *Trop. Agr. (Trinidad)* **42**:281–292.

Agarwal, A. S., B. R. Singh, and Y. Kanehiro. 1971. Soil nitrogen and carbon mineralization as affected by drying-rewetting cycles. *Soil Sci. Soc. Amer. Proc.* **35**:96–100.

Agarwal, A. S., B. R. Singh, and Y. Kanehiro. 1972. Differential effects of carbon sources on nitrogen transformations in Hawaiian soils. *Plant and Soil* **36**:529–537.

Agboola, A. A. 1968. Increasing the efficiency of applied fertilizer on maize. I. Timing of application of nitrogenous fertilizer. *Nigerian Agr. J.* **5**:45–48.

Bartholomew, W. V. 1972a. Soil nitrogen and organic matter. Pp. 63–81. In *Soils of the Humid Tropics*. National Academy of Sciences, Washington, D. C.

Bartholomew, W. V. 1972b. Soil nitrogen: Supply processes and crop requirements. *Int. Soil Fert. Eval. Improvement Program Tech. Bull.* 6. North Carolina State University, Raleigh.

Birch, H. F. 1958. The effect of soil drying on humus decomposition and nitrogen availability. *Plant and Soil* **10**:9–31.

Birch, H. F. 1960a. Nitrification of soils after different periods of dryness. *Plant and Soil* **12**:81–96.

Birch, H. F. 1960b. Soil drying and soil fertility. *Trop. Agr.* (*Trinidad*) **37**:3–10.

Birch, H. F. 1964. Mineralization of plant nitrogen following alternate wet and dry conditions. *Plant and Soil* **20**:43–49.

Blondel, D. 1971a. Contribution à l'étude de la croissance-matière sèche et de l'alimentation azotée des céréales de culture sèche au Senegal. *Agron. Tropicale* (*France*) **26**:707–720.

Blondel, D. 1971b. Contribution à la connaissance de la dynamique de l'azote minéral en sol sableux (Dior) au Senegal. *Agron. Tropicale* (*France*) **26**:1303–1333.

Blondel, D. 1971c. Contribution à la connaissance de la dynamique de l'azote en sol ferrugineux tropical à Séfa (Senegal). *Agron. Tropicale* (*France*) **26**:1334–1353.

Blondel, D. 1971d. Contribution à la connaissance de l'azote minéral en sol ferrugineux tropical à Nioro-du-Rip (Senegal). *Agron. Tropicale* (*France*) **26**:1354–1361.

Blondel, D. 1971e. Rôle de la plante dans l'orientation de la dynamique de l'azote en sol sableux. *Agron. Tropicale* (*France*) **26**:1362–1371.

Blondel, D. 1971f. Rale de la matière organique libre dans la minéralisation en sol sableux; relation avec l'alimentation azotée du mil. *Agron. Tropicale* (*France*) **26**:1372–1377.

Bornemisza, E. and R. Pineda. 1969. Amorphous minerals and nitrogen mineralization in volcanic ash derived soils Pp. B7.1–B7.7. In *Panel on Soils Derived from Volcanic Ash in Latin America*. Inter-American Institute of Agricultural Sciences, Turrialba, Costa Rica.

Bruce, R. C. and E. H. Tyner. 1960. Timing nitrogen applications for maize in tropical regions characterized by wet and dry seasons. *Trans. 7th Int. Congr. Soil Sci.* (*Madison, Wisc.*) **3**:504–509.

Calder, E. A. 1957. Features of nitrate accumulation in Uganda. *J. Soil Sci.* **8**:60–72.

Chapman, R. 1965. Some investigations into factors limiting yields of White Lisbon yams (*Dioscorea alata*) under Trinidad conditions. *Trop. Agr.* (*Trinidad*) **42**:145–151.

Commonwealth Agricultural Bureau. 1962. "A Review of Nitrogen in the Tropics with Particular Reference to Pastures." *Commonwealth. Bur. Pastures Field Crops Bull.* 46. Harpenden, England.

Cornforth, I. S. 1971a. Seasonal changes in mineralizable nitrogen in Trinidad soils. *Trop. Agr.* (*Trinidad*) **2**:157–162.

Cornforth, I. S. 1971b. Nitrogen mineralization in West Indian soils. *Exptal. Agr.* **7**:345–349.

Cornforth, I. S. and J. B. Davies. 1968. Nitrogen transformations in tropical soils. I. Mineralization of nitrogen rich organic materials added to the soil. *Trop. Agr.* (*Trinidad*) **45**:211–221.

Cunningham, R. K. 1962. Mineral nitrogen in tropical forest soils. *J. Agr. Sci.* **59**:257–262.

DeDatta, S. K. and C. P. Magnaye. 1969. A survey of forms and sources of fertilizer nitrogen for flooded rice. *Soils and Fert.* **32**:103–109.

De la Peña, R. and D. L. Plucknett. 1972. Effect of nitrogen fertilization on the growth, composition and yield of upland and lowland taro (*Colocasia esculenta*). *Exptal. Agr.* **8**:187–194.

Delaune, R. D. and W. H. Patrick. 1970. Urea conversion to ammonia in water-logged soils. *Soils Sci. Soc. Amer. Proc.* **34**:603–607.

Diamond, R. B. and F. J. Myers. 1972. Crop responses and related benefits from sulfur-coated urea. *Sulphur Inst. J.* **8**:9–11.

Diamond, W. E. 1937. Fluctuation in nitrogen contents of some Nigerian soils. *Emp. J. Exptal. Agr.* **5**:264–280.

Döbereiner, J. 1961. Nitrogen fixing bacteria in the rhizosphere of sugar cane. *Plant and Soil* **14**:211–217.

Döbereiner, J. 1968. Non symbiotic fixation in tropical soils. *Pesq. Agropec. Bras.* **3**:1–6.

Döbereiner, J., J. M. Day, and P. J. Dart. 1972. Nitrogenase activity in the rhizosphere of sugar cane and some other tropical grasses. *Plant and Soil* **37**:191–196.

Döbereiner, J. and J. M. Day. 1974. Associative symbiosis in tropical grasses: characterization of microorganisms and dinitrogen fixing sites. Paper presented at the International Symposium on N_2 Fixation, June 3–7, Washington State University. 27 pp.

Döbereiner, J. and J. M. Day. 1975. Potential significance of nitrogen fixation in rhizosphere association of tropical grasses. Pp 197–210. In E. Bornemisza and A. Alvarado (eds.), *Soil Management in Tropical America*. North Carolina State University, Raleigh.

Dommergues, Y. 1960. Nitrogen mineralization at low moisture contents. *Trans. 7th Int. Congr. Soil Sci.* (*Madison, Wisc.*) **2**:672–678.

Enyi, B. A. C. 1965. The efficiency of urea as fertilizers under tropical conditions. *Plant and Soil* **23**:385–396.

Ezeta, F. N. and R. E. McCollum. 1972. Dry matter production, nutrient uptake and removal by *Solanum andigena* in the Peruvian Andes. *Amer. Potato J.* **49**:151–163.

FAO. 1971. *The State of Food and Agriculture*. Food and Agricultural Organization of the United Nations, Rome.

Fassbender, W. H. 1967. La fertilización del frijol (*Phaseolus* spp.). *Turrialba* **17**:46–52.

Fernandez, R. and R. J. Laird. 1958. Efectos de la sequía durante el espigamiento del maiz fertilizado con diferentes cantidades de nitrógeno. *Secretaria de Agricultura y Ganadería* (*Mexico*) *Folleto Tec.* 30.

Fox, R. H., H. Talleyrand, and D. R. Bouldin. 1974. Nitrogen fertilization of corn and sorghum grown in Oxisols and Ultisols in Puerto Rico. *Agron. J.* **66**:534–540.

Freire, J. R. S., C. P. Goepfert, and C. Vidor. 1969. Inoculation of legumes in Brazil. Pp. 101–113. In *Biology and Ecology of Nitrogen*. National Academy of Sciences, Washington.

Greenland, D. J. 1958. Nitrate fluctuations in tropical soils. *J. Agr. Sci.* **50**:82–91.

Greenland, D. J. and P. H. Nye. 1959. Increases in carbon and nitrogen contents of tropical soils under natural fallows. *J. Soil Sci.* **9**:284–299.

Hamid, A. 1972. Efficiency of N uptake by wheat as affected by time and rate of application, using N^{15} labelled amonium sulfate and sodium nitrate. *Plant and Soil* **37**:389–394.

Hardy, F. 1946a. Seasonal fluctuations of soil moisture and nitrate in a humid tropical climate. *Trop. Agr.* (*Trinidad*) **23**:40–49.

Hardy, F. 1946b. The significance of carbon–nitrogen ratios in soils growing cotton. III. Nitrate fluctuations in relation to planting date and soil manurial requirements in the British West Indies. *Trop Agr. (Trinidad)* 23:201–210.

Hendershott, C. H. et al. (eds.). 1972. *A Literature Review and Research Recommendation on Cassava.* University of Georgia, Athens.

Henzell, E. F. 1971. Recovery of nitrogen from four fertilizers applied to Rhodes grass in small plots. *Austr. J. Exptal. Agr. Anim. Husb.* 11:420–430.

Henzell, E. F. and D. O. Norris. 1962. Processes by which nitrogen is added to the soil-plant system. Pp. 1–18. In "A Review of Nitrogen in the Tropics with Particular Reference to Pastures." *Commonwealth Bur. Pastures Field Crops, Bull.* 46.

International Atomic Energy Agency. 1970a. Rice fertilization. *Tech. Repts. Ser.* 108, Vienna.

International Atomic Energy Agency. 1970b. Fertilizer management practices for maize: Results of experiments with radioisotopes. *Tech. Repts. Ser.* 121, Vienna.

Ishaque, M. and A. H. Cornfield. 1972. Effect of level of soil moisture on nitrogen mineralization and nitrification during incubation on East Pakistan tea soils. *Bangladesh J. Biol. Agr. Sci.* 1:52–58.

Jain, N. K., D. P. Maurya, and H. P. Singh. 1971. Effect of time and methods of applying nitrogen to dwarf wheats. *Exptal. Agr.* 7:21–26.

Jewitt, T. N. 1942. Loss of ammonia from ammonium sulfate applied to alkaline soils. *Soil Sci.* 54:401–409.

Jones, M. J. 1973. Time of application of nitrogen fertilizer to maize at Samaru, Nigeria. *Exptal. Agr.* 9:113–120.

Jones, M. J. and A. R. Bromfield. 1970. Nitrogen in the rainfall at Samaru, Nigeria. *Nature* 227:86.

Kanehiro, Y., L. K. Nagasako, and M. F. Hadano. 1960. Leaching losses of nitrogen fertilizers. *Hawaii Farm Sci.* October 1970:6–7.

Kass, D. L. and M. Drosdoff. 1970. Sources of nitrogen in tropical environments. *Agron. Mimeo* 70-9. Cornell University, Ithaca, N.Y.

Kass, D. L., M. Drosdoff, and M. Alexander. 1971. Nitrogen fixation by *Azotobacter paspali* in association with Bahiagrass (*Paspalum notatum*). *Soil Sci. Soc. Amer. Proc.* 35:286–289.

Khalifa, M. A. 1973. Effects of nitrogen on leaf area index, leaf area duration, net assimilation rate and yield of wheat. *Agron. J.* 65:253–256.

Kinjo, T. and P. F. Pratt. 1971. Nitrate adsorption. *Soil Sci. Soc. Amer. Proc.* 35:722–732.

Laird, R. J. and H. Lizárraga. 1959. Fertilizantes y población óptima de plantas para maiz de temporal en Jalisco. *Secretaría de Agricultura y Ganadería (Mexico) Folleto Tec.* 35.

Landrau, P. and G. Samuels. 1951. The effect of fertilizers on the yield and quality of sweet potatoes. *J. Agr. Univ. Puerto Rico* 35:71–87.

Lathwell, D. J., H. D. Dubey, and R. H. Fox. 1972. Nitrogen supplying power of some tropical soils of Puerto Rico and methods for its evaluation. *Agron. J.* 64:763–766.

Lugo, J. C. 1970. Determination of nitrogen use efficiency in wheat by the isotopically labelled fertilizer techniques. *Ceiba* 16:57–87.

Malavolta, E. 1955. Studies on the mineral nutrition of cassava. *Plant Physiol.* 30:81–82.

McCollum, R. E. and C. Valverde. 1968. The fertilization of potatoes in Peru. *North Carolina Agr. Exp. Sta. Tech. Bull.* 185.

Medina, H. (ed.). 1972. El uso del nitrógeno en el trópico. Segundo Coloquio de Suelos, Sociedad Colombiana de la Ciencia del Suelo. *Suelos Ecuatoriales* 3(1):1–464.

Meiklejohn, J. 1962. Microbiology of the nitrogen cycle in some Ghana soils. *Emp. J. Exptal. Agr.* 30:115–126.

Montojos, J. C. and A. C. Magalhaes. 1971. Growth analysis of dry beans (*Phaseolus vulgaris* L. var *Pintado*) under varying conditions of solar radiation and nitrogen application. *Plant and Soil* 35:217–223.

Normanha, E. S., A. S. Pereira, and E. S. Freire. 1968. Modo e época de aplicação de adubos minerais em cultura de mandioca. *Bragantia* 27:143–154.

Nye, P. H. 1954. Fertilizer responses in the Gold Coast in relation to time and method of application. *Emp. J. Exptal. Agr.* 22. 101–111.

Ochse, J. J., M. J. Soule, M. J. Dijkman, and C. Wehlburg. 1961. *Tropical and Subtropical Agriculture*. Macmillan, New York.

Oelsligle, D. D. 1975. Accumulation of dry matter, nitrogen, phosphorus and potassium in cassava (*Manihot esculenta* Crantz). *Turrialba* 25:85–87.

Orioli, G. A., I. Mogilner, and W. L. Bartra. 1967. Acumulación de materia seca, N, P, K, y Ca en *Manihot esculenta*. *Bonplandia* 2:175–182.

Passioura, J. B. and R. Wetselaar. 1972. Consequences of banding nitrogen fertilizers in soil. II. Effect on the growth of wheat roots. *Plant and Soil* 36:461–473.

Pearson, R. W., F. Abruña, and J. Vicente-Chandler. 1962. Effects of lime and nitrogen applications in the downward movement of calcium and magnesium in two humid tropical soils of Puerto Rico. *Soil Sci.* 93:77–82.

Prasad, R., G. B. Rajale, and B. A. Lakdive. 1971. Nitrification retarders and slow-release nitrogen fertilizers. *Adv. Agron.* 23:337–383.

Puente, F. F., D. N. Sanchez, S. Chavez, and R. J. Laird. 1963. Prácticas de fertilización y población óptima para siembras de maiz en las regiones tropicales de Veracruz. *Secretaría de Agricultura y Ganadería (Mexico) Folleto Tec.* 45.

Ramirez, A. and J. Lotero. 1969. Efecto de la dosis y frequencia de aplicación de nitrógeno en la fertilidad y propiedades quimícas del suelo. *Rev. Inst. Colomb. Agropec.* 4:227–254.

Richardson, H. L. 1966. The use of fertilizers. Pp. 137–154. In R. P. Moss (ed.), *The Soil Resources of Tropical Africa*. Cambridge University Press, Oxford.

Rockfeller Foundation. 1963–1964. Annual Report Program in the Agricultural Sciences, New York.

Roy, R. N. and B. C. Wright. 1973. Sorghum growth and nutrient uptake in relation to soil fertility. I. Dry matter accumulation patterns, yield and N content of the grain. *Agron. J.* 65:709–711.

Sanchez, P. A. 1972. Nitrogen fertilization and management in tropical rice. *North Carolina Agr. Exp. Sta. Tech. Bull.* 213.

Sanchez, P. A. 1973. Nitrogen fertilization. Pp. 90–125. In "A Review of Soils Research in Tropical Latin America." *North Carolina Agr. Exp. Sta. Tech. Bull.* 219.

Sanchez, P. A. and M. V. Calderon. 1971. Timing of nitrogen applications for rice grown under intermittent flooding in the Coast of Peru. *Proc. Int. Symp. Soil Fert. Eval. (New Delhi)* 1:595–602.

Sanchez, P. A., G. E. Ramirez, and M. V. de Calderon. 1973a. Rice responses to nitrogen under high solar radiation and intermittent flooding in Peru. *Agron. J.* 65:523–529.

Sanchez, P. A., A. Gavidia, G. E. Ramirez, R. Vergara, and F. Muinguillo. 1973b. Performance of sulfur-coated urea under intermittently flooded rice culture in Peru. *Soil Sci. Soc. Amer. Proc.* **37**:789–792.

Semb, G. and J. B. D. Robinson. 1969. The natural nitrogen flush in different arable soils and climates in East Africa. *East Afr. Agr. For. J.* **34**:350–370.

Shankaracharya, N. B. and B. V. Mehta. 1971. Note on the losses of nitrogen by volatilization of ammonia from loamy-sand soil of Anand treated with different N carriers under field conditions. *Indian J. Agr. Sci.* **41**:131–133.

Sharma, K. C., R. D. Misra, B. C. Wright, and B. Krantz. 1970. Response of some dwarfs and tall wheats to nitrogen. *Indian J. Agron.* **15**:97–105.

Singh, B. R. and Y. Kanehiro. 1969. Adsorption of nitrate in amorphous and kaolinitic soils. *Soil Sci. Soc. Amer. Proc.* **33**:681–683.

Smith, R. L., J. H. Bouton, S. C. Schank, and K. H. Queensberry. 1975. Yield increases in tropical grain and forage grasses after inoculation with *Spirillum lipoferum* in Florida. Paper presented at a Conference on Biological Nitrogen Fixation and Farming Systems of the Humid Tropics. International Institute for Tropical Agriculture, Ibadan, Nigeria.

Srivatsava, S. P. 1969. An appraisal of nitrogen fertilization practices for paddy, wheat, sugar cane and potato. *Plant and Soil* **33**:265–271.

Srivatsava, S. P. and A. Singh. 1971. Utilization of nitrogen by dwarf sorghum. *Indian J. Agr. Sci.* **41**:543–546.

Stanford, G. 1966. Nitrogen requirements of crops for maximum yields. Pp. 237–257. In M. H. MacVickar. et al. (eds.), *Agricultural Anhydrous Ammonia—Technology and Use.* American Society of Agronomy, Madison, Wisc.

Stanford, G. 1973. Rationale for optimum nitrogen fertilization in corn production. *J. Environ. Quality* **2**:159–166.

Stanford, G., A. S. Ayres, and M. Doi. 1965. Mineralizable soil nitrogen in relation to fertilizer needs of sugar cane in Hawaii. *Soil Sci.* **99**:132–137.

Stanford, G. and S. J. Smith. 1972. Nitrogen mineralization potentials of soils. *Soil Sci. Soc. Amer. Proc.* **36**:465–472.

Stephens, D. 1962. Upward movement of nitrate in a bare soil on Uganda. *J. Soil Sci.* **13**:52–59.

Takahashi, D. T. 1970. Fate of unrecovered fertilizer nitrogen in lysimeter studies with N^{15}. *Hawaii Sugar Planters Rec.* **58**:95–101.

Tamini, Y. N. and Y. Kanehiro. 1962. Urea transformations in Hawaiian soils. *Hawaii Farm Sci.* July 1962:6–7.

Terry, D. L. and C. B. McCants. 1970. Quantitative prediction of leaching in field soils. *Soil Sci. Soc. Amer. Proc.* **34**:271–276.

Vicente-Chandler, J., R. Caro-Costas, R. W. Pearson, et al. 1964. Intensive pasture management in Puerto Rico. *Univ. Puerto Rico Agr. Exp. Sta. Bull.* 202.

Villamízar, F. and J. Lotero. 1967. Respuesta del pasto Pangola a diferentes fuentes y dósis de nitrógeno. *Rev. Inst. Colomb. Agropec.* **2**:57–70.

Webster, C. C. and P. N. Wilson. 1966. *Agriculture in the Tropics.* Longmans, London.

Wetselaar, R. 1961. Nitrate distribution in tropical soils. I. Possible causes of nitrate accumulation near the surface after a long dry period. II. Extent of capillary accumulation of nitrate during a long dry period. *Plant and Soil* **15**:110–133.

Wetselaar, R. 1962a. Nitrate distribution in tropical soils. III. Downward movement and accumulation of nitrate in the subsoil. *Plant and Soil* **16**:19–31.

Wetselaar, R. 1962f. The fate of nitrogenous fertilizers in a monsoonal climate. Pp. 588–595. In *Trans. Comm. IV and V, Int. Soc. Soil Sci. (New Zealand).*

Wetselaar, R., J. B. Passioura, and B. R. Singh. 1972. Consequences of banding nitrogen fertilizers in soil. I. Effects of nitrification. *Plant and Soil* **36**:159–175.

Wetselaar, R., P. Jakobsen, and G. R. Chaplin. 1973a. Nitrogen balance in crop systems in tropical Australia. *Soil Biol. Biochem.* **5**:35–40.

Wetselaar, R., J. B. Passioura, D. A. Rose, and P. Jakobsen. 1973b. Banding nitrogen fertilizers in soil: principles and practice. *Chim. et Ind.* **106**:567–572.

Wild, A. 1972a. Mineralization of soil nitrogen at a savanna site in Nigeria. *Exptal. Agr.* **8**:91–97.

Wild, A. 1972b. Nitrate leaching under bare fallow at a site in northern Nigeria. *J. Soil Sci.* **23**:315–324.

Wrigley, G. 1961. *Tropical Agriculture.* Batsford, London.

7

SOIL ACIDITY AND LIMING

The vast majority of the soils of the humid tropics are acid. The vast majority of the cultivated soils of the humid tropics, however, are not acid. Throughout civilization man has tended to settle on high-base-status soils. When the population grows beyond these initial centers, however, acid soils are often brought into cultivation. It is usually safe to generalize that soils with pH values of less than 6 occupy a large proportion of the tropics. As a region tropical America has more acid soils than tropical Asia or Africa. For example, León (1970) estimates that 70 percent of the soils of Colombia have acidity problems.

The well-established practice of liming temperate-region soils to neutrality is not effective in most of the highly weathered soils of the tropics. More often than not, liming to pH 7 causes more harm than good. Also, many tropical crops are well adapted to acid soil conditions; frequently, they do not respond to lime as better-known crops do.

The purpose of this chapter is to examine the status of soil acidity and liming problems in the tropics in light of modern concepts and in terms of varietal and species differences.

THE NATURE OF SOIL ACIDITY

Exchangeable Aluminum and Percent Aluminum Saturation

Soil acidity is a poorly defined parameter. Before the late 1950s exchangeable hydrogen was believed to be the cause of soil acidity. The work of Coleman and others, however, proved that exchangeable aluminum is the dominant cation associated with soil acidity. Hydrogen ions produced by organic matter decomposition are unstable in mineral soils because they react with layer silicate clays, releasing exchangeable aluminum and siliceous acid (Coleman and Thomas, 1967). Exchangeable hydrogen is found in small amounts in acid mineral soils. In soils high in organic matter, exchangeable hydrogen is associated with the carboxyl groups of the organic matter.

Exchangeable aluminum is determined by extracting with unbuffered normal salts such as 1 N KCl and titrating the extract with a base (Lin and Coleman, 1960). Exchangeable aluminum is precipitated out at a pH of about 5.5 to 6.0. Thus little or no exchangeable aluminum is found at higher soil pH values. Figure 7.1 illustrates this relationship in Oxisols and Andepts of Panama.

In addition to actual exchangeable aluminum values, a useful measure of soil acidity is the percent aluminum saturation of the effective CEC. Aluminum saturation is calculated by dividing the exchangeable aluminum (and H^+ if present) extracted by normal unbuffered KCl by the sum of exchangeable bases plus exchangeable aluminum (and hydrogen). Percent aluminum saturation or its reciprocal, percent base saturation, is a useful parameter. An example of pH–aluminum saturation relationships appears in Fig. 7.2 for Oxisols and Ultisols of Puerto Rico. Since aluminum is precipitated at a pH of about 5.5 to 6.0, soils are essentially base saturated at such pH levels.

Other methods that ignore the concepts of pH-dependent charge are still widely used throughout the tropics. Titratable, total, or nonexchangeable acidity is the amount of acidity extracted with $BaCl_2$–TEA solution buffered at pH 8.2. This value is generally much (sometimes 10 times) greater than that for exchangeable aluminum because it also includes nonexchangeable hydrogen associated with carboxyl groups, iron, and aluminum hydrated oxides. These components have no detrimental effect on plant growth (Kamprath, 1972). Therefore titratable acidity is of no practical value. When percent base saturation is calculated as the sum of basic cations divided by the sum of basic cations plus titratable acidity, the values obtained exaggerate the actual acidity of soils that have pH-dependent

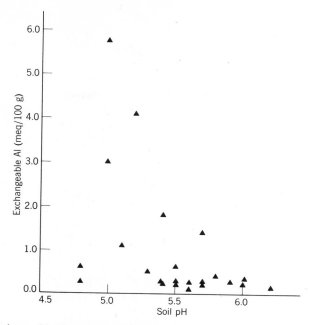

Fig. 7.1 Exchangeable aluminum at different pH values in nine Oxisols and Andepts from Panama. *Source:* Mendez-Lay (1973).

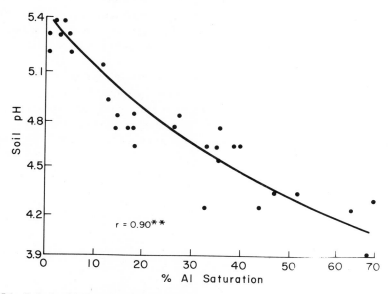

Fig. 7.2 Relationship between soil pH and aluminum saturation in eight Ultisols and Oxisols of Puerto Rico. *Source:* Abruña et al. (1975).

charge. Table 4.6 illustrated such differences. When base saturation is cal-culated as the sum of bases divided by CEC determined with ammonium acetate buffered at pH 7, the resulting value also exaggerates the acidity of soils that have pH values lower than 7. Unfortunately, base saturation values based on buffered extractions at pH 7 or 8.2 are still widely used by soil fertility specialists. The terms "aluminum saturation" and "base satura-tion" used in this book refer to the effective CEC method unless otherwise specified.

Aluminum in the Soil Solution

Exchangeable aluminum is held very tightly to the negative charges of layer silicate and layer silicate oxide-coated systems. Work in North Carolina (Nye et al., 1961; Evans and Kamprath, 1970) and in Guyana (Cate and Sukhai, 1964) showed that there is less than 1 ppm Al in the soil solution when the aluminum saturation is lower than 60 percent. The aluminum in the soil solution rises sharply beyond 60 percent aluminum saturation, however, as shown in Fig. 7.3.

In oxide systems the relationship between exchangeable and soil solution aluminum is a constant one, resembling the solubility of gibbsite. Ayres et al. (1965) studied this relationship on Andepts, Ultisols, and Oxisols of Hawaii, as shown in Fig. 7.3. The levels of aluminum saturation are too low to ascertain whether 60 percent aluminum saturation is a critical point in such soils.

The levels of aluminum in the soil solution also depend on the soil organic matter content and the salt content (Kamprath, 1972). Aluminum in the soil solution decreases as organic matter increases because organic matter forms very strong complexes with aluminum. Aluminum in the soil solution increases with increasing salt content because other cations then

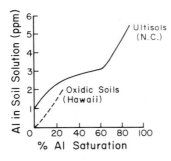

Fig. 7.3 Relationship between aluminum saturation and aluminum concentration in the soil solution in Ultisols of North Carolina and Oxisols, Ultisols, and Andepts of Hawaii. *Source:* Adapted from Ayers et al. (1965) and Evans and Kamprath (1970).

displace exchangeable aluminum by mass action (Brenes and Pearson, 1973).

Aluminum Saturation Levels in Tropical Soils

Table 7.1 shows the acidity status of some representative Oxisol, Ultisol, Alfisol, and Inceptisol profiles. Oxisols usually have a high percentage of aluminum saturation throughout their profiles (Guerrero, 1971). An exception is the Eutrustox or Eutrorthox, found in Brazil, Cuba, and other areas that are essentially 100 percent base saturated (Moura and Buol, 1972).

Some Ultisols have also high aluminum saturation, particularly in the subsoil. In Table 7.1 examples from the Amazon Jungle of Peru show very low pH values, probably because of the high aluminum contents. The very high exchangeable aluminum level in the subsoil of the Tropaquult is associated with a mottled montmorillonitic layer that releases large quantities of aluminum (Sanchez and Buol, 1974).

Most Alfisols have low aluminum saturation values. Since the requirement of more than 35 percent base saturation (at pH 8.2) refers to depths below the control section, some Alfisols may have low base saturation in the topsoil. Many Alfisols from West Africa have low aluminum saturation, but they are very poorly buffered soils (Juo, 1972). Consequently, their base status can change quickly with management.

Andepts usually have pH values above 5.5 and low aluminum saturation, except for the Dystrandepts and Hydrandepts, which can have over 60 percent aluminum saturation. Exchangeable hydrogen may be considerable in these soils because of their high organic matter contents (Igue and Fuentes, 1971). The examples in Table 7.1 range from 13 to 27 percent organic matter.

The aluminum status of other Inceptisols and Entisols is quite variable. The dystric (acid) and eutric (nonacid) terminology employed by the Soil Taxonomy and the FAO system is useful to distinguish acid soils. Most Vertisols, Mollisols, and Aridisols are essentially 100 percent base saturated. Most Spodosols and Histosols are acid. Some organic soils have very high exchangeable hydrogen contents.

CAUSES OF ACID SOIL INFERTILITY

Poor crop growth in acid soils can be directly correlated with aluminum saturation. Figure 7.4 shows a typical example. It is well known that pH per

Table 7.1 *Soil Acidity Levels in Some Tropical Soil Profiles*

Soil and Location	Horizon (cm)	pH (in H_2O)	Exch. Al[a] (meq/100 g)	Exch. H[a] (meq/100 g)	Effective CEC (meq/100 g)	Aluminum Saturation (%)
Oxisols						
Haplustox	0–8	4.8	3.1	1.0	6.5	63
(Carimagua, Colombia)	8–22	4.7	3.2	0.6	4.6	83
	22–46	4.4	1.9	0.5	2.9	83
	46–132	4.9	0.6	0.4	1.4	71
	132–140	5.4	0.3	0.4	1.6	44
Eutrustox	0–10	5.5	0.2	—	19.5	2
(Minas Gerais, Brazil)	20–30	4.9	0.8	—	13.0	6
	40–50	5.1	0.6	—	10.7	5
	100–110	5.3	0.3	—	9.0	4
	140–170	5.5	0.3	—	9.1	4
Acrustox	0–20	4.7	2.6	—	2.7	94
(Brasilia, Brazil)	20–40	4.8	1.9	—	2.0	94
Ultisols						
Paleudult	0–5	3.6	1.9	—	3.6	52
(Yurimaguas, Peru)	5–40	4.2	4.2	—	4.7	89
	40–60	4.1	4.5	—	6.2	72
	60–90	4.2	6.0	—	6.5	92
	90–140+	4.0	6.1	—	8.4	73
Tropaquult	0–5	5.6	0.0	—	16.7	0
(Iquitos, Peru)	5–10	4.7	5.5	—	19.3	28
	10–50	4.6	14.6	—	18.4	79
	50–90	4.7	29.3	—	35.3	83

Alfisols

Haplustalf	0–5	6.8	0.01	—	5.36	0.2
(Ibadan, Nigeria)	5–15	6.7	0.01	—	6.99	0.1
	15–45	7.1	0.03	—	5.26	0.6
	45–65	6.7	0.04	—	4.24	0.9
	65–95	6.3	0.06	—	4.06	1.5
Ustalf	0–27	5.8	0.20	—	4.41	4.5
(Zaria, Nigeria)	27–75	6.0	0.18	—	6.45	2.8
	75–116	6.1	0.17	—	8.43	2.0
Tropaqualf	0–5	5.5	0.55	—	21.13	3
(Yurimaguas, Peru)	5–25	4.9	9.20	—	27.55	33
	25–80	5.0	12.50	—	28.35	44
	80–100	5.2	14.55	—	33.37	44

Inceptisols

Vitrandept	A_1	4.5	0.52	0.23	10.57	7
(Turrialba, Costa Rica)	B_2	5.8	0.30	0.53	10.83	7
Dystrandept	A_p	5.4	2.45	1.25	5.39	69
(Turrialba, Costa Rica)	IIBi	5.4	0.38	0.53	1.96	46
Umbrandept	A	5.5	0.4	—	7.4	5
(Chinchiná, Colombia)	B	5.8	0.2	—	1.2	17
Dystropept	A_1	4.8	4.43	1.90	9.53	66
(Turrialba, Costa Rica)	B_{12}	5.3	3.95	1.16	6.61	77

Source: Guerrero (1971), Igue and Fuentes (1972), Juo (1972), León (1967), Moura and Buol (1972), and Sanchez and Buol (1974).
[a] Extracted by unbuffered $1 N$ KCl. Dashes in the "Exch. H" column indicate that no analysis was performed. In the samples from Peru, exch. Al determinations include exch. H.

229

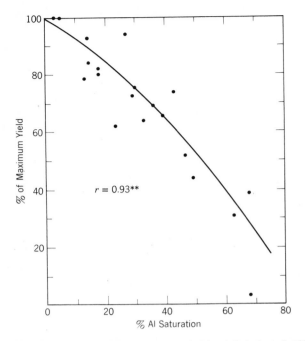

Fig. 7.4 Relationship between aluminum saturation levels of Oxisols and Alfisols of Puerto Rico and snap bean production. *Source:* Abruña et al. (1975).

se has no direct effect on plant growth, except at pH values below 4.2, where the hydrogen ion concentration may stop or even reverse cation uptake by roots (Black, 1967). Acid soil infertility is due to one or more of the following factors: aluminum toxicity, calcium or magnesium deficiency, and manganese toxicity. Jackson (1967) has summarized these concepts in a thorough review. The following examples illustrate the individual effects under tropical conditions.

Aluminum Toxicity

Concentrations of soil solution aluminum above 1 ppm often cause direct yield reduction. Studies by Abruña et al. (1970) on tobacco and by Villagarcia (1973) on potatoes show that the primary effect of aluminum toxicity is direct injury to the root system. Root development is restricted, and the roots become thicker and stubby and show dead spots. Table 7.2

shows the effect of aluminum on root growth. Corn root growth was not affected until 60 percent aluminum saturation was reached, whereas sorghum root growth was restricted at the first increment of aluminum. These studies, as well as many similar ones conducted in the temperate region, also indicate that aluminum tends to accumulate in the roots and impede the uptake and translocation of calcium and phosphorus to the tops (Foy, 1974). Thus aluminum toxicity may produce or accentuate calcium and phosphorus deficiencies.

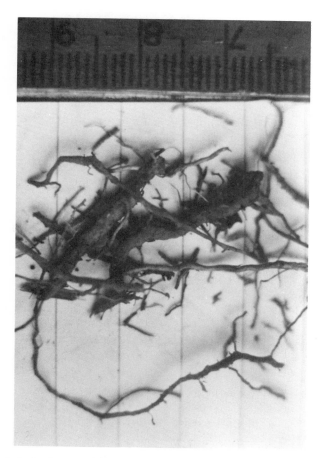

Plate 9 Typical aluminum toxicity symptoms in corn roots growing in an acid Oxisol. Note thickened, stubby roots. Courtesy of Mr. Enrique Gonzalez, North Carolina State University.

Table 7.2   Effects of Aluminum on Root Growth in Two Puerto Rican Soils

Soil	pH	Exch. Al (meq/100 g)	Al Saturation (%)	Root Dry Weights Corn	Sorghum (mg/pot)
Humatas	4.8	4	40	931	400
(Ultisol)	4.5	6	57	874	296
	3.9	11	87	209	19
Coto	4.8	3	52	687	345
(Oxisol)	4.5	4	70	630	126
	4.0	5	87	389	128

Source: Brenes and Pearson (1973).

Calcium and Magnesium Deficiencies

Although aluminum is the primary culprit, poor growth in acid soils may be due also to direct calcium or magnesium deficiencies. The work of Abruña and his coworkers on tobacco in Puerto Rico, reproduced in Fig. 7.5, also illustrates this point. Tobacco was grown in an Ultisol with a pH of 4.2 and 0.4 meq Ca/100 g. Without liming, this resulted in restricted root growth because of both aluminum toxicity and calcium deficiency. When aluminum was precipitated by liming to pH 5.6 with $MgCO_3$, maintaining calcium at the low level, root growth stopped at 60 hours. When both aluminum was precipitated and the calcium level raised to 4.4 meq by liming with $CaCO_3$, root growth progressed normally. Consequently, both aluminum toxicity

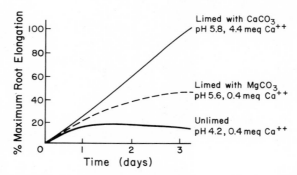

Fig. 7.5 Evidence of aluminum toxicity and calcium deficiency affecting elongation of tobacco roots growing on an Ultisol of Puerto Rico. *Source:* Abruña et al. (1970).

and calcium deficiency were limiting factors. Similarly, severe magnesium deficiencies also can restrict growth by themselves.

Many acid soils in the tropics are deficient in calcium without having aluminum toxicity problems. Examples of these are found in Hawaii, where many soils have pH values below 5 but little exchangeable aluminum (Ayres et al., 1965). Liming is essentially used as calcium fertilization for sugarcane in such soils. Likewise, certain Oxisols of the Cerrado of Brazil are very low in magnesium and respond directly to magnesium fertilization, according to Mikklesen et al. (1963).

Manganese Toxicity

Manganese is very soluble at pH values lower than 5.5 (Black, 1967). If this element is present in sufficient amounts, manganese toxicity can occur along with aluminum toxicity at pH values of up to about 5.5 to 6.0. The solubility of manganese also increases with soil reduction when the Mn^{4+} ions are converted to Mn^{2+}. Certain acid soils may be low in aluminum but high in manganese. Coto clay, a Puerto Rican Oxisol with a pH of 4.4, is an example of this situation (Pearson, 1975). Abruña and his coworkers (1970) found that yield responses to liming were correlated with a progressive decrease in toxic levels of manganese in tobacco leaves in Coto clay but not in Humatas clay, an Ultisol with higher levels of exchangeable aluminum. Their results are shown in Table 7.3. In soils like Coto clay, liming is aimed at decreasing the solubility of manganese. Unlike aluminum, manganese is a plant nutrient; consequently, the aim is not to eliminate soluble man-

Table 7.3   Effects of Liming on Yields and Manganese Contents of Tobacco Leaves in Two Puerto Rican Soils

Base Saturation Range (%)	Coto (Oxisol)		Humatas (Ultisol)	
	Leaf Yield (kg/ha)	Mn in Leaves (ppm)	Leaf Yield (kg/ha)	Mn in Leaves (ppm)
20–40	—	—	630	274
30–40	262	2380	1043	200
50–60	503	1170	1688	142
70–80	539	660	1686	140
80–90	682	580	—	—

Source: Abruña et al. (1970).

ganese but to keep it within a range between toxicity and deficiency. A soil solution concentration between 1 and 4 ppm Mn represents such a range, although there is considerable variability among soils.

CROP AND VARIETAL DIFFERENCES

The growth of different species in acid soils depends on their relative tolerances to aluminum and manganese levels and their relative requirements for calcium and magnesium. More recently, it has been found that substantial differences exist between and within crop species in relation to their tolerances to soil acidity factors.

Certain crops grown exclusively in the tropics grow normally at pH levels where corn or soybeans would die. Pineapple is perhaps the best-known example, but coffee, tea, rubber, and cassava also tolerate very high levels of exchangeable aluminum. Among the pasture species, several grasses and legumes are apparently very well adapted to acid soil conditions. Tropical grasses such as guinea grass (*Panicum maximum*), jaragua (*Hyparrhanea rufa*), molasses grass (*Melinis minutiflora*), and several species of the genera *Paspalum* and *Brachiaria* grow well in very acid soils.

Although legumes are considered very susceptible to soil acidity because of their high calcium requirements for nodulation, several tropical pasture legumes are strikingly well adapted to acid conditions. *Stylosanthes* spp., *Desmodium* spp., *Centrosema* spp., *Calopogonium* spp., and tropical Kudzu (*Pueraria phaseoloides*) are the principal ones. Among the grain legumes, cowpeas and pigeon peas seem to be more tolerant of acidity than field beans or soybeans.

Many of these species have evolved in acid soils and possess genes responsible for tolerating conditions associated with high aluminum levels. The mechanisms involved are not well understood, but the different responses of these species to liming can be tested in simple field trials for practical purposes.

Figure 7.6 shows the striking species differences among pasture legumes in tolerating aluminum in the soil solution. Andrew and Vanden Berg (1973) showed that Townsville stylo (*Stylosanthes humilis*) and *Desmodium intortum* are relatively insensitive to levels of soil solution aluminum as high as 2 ppm, whereas alfalfa practically dies at 1 ppm. Another tropical legume, *Glycine wightii* (perennial soybean), is also sensitive to low aluminum levels. It developed in the tropics but in calcareous areas.

Figure 7.6 also shows the different responses of these four species to high levels of manganese in solution. *Stylosanthes humilis* was relatively insensitive to manganese, but *Desmodium intortum* was as susceptible as alfalfa.

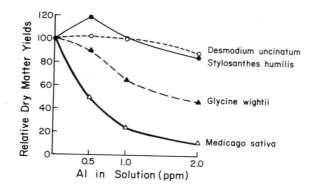

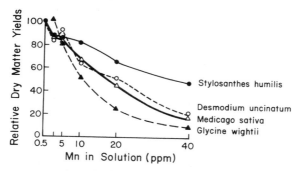

Fig. 7.6 Species differences in tolerance to aluminum and manganese among four pasture legumes. *Source:* Adapted from Andrew and Hegarty (1969) and Andrew and Vanden Berg (1973).

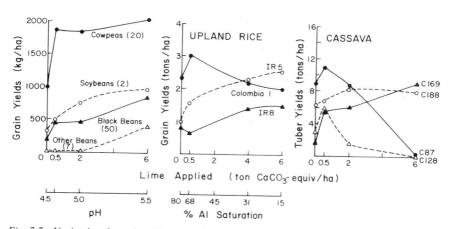

Fig. 7.7 Varietal and species differences in tolerance to acid soil conditions in an Oxisol from Carimagua, Colombia. Numbers in parentheses refer to the number of grain legume varieties tested. *Source:* Adapted from Spain et al. (1975).

235

Conseqently, tolerance or susceptibility to one cause of acid soil infertility does not necessarily imply a similar reaction to another cause.

Similar evidence has recently been obtained with tropical food crops by Spain et al. (1975) in Oxisols of the Llanos Orientales of Colombia. The soils on which the experiments were conducted had a pH of 4.3 and 3.5 meq/100 g of exchangeable aluminum, which amounted to 80 percent aluminum saturation. Figure 7.7 shows that substantial differences in aluminum tolerance and required liming exist among species. Cowpeas (*Vigna sinensis*) reach maximum yield with 0.5 ton of lime, whereas soybeans and field beans (*Phaseolus vulgaris*) require higher lime applications and even then do not approach the productivity of cowpeas under these conditions. Significant varietal differences exist within *Phaseolus vulgaris*. The black-skinned varieties as a group are much more tolerant to aluminum than are the white, yellow, or brown varieties. It is interesting to note that soybeans are somewhat better performers than field beans in these acid soils. Another grain legume that is well adapted to acid soil conditions is pigeon peas (*Cajanus cajan*). This information supports previous findings about the desirability of using cowpeas rather than field beans in acid soil conditions.

Spain and his coworkers have also tested several hundred rice varieties under the same conditions. Several local varieties responded to the first lime increment and lodged afterwards. This behavior is illustrated in Fig. 7.7 by the Colombia 1 variety. Most varieties developed at the International Rice Research Institute do not exhibit such tolerance because they were selected in soils with high base status. Consequently, they require at least 4 tons/ha of lime to approach good yields, as is illustrated with the variety IR5 in Fig. 7.7. Unfortunately most aluminum-tolerant varieties like Colombia 1 have many undesirable characteristics; most are tall-statured and therefore susceptible to lodging. Breeding efforts are required to incorporate aluminum tolerance into short-statured varieties with a high yield potential.

Varietal differences also exist in cassava, a species considered to be well adapted to acid soil conditions. Figure 7.7 indicates that some varieties like C169 show a direct lime response with highest yields at the highest lime rates, whereas another shows no response and two show just a response to the first increment and a sharp decrease afterwards. The yield level in this trial was quite low because of disease problems. Other investigations have shown a positive response to lime by this crop at higher yield levels.

The selection of varieties or species that perform well at high aluminum saturation levels and thus need only a fraction of the normal lime requirement is of great practical importance. On the basis of their observations, Spain et al. (1975) have produced a list of species adapted to such conditions (Table 7.4).

Several physiological mechanisms have been identified as associated with

Table 7.4   Crops and Pasture Species Suitable for Acid Soils with Minimum Lime Requirements

Lime Requirement (tons/ha)	Al Saturation (%)	pH	Crops (Using Tolerant Varieties)
0.25 to 0.5	68 to 75	4.5 to 4.7	Upland rice, cassava, mango, cashew, citrus, pineapple, *Stylosanthes, Desmodium,* kudzu, *Centrosema,* molasses grass, jaragua, *Brachiaria decumbens, Paspalum plicatulum*
0.5 to 1.0	45 to 58	4.7 to 5.0	Cowpeas, plantains (?)
1.0 to 2.0	31 to 45	5.0 to 5.3	Corn, black beans

Source: Spain et al. (1975).

tolerance or sensitivity to aluminum among or within species, but none has been found to apply in all cases. In a review of the subject, Foy (1974) cites the following:

1. Differences in root morphology. Some aluminum-tolerant varieties keep developing and are not injured in the root tips or laterals in acid soils.

2. Changes in pH of the root rhizosphere. Some aluminum-tolerant varieties increase the pH of the growth medium, whereas sensitive ones decrease it. Such changes are believed to be a result of differential cation–anion uptake, secretion of organic acids, carbon dioxide, and bicarbonate.

3. Lower translocation of aluminum to plant tops. Several tolerant species and varieties accumulate aluminum in the roots, but translocate it to the tops at a lower rate than sensitive varieties. However, several tree and fern species adapted to acid conditions accumulate large amounts of aluminum in their tops.

4. Aluminum in the roots does not inhibit the uptake and translocation of calcium, magnesium, and potassium in tolerant varieties, whereas it does so in sensitive varieties. Varietal tolerance of aluminum in soybeans, wheat, and barley is related to calcium uptake and translocation; in sorghum, to potassium; and in potatoes, to both magnesium and potassium translocation.

5. High plant silicon content is associated with aluminum tolerance in certain rice varieties.

6. Aluminum-tolerant varieties do not inhibit phosphorus uptake and translocation as much as susceptible varieties or species. Also many aluminum-tolerant species or varieties are also tolerant of low phosphorus levels.

Subsequent work in Brazil (North Carolina State University, 1974; Salinas and Sanchez, 1976) suggests that varietal tolerances to both high exchangeable aluminum and low available phosphorus levels are related, and depend on the plant's ability to translocate phosphorus from the roots to the shoot in the presence of high levels of aluminum in solution. These results were obtained with rice, corn, wheat, sorghum, and *Phaseolus vulgaris*.

Andrew and his associates in Australia have shown that several of these factors are responsible for the species differences in tropical pasture legumes presented in Fig. 7.6. They found that aluminum-tolerant species nodulated abundantly at low pH levels, regardless of calcium level. In the

Plate 10 Example of varietal differences in aluminum tolerance. Two sorghum varieties are shown growing on an Oxisol with pH 4.5 and 75 percent aluminum saturation in Brasília, Brazil. The variety on the left is suffering severely from aluminum toxicity, whereas the variety on the right is growing adequately.

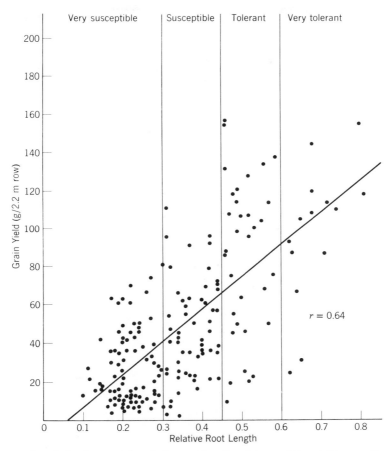

Fig. 7.8 Correlation between the relative root lengths of rice varieties at 30 *versus* 3 ppm Al
in culture solution and their grain yields when grown in an Oxisol from Carimagua, Colombia.
Source: Spain et al. (1975).

presence of aluminum, the calcium content of roots did not decrease
sharply in tolerant species, whereas it decreased from 1.0 to 0.1 percent in
alfalfa. Aluminum-tolerant species accumulated aluminum in the cortical
layer root cells, but in the tops the aluminum concentration was much lower
than in alfalfa. Phosphorus translocation to the tops drastically decreased at
high aluminum levels in alfalfa and *Glycine wightii*, but it actually increased
in the aluminum-tolerant species at high levels of soil aluminum.
Stylosanthes in addition has a very low requirement for phosphorus.
Maximum yields were attained with a phosphorus concentration in the

plant of 0.17 percent, whereas other tropical legumes such as *Phaseolus lathyroides* needed 0.24% P to produce maximum yields.

Evidence which indicates that one or several genes control aluminum tolerance within a species is beginning to accumulate. One dominant gene was found for winter barley, and three genes for corn (Reid, 1971; Gorsline et al., 1968). Greenhouse or nutrient solution techniques have been developed to screen rapidly large number of lines for aluminum tolerance. Several hundred rice varieties were grown at 3 and 30 ppm Al in solution for 3 weeks at CIAT. When the roots lengths were measured, relative root length was found to be well correlated with field yield at low lime levels. The results are shown in Fig. 7.8. With such techniques, varieties exhibiting both a high yield potential and aluminum tolerance could be selected and tested in the field.

LIMING

From the above considerations it should be obvious that the purpose of liming is primarily to neutralize the exchangeable aluminum, and that this is normally accomplished by raising the pH to 5.5. When manganese toxicity is suspected, the pH should be raised to 6.0. The factors to be considered are (1) the amount of lime needed to decrease the percent aluminum saturation to a level at which the particular crop and variety will grow well, (2) the quality of lime, and (3) the placement method.

Determination of Lime Requirement

A large amount of effort has been devoted to finding the best methods for estimating lime needs in the tropics. Part of the confusion was certainly caused by attempts to lime soils to neutrality and the use of titratable acidity as the criterion. Kamprath (1970) suggested that lime recommendations be based on the amount of exchangeable aluminum in the topsoil and that lime rates be calculated by multiplying the milliequivalents of aluminum by 1.5. The result is the milliequivalents of calcium needed to be applied as lime. Lime rates calculated by this method neutralize 85 to 90 percent of the exchangeable aluminum in soils containing 2 to 7 percent organic matter. The reason for 1.5 as a factor is the need to neutralize the hydrogen ions released by organic matter or iron and aluminum hydroxides as the pH increases. In soils with higher organic matter, the factor has to be raised to 2 or 3 because of the presence of exchangeable hydrogen. This method has been used successfully in Brazil since 1965 (Cate, 1965) and is employed in most Latin American countries at the time of this writing.

For every milliequivalent of exchangeable aluminum present, 1.5 meq of calcium or 1.65 tons/ha of $CaCO_3$-equivalent should be applied. This procedure has been found successful in the tropics and eliminates the need for time-consuming neutralization tests in the laboratory.

The application of this concept has reduced the rates of liming substantially, particularly in acid Oxisols and Ultisols low in effective CEC. In most cases where 1 to 3 meq of exchangeable aluminum is present, lime applications are now on the order of 1.6 to 5 tons/ha. In the past, rates on the order of 10 to 30 tons/ha were frequently recommended and applied with mixed results. Farmers accustomed to the high exchangeable aluminum contents of certain Oxisols and Ultisols in southern Brazil are pleasantly surprised when they move north to the Cerrado of central Brazil, where the more highly weathered Oxisols have about 1 meq of exchangeable aluminum. Their lime application rates are reduced to one-half or one-third of what they were used to applying. Although in both areas soils have high aluminum saturation, the lower exchangeable aluminum levels in the more weathered Oxisols decrease the lime requirement.

Another important component of lime requirement determination is the level of exchangeable aluminum that specific crops or varieties can tolerate. Crops originally developed in calcareous soils, such as cotton, sorghum, and alfalfa, are susceptible to levels of 10 to 20 percent aluminum saturation. For them, liming should be aimed at zero aluminum saturation in order for the application to last for a few years. Corn is sensitive to 40 to 60 percent aluminum saturation. Although liming to zero aluminum saturation might be beneficial, lowering the aluminum saturation level to 20 percent could be more economical. Other crops such as rice and cowpeas are more tolerant. Coffee, pineapple, and some pasture species seldom respond to lime, even in soils with high aluminum saturation.

An example of these relationships is illustrated in Fig. 7.9, where the relative yields of sorghum, corn, elephant grass (*Pennisetum purpureum*), and coffee from liming trials are plotted as a function of aluminum saturation. Sorghum yields began to drop dramatically at about 15 percent aluminum saturation, and corn at about 40 percent. Elephant grass yields dropped sharply at about 60 percent aluminum saturation, while coffee suffered a yield decrease only at 80 percent.

Liming may be needed for some of these tolerant crops, however, to counteract calcium or magnesium deficiencies or to correct manganese toxicity. In Puerto Rico high coffee yields are produced without lime in soils at pH 3.8, but liming is necessary in certain soils high in manganese to

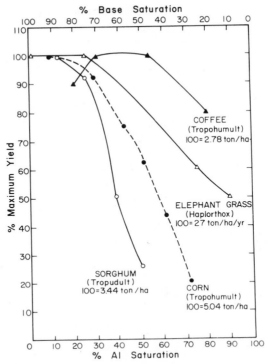

Fig. 7.9 Yield responses to liming in Puerto Rican Oxisols and Ultisols. *Source:* Compiled from Abruña et al. (1964, 1965, 1975).

prevent manganese toxicity. Much of the response to 0.5 ton of lime in the Colombian Oxisols shown in Fig. 7.7 may simply represent the response to calcium and magnesium fertilization.

Sources of Lime

To find lime sources of sufficient fineness and purity is a major practical problem in the tropics. The selection of sources must take into account the calcium and magnesium contents of the liming material and the calcium and magnesium status of the soil. Fineness is crucial. Lime that does not pass through a 20 mesh sieve will have very little reactivity; what does not pass through a 60 mesh sieve will react very slowly. Lime that passes a 100 mesh sieve will react quickly. A good grade of fineness is more than 60 mesh; a better grade, 100 mesh.

Placement

Lime is commonly incorporated into the top 15 cm several days before planting the crop. Although this is usually the best way, there are instances in which plowing in is not possible and others in which deeper incorporation is beneficial.

Liming established pastures precludes incorporation unless reseeding is contemplated. Usually basal applications are well incorporated, but smaller applications aimed at correcting the acidifying effect of nitrogen fertilizers are effective when broadcast to the pasture surface, according to the work of Vicente-Chandler et al. (1964). In a Puerto Rican Oxisol, Abruña et al. (1964) observed no differences in pasture yields between surface-applied and incorporated lime applications.

When extremely acid Oxisols have their topsoil limed to pH 5.5, most of the root development of corn occurs in the topsoil. The high subsoil aluminum saturation prevents deeper root development. When short-term droughts occur during the rainy season, plants may suffer from water stress while the subsoil is still moist. For this reason, Gonzalez and Kamprath (North Carolina State University, 1973) compared incorporating lime at two depths, 0 to 15 cm and 0 to 30 cm, in an Oxisol from Brasilia. This soil has an excellent granular structure that permitted deep incorporation with a rototiller. The results of the first corn crop are shown in Fig. 7.10. Deeper applications produced higher yields. Root studies showed that higher yields were associated with deeper root development in the 0 to 30 cm layer, which diminished water stress during short-term droughts.

The feasibility of deep lime incorporation depends largely on soil structural properties and available equipment. It seems reasonable to assume that this would be possible in sandy soils and highly aggregated Oxisols and Andepts, but doubtful in Ultisols with clayey argillic horizons.

Downward Movement of Calcium and Magnesium

Reducing aluminum toxicity in the subsoil is a major but difficult management objective in many areas of the tropics. When deeper lime incorporation is not feasible, other ways must be sought. The use of aluminum-tolerant varieties is one alternative. The possibility of downward movement of calcium and magnesium is another. Pearson and his coworkers (1962) found that under intensive pasture management an application of 12 tons/ ha of lime caused calcium and magnesium to move from the top 15 cm into the subsoil when accompanied by high annual ammonium sulfate applications equivalent to 800 kg N/ha in an Ultisol from Puerto Rico. Movement

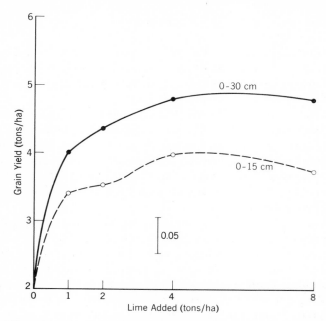

Fig. 7.10   Effect of lime incorporation on corn yields in an Oxisol from Brasilia, Brazil. *Source:* E. Gonzalez (North Carolina State University, 1973).

occurred in the form of exchangeable calcium and magnesium but also as calcium and magnesium sulfate. The high rates of lime and ammonium sulfate used limit the application of these findings to capital-intensive systems.

Similar downward calcium and magnesium movement was observed in an Oxic Dystrandept of Costa Rica by Morelli et al. (1971). Four years after liming, they measured changes in pH and aluminum saturation in the profile. The results shown in Fig. 7.11 indicate substantial calcium and magnesium movement from the topsoil (0 to 20 cm) down to 60 or 80 cm. The effect achieved with the 3.6 tons/ha rate has practical significance. Before liming, this Andept had positive ΔpH values indicative of an oxidic mineralogy throughout the profile. The ΔpH changed to negative values upon liming as deep as the downward movement of calcium and magnesium went. Mahilum et al. (1970) have made similar but less well-quantified observations in a Hydrandept from Hawaii.

In a laboratory study with a South African Oxisol, Reeve and Sumner (1972) showed much faster movement of calcium and magnesium to the subsoil when gypsum was used instead of calcium hydroxide. At a lime rate that precipitated the exchangeable aluminum in the topsoil, aluminum saturation in the subsoil decreased from 57 to 53 percent with Ca(OH)$_2$; but when the same amount of calcium was applied as CaSO$_4$, the subsoil aluminum saturation decreased from 53 to 43 percent. Overliming the topsoil reduced aluminum saturation in the subsoil even more. This is illustrated in Table 7.5. Amaral et al. (1965) also obtained evidence of downward calcium and magnesium movement in Oxisols of São Paulo, Brazil.

These results show that downward movement of calcium and magnesium occurs in well-aggregated tropical soils when they are limed at the rates needed to neutralize exchangeable aluminum or at higher rates. One possible explanation of this phenomenon is that, once the permanent charge sites are saturated, exchangeable calcium and magnesium held on the pH-dependent charge sites is likely to move down fairly easily. The porous nature of well-aggregated soils with oxidic mineralogy and the high annual rainfall characteristic of these sites favor such a movement. This situation is certainly different from that prevailing in soils with layer silicate systems. There is no information available on the downward movement of unreacted lime particles in oxide or oxide-coated soils. Research in this direction is important in areas where subsoil acidity is a major limiting factor.

Residual Effect of Liming

The residual effect of liming depends on how fast calcium and magnesium are being displaced by the residual acidity of nitrogen fertilizers. Soil tests

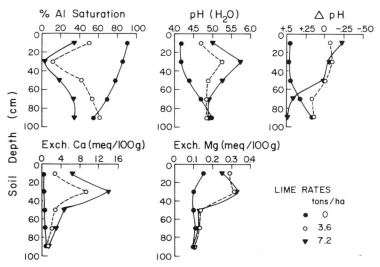

Fig. 7.11 Effects of lime applied 4 years before sampling on the profile properties of an Oxic Dystrandept from Costa Rica. *Source:* Calculated from data by Morelli et al. (1971).

Table 7.5   Effects of Liming Topsoils on the % Aluminum Saturation of Subsoils of an Oxisol from South Africa in Laboratory Conditions

Rate Incorporated into Topsoil (meq Ca/100 g)	Liming Source	
	$Ca(OH)_2$	$CaSO_4$
0	57	57
3[a]	53	43
6	49	34
9	44	30

[a] Rate recommended to neutralize topsoil exchangable Al.
Source: Adapted from Reeve and Sumner (1972).

are an adequate tool for determining when additional lime should be applied.

The residual effect of liming elephant grass pasture fertilized annually with 800 kg N/ha as ammonium sulfate (Abruña et al., 1964) is shown in Fig. 7.12. There was little response during the first year; but as the residual acidity of the ammonium sulfate applications became effective, optimum yields were obtained with 4 tons/ha of lime. The response curves were sharper every year because of the decreases of the unlimed treatments.

Mahilum et al. (1970) studied the residual effect of liming on a Hawaiian Hydrandept. They found that after 5 years a rate of 2 tons lime/ha kept the aluminum level at about 1 meq (from an original value of 3 meq), even though most of the calcium was leached to lower levels. Apparently alu-

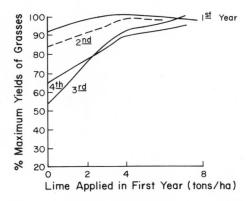

Fig. 7.12 Residual effect of lime applications on the relative yields of pasture grasses in Puerto Rican Oxisols and Ultisols. *Source:* Abruña et al. (1964).

minum ions did not readily reoccupy the exchange sites even when calcium leached below. After 5 years the residual effect of liming at the rate of 5 tons/ha completely disappeared.

In sharp contrast De Freitas and Van Raij (1975) obtained positive corn and soybean responses to lime in a sandy Oxisol of São Paulo, Brazil, 6 years after application. They observed increasing yield responses with time and attributed them to the dissolution of the coarser lime particles.

OVERLIMING

The tropical soil literature is full of reports citing the lack of response or the negative response when tropical soils are limed. This has created the generalized idea that liming does not work in the tropics (Richardson, 1951). In many cases, soils were limed to neutrality. This concept originated in the United States Midwest in corn–legume rotations because alfalfa and clover grew best at a pH of 6.5 to 7.0 because of their high calcium requirements. This practice implied that liming to neutrality was also best for corn and small grains in rotation with alfalfa. The soils in question were Mollisols or Alfisols essentially devoid of pH-dependent charge. Kamprath (1971) reviewed the reasons for the lack of positive lime responses when highly leached soils are limed to neutrality. The consequences of overliming are yield reduction, soil structure deterioration, and decreased availability of phosphorus, boron, zinc, and manganese. Overliming can be defined as liming at rates higher than necessary to neutralize the exchangeable aluminum or eliminate manganese toxicity.

In his review Kamprath cites several cases in which crop yields decreased when Oxisols and Ultisols were limed to pH 7. An example is illustrated with corn growing on a Hawaiian Ultisol in Fig. 7.13. Some evidence from Indonesia and South Carolina indicates that liming to neutrality promotes

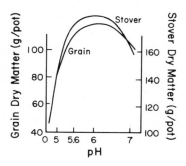

Fig. 7.13 Effects of liming to neutrality on corn growth in an Ultisol from Hawaii. *Source:* Younge and Plucknett (1964).

the formation of smaller aggregates and thus reduces infiltration rates and makes some Oxisols and Ultisols more susceptible to erosion (Peele, 1936; Schuffelen and Middleburg, 1954.) This may be a consequence of increased microbial activity due to liming. Ghani et al. (1955) found that liming with MgO decreased infiltration rates and noncapillary porosity in an acid lateritic soil of Bangladesh, while $CaCO_3$ did not produce a consistent effect. No correlations between such changes in physical properties and crop yields have been recorded.

Overliming induces phosphorus deficiency in soils with high phosphorus fixation capacity. When a Gibbsihumox from Hawaii was limed from pH 5.3 to 6.1, the uptake of phosphorus by sorghum and *Desmodium intortum* increased dramatically (Fox et al., 1964). When this soil was limed to pH 7, however, phosphorus uptake decreased and severe phosphorus deficiency was observed, apparently because of the formation of insoluble calcium phosphates.

Overliming soils high in oxide coatings greatly increases the adsorption of boron by clays and reduces the availability of boron. Kamprath (1971) cites several examples from the southeastern United States of such problems arising when the soil pH increases beyond 6.5 in Ultisols. Lime-induced manganese deficiencies can also be observed at pH levels above 6.2, because manganese tends to precipitate at this pH range in the presence of iron and aluminum oxides.

Overliming may also induce zinc deficiencies. The solubility of zinc decreases rapidly at a pH of 6 to 7. In soils naturally low in zinc excessive liming does decrease the availability of zinc, as demonstrated when Cerrado Oxisols are overlimed (North Carolina State University, 1973).

The bulk of the evidence suggests that highly weathered soils should not be limed to pH values greater than 5.5. Beyond that level, yield decreases can occur. However, in many cases overliming produces no yield decreases, just a flat plateau. Detrimental effects are most commonly noted in oxide-coated layer silicate systems or oxide systems in soils low in available phosphorus, boron, and zinc and high in phosphorus fixation capacity. According to McLean (1971), overliming layer silicate systems with little pH-dependent charge causes hardly any detrimental effects.

SUMMARY AND CONCLUSIONS

1. Soil acidity problems are associated with pH levels lower than 5.5 and the presence of exchangeable aluminum in the soil. Percent aluminum saturation, calculated on the basis of effective cation exchange capacity, is a useful measure of soil acidity. When total extractable acidity or aluminum

saturation is calculated using buffered solutions at pH 7 or 8.2, the result tends to grossly overestimate the acidity problems of most tropical soils.

2. Aluminum toxicity is the most common cause of acid soil infertility. High aluminum levels in the soil solution do direct harm to roots and decrease root growth and the translocation of calcium and phosphorus to tops. Aluminum toxicity can be corrected by liming to pH 5.5 to 6.0, to precipitate the exchangeable aluminum as aluminum hydroxide. Manganese toxicity can occur in certain soils high in soluble manganese. This problem can also be corrected by liming to pH levels of 5.5 to 6.0, where the solubility of manganese decreases enough to eliminate toxicity but not enough to prevent deficiencies. Calcium and magnesium deficiencies are also important causes of acid soil infertility. Often they occur together with aluminum or manganese toxicity, but in certain soils, low in all these elements, liming serves as calcium and/or magnesium fertilizer.

3. Tropical crops differ widely in the ability to tolerate acid soil conditions. Coffee, rubber, pineapple, certain pasture grasses and legumes are very tolerant of high levels of aluminum saturation. Rice and black beans are also fairly tolerant, whereas sorghum and cotton are very intolerant. Important varietal differences in relation to aluminum tolerance exist in rice, corn, wheat, beans, and soybeans.

4. Liming rates can be calculated on the basis of 1.65 tons/ha of $CaCO_3$-equivalent per milliequivalent of exchangeable aluminum. Such rates raise the soil pH to about 5.5 to 6.0 and virtually eliminate aluminum saturation in most mineral soils. For the more tolerant species or varieties, liming should be aimed at decreasing aluminum saturation to 20 or 40 percent. Achieving fineness and the proper magnesium content of liming materials is one of the most critical management problems.

5. In well-aggregated Oxisols with very acid subsoils, liming the topsoil does not solve the problem of impeded root development in the subsoil to take advantage of stored moisture. Deep liming applications are feasible in certain soils and result in substantially increased yields because of deeper root development and improved water relationships. In well-aggregated oxidic soils downward movement of calcium and magnesium into the subsoil takes place.

6. Adequate lime applications can have a long-term residual effect in some soils, and relatively short ones in others. Periodic maintenance lime applications can be calculated according to the exchangeable aluminum criterion.

7. Overliming to pH values greater than 6 or 7 can seriously decrease yields, particularly in soils high in iron and aluminum oxides. Liming to neutrality may cause structural deterioration, reduce phosphorus avail-

ability, and induce zinc, boron, and manganese deficiency. Liming to pH 7 is usually not harmful in layer silicate soils but is either detrimental or neutral in oxides and oxide-coated layer silicate systems.

8. Liming management in the tropics should be aimed at determining the minimum level of lime needed, selecting species and varieties more tolerant to aluminum, following practices that promote deeper root development in acid subsoils.

REFERENCES

Abruña, F., J. Vicente-Chandler, and R. W. Pearson. 1964. Effects of liming on yields and composition of heavily fertilized grasses and on soil properties under humid tropical conditions. *Soil Sci. Soc. Amer. Proc.* **28**:657–661.

Abruña, F., J. Vicente-Chandler, L. Becerra, and R. Bosque-Lugo. 1965. Effects of liming and fertilization on yields and foliar composition of high yielding sun-grown coffee in Puerto Rico. *J. Agr. Univ. Puerto Rico* **49**:413–428.

Abruña, F. and J. Vicente-Chandler. 1967. Sugar cane yields as related to the acidity of a humid tropic Ultisol. *Agron. J.* **59**:330–332.

Abruña, F., J. Vicente-Chandler, R. W. Pearson, et al. 1970. Crop response to soil acidity factors in Ultisols and Oxisols. I. Tobacco. *Soil Sci. Soc. Amer. Proc.* **34**:629–635.

Abruña, F., R. W. Pearson, and R. Perez-Escolar. 1975. Lime responses of corn and beans grown on typical Ultisols and Oxisols of Puerto Rico. Pp 261–281. In E. Bornemisza and A. Alvarado (eds.), *Soil Management in Tropical America*. North Carolina State University, Raleigh.

Amaral, A. Z., F. C. Verdade, N. C. Schmidt, A. C. P. Wutke, and K. Igue. 1965. Parcelamento e intervalo de aplicação de calcário. *Bragantia* **24**:83–96.

Andrew, C. S. and M. P. Hegarty. 1969. Comparative responses to manganese excess of eight tropical and four temperate pasture legume species. *Aust. J. Agr. Res.* **20**:687–696.

Andrew, C. S. and P. J. Vanden Berg. 1973. The influence of aluminum on phosphate sorption by whole plants and exised roots of some pasture legumes. *Aust. J. Agr. Res.* **24**:341–351.

Andrew, C. S., A. D. Johnson, and R. L. Sandland. 1973. Effect of aluminum on the growth and chemical composition of some tropical pasture legumes. *Aust. J. Agr. Res.* **24**:325–339.

Ayres, A. S., H. H. Hagihara, and G. Stanford, 1965. Significance of extractable aluminum in Hawaiian sugar cane soils. *Soil Sci. Soc. Amer. Proc.* **29**:387–392.

Black, C. A. 1967. *Soil-Plant Relationships*, 2nd ed. Wiley, N.Y.

Brenes, E. and R. W. Pearson. 1973. Root responses of three gramineae species to soil acidity in an Oxisol and an Ultisol. *Soil Sci.* **116**:295–302.

Cate, R. B. 1965. *Sugestões para Adubação na Base de Análise de Solo: Primera Aproximação.* Duarte Coelho, Recife, Brasil.

Cate, R. B., Jr., and A. P. Sukhai. 1964. A study of aluminum in rice soils. *Soil Sci.* **98**:85–93.

Cervantes, O., L. A. León, and G. Marín. 1970. Relaciones entre pH, aluminio y materia orgánica en algunos suelos de Colombia. *Rev. Inst. Colomb. Agropec.* **5**:43–64.

Chakraborty, M., B. Chakravarti, and S. K. Mukherjee. 1961. Liming in crop production in India. *Indian Soc. Soil Sci. Bull.* 7.

Coleman, N. T. and G. W. Thomas. 1967. The basic chemistry of soil acidity. *Agron. Monogr.* 12:1–41.

De Freitas, L. M. M. and P. F. Pratt. 1969. Response of three legumes to lime in various acid soils in São Paulo. *Pesq. Agropec. Bras.* 4:89–95.

De Freitas, L. M. M., E. Lobato, and W. V. Soares. 1971. Experimentos de calagem em solos sob vegetação de cerrado de Distrito Federal. *Pesq. Agropec. Bras.* 6:81–90.

De Freitas, L. M. M. and B. Van Raij. 1975. Residual effects of liming a sandy clay loam Latosol. Pp. 300–307. In E. Bornemisza and A. Alvarado (eds.), *Soil Management in Tropical America.* North Carolina State University, Raleigh.

Ekpete, D. M. 1972. Assessment of lime requirements of Eastern Nigeria soils. *Soil Sci.* 113:363–372.

Evans, C. E. and E. J. Kamprath. 1970. Lime response as related to percent aluminum saturation, soil solution aluminum and organic matter content. *Soil Sci. Soc. Amer. Proc.* 34:893–896.

Fassbender, H. W. and R. Molina. 1969. The influence of silicates and liming on phosphate fertilization on volcanic ash soil in Costa Rica. Pp. C2.1–C2.12. In *Panel on Soils Derived from Volcanic Ash in Latin America.* Inter-American Institute of Agricultural Sciences, Turrialba, Costa Rica.

Fox, R. L., S. K. DeDatta, and G. D. Sherman. 1962. Phosphorus solubility and availability to plants and aluminum status of Hawaiian soils as influenced by liming. Pp. 574–583. In *Trans. Comm. IV and V, Int. Soc. Soil Sci. (New Zealand).*

Fox, R. L., S. K. DeDatta, and J. M. Wang. 1964. Phosphorus and aluminum uptake by latosols in relation to liming. *Trans. 8th Int. Congr. Soil Sci. (Bucharest)* 4:595–603.

Foy, C. D. 1974. Effects of aluminum on plant growth. Pp. 601–642. In E. W. Carson (Ed.), *The Plant Root and Its Environment.* University Press of Virginia, Charlottesville.

Foy, C. D. and J. C. Brown. 1964. Toxic factors in acid soils. II. Differential aluminum tolerance of plant species. *Soil Sci. Soc. Amer. Proc.* 28:27–32.

Ghani, M. O., K. A. Hassan, and M. F. A. Kahn. 1955. Effects of liming on aggregation, non-capillary pore space and permeability of laterite soils. *Soil Sci.* 80:469–478.

Gorsline, G. W., W. I. Thomas, and D. E. Baker. 1968. Major gene inheritance of Sr-Ca, Mg, K, P, Zn, Cu, B, Al-Fe, and Mn concentrations in corn. *Pennsylvania Agr. Exp. Sta. Bull.* 746.

Guerrero, R. 1971. Soils of the Colombian Llanos Orientales—composition and classification of selected soil properties. Ph.D. Thesis, North Carolina State University, Raleigh.

Hardy, F. 1926. The role of aluminum in soil in fertility and toxicity. *J. Agr. Sci.* 16:616–631.

Heylar, K. R. and A. J. Anderson. 1971. Effects of lime on the growth of five species, on aluminum toxicity, and on phosphorus availability. *Aust. J. Agr. Res.* 22:707–721.

Igue, K. and R. Fuentes. 1972. Characterization of aluminum in volcanic ash soils. *Soil Sci. Soc. Amer. Proc.* 36:292–296.

Jackson, W. A. 1967. Physiological effects of soil acidity. *Agron. Monogr.* 12:43–124.

Juo, A. S. R. 1972. The problems of soil acidity and crop growth in acid tropical soils. Paper presented at a Tropical Soils Research Seminar, International Institute for Tropical Agriculture, Ibadan, Nigeria.

Kamprath, E. J. 1967. Soil acidity and response to liming. *Int. Soil Testing Program Tech. Bull.* 4. North Carolina State University, Raleigh.

Kamprath, E. J. 1970. Exchangeable aluminum as a criterion for liming leached mineral soils. *Soil Sci. Soc. Amer. Proc.* **34**:252–254.

Kamprath, E. J. 1971. Potential detrimental effects from liming highly weathered soils to neutrality. *Proc. Soil Crop Sci. Soc. Fla.* **31**:200–203.

Kamprath, E. J. 1972. Soil acidity and liming. Pp. 136–149. *Soils of the Humid Tropics.* National Academy of Sciences, Washington.

Kamprath, E. J. 1973. Soil acidity and liming. Pp. 126–128. In P. A. Sanchez (ed.), "A Review of Soils Research in Tropical Latin America." *North Carolina Agr. Exp. Sta. Tech. Bull.* 219.

León, L. A. 1967. Chemistry of some acid tropical soils of Colombia. Ph.D. Thesis, University of California at Riverside.

León, L. A. 1970. Teorías modernas sobre la naturaleza de la acidez del suelo. *Suelos Ecuatoriales* **3**:1–23.

Lin, C. and N. T. Coleman. 1960. The measurement of exchangeable aluminum in soil and clays. *Soil Sci. Soc. Amer. Proc.* **23**:12–18.

Lugo-Lopez, M. A., E. Hernandez, and G. Acevedo. 1959. Response of some tropical soils and crops of Puerto Rico to applications of lime. *Puerto Rico Agr. Exp. Sta. Tech. Paper* 28.

Mahilum, B. C., R. L. Fox, and J. A. Silva. 1970. Residual effects of liming volcanic ash soils in the tropics. *Soil Sci.* **109**:102–109.

McLean, E. O. 1971. Potentially beneficial effects of liming: chemical and physical. *Proc. Soil Crop Sci. Soc. Fla.* **31**:189–199.

Mendez-Lay, J. 1973. Effects of lime on phosphorus fixation and plant growth in various soils of Panama. M.S. Thesis, Soil Science Department, North Carolina State University, Raleigh. 90 pp.

Mikklesen, D. S., L. M. M. De Freitas, and A. C. McClung. 1963. Effects of liming and fertilizing cotton, corn and soybeans on Campo Cerrado soils, State of São Paulo, Brazil. *IRI Res. Inst. Bull.* 29.

Monteith, N. H. and G. D. Sherman. 1962. A comparison of the use of liming materials on a hydrol humic latosol and a humic ferruginous latosol in Hawaii. Pp. 584–587. In *Trans. Comm. IV and V, Int. Soc. Soil Sci.* (*New Zealand*).

Morelli, M., K. Igue, and R. Fuentes. 1971. Effect of liming on the exchange complex and on the movement of Ca and Mg. *Turrialba* **21**:317–322.

Moura, W. and S. W. Buol. 1972. Studies on a Latosol Roxo (Eutrustox) in Brazil. *Experientiae* **13**:201–212.

Navas, J. A. and F. Silva (eds.). 1971. Acidez y Encalamiento en el Trópico. Primer Coloquio de Suelos. *Suelos Ecuatoriales* **3**:1–309.

Norris, D. O. 1959. The role of calcium and magnesium in the nutrition of *Rhizobium. Aust. J. Agr. Res.* **10**:651–698.

North Carolina State University. 1973, 1974. *Agronomic–Economic Research on Tropical Soils.* Annual Reports, 1973 and 1974. Soil Science Department, North Carolina State University, Raleigh.

Nye, P., D. Craig, N. T. Coleman, and J. L. Ragland. 1961. Ion exchange equilibrium involving aluminum. *Soil Sci. Soc. Amer. Proc.* **25**:14–17.

Pearson, R. W. 1975. Soil acidity and liming in the humid tropics. *Cornell Int. Agr. Bull.* 30.

Pearson, R. W., F. Abruña, and J. Vicente-Chandler. 1962. Effects of lime and nitrogen applications on the downward movement of Ca and Mg in two humid tropical soils of Puerto Rico. *Soil Sci.* **93**:77–82.

Peele, T. C. 1936. The effect of calcium on the erodibility of soils. *Soil Sci. Soc. Amer. Proc.* **1**:47–58.

Pratt, P. F. and R. Alvahydo. 1966. Cation exchange characteristics of soils from São Paulo, Brazil. *IRI Res. Inst. Bull.* 31.

Rana, S. K. and G. D. Sherman. 1971. Effect of liming and air drying on plant nutrition in two Hawaiian latosols. *J. Indian Soc. Soil Sci.* **19**:203–208.

Reeve, N. G. and M. E. Sumner. 1970. Lime requirements of Natal Oxisols based on exchangeable aluminum. *Soil Sci. Soc. Amer. Proc.* **34**:595–598.

Reeve, N. G. and M. E. Sumner. 1972. Amelioration of subsoil acidity in Natal Oxisols by leaching of surface applied amendments. *Agrochemophysica* **4**:1–5.

Reid, D. A. 1971. Genetic control of reaction to aluminum in winter barley. Pp. 409–413. In R. A. Nilan (ed.), *Barley Genetics*, Vol. II. Washington State University Press.

Richardson, H. L. 1951. Soil acidity and liming with tropical crops. *World Crops* **3**:339–340.

Rixon, A. J. and G. D. Sherman. 1962. Effects of heavy lime applications to volcanic ash soils in the humid tropics. *Soil Sci.* **94**:119–127.

Salinas, J. G. and P. A. Sanchez. 1976. Soil-plant relationships affecting varietal and species differences in tolerance to low available soil phosphorus. *Ciencia e Cultura (Brazil)* **28**:156–168.

Samuels, G. 1962. The pH of Puerto Rican soils used for principal crops. *J. Agr. Univ. Puerto Rico.* **46**:107–119.

Sanchez, P. A. and S. W. Buol. 1974. Properties of some soils of the upper Amazon Basin of Peru. *Soil Sci. Soc. Amer. Proc.* **38**:117–121.

Schuffelen, A. C. and H. A. Middleburg. 1954. Structural deterioration of lateritic soils through liming. *Trans. Fifth Int. Congr. Soil Sci.* **2**:158–165.

Soares, W. V., E. Lobato, E. Gonzalez, and G. C. Naderman, Jr. 1975. Pp. 283–299. Liming soils of the Brazilian Cerrado. In E. Bornemisza and A. Alvarado (eds.), *Soil Management in Tropical America*. North Carolina State University, Raleigh.

Spain, J. M., C. A. Francis, R. H. Howeler, and F. Calvo. 1975. Differential species and varietal tolerance to soil acidity in tropical crops and pastures. Pp. 308–324. In E. Bornemisza and A. Alvarado (eds.), *Soil Management in Tropical America*. North Carolina State University, Raleigh.

Sumner, M. E. 1970. Aluminum toxicity—a growth-limiting factor in some Natal sands. *Proc. Ann. Congr. South Afr. Sugar Technol. Assoc.* **44**:176–182.

Van Raij, B., M. Sacchetto, and Th. Dovichi. 1968. Correlações entre o pH e o grão de saturação em bases nes solos com horizonte B textural e horizonte B latossólico. *Bragantia* **27**:193–200.

Vicente-Chandler, J., R. Caro-Costas, R. W. Pearson, F. Abruña, J. Figarella, and S. Silva. 1964. The intensive management of tropical forages in Puerto Rico. *Univ. Puerto Rico Agr. Exp. Sta. Bull.* 187.

Villagarcía, S. 1973. Aluminum tolerance in the Irish potato and the influence of substrate aluminum on growth and mineral nutrition of potatoes. Ph.D. Thesis, North Carolina State University, Raleigh. 200 pp.

Younge, O. R. and D. L. Plucknett. 1964. Liming materials. *Hawaii Farm Sci.* **13**(3):7–8.

8

PHOSPHORUS, SILICON, AND SULFUR

After water and nitrogen, the two most commonly limiting nutrient elements in the tropics are probably phosphorus and sulfur. Phosphorus deficiencies are very common in highly weathered Oxisols and Ultisols and in slightly weathered Andepts and Vertisols. Many tropical soils have extremely high capacities to immobilize phosphorus. Considerable modification of phosphorus management practices developed for soils with moderate fixation capacity is required for such soils. In addition, significant differences exist between and within species in the ability to tolerate low levels of available soil phosphorus.

Silicate applications often have a beneficial effect in decreasing phosphorus fixation. Although silicon is generally not considered an essential nutrient for plant growth, nutritional responses to silicon applications have been found in highly weathered tropical soils.

Rapidly accumulating evidence indicates that sulfur deficiencies may be more widespread in the tropics than potassium deficiencies. Sulfur deficiencies have been unconsciously corrected by the use of sulfur-bearing fertilizers such as ammonium sulfate and ordinary superphosphate. With the use of high-analysis sources that contain no sulfur, such as urea and triple superphosphate, reports of severe sulfur deficiencies have become widespread.

The purpose of this chapter is to summarize the present knowledge of the management of these nutrients in tropical conditions.

PHOSPHORUS CONTENTS AND FORMS

Total Phosphorus

The total phosphorus content of the soil is of no direct practical importance, but it has often been used as a weathering index. Total phosphorus in the topsoil decreases with increasing weathering intensity. Olson and Englestad (1972) stated that representative soils from the United States Midwest average about 3000 ppm P in the topsoil, the more weathered soils of the United States Southeast about 500 ppm, and "tropical soils" about 200 ppm. Because of the greater areal extent of highly weathered soils in the tropics, this is generally true, but it ignores the sharp differences among tropical soils. In Venezuela, for example, total phosphorus contents are correlated with increasing weathering, as measured by the cation exchange capacity of the clays. These relationships are shown in Table 8.1.

Although Oxisols are generally low in total phosphorus (less than 200 ppm), some are extremely high. Moura et al. (1972) reported 3760 ppm total P in a Eutrustox from Brazil. Ultisols and Alfisols are also generally low in total phosphorus, with values mostly below 200 ppm (Nye and Bertheux, 1957; Westin and de Brito, 1969; Cabala and Fassbender, 1970). Andepts are generally high in total phosphorus, with ranges between 1000

Table 8.1 Distributions of Topsoil Phosphorus Fractions in Some Venezuelan Soils as Related to Their Degrees of Weathering (ppm)

Series	Order	CEC of Clay (meq/100 g)	pH	Total P	Organic P	Ca-P	Al-P	Fe-P
Chispa	Mollisol	100	6.9	692	235	70	33	43
Maracay	Entisol	127	5.9	298	79	88	20	33
Paya	Alfisol	50	5.0	144	85	3	14	19
Guataparo	Oxisol	18	4.8	59	11	0	2	17

Source: Adapted from Westin and de Brito (1969).

and 3000 ppm, according to Fassbender (1969). Many Vertisols from India and Central America are extremely low in total phosphorus, ranging from 20 to 90 ppm (Mehta and Patel, 1963; Danke et al. 1964). The total phosphorus contents of Entisols and other Inceptisols vary with parent materials.

Organic Phosphorus

Organic phosphorus normally accounts for 20 to 50 percent of the total topsoil phosphorus. In the more highly weathered Oxisols, Ultisols, and Alfisols it often represents 60 to 80 percent of the total soil phosphorus. African workers consider organic phosphorus to be the main source of phosphorus for plants in no-fertilizer agriculture. Its maintenance, therefore, is of great practical significance in traditional agricultural systems. For example, Smith and Acquaye (1963) observed a high correlation between cacao yield and soil organic phosphorus content. Friend and Birch (1960) obtained a negative correlation between responses to applied phosphorus and organic phosphorus contents in East Africa. Omatoso (1971) found sharp yield increases with phosphorus applications only when the soil organic phosphorus content was less than 150 ppm in cacao plantations of Ghana.

The C:P ratios of organic matter in representative soils of Ghana are on the order of 240:1, according to Nye and Bertheux (1957). This value represents a much lower phosphorus content than in the United States, where the average C:P ratio is on the order of 110:1. The N:P ratios in Ghana are on the order of 20:1; in the United States, about 9:1. These wide C:N and N:P ratios are indicative of phosphorus deficiency. The fact that Nye and Bertheux found no differences in C:P ratios with cultivation suggests that phosphorus is mineralized at about the same rate as organic carbon. The N:P ratios decrease with cultivation, implying that organic phosphorus is mineralized at a faster rate than organic nitrogen. This view is confirmed by evidence that recently cleared savanna soils are deficient in nitrogen but not in phosphorus during the first year in West African Alfisols (Nye and Greenland, 1960).

Organic phosphorus mineralization is difficult to quantify because the released $H_2PO_4^-$ ions may be quickly fixed into inorganic forms. Laboratory experiments in Ghana suggest that organic phosphorus may release from 2 to 27 ppm P to the soil, according to Acquaye (1963). He also found that nitrogen and phosphorus applications increased organic phosphorus mineralization. Awan (1964), working in Honduras, also found that increased yields of corn, sorghum, and cowpeas due to liming were associated with increased mineralization of soil organic phosphorus. These effects can probably be attributed to the establishment, with fertilization and liming, of a more favorable environment for the microorganisms responsible for mineralization.

The importance of maintaining organic matter is also a function of the maintenance of organic phosphorus, particularly in soils where most of the phosphorus is in this form. Where fertilizers are available and economical, organic phosphorus deserves less consideration.

Inorganic Phosphorus Fractions

The solid inorganic forms of phosphorus are usually divided into three active fractions and two relatively inactive fractions. The active fractions can be grouped into calcium-bonded phosphates (Ca-P), aluminum-bonded phosphates (Al-P), and iron-bonded phosphates (Fe-P). Calcium phosphates are present as films or as discrete particles, while Al-P and Fe-P occur as films or are simply adsorbed on clay or silt surfaces. The relatively inactive fractions are the occluded and reductant-soluble forms. Occluded phosphorus consists of Fe-P and Al-P compounds surrounded by an inert coat of another material that prevents the reaction of these phosphates with the soil solution. Reductant-soluble forms are covered by a coat that may be partially or totally dissolved under anaerobic conditions.

The forms of inorganic phosphorus present in a soil depend on its chemical stage of weathering. This is clearly shown in Table 8.1, where the proportion of calcium phosphates decreases with weathering and the proportion of iron phosphates increases. Calcium phosphates are more soluble than aluminum phosphates, which are in turn more soluble than iron phosphates.

The transformation of one form of phosphate into another is controlled mainly by soil pH. As soils become more acid, the activity of iron and aluminum increases and the relatively soluble calcium phosphates are converted into less soluble aluminum and iron phosphates (Kamprath, 1973). These processes are slow enough to permit considerable quantities of calcium phosphates to be present in acid soils with pH values below 5.5.

In highly weathered soils most of the inorganic phosphorus is in the occluded or reductant-soluble form because of the formation of iron and aluminum oxide coatings. Table 8.2 shows that reductant-soluble iron phosphates are the dominant inorganic form of phosphorus in an Oxisol from the Llanos Orientales of Colombia. Most of the total phosphorus found in Moura's Brazilian Eutrustox is also in the reductant-soluble form. In Ultisols both iron and aluminum phosphates predominate. In Andepts calcium phosphates are most common at pH values higher than 6, and aluminum

Table 8.2 Soil Phosphorus Fractions in the Profile of an Oxisol of Carimagua, Llanos Orientales, Colombia

Horizon (cm)	(pH)	Organic C (%)	Base Saturation (%)	Total P (ppm)	Percent of total P					
					Organic P	Ca-P	Al-P	Fe-P	Reductant-Sol Fe-P	Occluded Al-P
0–6	4.5	2.26	7	185	77	0.9	0.8	10	9	1
6–15	4.6	1.84	7	151	75	0.6	0.9	11	11	1
15–40	4.6	1.13	13	126	73	0.7	1.2	6	17	1
40–70	4.9	0.53	15	114	55	0.8	1.3	7	34	1
70–100	5.1	0.43	29	90	47	0.6	1.0	9	41	1
100–150	5.1	0.24	21	84	35	0.7	1.2	4	53	4

Source: Benavides (1963).

phosphates at lower pH values. Moisture regimes also play a significant part in phosphorus transformations. Aluminum phosphates tend to accumulate in aquic soil moisture regimes, whereas iron phosphates accumulate in ustic soil moisture regimes.

PHOSPHORUS FIXATION* AND RELEASE

Phosphorus Fixation Processes

"Phosphorus fixation" is another poorly defined term in soil science. This writer prefers to consider fixation as the process that alters the availability of phosphate compounds as they are measured by plant growth. The term "fixation" has an erroneous connotation of irreversibility. The process consists of transforming soluble monocalcium phosphates (superphosphates) into less soluble calcium, aluminum, or iron phosphates. These compounds release phosphorus into the soil solution over a period of years.

When a superphosphate granule is placed in the soil, water initially moves into the granule and dissolves some of the monocalcium phosphate into dicalcium phosphate and free phosphoric acid. The solution coming out of the granule has a pH of 1 to 1.5. It dissolves aluminum, iron, potassium, and magnesium compounds in soil particles. In acid soils aluminum and iron are most abundant and react with phosphorus to form relatively insoluble aluminum and iron phosphates. In calcareous soils the phosphate ions are precipitated by calcium and magnesium as relatively insoluble compounds.

In acid soils there is an additional fixation mechanism. Exchangeable aluminum reacts with monocalcium phosphate and forms compounds having the general formula $Al(OH)_2H_2PO_4$. They resemble the crystalline form, variscite, but are more soluble (Coleman et al., 1960). An indirect effect of this mechanism is the precipitation of exchangeable aluminum with phosphorus. According to the calculations of Coleman and his coworkers, 1 meq of exchangeable aluminum can fix about 70 ppm P because aluminum is precipitated in this fashion. This is often referred to as "liming with phosphorus."

The higher the content of iron and aluminum oxides, the larger is the phosphorus-fixing capacity of the soil. Also, the higher the exchangeable aluminum content, the larger the phosphorus-fixing capacity will be. Therefore highly acid and weathered Oxisols and Ultisols generally have

---

* The terms "sorption" and "adsorption" are also used to express the same concept. For clarity, "fixation" will be used as defined in this section.

high phosphorus fixation capacities, whereas less acid soils with layer silicate mineralogy have much lower ones. Because of the high contents of aluminum oxides, Andepts also have a very high fixing capacity.

In oxide or oxide-coated layer silicate systems, phosphorus fixation increases with increasing clay content. This is an indirect effect of the iron and aluminum oxide contents found in the clay fraction. Woodruff and Kamprath (1965) found that sandy Ultisols retained much less phosphorus than clayey Ultisols of similar mineralogy. In addition to the iron and aluminum oxide contents, the degree of crystallinity affects the magnitude of phosphorus fixation. Pratt et al. (1969) found that Brazilian Ultisols fixed more phosphorus per unit of iron oxide content than did Brazilian Oxisols. They associated this effect with the less crystalline iron oxide forms found in these Ultisols, in contrast to the more crystalline forms in the Oxisols. Because of their larger surface area, amorphous oxides retain more phosphorus than crystalline species. However, since the iron oxide content is generally higher in Oxisols, as an order they probably fix more phosphorus than Ultisols. Among the amorphous materials Syers et al. (1971) found that amorphous aluminum compounds were more active fixers than amorphous iron oxides in Ultisols and Oxisols of southern Brazil.

The mineralogy of the soil defines the product of phosphorus fixation. In soils with oxidic mineralogy, most of the phosphorus will be fixed as iron phosphates. In kaolinitic families aluminum phosphates will be the predominant form, but with age they are transformed into iron phosphates. Shelton and Coleman (1968) in a long-term experiment on an Ultisol found that the added phosphorus was converted into equal proportions of iron and aluminum phosphates during the first year. Within 3 years two-thirds of the phosphorus was detected as iron phosphate, and during the sixth year the proportion rose to 75 percent. Aluminum phosphates, being more soluble, are gradually transformed into iron phosphates. In allophanic families fixation is primarily in the form of aluminum phosphates because of the high content of aluminum oxides and the low content of iron oxides in acid Andepts.

The intensity of fixation according to mineralogy is as follows:

Amorphous oxides > Crystalline oxides > 1 : 1 clays > 2 : 1 clays

(including   (gibbsite, goethite,
allophane)   etc.)

Magnitude of Phosphorus Fixation

Some idea of the maximum amounts of phosphorus that tropical soils can fix is given in Tables 8.3 and 8.4. As a group, Andepts and other soils high

Table 8.3   Amounts of Phosphorus Fixation and Amounts of Added Phosphorus Required to Give Soil Solution Concentrations of 0.1 and 0.2 ppm P in Soils from Mexico, Central America, and Brazil

			Fixed P (ppm)		
Soil	Dominant Clay Mineral	Clay (%)	Adsorption Maxima	At 0.1 ppm P Soil Solution	At 0.2 ppm P Soil Solution
Inceptisol	Montmorillonite	27	106	65	83
Ultisol	Kaolinite	38	480	285	360
Oxisol	Kaolinite	36	531	310	395
Oxisol	Kaolinite	78	—	720	900
Andept	Allophane	11	1050	500	670

Source: Adapted from Rivera-House (1971) and North Carolina State University (1973).

in allophane are the highest fixers with over 1000 ppm added P. Kaolinitic soils, including Oxisols and Ultisols, follow with values ranging from 500 and 1000 ppm P except for coarse-textured ones. Soils high in montmorillonite and on the calcareous side may fix on the order of 100 ppm added P.

Phosphorus Release

Phosphate ions in the soil solution are what the plants actually use. The concentration of $H_2PO_4^-$ ions is on the order of a fraction of a part per

Table 8.4   Phosphorus Fixation and Amount Needed to Maintain 0.2 ppm P in the Soil Solution on a "Weathering Transect" in Oahu, Hawaii

				Fixed P (ppm)	
Soil	Annual Rainfall (mm)	Clay Minerals	pH	Adsorption Maxima	At 0.2 ppm P Soil Solution
Chromustert	500	Montmorillonite	7.6	300	85
Haplustox	600	Kaolinite	7.8	500	145
Haplustox	950	Kaolinite	6.8	525	145
Eutrorthox	1200	Kaolinite	5.1	725	235
Tropohumult	2200	Kaolinite, Gibbsite	4.5	670	335
Tropohumult	2300	Gibbsite, Goethite	6.6	1320	435

Source: Fox et al. (1971).

million. Plants accumulate in their tissue about 2000 ppm P. Therefore the soil solution must be renewed several times a day for the plants to create such a concentration gradient. Fixed phosphorus is released slowly to the soil solution as a function of the solubility of the species. Also, organic phosphorus mineralization adds $H_2PO_4^-$ ions to the soil solution, but most of them are fixed immediately.

It has been found that there is an optimum concentration of phosphorus in the soil solution associated with maximum crop growth in some species. Baldovinos and Thomas (1967) in Virginia and Fox et al. (1971, 1974) in Hawaii have shown that maximum bean and corn growth is associated with a level of 0.07 ppm P in the soil solution of clayey Ultisols, Oxisols, and Andepts. In sandy soils the optimum concentration is about 0.2 ppm P in the soil solution. This difference is probably due to the slower diffusion rate of soil solution in sandy soils because of discontinuous water films around the particles. Even at 15 bars, water films around clays can be continuous.

The relationship between the amount of inorganic phosphorus added to the soil and the equilibrium concentration of phosphorus in the soil solution is a good parameter for determining how much fertilizer phosphorus should be added to arrive at a desired level of soil solution phosphorus. These relationships, obtained in the laboratory by adding various amounts of phosphorus to the soil, shaking for 6 days, and determining how much remains in solution, are known as "phosphorus fixation isotherms" (Fox and Kamprath, 1970). Figure 8.1 shows examples of such relationships in several tropical soils. Phosphorus fixation isotherms evaluate the degree of fixation and release at one time. They are essentially intensity-capacity functions.

Figure 8.1 shows that a montmorillonitic soil fixes little added phosphorus and requires very little phosphorus in the soil to supply 0.1 ppm in solution. Clayey Oxisols and Ultisols need close to 300 ppm added P to supply the same amount. The Andept, because of its high contents of amorphous aluminosilicates, needs about 450 ppm added P to reach that level. A very clayey Oxisol from Brazil requires over 700 ppm P to reach 0.1 ppm in the soil solution. Relatively little additional fertilizer phosphorus is required to raise the soil solution level to 0.2 ppm.

Fox and his coworkers (1971) have shown similar relationships with other tropical soils. They have also demonstrated a relationship between clay mineralogy and phosphate fixation (Table 8.4). The highest fixers are oxidic and allophanic families, followed by soils with mixtures of kaolinite and oxides; next come kaolinitic and montmorillonitic families.

Phosphorus soil tests are empirical determinations which attempt to extract an amount that will correlate with plant response to phosphorus fertilization. When used and interpreted properly, soil tests can separate soils that will probably respond to phosphorus from those that are not likely

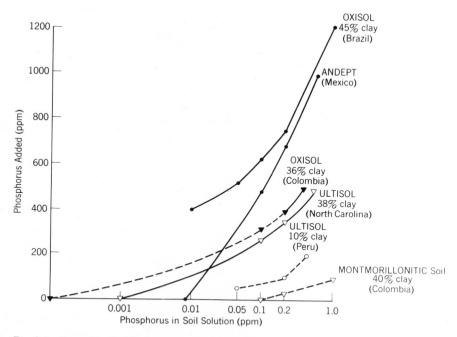

Fig. 8.1 Examples of phosphorus fixation isotherms in tropical soils. *Sources:* Rivera-House (1971), North Carolina State University (1973), and Fox (1974).

to do so. When the fixing capacity of the soils is also known, the meaning of soil tests assumes a sharper focus. Examples of such calibrations are presented in Table 8.5 for two commonly used soil tests. Soil test critical levels can then be defined as those that predict 0.07 to 0.2 ppm P in the soil solution. Critical levels are quite variable with major soil types.

PHOSPHORUS REQUIREMENTS OF TROPICAL CROPS

Uptake

At present low yield levels of about 1 ton/ha, the phosphorus uptake of cereals and grain legumes is less than 10 kg P/ha (see Table 6.7). At high yield levels of 4 to 8 tons/ha, corn, rice, and wheat remove from 14 to 35 kg P/ha.

Root crops are high extractors of phosphorus. At average yield levels of about 8 tons/ha, the phosphorus contents of cassava and potato tubers are

Table 8.5   Relationship between North Carolina and Olsen Extraction Values Needed to Predict Equilibrium Soil Solution Phosphorus Values in Soils of Mexico, Colombia, and North Carolina

Dominant Clay Mineral and Soil	N.C. Extraction (ppm P)			Olsen Extraction (ppm P)		
	0.07	0.1	0.2	0.07	0.1	0.2
Allophane						
Olotepec	14	20	25	—	—	—
Kaolinite						
Acayucan	9	11	20	4	5	9
Carimagua	60	72	100	21	26	34
Georgeville	57	70	87	17	21	28
Montmorillonite or 14 Å						
Escorcega	—	—	—	4	4	6
Cardenas	—	—	—	15	15	19

Source: Rivera-House (1971).

on the order of 35 to 40 kg P/ha. When the tops are considered, and at higher yield levels, the phosphorus uptake is of course much higher. Possible exceptions are sweet potatoes and yams, which have low phosphorus requirements.

Intensively managed pasture grasses and sugarcane may remove from 20 to 70 kg P/ha a year, depending on yeild levels. Some pasture legumes respond dramatically to phosphorus fertilization but do not require as much phosphorus as the grasses because of their generally lower dry matter production. Bananas are relatively low phosphorus extractors; the bunches may remove annually less than 10 kg P/ha. Tree crops remove little phosphorus in the harvested parts. Coffee, rubber, and cacao usually remove less than 2 kg P/ha in the harvested portions. Nutrient cycling in these crops probably provides for a large part of the tree's phosphorus requirements. Tropical rainforests, for example, add annually an average of 15 kg P/ha to the soil in the form of biomass additions.

Varietal and Species Differences

Significant species differences in tolerating low available phosphorus exist. At similar yield levels, upland rice usually requires less phosphorus than corn. The general recommendation for these crops in Latin America ranges from 100 to 150 kg P_2O_5/ha for corn from 0 to 60 for upland rice (Kamprath, 1973). An example of this relationship is illustrated in Fig. 8.2. Corn and upland rice were grown in identical adjacent experiments on an acid Oxisol in Colombia. In the limed plots corn showed a marked response to 50 kg P_2O_5/ha with a yield increase from 0.8 to 3.2 tons/ha. In the limed plots rice did not respons to phosphorus and attained a yield of 3.2 tons/ha without added phosphorus. The reasons for such differences are not completely understood. Recent evidence establishes differences in the external (ppm P in the soil solution) and internal (percent P in the plant tissue) phosphorus requirements associated with maximum yields.

The only form of phosphorus available to plants is the phosphate ions in the soil solution. A report by Fox et al. (1974) demonstrated that there is an optimum concentration of phosphorus in the soil solution correlated with good growth and that it varies with species. The results illustrated in Fig. 8.3 show that sweet potatoes (*Ipomoea batatas*) are more tolerant of low levels of soil solution phosphorus than is lettuce (*Lactuca sativa*), while corn (*Zea mays*) and Chinese cabbage (*Brassica pekinensis*) occupy intermediate positions.

If the phosphorus concentration in the soil solution that produces 95 percent of the maximum yield is considered to be the external phosphorus requirement, there are tremendous species differences. Table 8.6 indicates a tenfold difference in the external requirement between two similar vegetables, lettuce and cabbage. This table also indicates that the external require-

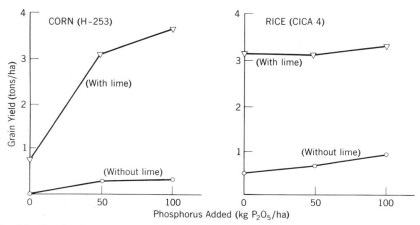

Fig. 8.2 Species differences in phosphorus response between corn and upland rice grown in adjacent plots in an Oxisol from the Llanos Orientales of Colombia. *Source:* Adapted from CIAT (1971).

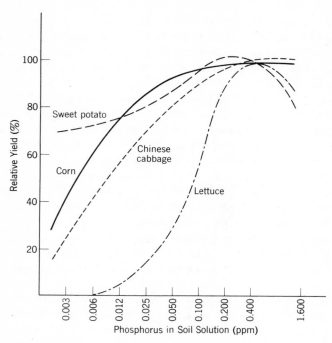

Fig. 8.3 Differential phosphorus response between sweet potato, corn, Chinese cabbage, and lettuce grown in a Eutrustox of Hawaii. *Source:* Fox et al. (1974).

Table 8.6 *External Critical Phosphorus Levels of Several Tropical Crops*

Crop	Soil Solution P Associated with 95% Maximum Yield (ppm)
Lettuce	0.40
Tomato	0.25
Cucumber	0.20
Soybean (vegetable)	0.20
Desmodium aparines	
Establishment	0.20
Second cut	0.01
Sweet potato	0.10
75% maximum yield	0.003
Corn	0.60
Sorghum	0.50
Cabbage	0.04

Source: Fox et al. (1974).

ment for a forage legume of the genus *Desmodium* is high during the establishment period (0.20 ppm P) but decreases to 0.01 ppm P after the second cutting. This information reflects the need for applying phosphorus to tropical legumes during the establishment period in soils low in this element.

The low external phosphorus requirement of sweet potatoes is of considerable interest. The fact that 75 percent maximum yield has been obtained with levels as low as 0.003 ppm P has great economic implications. Unfortunately, there are no similar data for other important crops such as rice, cassava, wheat, cowpeas, yams, and also many grasses and forage legumes. Nor do there exist data similar to those on varietal differences within the same species.

The work of Andrew and Robins (1969, 1971) in Australia shows the existence of species differences in internal critical levels of phosphorus. They determined the critical concentrations of phosphorus in the tops of several tropical grass and legume species that were correlated with maximum yields. The percent phosphorus level in the plant tops associated with maximum growth was considered to be the internal critical level of phosphorus. Some of their results (Table 8.7) show that legume species such as *Stylosanthes humilis* and *Centrosema pubescens* have internal critical

Table 8.7   Plant Phosphorus Contents Associated with Maximum Growth of Several Pasture Species in Queensland, Australia

Species	Internal P Requirement (%P)
Legumes	
Stylosanthes humilis	0.17
Centrosema pubescens	0.16
Desmodium intortum	0.22
Glycine wightii	0.23
Medicago sativa	0.25
Grasses	
Digitaria decumbens	0.16
Melinis minutiflora	0.18
Panicum maximum	0.19
Pennisetum clandestinum	0.22
Chloris gayana	0.23
Paspalum dilatatum	0.25

Source: Andrew and Robins (1969, 1971).

levels lower than those of species such as *Glycine wightii* and *Medicago sativa*. The first two species are native of regions with soils low in available phosphorus, while the last two originated in regions with soils high in available phosphorus and other nutrients. The same situation occurs with the forage grasses. Grasses such as *Melinis minutiflora* and *Panicum maximum* are very common in acid soils with low phosphorus availability, whereas *Chloris gayana* and *Paspalum dilatatum* have higher internal critical levels.

The International Rice Research Institute in the Philippines is conducting a comprehensive program to select rice varieties tolerant of various soil problems, including low available soil phosphorus. Through preliminary greenhouse experiments (IRRI, 1971; Ponnamperuma and Castro, 1971), followed by field experiments at two rates of applied phosphorus (IRRI, 1972), the institute has classified a large group of varieties according to their tolerance of low soil phosphorus levels. Some examples are illustrated in Table 8.8. The classification by degree of tolerance is based on the relative response to phosphorus applications in the field.

An interesting point of this study is the possible selection of varieties not only for phosphorus tolerance but also for tolerance of other soil problems, such as iron deficiency or toxicity and the presence of toxic soil reduction products. The varieties with more ample spectra of resistance to adverse soil conditions are IR24, CAS209, and BG79. The varieties Pelita I/1 and

Table 8.8 Classification of Rice Varieties According to Low Phosphorus Tolerance in an Ultisol from Luisiana, Philippines

Tolerant	Moderately Tolerant	Moderately Sensitive	Sensitive
IR4-11	IR5	IR442-2-58	IR579-48-1
Bahagia	IR8	IR1008-14-1	IR747B-26-3
BG 79	IR20	IR1154-233-2	IR878B-4-220
CAS 209	IR22	Taichung (N) 1	Bala
Engkatok	IR24	TKM6	C 22
Pelita I/1	IR661-1-70		Colombia 1
Pelita I/2	IR1154-68-2		CP231xSL017
RD 1	CICA 4		ICA 10
SR 26 B	Peta		Ml-48
	T442-35		Ml-273
			OS 4
			SML Acorni

Source: IRRI (1972).

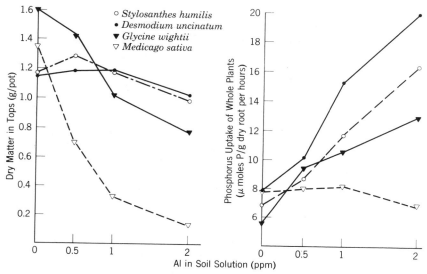

Fig. 8.4   Effects of aluminum on growth and phosphorus uptake of four pasture legumes. *Source:* Adapted from Andrew and Vanden Berg (1973).

Pelita I/2 are considered genetic sources for tolerance to low phosphorus levels. Although the genetics of varietal differences in low phosphorus tolerance are not as advanced as those of aluminum tolerance, it seems reasonable to assume that species or varieties developed in soils low in available phosphorus will show tolerance, whereas those developed in soils well supplied with phosphorus, either naturally or heavily fertilized, are not likely to show such tolerance.

The physiological mechanisms responsible for these varietal and species differences are not completely understood. In a recent review of this subject, Salinas and Sanchez (1976) observed that varieties and species more tolerant of low available soil phosphorus can either absorb phosphorus as a faster rate or translocate it to the tops at a faster rate than sensitive varieties or species. In some cases tolerant varieties have a slower relative growth rate that permits them to operate at a low level of external phosphorus supply.

In many acid soils it is difficult to separate the effects of high aluminum saturation from those of low phosphorus availability. Aluminum tolerance has been related to the plant's ability to absorb and translocate phosphorus in the presence of aluminum. Evidence of this was supplied by Andrew and Vanden Berg (1973) for the same group of pasture legume species discussed in Chapter 7. Figure 8.4 shows that increasing aluminum levels in solution decreased the top growth of the two aluminum-sensitive species, *Glycine wightii* and alfalfa, but did not have much effect on the two tolerant genera, *Stylosanthes* and *Desmodium*. Increasing the aluminum concentration increased the phosphorus uptake of three species, but not alfalfa. The phosphorus taken up, however, was not translocated to the tops at equal rates. The translocation rate of *Stylosanthes* and *Desmodium* was about twice as fast as that of *Glycine wightii* and alfalfa. Similar results have recently been obtained with rice, corn, and bean varieties (North Carolina State University, 1974).

Aluminum is known to precipitate phosphorus in root cell walls and cytoplasmic membranes as aluminum phosphates (Rasmussen, 1968; McCormick and Borden, 1972). Consequently, species or varieties that can translocate phosphorus to the top faster may escape this precipitation and be more tolerant of high aluminum levels.

Varietal and species differences in tolerance of low available soil phosphorus are of great practical significance, and application of such information may reduce phosphorus fertilization significantly in many cases. Unfortunately, little is known about the nature of such differences in most tropical crops. Research in this direction is a high-priority concern.

MANAGEMENT OF PHOSPHORUS FERTILIZERS

Rates

Phosphorus responses are common in Oxisols, Ultisols, Andepts, and Vertisols. Well-calibrated soil test procedures easily identify the soils that have a high probability of phosphorus response and those that do not (see the next chapter). In tropical America the generally recommended rates for corn, soybeans, forages, and sugarcane are on the order of 100 to 150 kg P_2O_5/ha, usually in banded applications. The recommended rates for potatoes and wheat are on the order of 120 to 240 kg P_2O_5/ha, but for upland rice only from 0 to 60 kg P_2O_5/ha is recommended (Kamprath, 1973). In tropical Africa, where the research has been conducted at lower application rates and lower yield levels, FAO simple fertilizer trials show good crop response to rates of 20 kg P_2O_5/ha (Richardson, 1968).

Phosphorus management in soils with moderate fixation capacity is usually a simple proposition. Small annual rates of superphosphate can be broadcast and incorporated into the topsoil or banded once a year.

In soils with high phosphorus fixation capacity, like certain clayey Ultisols and Oxisols and most Andepts, economically sound phosphorus management involves several approaches. In many cases no response to phosphorus applications will be noticed at low to moderate rates. In volcanic ash soils of Japan, for example, it was believed that paddy rice did not respond to phosphorus applications until an investigator added about 1000 kg P_2O_5/ha in one experiment and obtained a dramatic increase in yield.

Massive Initial Applications versus Moderate Banded Applications

In soils with high phosphorus fixation capacity two strategies of phosphorus fertilization are presently in use. One is to apply moderate rates placed in bands to every crop. The other is to apply phosphorus at a rate large enough to saturate the fixation capacity of the soil at once and to count on an adequate release for many years. The high initial cost is then considered an investment that can be amortized in several years with the residual effects.

An example of this approach is illustrated by the work of Younge and Plucknett (1966) in Hawaii. They applied rates of 330 to 1320 kg P/ha to a Gibbsihumox and measured the residual effect to a grass–legume pasture for several years. After 12 years these massive applications continued to maintain forage yields, as shown in Fig. 8.5. During the first 3 years yield responses were obtained only at the lowest rate, 300 kg P/ha. With time,

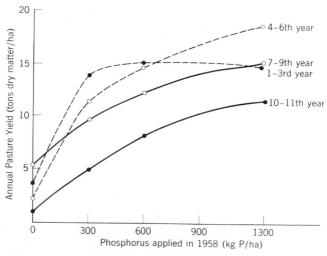

Fig. 8.5 Residual effects of massive phosphorus applications to a Hawaiian Oxisol on grass–legume pasture yields. *Source:* Adapted from Fox et al. (1971).

maximum yields were obtained at progressively higher rates, but during the first 9 years the highest yields did not vary.

The advantages of "quenching" the phosphorus fixation capacity of such soils is that the problem is eliminated right away. Also, these massive applications may increase soil pH and cation exchange capacity in oxide systems. Organic matter content also increases with sustained crop growth for many years. The disadvantages are the very high initial investment and need for adequate financing, a stable land tenure pattern, and high-value crops. Economic considerations probably limit this approach to highly developed areas such as Hawaii. The sharp increases in phosphorus fertilizer prices since 1972 have made many farmers wish they had adopted this procedure before that date. In a way, massive phosphorus applications may be a hedge against inflation.

In another high-fixing soil, an Ultisol from North Carolina, Kamprath (1967) studied the residual effect of massive initial applications versus small annual maintenance rates applied in bands. The corn yields obtained during the seventh year are shown in Table 8.9. The results show that small annual banded applications are superior to a high initial investment. An initial rate of 337 kg P/ha gave results similar to those obtained with an annual rate of 22 kg P/ha (as a total of 154 kg P/ha) in 7 years. This Ultisol, however, fixed less phosphorus than the Hawaiian Oxisol of Fig. 8.5. The Georgeville series of North Carolina needed 360 ppm P to provide 0.2 ppm to the soil

Table 8.9   Effects of Massive Initial Phosphorus Application versus Small Annual Maintenance Rates on Corn Yields in a North Carolina Ultisol with High Phosphorus Fixation Capacity 7 Years after the Initial Application[a]

| Initial Application (kg P/ha) | Yield (tons/ha) | |
| | Annual Maintenance Applications (banded) | |
	None	22 kg P/ha
0	1.7	5.7
168	3.9	6.2
337	5.5	6.8
673	7.0	7.1

Source: Kamprath (1967).
[a] $LSD_{0.05}$ = 1.3 tons/ha.

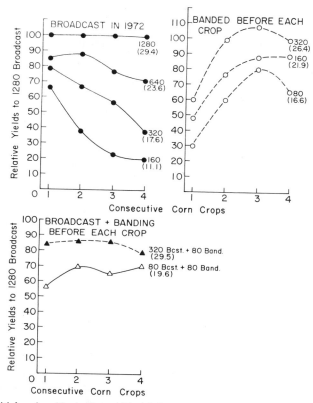

Fig. 8.6 Initial and residual effects of banded and/or broadcast superphosphate applications on corn yields in an Oxisol from Brasilia, Brazil. Figures in parentheses are the cumulative corn yields (tons/ha). *Source:* R. S. Yost (North Carolina State University, 1974).

solution, whereas the Kapaa series of Hawaii required 1000 ppm P to provide the same soil solution level.

The choice of strategy, therefore, depends on both soil properties and economic considerations. Unfortunately there are few additional data from tropical areas that measure the residual effect of phosphorus applications during a long enough period of time to supplement the Hawaii study.

Applying phosphorus fertilizers in bands is a simple practice that satisfies the phosphorus fixation capacity of a small soil volume and thus makes much of the fertilizer applied directly available to plants. The effectiveness of banding depends on other soil factors. For example, banded versus broadcast applications were compared in an Oxisol from Brasilia with an extremely high phosphorus fixation capacity (about 700 ppm P added to provide 0.2 ppm P in solution). The results (Fig. 8.6) show that broadcast and incorporated applications were definitely superior to banded applications in the first crop. Yost and his coworkers found that banded applications concentrated corn root development around the band and that, when a temporary drought struck, these plants suffered more than those of the broadcast plots, which showed more extensive root development. In time, however, the effectiveness of the banded treatments increased, whereas that of the broadcast treatments decreased. The best combination was attained by broadcasting and incorporating a rate of 320 kg P_2O_5/ha before the first crop, with a banded application of 80 kg P_2O_5/ha before planting each crop.

With continuous cropping, banded applications are mixed with a larger volume of soil after each tillage operation. In a sense, annual banding begins to approach broadcast applications in terms of phosphorus distribution with time. For soils with high fixation capacities and low available moisture ranges, the most appropriate scheme seems to be a moderately high initial broadcast application, followed by small banded applications to each subsequent planting.

Sources of Phosphorus

Research in the temperate region indicates that phosphorus fertilizers should have at least 40 to 50 percent P in water-soluble form to insure an adequate supply at early growth stages (Englestad, 1972). Ordinary and triple superphosphates and monoammonium and diammonium phosphates meet this requirement and can be used effectively in soils with low to moderate fixation capacities.

In acid soils that fix large quantities of phosphorus, applications of less soluble phosphorus sources such as rock phosphates may be more effective and economical than the highly soluble forms. Rock phosphates are more reactive in acid soils and usually cost one-third to one-fifth as much as superphosphate per unit of phosphorus.

The tropical literature is full of examples that indicate the desirability of using high-quality rock phosphate sources instead of superphosphates in acid soils (Motsara and Datta, 1971; Awan et al., 1971; Englestad, 1972) and the poor performance of low-citrate-solubility rock phosphate sources in acid soils (Alvarez et al. 1965; Viegas et al. 1970; Miranda et al. 1970). Studies by Lehr and McClellan (1972) at the Tennessee Valley Authority indicate that, when rock phosphates are classified according to absolute citrate solubility, their agronomic effectiveness can be predicted. If the solubility of the best rock phosphate deposits (North Carolina and Tunisia) is given an index of 100, rock phosphates with a solubility index of 70 percent or greater can be recommended for direct application without agronomic research. These are largely concentrated in North Africa, the Soviet Union, and the southeastern United States.

The effects of rock phosphates of varying citrate solubility on flooded rice yields in an acid sulfate soil from Thailand are illustrated in Fig. 8.7. The initial and residual effects of the rock phosphates were highly dependent on their absolute citrate solubilities. The yield responses of the North Carolina and Florida rocks approximated those of triple superphosphate. The results were then interpreted in terms of the relative agronomic effectiveness, which is the ratio of the yield response with rock phosphate

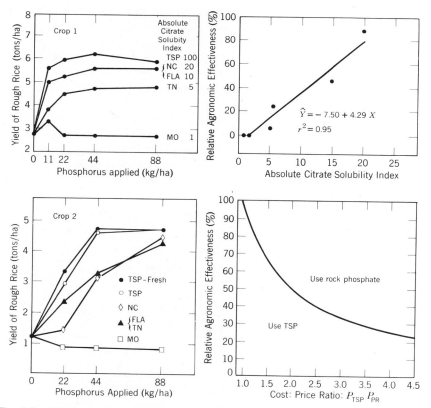

Fig. 8.7 Efficiencies of rock phosphate sources of different citrate solubilities on the yield responses of two flooded rice crops in an acid sulfate soil of Klong Luang, Thailand, and the relationship between absolute citrate solubility, relative agronomic effectiveness, and price ratio. *Source:* Englestad et al. (1974). TSP = Triple superphosphate. Rock phosphates: NC = North Carolina; FLA = Florida; TN = Tennessee; MO = Missouri.

divided by the yield response with triple superphosphate. The economic choice of using rock phosphate or superphosphate can be made by using the bottom right diagram of Fig. 8.7, in which the relative agronomic effectiveness is plotted against the cost:price ratios of the sources.

In the tropics high-citrate-solubility deposits are limited to relatively small areas in Peru and India. The majority of the deposits in most tropical areas, including significant ones in Brazil, Colombia, Venezuela, Togo, and India, have relative solubilities lower than 40 percent. Most are unsuitable for direct applications, but their reactivity can be increased by fine grinding or by thermal alteration and fusion with silica sand, sodium, or magnesium carbonates. These silicophosphates, called "Rhenania" or "thermophosphates," appear to be promising for acid soils that fix large quantities of phosphorus, because of the blocking effect of silicon on phosphorus fixation sites (Olson and Englestad, 1972; Fassbender and Molina, 1969).

The potential effectiveness of these cheaper forms of phosphorus in acid, high-fixing soils is illustrated in Table 8.10, where the various phosphorus sources are rated for relative effectiveness on the basis of a 5 year field experiment in southern Brazil. The low-citrate-solubility Olinda rock phosphate was inferior to ordinary superphosphate; but when this rock or a similar one was thermally treated with silicates and carbonates to produce the thermophosphate, its effectiveness was superior to that of ordinary superphosphate. In view of the substantially lower costs of the rock phosphates and some thermophosphates, both seem desirable alternatives for soils with high fixation capacities.

Effects of Lime and Silicate Applications for Decreasing Phosphorus Fixation

An additional strategy is sometimes feasible for managing soils with high phosphorus fixation capacities: reduce their fixation through amendments

Table 8.10   Behavior of Different Fertilizer Sources on Wheat Grown in Oxisols of Southern Brazil

Phosphorus Source	Relative yield (5 year average)
No phosphorus	100
Olinda rock phosphate[a]	179
Simple superphosphate	206
Thermophosphate	218

Source: W. J. Goedert (personal communication).
[a] Low citrate solubility.

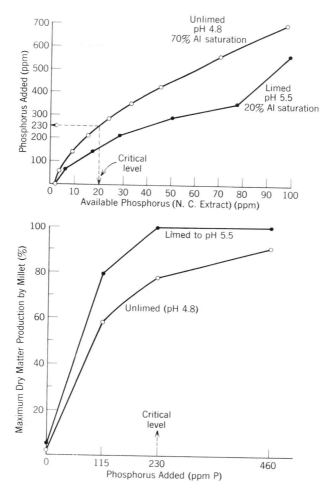

Fig. 8.8 Effects of liming on phosphorus fixation and response in an Oxisol from Panama. *Source:* Adapted from Mendez-Lay (1973).

that will block some of the fixing sites in the soil. This can be accomplished in some soils through liming or silicate additions.

Liming soils to pH 5.5 generally increases the availability of phosphorus. Liming precipitates the exchangeable aluminum and some hydroxy aluminum as aluminum hydroxides, which fix considerably less phosphorus. The top half of Fig. 8.8 shows the decrease in phosphorus fixation when an Oxisol from Panama was limed to pH 5.5. In this case less than half the phosphorus application rate was needed in the limed soil to approach maximum yields in relation to the unlimed soil, as illustrated in the bottom half of Fig. 8.8.

Large increases in the phosphorus uptake derived from fertilizers by several pasture species have been reported by Fox et al. (1964b) in Hawaii when several soils were limed to pH values between 5 and 6. The effect was more pronounced in soils with high phosphorus fixation capacity and high exchangeable aluminum levels. Fox and his coworkers also found that liming to pH levels beyond 6 decreases phosphorus uptake, perhaps through the formation of insoluble calcium phosphates.

This evidence indicates that liming acid soils to pH levels between 5.5 and 6.0 decreases phosphorus fixation but does not eliminate it. Fixation by iron and aluminum oxides and hydroxides is not affected by liming. In addition, many soils that fix large quantities of phosphorus have very low levels of exchangeable aluminum; the most common case is the Andepts. Liming such soils will probably result in the detrimental effects of overliming.

Silicon applications, usually as calcium silicate, sodium silicate, or basic slag, are known to decrease phosphorus fixation and increase phosphorus uptake by crops. For example, Suehisa et al. (1963) reported that Sudan grass yields increased from 2 to 7.6 tons/ha and phosphorus uptake rose from 4 to 15 kg P/ha when 1 ton Si/ha was applied to an Ultisol without added phosphorus. Silva (1971) reviewed most of the Hawaii work and concluded that the silicate anions may replace phosphate anions in the fixation sites and that silicate applications increase the solubility of fixed phosphorus. Roy et al. (1971) found that applications of 500 ppm Si as calcium metasilicate dramatically decreased the phosphorus fixed by several oxidic soils (Table 8.11). The decrease was greater in the soils that had lower phosphorus fixation capacities, the Tropohumult and the Eutrorthox.

The combined effect of these two amendments on decreasing phosphorus

Table 8.11   Effects of Calcium Silicate Applications on Phosphorus Fixation in Hawaiian Soils (expressed as ppm P fixed to give 0.2 ppm P in soil solution)

Soil	Si Added (ppm)		Decrease (%)
	0	500	
Tropohumult	187	100	47
Eutrorthox	425	250	41
Gibbsihumox	725	550	24
Hydrandept	1150	1050	9

Source: Roy et al. (1971).

Table 8.12   Combined Effect of Lime and
Calcium Silicate Applications on
Phosphorus Fixation in a Gibbsihumox from
Hawaii (expressed as ppm P fixed to give 0.2
ppm P in soil solution)

Silicate Applied (tons Si/ha)	No Lime (pH 5.5)	Limed to pH 6.2
0	910	800
1.6	580	675

Source: Silva (1971).

fixation is shown in Table 8.12 for a Humox with an original pH of 5.5 and a phosphorus requirement of 900 ppm P to reach the adequate level of 0.2 ppm in the soil solution. Liming to pH 6.2 decreased the phosphorus requirement to 800 ppm P, while applying silicate as TVA slag decreased it to 580 ppm P without liming. In this soil, liming is of questionable value because of initially high pH.

Phosphorus management of acid tropical soils with very high fixation capacities should include the following possibilities: (1) a combination of banding and broadcast applications, (2) the use of citrate-soluble rock phosphates or thermally altered rock phosphates with initially low citrate solubilities, (3) decreasing the fixing capacity by liming or silicate applications, and (4) the use of species and varieties tolerant to low levels of available soil phosphorus. Economic considerations, such as the relative costs of heavy initial broadcast and of banded applications and the relative costs of the different sources of fertilizer phosphorus, will largely dictate the best alternatives.

SILICON AS A NUTRIENT

In addition to the beneficial effect of silicon applications as a means of decreasing phosphorus fixation and increasing the availability of phosphorus, straight nutritional responses to silicon have been obtained in the tropics. Although silicon is not generally considered an essential element for plant growth, definite silicon deficiencies and responses have been observed in highly leached soils of the tropics under intense cultivation of certain grass species, particularly sugarcane and rice.

Soils having low contents of soluble silicon are the ones likely to show response to silicon applications. Fox et al. (1967) suggested that the critical level is 0.9 ppm Si in water extracts. Responses have been obtained in Oxisols and Ultisols derived from basalt in Hawaii and Mauritius and in rice soils in Japan, Korea, and Sri Lanka.

In these rice soils, silicon applications increased yields because of a more erect leaf habit, greater tolerance of insects and disease attacks, lower uptake of iron and manganese when present in toxic concentrations in the soil, and perhaps a rise in the oxidizing power of rice roots (Okuda and Takahashi, 1965). Also, the beneficial effects of silicon on phosphorus uptake have been observed in paddy rice soils.

Reviews by Silva (1971) and Plucknett (1972) summarize the extensive research conducted in Hawaii on silica fertilization of sugarcane. Dramatic responses to silicate applications, such as the one illustrated in Table 8.13, are common in the absence of phosphorus fixation problems. Yield responses to silica applications result in greater dry matter and sugar contents, larger stalk size, greater elongation and number of green and functioning leaves, and the disappearance of "freckling" leaf symptoms. The freckling problem seems to be associated with high manganese contents but is not completely understood. Calcium silicate applications have a significant residual effect, although small maintenance applications every year or two may be needed (Wong, 1971; Plucknett, 1972).

Table 8.13 Direct Response of Sugarcane to Calcium Metasilicate Fertilization in an Oxisol from Hawaii

CaSiO₃ Applied (tons/ha)	Sugar Production (tons sugar (Pol)/ha)	
	First Crop	Ratoon
0	34	27
4.5	40	28
9.0	42	29
13.5	43	31
18.0	45	38

Source: Plucknett (1972).

SULFUR DEFICIENCY

Widespread sulfur deficiencies and responses have been reported all over the tropics. One of the first reports was from the Cerrado of Brazil, where McClung et al. (1959) observed responses not only in savannas but also in soils recently cleared of virgin forests. In Central America sulfur deficiencies are also widespread, particularly in Andepts, where their volcanic origin would lead one to assume that such soils whould be high in this element (Muller, 1965; Fitts, 1970). In sub-Saharan Africa sulfur is limiting in Alfisols and Oxisols with annual rainfall greater than 600 mm and in sandy soils of central Africa, according to a review by Bolle-Jones (1964). Sulfur deficiencies also occur in Asia, particularly the Punjab of India and in Malaysia (Olson and Englestad, 1972). They also have been reported in Australia and Hawaii (Williams, 1972; Fox et al., 1971).

In general, sulfur-deficient soils have one or more of the following properties. They are high in allophane or oxides. They are also low in organic matter, and often sandy. Soils subject to repeated annual burning are often sulfur deficient since about 75 percent of the sulfur is volatilized by fire. Sulfur-deficient soils occur in unpolluted, inland areas where the atmosphere is low in sulfur.

SULFUR CONTENTS AND FORMS

Total Sulfur

In the temperate region total soil sulfur is positively correlated with organic matter content and inversely correlated with degree of weathering. Olson and Englestad (1972) provided the following topsoil average values for total sulfur in the temperate region: 500 ppm for Mollisols, 400 ppm for Alfisols, and 200 ppm for Ultisols. Following this reasoning, they assumed that tropical soils would average about 100 ppms. A summary of eastern Australia soils by Williams and Steinbergs (1958) shows an average content of 167 ppm S, of which only 7 ppm is inorganic sulfur. Tropical soils high in organic matter and allophane, however, have large quantities of total sulfur. A Eutrandept from Hawaii had 1280 ppm S in its topsoil (Fox et al., 1971).

Organic Sulfur

Most of the sulfur contents of unfertilized tropical soils are in the organic form. The C:N:S or organic matter ratio is on the order of 126:10:1 for soils of Nigeria and eastern Australia, according to surveys by Oke (1971) and Williams and Steinbergs (1958).

Organic sulfur is mineralized like organic nitrogen. There is a great similarity between nitrogen and sulfur cycles. Sulfur mineralization rates range from 1 to 10 percent per year. Barrow (1961) found that immobilization occurs at C:S ratios greater than 200 and that materials containing less than 0.15 percent S are immobilized. Andepts and other soils high in allophane are also high in organic sulfur, but plants growing on these soils are usually very deficient in sulfur because the mineralization rate of this element is slow when organic matter is intimately associated with allophane. Flushes of sulfur mineralization upon wetting previously dried soils have also been observed by Barrow. It seems logical to assume that the processes involved are similar to those governing nitrogen flushes. The fate of the mineralized sulfur, however, may be different from that of nitrates because of the sulfur fixation capacity of many soils.

As in the case of nitrogen, cultivation reduces the organic sulfur contents because of the increased organic matter decomposition rates. McClung et al. (1959) observed an organic sulfur range of 79 to 540 kg S/ha in virgin soils of central Brazil (Table 8.14). After 20 to 30 years of cropping, organic sulfur decreased to 44 to 130 kg S/ha.

Inorganic Sulfur

Sulfates are the main inorganic form of sulfur in aerobic soils. In tropical soils susceptible to leaching, SO_4^{2-} ions are found in small quantities, associated with Ca^{2+}, Mg^{2+}, and K^+. In addition to organic sulfur

Table 8.14   Effect of 20 to 30 Years of Cropping on the Contents and Responses to Sulfur in Two Cerrado Soils of Brazil in Pot Conditions

Soil	Cropping History	Topsoils (ppm)		Subsoils (ppm)	
		Organic S	Inorganic SO_4-S	Organic S	Inorganic SO_4-S
Ultisol	Virgin	36	4.0	12	6.4
(Baurú)	Cropped	24	2.5	10	12.0
Oxisol	Virgin	247	3.7	11	3.3
(Riberão Preto)	Cropped	60	7.2	10	12.3

Source: McClung et al. (1959).

Table 8.15   Inorganic Sulfur Contents Extracted from Hawaiian Topsoils (ppm S)

Great Soil Group	H_2O- Extractable	$Ca(H_2PO_4)$- Extractable
Red Desert (Aridisol)	8	16
Dark Magnesium Clay (Vertisol)	18	11
Low Humic Latosol (Ustox)	20	55
Humic Latosol (Ultisol)	23	33
Humic Ferruginous Latosol (Humult)	22	53
Hydrol Humic Latosol (Hydrandept)	11	134

Source: Fox et al. (1965).

mineralization, SO_4^{2-} ions may be added to the soil through rainfall, fertilizers, pesticides, irrigation water, and gases from volcanic eruptions. Rainfall usually contains less than 1 ppm S; this is equivalent to about 10 kg S/ha a year, a quantity that may be sufficient to prevent sulfur deficiencies. Rainfall contributes high amounts of sulfur in polluted areas and near the oceans. The contributions of fertilizers and other farm chemicals may be considerable, particularily when ammonium sulfate and ordinary superphosphates are used. Sulfurous gases from volcanic eruptions may be adsorbed directly by plants or dissolved in the soil solution.

Soils high in iron and aluminum oxides or allophane have considerable capacity to sorb sulfate ions through anion exchange sites. This process results in an accumulation of inorganic SO_4^{2-} in many Ultisols, Oxisols, and Andepts with oxidic mineralogy. Water-soluble SO_4^{2-} and phosphate-extractable SO_4^{2-} contents are generally higher in such soils than in those with pure layer silicate systems, as Table 8.15 shows.

When soils of oxide-coated or oxide mineralogy are heavily fertilized and leaching occurs, relatively large quantities of SO_4^{2-} accumulate in the subsoils because of their sulfur sorbing capacity. In the Brazilian subsoils of Table 8.14 inorganic sulfur became the dominant fraction. The same occurred in a heavily fertilized Eutrorthox of Hawaii, where large quantities of SO_4^{2-} accumulated in the subsoil. Similar relationships have been observed in Ultisols of North Carolina by Kamprath et al. (1957). Under such conditions certain crops may show sulfur deficiency symptoms at eary growth stages when the root systems have not penetrated the subsoil. This deficiency gradually disappears as the roots come in contact with the SO_4^{2-}-rich subsoil.

SULFUR SORPTION* AND RELEASE

Two mechanisms are recognized as responsible for the sulfur sorption in soil. One is the exchange of SO_4^{2-} for OH^- ions on iron and aluminum oxide surfaces. This reaction also occurs in anion exchange sites along the edges of kaolinite particles and positively charged organic radicals. In subsoils it may result in pH increases of 0.3 to 0.9 unit.

The second mechanism is the formation of complexes with hydroxy aluminum as follows:

$$Al(OH)^{2+} + SO_4^{2-} \rightarrow Al(OH)SO_4$$

The sorption and release properties of sulfur can be characterized by adsorption isotherms in a way very similar to that used for phosphorus. Figure 8.9 shows examples of these relationships in some tropical soils. The montmorillonitic Mollisol from the Cauca Valley of Colombia does not exhibit any sulfur sorption, a finding typical also of other soils with similar mineralogy from Hawaii and Australia, according to Fox et al. (1971). Several Oxisols show strong sulfur sorption capacity, particularly in the subsoil.

In addition to mineralogy, other factors affect the intensity of sulfur sorption. In oxide systems sorption increases as pH decreases since OH^- ions tend to replace SO_4^{2-} ions at high pH values. Sulfur sorption is decreased by phosphorus sorption since the $H_2PO_4^-$ ion replaces SO_4^{2-} ions. This reaction takes place in the topsoil and may cause sulfate movement into the subsoil, where the sulfur can be sorbed.

Sorbed sulfur is held much less tightly by clay particles than is fixed (or sorbed) phosphorus. Therefore the availability of sorbed sulfur is generally higher than that of fixed phosphorus. The SO_4^{2-} ions have to be released in the soil solution before they can be taken up by plants. In Hawaii 8 to 10 ppm of phosphate-extractable sulfur is considered to be the critical level (Hasan et al., 1970). No critical levels for water-soluble sulfur have been developed.

* As in the case of phosphorus, "fixation" and "sorption" are used interchangeably to mean the same process: the transformation of inorganic sulfur into slowly available forms. The term "sorption" will be employed here because of its widespread usage.

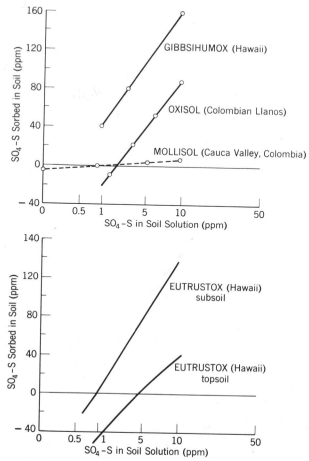

Fig. 8.9 Examples of sulfur sorption isotherms. *Source:* Fox et al. (1971) and Fox (1974).

SULFUR REQUIREMENTS BY TROPICAL CROPS

Sulfur uptake values are similar to those of phosphorus in tropical crops for which data are available. Sulfur concentrations in plant tissue are also similar to phosphorus concentrations, ranging from 0.1 to 0.3 percent S. A good indication of sulfur deficiency is a decline of the N:S ratio of tissues below 17:1 because the ratio of protein nitrogen to protein sulfur is 15:1. Among species, grain and pasture legumes as well as cotton are especially sensitive to sulfur deficiencies.

A sulfur deficiency at eary growth stages may disappear later when the

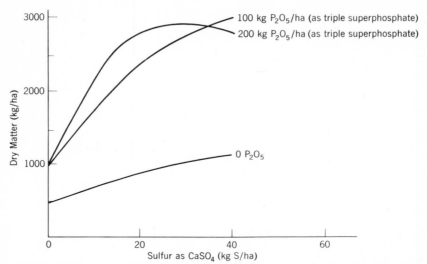

Fig. 8.10 Effects of sulfur and phosphorus fertilization on the growth of *Paspalum notatum* (Batatais grass) in a Brazilian Ultisol. *Source:* McClung and Quinn (1959).

soil comes in contact with the subsoil. In such cases small starter fertilizer applications may be needed.

SULFUR FERTILIZATION

In general, rates on the order of 10 to 40 kg S/ha are sufficient to overcome sulfur deficiencies. Figure 8.10 shows how such rates tripled the yield of pasture in the presence of phosphorus. Jones and Quagliato (1970) showed that 10 to 20 kg S/ha is sufficient for most tropical pasture legumes and

Table 8.16   Effect of Sulfur Fertilization on Annual Coffee Yields on a Red–Yellow Latosol (Oxisol) of Brazil

Annual Application (kg S/ha)	Coffee Yield (average of 10 years) (kg/ha)
0	1320
16	2040
33	2341
66	2400
132	2172

Source: De Freitas et al. (1972).

alfalfa. For coffee an annual rate of 30 kg S/ha increased yields by 82 percent over a 10 year period in Brazil (Table 8.16). Sulfur fertilization is an established practice in this area.

Most of these requirements are small enough for some nitrogen and phosphorus carriers to supply them. When urea and triple superphosphate are used, only small applications are needed to achieve high yields.

SUMMARY AND CONCLUSIONS

1. The total soil phosphorus content, as well as the predominant forms of phosphorus found in the soil, reflects the intensity of weathering. Most high-base-status soils of the orders Entisols, Vertisols, Inceptisols, and Mollisols have calcium-bonded phosphorus as the dominant form. Andepts are very high in organic phosphorus and calcium or aluminum phosphates, depending on pH. In Ultisols and Oxisols, iron and aluminum phosphates predominate, often in occluded or reductant-soluble forms.

2. In highly weathered soils and in Andepts, organic phosphorus accounts for over 50 percent of the total soil phosphorus. The maintenance of organic phosphorus is an important management practice for no-fertilizer agriculture.

3. Phosphorus sorption isotherms provide an adequate means for quantifying the intensity–capacity relationships. The amount of fertilizer phosphorus that must be added to provide a concentration of 0.05 to 0.2 ppm P in the supernatant soil solution provides an indication of the magnitude of phosphorus fixation. Clayey Oxisols, clayey Ultisols, and most Andepts have extremely high phosphorus fixation capacities, ranging from 300 to 1000 ppm P required to reach the desired concentration in the soil solution. Other tropical soils with predominant layer silicate mineralogy, as well as coarse-textured Ultisols and Oxisols, have moderate to weak phosphorus fixation capacity.

4. Tropical crops differ widely in their phosphorus requirements. Rice, sweet potatoes, and many tropical pastures have substantially lower phosphorus requirements than corn, potatoes, and certain pasture species developed under high-fertility conditions. Varietal differences in tolerance of low available phosphorus exist. Varieties requiring less phosphorus should be selected in order to lower the required phosphorus application rates.

5. Most Oxisols, Ultisols, Andepts, and Vertisols in the tropics are deficient in phosphorus. Phosphorus deficiency can be predicted by soil test procedures. The correction of such deficiency depends on the magnitude of phosphorus fixation in the soil. For soils that fix low to moderate amounts,

phosphorus management is a relatively simple proposition. The recommended rates can be broadcast as water-soluble superphosphates or ammonium phosphates and incorporated into the plowed layer. Small maintenance applications can be broadcast or banded once a year.

6. Phosphorus management in high-fixing soils is a more complicated procedure. When capital is available, one approach is to broadcast a high rate sufficient to satisfy the fixing capacity of the soil. The fixed phosphorus will be released gradually over the years. The extremely high investment required must be amortized over a considerable period of time. This approach, however, is probably beyond the reach of most farmers in the tropics.

In such cases another strategy is required. This consists of using low-cost phosphorus sources such as rock phosphates of high citrate solubility or thermally modified rock phosphates of low citrate solubility. A combination of an initial broadcast application followed by banded applications of superphosphate before each successive planting seems promising. Banding alone may not be sufficient in Oxisols subject to serious moisture stress because it limits root development around the band. A third component of this strategy consists of reducing the phosphorus fixation capacity of the soil by lime and silicate applications. Liming to pH 5.5 precipitates the exchangeable aluminum, which fixes phosphorus, while silica seems to block phosphorus fixation sites because of the similar size of the phosphate and silicate anions. The fourth component of the strategy is to use species and varieties most tolerant of low available phosphorus levels so that the application rates can be minimized.

7. In addition to the effect of silica in decreasing phosphorus fixation, direct nutritional responses to silica are commonly observed in some Oxisols and Ultisols in the tropics having less than 1 ppm Si in the soil solution. Silica fertilization is a common practice in such soils when they are used for paddy rice in Asia and sugarcane in Hawaii and Mauritius. The possibility of silicon deficiencies in other crops grown in such soils should be investigated.

8. Sulfur deficiencies are widespread throughout the tropics. They are found in highly weathered Oxisols, Ultisols, and Alfisols, in young soils developed from volcanic ash, in sandy soils, in savanna areas subject to annual burning, and in inland, unpolluted areas.

9. Sulfur sorption and release can be characterized by isotherms similar to those used for phosphorus. Sulfur is less tightly held by oxide particles than is phosphorus. In many Ultisol and Oxisols sulfate-sulfur tends to accumulate in the subsoil, where the sorption capacity or anion exchange is greater. This mechanism in effect prevents leaching.

10. The requirements of plants for sulfur are similar to those for phosphorus. Several species—cotton and certain pasture legumes—are more susceptible to sulfur deficiency than others. Application rates of 10 to 40 kg S/ha are usually sufficient to correct sulfur deficiencies. Such rates may be supplied by ammonium sulfate or ordinary superphosphate instead of other fertilizer sources that do not contain sulfur.

REFERENCES

Acquaye, D. K. 1963. Some significance of soil organic matter on soil organic phosphorus mineralization in phosphorus nutrition of cocoa in Ghana. *Plant and Soil* **19**:65–80.

Alvarez, R., J. C. Ometto, J. Ometto, A. Wutke, et al 1965. Adubação de cana-de-açúcar. *Bragantia* **24**:97–107.

Andrew, C. S. and M. F. Robins. 1969. The effect of phosphorus on the growth and chemical composition of some tropical pasture legumes. I. Growth and critical percentages of phosphorus. *Aust. J. Agr. Res.* **20**:655–674.

Andrew, C. S. and M. F. Robins. 1971. The effect of phosphorus on the growth, chemical composition and critical phosphorus percentages of some tropical pasture grasses. *Aust. J. Agr. Res.* **22**:693–703.

Andrew, C. S. and P. J. Vanden Berg. 1973. Influence of aluminum on phosphate sorption by whole plants and excised roots of some pasture legumes. *Aust. J. Agr. Res.* **24**(3):341–351.

Awan, A. B. 1964. Effect of lime on the availability of phosphorus in Zamorano soils. *Soil Sci. Soc. Amer. Proc.* **28**:672–673.

Awan, A. B. et al. 1971. Estudio comparativo de roca fosfatada y superfosfato triple como fuerte de fósforo para los cultivos. *Rev. Agr. Cuba* **4**:55–61.

Bache, B. W. 1964. Aluminum and iron phosphate studies related to soils. II. Reactions between phosphates and hydrous oxides. *J. Soil Sci.* **15**:110–116.

Bache, B. W. and N. E. Rogers. 1970. Soil phosphate values in relation to phosphate supply to plants from Nigerian soils. *Samaru Res. Bull.* **123**:383–390.

Baldovinos, S. F. and G. W. Thomas. 1967. The effect of soil clay on phosphorus uptake. *Soil Sci. Soc. Amer. Proc.* **31**:680–682.

Barrow, N. J. 1961. Studies on the mineralization of sulfur from soil organic matter. *Aust. J. Agr. Res.* **12**:306–391.

Benavides, S. T. 1963. Fractionation of phosphorus from soils of the Llanos Orientales of Colombia. M.S. Thesis, Oklahoma State University, Stillwater.

Bolle-Jones, E. W. 1964. Incidence of sulfur deficiency in Africa—a review. *Emp. J. Exptal. Agr.* **32**:241–248.

Bornemisza, E. and R. Llanos. 1967. Sulfate movement, adsorption and desorption in three Costa Rican soils. *Soil Sci. Soc. Amer. Proc.* **31**:356–360.

Brams, E. 1973. Soil organic matter and phosphorus relationships under tropical forests. *Plant and Soil* **39**:465–468.

Bromfield, A. R. 1972. Sulfur in northern Nigerian soils. I. Effects of cultivation and fertilizers on total sulfur and sulfate pattern in soil profiles. *J. Agr. Sci.* **78**:465–470.

Cabala, P. and H. W. Fassbender. 1970. Formas de fósforo en la región cacaotera del Brasil. *Turrialba* **20**:439–444.

Chang, C. A. and G. D. Sherman. 1961. Differential fixation of phosphate by typical soils of the Hawaiian great soil groups. *Hawaii Agr. Exp. Sta. Tech. Bull.* 16.

CIAT. 1971. Annual Report. Pp. 119–121. Centro Internacional de Agricultura Tropical, Cali, Colombia.

Cock, J. H. and S. Yoshida. 1970. Effects of silicate applications on rice by the simulation method. *Soil Sci. Plant Nutr.* **16**:212–214.

Coleman, N. T., J. T. Thorup, and W. A. Jackson. 1960. Phosphate sorption reactions that involve exchangeable aluminum. *Soil Sci.* **90**:1–7.

Dabin, B. 1971. The use of phosphate fertilizer in a long-term experiment on a ferralitic soil at Bambari, Central African Republic. *Phosphorus in Agr.* **58**:1–11.

Dabin, B., and E. Gavinelli. 1970. Preliminary results of a survey showing the sulfur contents of soils of tropical Africa. Pp. 113–136. In *Proceedings of the International Symposium on Sulphur in Agriculture.*

Dankhe, W. C., J. L. Malcolm, and M. E. Menendez. 1964. Phosphorus fractions in selected soil profiles of El Salvador as related to their development. *Soil Sci.* **98**:33–38.

De Freitas, L. M. M., F. P. Grunes, and W. L. Lott. 1972. Effect of sulphur fertilizer on coffee. *IRI Res. Inst. Bull.* 41.

D'Hoore, J. L. and J. K. Coulter. 1972. Soil silicon and plant nutrition pp. 163–173. In *Soils of the Humid Tropics*. National Academy of Sciences, Washington.

Englestad, O. P. 1972. Fertilizers. Pp. 174–188. In *Soils of the Humid Tropics*. National Academy of Sciences, Washington.

Englestad, O. P., A. Jugsujinda, and S. K. DeDatta. 1974. Response by flooded rice to phosphate rocks varying in citrate solubility. *Soil Sci. Soc. Amer. Proc.* **38**:524–529.

Ensminger, L. E. 1954. Some factors affecting adsorption of sulfate by Alabama soils. *Soil Sci. Soc. Amer. Proc.* **18**:259–264.

Fassbender, H. W. 1969. Estudio del fósforo en suelos de América Central. IV. Capacidad de fijación del fósforo y su relación con características edáficas. *Turrialba* **19**:497–505.

Fassbender, H. W. and R. Molina. 1969. Influencia de enmiendas calcáreas y silicatadas sobre el efecto de fertilizantes fosfatados en suelos derivados de cenizas volcánicas en Costa Rica. Pp. C2.1–C2.12. In *Panel on Soils Derived from Volcanic Ash in Latin America*. Inter-American Institute of Agricultural Sciences, Turrialba, Costa Rica.

Fitts, J. W. 1970. Sulfur deficiency in Latin America. *Sulphur Inst. J.* **6**:14–16.

Fox, R. L. 1974. Examples of anion and cation adsorption by soils of tropical America. *Trop. Agr.* (*Trinidad*) **51**:200–210.

Fox, R. L., R. A. Olson, and J. F. Rhodes. 1964a. Evaluating the sulphur status of soils by plant and soil tests. *Soil Sci. Soc. Amer. Proc.* **28**:243–246.

Fox, R. L., S. K. DeDatta, and J. M. Wang. 1964b. Phosphorus and aluminum uptake by plants from Latosols in relation to liming. *Trans. 8th Int. Congr. Soil Sci.* **4**:595–603.

Fox, R. L., D. G. Moore, J. M. Wang, D. L. Plucknett, and R. D. Furr. 1965. Sulfur in soils, rainwater and forage plants in Hawaii. *Hawaii Farm Sci.* **14**:9–12.

Fox, R. L., J. A. Silva, O. R. Young, D. L. Plucknett, and G. D. Sherman. 1967. Soil and plant silicon and silicate response to sugar cane. *Soil Sci. Soc. Amer. Proc.* **31**:775–779.

Fox, R. L., D. L. Plucknett, and A. S. Whitney. 1968. Phosphate requirements of Hawaiian Latosols and residual effects of fertilizer phosphorus. *9th Int. Congr. Soil Sci.* **2**:301–310.

Fox, R. L. and E. J. Kamprath. 1970. Phosphate sorption isotherms for evaluating the phosphate requirements of soils. *Soil Sci. Soc. Amer. Proc.* **34**:902–906.

Fox, R. L., S. M. Hasan, and R. C. Jones. 1971. Phosphate and sulfate sorption by Latosols. *Proc. Int. Symp. Soil Fert. Eval. (New Delhi)* **1**:857–864.

Fox, R. L., R. K. Hashimoto, J. R. Thompson, and R. S. de la Peña. 1974. Comparative external phosphorus requirements of plants growing in tropical soils. *Tenth Int. Congr. Soil Sci. (Moscow)* **4**:232–239.

Friend, M. T. and H. F. Birch. 1960. Phosphate responses in relation to soil tests and organic phosphorus. *J. Agr. Sci.* **54**:341–347.

Greenwood, M. 1951. Fertilizer trials with groundnuts in Northern Nigeria. *Emp. J. Exptal. Agr.* **19**:225–241.

Hasan, S. M., R. L. Fox, and C. C. Boyd. 1970. Solubility and availability of sorbed sulfate in Hawaiian soils. *Soil Sci. Soc. Amer. Proc.* **34**:897–901.

Hesse, P. R. 1958. Sulfur and nitrogen changes in forest soils of East Africa. *Plant and Soil* **9**:86–96.

Hossner, L. R., J. A. Freehouf, and B. L. Folsom. 1973. Solution phosphorus concentration and the growth of rice in flooded soils. *Soil Sci. Soc. Amer. Proc.* **37**:405–408.

Igue, K. et al. 1971. Mineralización de P orgánico en suelos acidos de Costa Rica. *Turrialba* **21**:47–52.

Igue, R. and R. Fuentes. 1971. Retención y solubilización de fósforo-32 en suelos acidos de las regiones tropicales. *Turrialba* **21**:429–434.

IRRI. 1971, 1972. Annual Reports. Soil Chemistry Section, International Rice Research Institute, Los Baños, Philippines.

Jones, M. B. and L. M. M. De Freitas. 1970. Response of four tropical legumes to phosphorus, potassium and lime when grown in Red-Yellow Latosols from a dense forest. *Pesq. Agrop. Bras.* **5**:91–99.

Jones, M. B. and S. L. Quagliato. 1970. Respostas de quatro leguminosas tropicais e da alfalfa a varios niveles de enxofre. *Pesq. Agropec. Bras.* **5**:359–363.

Jones, R. K., P. J. Robinson, F. P. Haydock, and R. G. Megarrity. 1971. Sulfur–nitrogen relations in the tropical legume *Stylosanthes humilis*. *Aust. J. Agr. Res.* **22**:885–894.

Juo, A. S. R. and H. O. Maduakor. 1977. Phosphate sorption of some Nigerian soils and its effect on cation exchange capacity. *Commun. Soil Sci. Plant Anal.* **5**:479–497.

Kamath, M. B. and B. V. Subbiah. 1971. Phosphorus uptake pattern by crops from different soil depths. *Proc. Int. Symp. Soil Fert. Eval. (New Delhi)* **1**:281–291.

Kamprath, E. J. 1967. Residual effects of large applications of phosphorus on high fixing soils. *Agron. J.* **59**:25–27.

Kamprath, E. J. 1968. Sulfur reactions and availability in highly weathered soils. *Sulphur Inst. J.* August 1968:4–6.

Kamprath, E. J. 1973a. Phosphorus. Pp. 138–161. In P. A. Sanchez (ed.), "A Review of Soils Research in Tropical Latin America." *North Carolina Agr. Exp. Sta. Bull.* 219.

Kamprath, E. J. 1973b. Sulfur. Pp. 179–181. In P. A. Sanchez (ed.), "A Review of Soils Research in Tropical Latin America." *North Carolina Agr. Exp. Sta. Tech. Bull.* 219.

Kamprath, E. J. 1974. Aspectos quimícos y formas minerales del fósforo del suelo en regiones tropicales. *Suelos Ecuatoriales* **4**:1–18.

Kamprath, E. J., W. L. Nelson, and J. W. Fitts. 1957. Sulfur removed from soils by field crops. *Agron. J.* **49**:289–293.

Koyama, T. and C. Chammek. 1971. Soil–plant nutrition studies on tropical rice. I. Studies on the varietal differences in absorbing soil phosphorus from soils low in available phosphorus. *Soil Sci. Plant Nutr.* **17**:115–126.

Koyama, T. and P. Snitwongse. 1971. Soil–plant nutrition studies in tropical rice. II. Varietal differences in absorbing phosphorus from soils low in available phosphorus. *Soil Sci. Plant Nutr.* **17**:186–194.

Lehr, J. A. and G. H. McClellan. 1972. A revised laboratory scale for evaluating phosphate rocks for direct application. *Tennessee Valley Authority Bull.* Y-43.

Le Mare, P. H. 1968. Experiments on the effects of phosphate applied to a Buganda soil. II. Field experiments on the response curve to superphosphate. *J. Agr. Sci.* **70**:271–279.

Lott, W. L., A. C. McClung, and J. C. Medcalf. 1960. Sulfur deficiency in coffee. *IBEC Res. Inst. Bull.* 22.

McClung, A. C., L. M. M. De Freitas, and W. L. Lott. 1959. Analysis of several Brazilian soils in relation to plant responses to sulfur. *Soil Sci. Soc. Amer. Proc.* **23**:221–224.

McClung, A. C. and L. R. Quinn. 1959. Sulfur and phosphorus responses in Batatais grass (*Paspalum notatum*). *IBEC Res. Inst. Bull.* 18.

McCormick, L. H. and F. Y. Borden. 1972. Phosphate fixation by aluminum in plant roots. *Soil Sci. Soc. Amer. Proc.* **36**:799–802.

Mehta B. V. and J. M. Patel. 1963. Some aspects of phosphorus availability in Gurajat soils. *J. Indian Soc. Sci.* **11**:151–158.

Mendez-Lay, J. M. 1973. Effect of lime on P fixation and plant growth in various soils of Panama. M. S. Thesis, Soil Science Department, North Carolina State University, Raleigh.

Miranda, L. T. de, G. P. Viegas, E. S. Freire, and T. Igue. 1970. Adubação do milho. XXVII. Ensaios com diversos fosfatos. *Bragantia* **29**:301–308.

Monteith, N. H. and G. D. Sherman. 1963. The comparative effects of calcium carbonate and of calcium silicate on the yield of Sudan grass grown in a Ferruginous Latosol and a Hydrol Humic Latosol. *Hawaii Agr. Exp. Sta. Tech. Bull.* 53.

Moura Filho, W., S. W. Buol, and E. J. Kamprath. 1972. Studies on a Latosol Roxo of Brazil: Phosphate reactions. *Experientiae* **13**:235–247.

Motsara, M. R. and N. P. Datta. 1971. Rock phosphate as a fertilizer for direct application in acid soils. *J. Indian Soc. Soil Sci.* **19**:107–113.

Muller, L. E. 1965. Deficiencia de azufre en algunos suelos de Centro America. *Turrialba* **15**:208–215.

North Carolina State University. 1973, 1974. *Agronomic–Economic Research on Tropical Soils.* Annual Reports. Soil Science Department, North Carolina State University, Raleigh.

Nye, P. H. and M. H. Bertheux. 1957. The distribution of phosphorus in forest and savanna soils of the Gold Coast and its agricultural significance. *J. Agr. Sci.* **49**:141–149.

Nye, P. H. and D. J. Greenland. 1960. The soil under shifting cultivation. *Commonwealth Bur. Soils Tech. Commun.* 51.

Oke, O. L. 1971. Sulfur in relation to soil organic matter in Nigerian soils. *J. Indian Soc. Soil Sci.* **19**:309–311.

Okuda, A. and E. Takahashi. 1965. The role of silicon. Pp. 123–146. In *The Mineral Nutrition of the Rice Plant.* Johns Hopkins Press, Baltimore.

Olson, R. A. and O. P. Englestad. 1972. Soil phosphorus and sulfur. Pp. 82–101. In *Soils of the Humid Tropics*. National Academy of Sciences, Washington.

Omotoso, T. I. 1971. Organic phosphorus contents of some cocoa growing soils of Southern Nigeria. *Soil Sci.* 112:195–199.

Pichot, J. and P. Roche. 1972. Phosphore dans les sols tropicaux. *Agron. Tropicale (France)* 27:939–965.

Plucknett, D. L. 1972. The use of soluble silicates in Hawaiian agriculture. *Univ. Queensl. Papers, Dept. Agri.* 1:203–223.

Ponnamperuma, F. N. and R. U. Castro. 1971. Varietal differences in resistance to adverse soil conditions. Pp. 677–684. In *Rice Breeding*. International Rice Research Institute, Los Baños, Philippines.

Prasad, R., M. L. Bhendiu, and B. B. Turkhede. 1971. Relative efficiency of phosphate fertilizers in different soils of India. *Proc. Int. Soil Fert. Eval. (New Delhi)* 1:747–756.

Pratt, P. F., F. F. Peterson, and C. S. Holzley, 1969. Qualitative mineralogy and chemical properties of a few soils from São Paulo, Brazil. *Turrialba* 19:491–496.

Rajan, S. S. S. and R. L. Fox. 1972. Phosphate adsorption by soils. I. Influence of time and ionic environment on phosphate adsorption. *Commun. Soil Sci. Plant Anal.* 3:493–504.

Rasmussen, H. P. 1968. Entry and distribution of aluminum in *Zea mays*: Electron microprobe x-ray analysis. *Planta* 81:28–37.

Richardson, H. L. 1968. The use of fertilizers. Pp. 135–154. In R. P. Moss (ed.), *The Soil Resources of Tropical Africa*. Cambridge University Press, Oxford.

Rivera-House, C. 1971. Phosphate fixation by tropical soils. M. S. Thesis, North Carolina State University, Raleigh. 112 pp.

Roberts, K. J. and R. M. Weaver. 1973. Organic phosphorus in soils with special interest in soils of the tropics. *Agron. Mimeo* 73-1. Cornell University, Ithaca, N.Y.

Roy, A. C., M. Y. Ali, R. L. Fox, and J. A. Silva. 1971. Influence of calcium silicate on phosphate solubility and availability in Hawaiian Latosols. *Proc. Int. Symp. Soil Fert. Eval. (New Delhi)* 1:757–768.

Salinas, J. G. and P. A. Sanchez. 1976. Soil–plant relationships affecting varietal and species differences in tolerance to low available soil phosphorus. *Ciência e Cultura* 28:156–168.

Shelton, J. E. and N. T. Coleman. 1968. Inorganic phosphorus fractions and their relationship to residual value of large applications of phosphorus on high phosphorus fixing soils. *Soil Sci. Soc. Amer. Proc.* 32:91–94.

Silva, J. A. 1971. Possible mechanisms for crop response to silicate applications. *Proc. Int. Symp. Soil Fert. Eval. (New Delhi)* 1:805–814.

Smith, R. W. and D. K. Acquaye. 1963. Fertilizer responses in peasant cocoa farms in Ghana: A factorial experiment. *Emp. J. Exptal. Agr.* 31:115–123.

Soni, S. L., P. B. Kaufman, and W. C. Bigelow. 1972. Electron probe analysis of silicon and other elements in leaf epidermal cells of the rice plant. *Amer. J. Bot.* 59:38–42.

Suehisa, R. H., O. R. Younge, and G. D. Sherman. 1963. Effects of silicate on phosphorus availability to Sudan grass grown on Hawaiian soils. *Hawaii Agr. Exp. Sta. Tech. Bull.* 51.

Syers, J. K. et al. 1971. Phosphate sorption parameters of representative soils from Rio Grande do Sul, Brazil. *Soil Sci.* 112:267–275.

Takijima, Y., H. M. S. Wijayaratna, and E. J. Senewiratne. 1970. Nutrient deficiency and physiological disease of lowland rice in Ceylon. 3. Effect of silicate fertilizers and dolomite for increasing rice yield. *Soil Sci. Plant Nutr.* **16**:11–16.

Tennessee Valley Authority. 1972. *Tailoring Fertilizers for Rice.* Muscle Shoals, Ala.

Truong, B., S. Burdin, and J. Pichot. 1973. Etudes des effects residuels du phosphore dans deux sol ferralitíques par diverses methods analitíques. *Agron. Tropicale (France)* **28**:147–155.

Viegas, G. P., L. T. de Miranda, and E. S. Freire. 1970. Adubação do milho. XXVI. Ensaios com diversos fosfatos. *Bragantia* **29**:191–198.

Vogt, J. B. M. 1966. Responses to sulfur fertilization in northern Rhodesia. *Agrochimica* **10**:105–113.

Westin, F. C. and J. G. de Brito. 1969. Phosphorus fractions of some Venezuelan soils as related to their stage of weathering. *Soil Sci.* **107**:194–202.

Williams, C. H. 1971. Reaction of surface-applied superphosphate with soil. I. The fertilizer solution and its initial reaction with soil. II. Movement of phosphorus and sulfur into the soil. *Aust. J. Soils Res.* **9**:83–106.

Williams, C. H. 1972. Sulfur deficiency in Australia. *Sulphur Inst. J.* **8**:5–8.

Williams, C. H. and A. Steinbergs. 1958. Sulphur and phosphorus in Eastern Australian soils. *Aust. J. Agr. Res.* **9**:483–491.

Wong, Y. C. 1971. The residual effect of calcium silicate applications on sugarcane growth. Pp. 63–68. Annual Report Mauritius Sugar Industry Research Institute.

Wong, Y. C. and P. Halais. 1970. Needs of sugarcane for silicon when growing in highly weathered Latosol. *Exptal. Agr.* **6**:99–106.

Woodruff, J. R. and E. J. Kamprath. 1965. Phosphorus adsorption maximum as measured by the Langmuir isotherm and its relationship to phosphorus availability. *Soil Sci. Soc. Amer. Proc.* **29**:148–150.

Younge, O. R. and D. L. Plucknett. 1966. Quenching the high phosphorus fixation of Hawaiian Latosols. *Soil Sci. Soc. Amer. Proc.* **30**:653–655.

9

SOIL FERTILITY EVALUATION

Soil fertility evaluation is the process by which nutritional problems are diagnosed and fertilizer recommendations made. Several approaches are presently in use in the tropics as well as in the rest of the world. The most widespread ones are based on soil testing, plant analysis, missing element techniques, simple fertilizer trials, and frequently a combination of these. The advantages and disadvantages of the different approaches are a subject of much discussion, as can be appreciated by the large number of references at the end of this chapter.

In the tropics, effective soil fertility evaluation methods exist in many Latin American countries and in India, but few in tropical Africa and Southeast Asia. The main reason for the limited development in these areas is probably the lack of widespread fertilizer use in Africa, and the preponderance of flooded rice culture in Southeast Asia. This chapter examines the various approaches to soil fertility evaluation used in the tropics. It also describes some of the efforts to correlate soil fertility parameters with soil classification.

FERTILITY EVALUATION SYSTEMS BASED ON SOIL TESTING

One of the most popular approaches in use by a large number of tropical countries is that developed by the cooperative International Soil Fertility

Evaluation and Improvement Program (ISFEIP). Its concept and scope are best described as follows:

A soil fertility evaluation program involves several parts. Soil fertility has to do principally with plant nutrient elements and soil conditions. Evaluation is concerned with levels of availability and nutrient balance in the soil, including appropriate methods for assessing these factors (soil tests, plant analysis, soil survey, climatic conditions). Improvement involves the addition to the soil of fertilizers, lime, manures, and other amendments in such quantities, at such times in the season, and in such ways as to provide the optimum nutritional environment for crop production. Thus, a soil fertility evaluation and improvement program is site-specific and situation-specific. It is the judicious use of information for a specific field in which consideration is given to factors that will influence yield, the capability of the farmers, and the availability of capital (ISFEIP, 1974).

According to Fitts (1974) such a program involves six interrelated facets:

1. Sampling (soil and plant).
2. Laboratory analyses (soil and plant).
3. Correlation between analysis and yield response.
4. Interpretation and recommendations.
5. Putting information to use.
6. Research.

This broader philosophy has gradually evolved from the concept of soil testing, that is, the use of soil analysis as the means for obtaining fertilizer recommendations. The use of soil tests exclusively is not considered a satisfactory approach. The overall program has laboratory, greenhouse, and field components.

Soil Sampling

Taking a representative soil sample is both the first step and the largest source of error in the soil fertility evaluation program. Soil scientists assume that distribution of printed instructions and diagrams on how to take soil samples is sufficient to assure representative specimens. Unfortunately, this is seldom the case. On the other hand, detailed instructions to extension workers and farmers such as the ones included in a recent bulletin by Perur et al. (1974) from India are very valuable.

A representative soil sample is composed of 10 to 20 subsamples from the rooting zone of a field with no major variation in slope, drainage, color, or past fertilizer history. Nonrepresentative areas such as fence rows and manure piles must be avoided. Appropriate information is also needed, including the name and address of the farmer, field number, previous crop, and fertilizer practices. These are best provided by printing the desired information on the soil boxes.

How deep the sample should be depends on the crop to be grown and the depth at which amendments are likely to be incorporated. This is almost universally the plowed layer or the top 15 cm. In the case of deep-rooted crops such as sugarcane and tea, a subsoil sample is often taken (Wong, 1971). For established pastures or permanent crops where fertilizers are not likely to be incorporated, the top 15 cm is usually taken.

Another important question is where to sample when phosphorus has been applied in bands. The answer is between the bands if their locations are known. A third consideration is the time of the year. Soil samples should be taken substantially before planting so that the results will be available when the decision on how much fertilizer to apply is made. In ustic soil moisture regimes, this means sampling during the dry season. In the preceding chapters, upward ion movement during the dry season has been discussed. This may change some of the soil test results, principally in regard to soluble salts.

How often a field should be sampled depends primarily on the intensity of fertilizer use and the economic value of the crop. For average management intensity once every 3 years is recommended by ISFEIP (1971), while for very intensively managed areas annual sampling is necessary.

The reason why soil sampling is the largest source of error in evaluating soil fertility is the magnitude of extrapolation of the analytical results. When a 5 g subsample taken from a 500 g composite sample of a 2.5 ha field at 15 cm depth is used for phosphorus determination, it represents one-billionth (10^{-9}) of the total soil volume for which analysis is made (Perur et al., 1974).

This being a service program, the time between soil sampling and its analysis should be minimized. Delays in shipping, power failures, and absence of some reagents in many tropical laboratories are unfortunately very common. Several studies have been made on the effect of time of storage and drying on analytical results. Bouldin et al. (1971) studied the effect of drying versus keeping several Oxisols and Ultisols from Puerto Rico and Colombia moist for a period of months. Drying decreased the exchangeable aluminum content by 20 percent and increased the exchangeable NH_4 content when both were extracted by 1 N KCl. No changes in pH, potassium, calcium, or magnesium were observed. It is well established that storage increases the inorganic nitrogen content of the soil. Nelson and Bremner (1972) recommend storing air-dried soils in stoppered bottles if inorganic nitrogen analysis is to be delayed. The best procedure, however, is to minimize delays between sampling and analysis.

With increasing international cooperation, many soil samples are shipped to other countries for research purposes and quality control. All soil samples entering the United States are treated by the U.S. Plant Quarantine Service at the point of entry. Samples intended for laboratory use are fumigated with methyl bromide, while those for greenhouse tests are heat-treated in an autoclave. The effects of these treatments by quarantine authorities on soil test values have been evaluated by Lopes (1975). The methyl bromide treatment induces no major changes in soil properties. Autoclaving, on the other hand, increases NH_4^+ and manganese contents drastically.

After arrival at the laboratory, the soil sample is dried, ground either by pounding or by electric-powered grinders, passed through a 2 mm sieve, assigned a laboratory number, and stored until ready for analysis. In most cases the original sample box, which has the source information and now a laboratory number, is kept.

Laboratory Organization

The soil-testing laboratory is the backbone of the fertility evaluation program (Fitts, 1974). Unlike research laboratories, service laboratories must be geared to handle large numbers of samples *rapidly* and *accurately*. A complete system of semiautomated apparatus has been developed by the ISFEIP program especially for tropical laboratories. This system has been described by Vettori (1969), ISFEIP (1973), Perur et al. (1974), and Hunter (1975) and is summarized in the following paragraph.

Soil samples are measured volumetrically, eliminating the time-consuming weighing process, and then placed in multiple-unit trays of 30, where the extracting and diluting solutions are added or transferred by specially designed diluter-dispensers. Other apparatus transfers the aliquots to spectrophotometers and pH-measuring units automatically. Shaking, stirring, and cleaning apparatus are also semiautomatic and capable of handling 30 units at a time. A single technician can run 100 soil samples a day, measuring 10 determinations per sample (Hunter, 1975). The apparatus are based on manual operation and use a minimum of electrical power. Expensive electronic equipment such as pH meters, spectrophotometers, and atomic absorption apparatus are used only in the last phases. An additional advantage is that the same units can be used for plant tissue as for soil analysis. The bulk of the work is done with two extractants: the dilute double acid or modified Olsen extractants for available P, K, Ca, Mg, Na, Fe, Mn, Zn, and Cu, and the 1 N KCl extraction for Al. Tests for N, B, S, and Mo require other methods that have not yet been adapted to the system. Soil pH measurements are also adapted to routine analysis.

An interesting result of this process is that the analytical accuracy increases with the size of the operation. Every tenth sample is a control that is monitored periodically by a control laboratory. On the basis of these records, the larger the laboratory's size, the greater is its accuracy (Fitts, 1974). Investigations on the total error variation of routine chemical analyses by Ng et al. (1974) show that instrument and operator errors are small, whereas chemical treatment is the main source of error in soil-testing laboratories.

At present approximately 50 laboratories in Latin America are equipped with this system or modifications thereof. They analyze approximately 400,000 soil samples submitted by farmers per year (ISFEIP, 1970, 1973). Each of the 50 laboratories analyzes more than 100 samples per day, and 12 analyze more than 500 samples a day. The operation of these laboratories has had a measurable impact on increasing average yields by increasing fertilizer use (ISFEIP, 1972).

Selection of Extraction Methods

The selection of extractants for available nutrients has received major attention in soils research and is a subject of considerable discussion. According to Bray (1948), a successful soil test laboratory method should meet three criteria:

1. It must extract all or a proportionate part of the available forms of nutrients in soils with widely different properties.

2. The procedure must be rapid and accurate.

3. The amounts extracted should be correlated with the growth and response of each crop to the nutrient in question under various conditions.

If the voluminous literature on soil-testing methods is evaluated according to these criteria, only a few methods meet them. Excluded immediately are the biological (Neubauer) tests, exchange resins, and other time-consuming analytical methods such as those outlined by Muller and Van Baren (1971). These are good research techniques but not routine soil tests. The following is a summary of tests for each nutrient or adverse soil condition.

Nitrogen.   Bartholomew (1972) classified soil tests for nitrogen into three categories: (1) determination of organic nitrogen or some chemically extracted fraction of organic nitrogen, including organic matter, as an indicator of organic nitrogen; (2) incubation methods to evaluate mineralization rates, and (3) direct measurement of inorganic nitrogen. Unfortunately none of these methods meets the three criteria given above in most cases.

Total soil nitrogen or organic matter content is sometimes correlated with nitrogen response in soils with similar properties and climatic conditions. In such cases the tests meet the three criteria except the one for wide applicability, and can be useful at the local level. Although good correlations between total nitrogen and yield response in the greenhouse are common (Cornforth and Walmsley, 1971), they become less frequent under field conditions. Baynes and Walmsley (1973), for example, correlated total soil nitrogen with nitrogen response in corn at several locations in the West Indies. They were able to predict the presence or absence of nitrogen response with a critical level of 0.17 percent N in two-thirds of the cases.

Usually, total nitrogen and organic carbon are poorly correlated with nitrogen response in the field. Because of the lack of a better method, however, many laboratories use these measures to provide some information to farmers.

Methods in the second category consist of incubating a sizable amount of sample under aerobic or anaerobic conditions and measuring the mineralization rate, calculated from the inorganic nitrogen contents before and after incubation. Sometimes the total nitrogen content is also determined. Mineralization rates more nearly reflect nitrogen availability from soil organic matter than any other analysis, according to a review by Bartholomew (1972). Their time requirement, however, disqualifies them as a successful soil test. Although good correlations between incubation value and yield response exist (Cornforth and Walmsley, 1971; Lathwell et al., 1972; Baynes and Walmsley, 1973), these methods are too time consuming for routine purposes. However, they are excellent research tools.

Methods evaluating inorganic nitrogen directly have been successful in areas of limited rainfall when sampling is made at definite times. According to Bartholomew (1972), they are useful in soils high in organic matter and in irrigated fields where inorganic nitrogen accumulates. Application of these methods to humid tropical conditions was attempted by Lathwell et al. (1972). They were successful in correlating inorganic nitrogen extracted by 0.01 M CaCl$_2$ with greenhouse crop uptake in Puerto Rico. A 1 N KCl extraction is presently being studied by ISFEIP. To this writer's knowledge no correlations with field response have been achieved with such extractions in humid tropical areas.

Most nitrogen soil tests, therefore, do not meet the three criteria for a successful soil test. With few exceptions nitrogen soil tests are not reliable enough to predict nitrogen response. Other means, principally field experiments and crop uptake, are used to evaluate nitrogen fertilization.

Phosphorus. The situation with phosphorus is the opposite of that with nitrogen. Several effective soil tests for estimating available phosphorus meet all three of Bray's criteria and are used throughout the world. A list of the main procedures used in the tropics appears in Table 9.1. Of these the most common are the Olsen, the Bray 2, and the North Carolina (also called the "Mehlich" or "dilute double acid") methods.

The effectiveness of each of these methods is related to its ability to extract different forms of inorganic phosphorus. A recent comparison on widely different soils from Bangladesh illustrates the relative abilities of the common soil phosphorus tests to identify different phosphorus fractions (Table 9.1). Most acid extractants are effective in determining Ca-P. The North Carolina, Bray 1, Bray 2, 0.3 N HCl, and the Na-EDTA methods are effective in extracting both Ca-P and Al-P. The alkaline Olsen method, however, is the only one sensitive to Fe-P. No methods are capable of extracting reductant-soluble, occluded, and organic forms of phosphorus. Results similar to those presented in Table 9.1 have been obtained by Tyner and Davide (1962) in the Philippines, Chang and Juo (1963) in Taiwan, Balerdi et al. (1968) in Central America, Srivastsava and Pathak (1971) and John (1972) in India, Cholitkul and Tyner (1971) in Thailand, and Pagel (1972) throughout the tropics. The obvious conclusion from such studies— that there is no universal extractant effective in all soils—is logical in view of the variability in soil properties related to phosphorus availability. Consequently, several studies recommend different extractants for different broad soil groups (Goswami et al., 1971; Oko and Agboola, 1974).

The extraction that seems to have the widest applicability is the Olsen sodium bicarbonate method, as is evident in Table 9.1 by the highest correlation coefficients. The bothersome carbon filtration step caused by color contamination of alkaline extractants has been eliminated by treating soils with Superfloc 127, a commercial flocculant (Hunter, 1975). A further modification, adding EDTA, improved the effectiveness of the Olsen method. The "modified Olsen," as it is now called, is an effective extractant not only for available P but also for K, Ca, Mg, Zn, Mn, Fe, Cu, and NH_4–N (ISFEIP, 1971, 1972).

Phosphorus tests for flooded rice soils are not as successful as those for aerobic crops. This problem and the reasons for it are discussed in Chapter 11. Various references listed in this chapter indicate that the Olsen method is again the most effective, although none is satisfactory.

Table 9.1 Correlation between Common Phosphorus Soil Tests and Inorganic Phosphorus Fractions in Various Soils of Bangladesh

Soil Test	Extractant	Ca-P	Al-P	Fe-P	Reductant-Soluble Fe-P	Occluded Fe-P and Al-P
Olsen	0.5 M NaHCO$_3$ at pH 8.5	.55	.62	.78*	.01	−.17
Truog	0.002 N H$_2$SO$_4$ at pH 3	.90*[a]	.59	.09	−.32	−.49
North Carolina	0.025 N H$_2$SO$_4$ + 0.05 N HCl	.88*	.65*	.06	−.39	−.62
HCl	0.3 N HCl	.95*	.70*	.23	−.38	−.61
Bray 1	0.03 N HN$_4$F + 0.025 N HCl	.72*	.73*	.46	−.17	−.48
Bray 2	0.3 N NH$_4$F + 0.025 N HCl	.78*	.74*	.38	−.24	−.50
Schoefield	0.01 M CaCl$_2$	.06	.05	.03	−.30	−.60
Morgan	NaOAc + HOAc	.79*	.56	.18	−.47	−.60
EDTA	0.02 N Na$_2$-EDTA	.77*	.95*	.41	−.23	.53

Source: Ahmed and Islam (1975).

[a] Asterisks indicate statistical significance.

Potassium.   The available forms of potassium in most soils are exchange-able and soluble potassium. A series of extractants is able to account for both forms in ways that satisfy Bray's three criteria. The most commonly used method is the 1 N ammonium acetate extraction; it is usually the best or at least as good as others (Datta and Kalbande, 1967; Boyer, 1972; Ekpete, 1972; Ahmad et al., 1973; Rodriguez, 1974). Attempts to use the same extractant for estimating available potassium as is used for phos-phorus have also been successful. The North Carolina extraction is now used for both these elements in Brazil (PIPAEMG, 1972); and, as men-tioned before, the modified Olsen extraction is also effective for potassium. The interpretation varies somewhat with soil properties. Critical levels are higher for montmorillonitic soils with high potassium fixation capacities than for other soils (Boyer, 1972).

Lime requirement.   The many recommendations for determining the lime requirements of acid soils have been summarized by McLean (1973). Of these, only two meet Bray's soil test criteria: direct pH determination and the use of exchangeable aluminum extracted by 1 N KCl. Liming to a desired pH may not correlate well with crop responses because of the imperfect correlation between pH and aluminum, overliming problems, and differences in plant tolerance of acidity, as discussed in Chapter 7. The exchangeable aluminum determination is a simple soil test that has corre-lated well with lime response in acid soils of the tropics (Kamprath, 1967; Reeve and Sumner, 1970; Tobón and León, 1970). The percent aluminum saturation is often used as the critical level (Muzilli and Kalckmann, 1971), and this practice is recommended.

Secondary nutrients.   Exchangeable calcium and magnesium can be determined by atomic absorption from the same extracts as potassium with little complication (Doll and Lucas, 1973). When the necessary instruments are not available, EDTA extractions and filtrations are adequate. Testing for calcium and magnesium may facilitate the interpretation of potassium tests if a cation balance problem is suspected. Calcium and magnesium tests are needed to estimate effective cation exchange capacity and percent alu-minum saturation in conjunction with potassium and aluminum tests. Although not soil tests in the strict sense, these parameters are very useful in the interpretation process.

   Soil tests for sulfur are still a problem because current methods are cum-bersome and often inaccurate. The main problem is the turbidimetric determination of sulfate as $BaSO_4$. The present methods and their problems are discussed in articles by Tisdale (1971), Pal and Motiranami (1971), Reisenauer et al. (1973), and others. The water and $Ca(H_2PO_4)_2$ extractions are the most common ones. Efforts at simplifying sulfur determinations to the level of other soil tests are currently under way.

Micronutrients.   A variety of methods are presently in use for determining the availability of B, Cu, Fe, Mn, Mo, and Zn in soils. The literature on the subject has been reviewed by Cox and Kamprath (1971). The critical levels for the various routine soil tests and the interacting factors are shown in Table 9.2. As mentioned before, the modified Olsen extraction can now be used for determining Cu, Fe, Mn, and Zn from the same extraction as P, K, Ca, and Mg (ISFEIP, 1972). Boron and molybdenum, however, require separate extractions. In general, micronutrient soil tests are effective when sufficient efforts are made in regard to correlation and interpretation.

Table 9.2   Soil Test Methods, Soil Factors Influencing Their Interpretation, and Typical Ranges in Critical Level for Micronutrients

| Element | Interacting Factors[a] | | Method | Range in Critical Level (ppm) |
	Essential	Probable		
B	Texture, pH	Lime	Hot H_2O	0.1–0.7
Cu	—	O.M., Fe	$NH_4C_2H_3O_2$ (pH 4.8)	0.2
			0.5 M EDTA	0.75
			0.43 N HNO_3	3–4
			Biological assay	2–3
Fe	—	pH, lime	$NH_4C_2H_3O_2$ (pH 4.8)	2
			DTPA + $CaCl_2$ (pH 7.3)	2.5–4.5
Mn	pH	O.M.	0.05 N HCl + 0.025 N H_2SO_4	5–9
			0.1 N H_3PO_4 and 3 N $NH_4H_2PO_4$	15–20
			Hydroquinone + $NH_4C_2H_3O_2$	25–65
			H_2O	2
Mo	pH	Fe, P, S	$(NH_4)_2C_2O_4$ (pH 3.3)	0.04–0.2
Zn	pH, lime	P	0.1 N HCl	1.0–7.5
			Dithizone + $NH_4C_2H_3O_2$	0.3–2.3
			EDTA + $(NH_4)_2CO_3$	1.4–3.0
			DTPA + $CaCl_2$ (pH 7.3)	0.5–1.0

Source: Cox and Kamprath (1971).
[a] Climatic and crop factors, although highly important, are not considered here.

Salinity and Alkalinity. Electrical conductivity on a saturated soil paste is almost universally accepted as the best method for salinity determinations. Care must be taken concerning the time and the depth of samples, as soluble salts are extremely mobile in irrigated systems. Alkalinity is best determined by calculating exchangeable sodium saturation. Sodium is easily determined in the same extracting solution as potassium, provided that the extractants are sodium free. The details for these methods and their interpretation are explained in *Handbook* 60 of the U.S. Soil Salinity Laboratory (1954).

Soil Test Correlation

The most difficult aspects of the soil fertility evaluation process are the correlation, interpretation, and recommendation phases, because of the complex phenomena involved. A soil test value per se is worthless; it is an empirical number that may or may not indirectly reflect nutrient availability. Soil test values become useful only when they are correlated with crop responses. Such correlations are usually conducted at two levels: an exploratory one in the greenhouse with a large number of widely diverging soils, and a more definite one in the field with fewer, but carefully selected, soils.

The primary purpose of greenhouse correlation is to compare different extraction methods and determine *tentative* critical levels. The purpose of field test correlations is to establish *definite* critical levels for a selected extraction method. Good greenhouse correlations, however, do not prove the effectiveness of a particular soil test; this can be demonstrated only in the field. The bulk of soil-testing research in the tropics, unfortunately, stops at the greenhouse stage.

Plant growth and yields are functions of many variables beyond the single nutrient under consideration. Actual yields are functions of over a hundred variables, which can be grouped into soil, crop, climate, and management categories (Fitts, 1955). Consequently, when yields are correlated with *one* variable such as available phosphorus, this means that available phosphorus is a more important limiting factor than the numerous uncontrolled variables in any correlation study. Better correlations are normally obtained in greenhouse studies, where the uncontrolled variables are more uniform. Nevertheless a considerable scatter of points occurs when absolute yield, yield response, nutrient uptake, or other growth parameters are plotted as functions of soil test values. An example is shown in Fig. 9.1. The continuum of points is commonly split into several arbitrary categories, such as

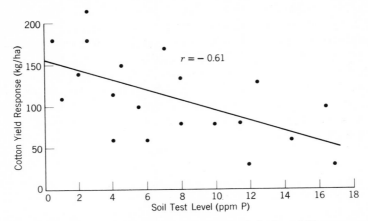

Fig. 9.1 Correlation between phosphorus soil test levels and absolute yield response with cotton in Iran. *Source:* Hauser (1973).

"low," "medium," and "high" (Fig. 9.2). Recommendation rates are based on the fertilizer needed to raise the soil test level to "high."

When relative yields rather than absolute yields are used, the variability is considerably reduced. Relative yield calculations are of two kinds: percent yield response, in which the actual yield response is divided by the absolute yield at the zero level, and the value obtained when the yield at a certain fertility level is divided by the maximum yield attained. In both cases, relative yield values reduce the variability of results and direct soil test correla-

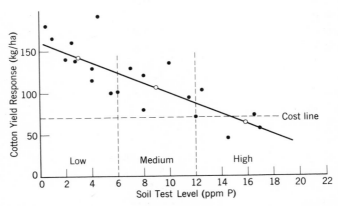

Fig. 9.2 Arbitrary grouping of soil test categories from Fig. 9.1. *Source:* Hauser (1973).

tion toward its realistic goal: the prediction of soil–crop situations for which there is a likelihood of significant fertilizer response (Waugh et al., 1973). Soil test correlations cannot predict yields or even absolute yield responses because of the many other variables involved.

A major breakthrough in soil test correlations occurred with the development of the Cate–Nelson method (Cate and Nelson, 1965). The simple graphic method consists of plotting relative yields (percents of maximum) as a function of soil test values, as Fig. 9.3 illustrates. Instead of attempting to fit a continuous mathematical function through the scattered points, a plastic overlay sheet divided into quadrants by a horizontal and a vertical line is used. This sheet is superimposed on the data in such a way that the maximum number of points falls in the lower left and upper right quadrants, and the minimum number of points is left in the upper left and lower right quadrants. The point at which the vertical line intersects the *x*-axis is considered to be at the *critical level* for the soil test in question. The point at which the horizontal line intersects the *y*-axis separates soils with high response from those with low response. The critical level, therefore,

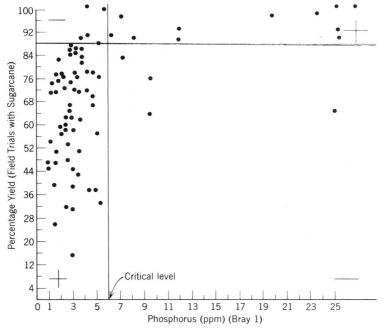

Fig. 9.3 Analysis of sugarcane data from Pernambuco, Brazil, by the method of Cate and Nelson. Each dot represents a field plot. *Source:* ISFEIP (1967).

divides the data points with a likelihood of a large yield response from those with little probability of obtaining a response.

In comparison with the conventional correlation techniques, the Cate–Nelson method presents some fundamental and practical advantages. By separating the data points into two populations, it follows Leibig's law of the minimum, because the critical level is the point beyond which the nutrient in question is no longer a limiting factor. In contrast, continuous regression models show no inflection points and force the groupings to be arbitrary. The main advantage of the Cate-Nelson approach is that it recognizes the basic limitation of soil tests: they are able only to separate the soils that are likely to respond to the added nutrient from those unlikely to respond. Thus the population is split into two soil–crop categories.

Another advantage is that the method identifies the soils in which the extractant does not work well (the points lying in the top left and bottom right quadrangles). They can be subjected to further study and refinement.

The simplicity of this approach has a major practical advantage, particularly in tropical laboratories without ready access to digital computers. All that is needed is a piece of transparent paper instead of complex mathematical calculations that must be handled by computers. The statistical soundness of the technique, however, has also been proved by fitting a discontinuous linear regression model and comparing it with the conventional curvilinear models (Cate and Nelson, 1971). Table 9.3 compares this model with the conventional quadratic, logarithmic, and Mitscherlich models on a worldwide collection of soil test–relative yield correlation data. The coefficients of determination (R^2 values) were somewhat higher with the new model.

The use of this technique is shown in Figs. 9.4 to 9.6. The first step, a

Table 9.3   Performance of Different Mathematical Models on a Worldwide Sample of Correlations between Relative Yields and Soil Test Levels

Regression Model	Equation	R^2
Quadratic	$Y = b_0 + b_1 x - b_2 x^2$	.58
Logarithmic	$Y = b_0 + b_1 (\log x)$	.59
Mitscherlich	$Y = A(1 - e^{cx})$	.66
Cate–Nelson	$Y = b_0 + b_1 x_1$	.73
	where $x = 0$ if below critical level	
	$x = 1$ if above critical level	

Source: Cate and Nelson (1971).

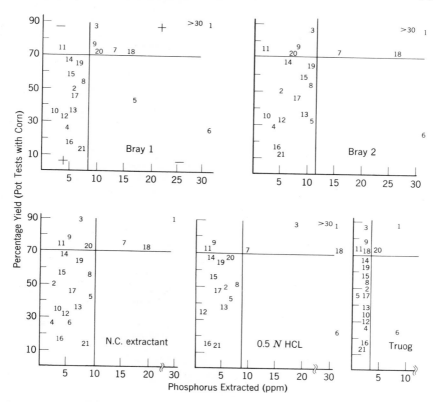

Fig. 9.4 Correlations between soil test levels and relative yields, using several soil test methods, in a greenhouse experiment with corn conducted in Recife, Brazil. Numbers indicate the quantity of data points in each position. *Source:* ISFEIP (1967).

comparison of different soil test methods, was conducted in the greenhouse (Fig. 9.4). In this case all soil test methods correlated well, and each one produced a critical level. The results were then taken to the field, using the North Carolina extractant. The results confirmed 10 ppm as the critical level under field conditions (Fig. 9.5).

The critical level is specific to certain soils–crop situations, even with the same extractants. Each laboratory, therefore, should establish its own critical levels for its major soils and crops. For example, two different critical levels were found for zinc in Peru when the soil population was split between calcareous and acid soils (Fig. 9.6). This example confirms the previously described strong influence of pH level on available zinc, and shows how the correlations can be refined by such groupings. Crops with different

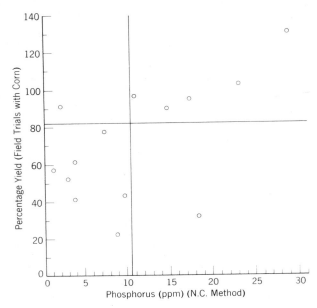

Fig. 9.5 Correlation between the chosen soil test method and corn response in the field in Recife, Brazil, according to the results of Fig. 9.4. *Source:* ISFEIP (1967).

nutritional requirements are likely to have different critical levels. Goswami et al. (1971), for example, compiled a large number of correlation studies and found a critical level of 30 ppm P (Olsen) for lowland rice and 49 ppm P (Olsen) for wheat in Vertisols. This difference is probably due to the lower phosphorous requirement of rice and perhaps also to the increased availability of phosphorus under flooded conditions.

The Cate–Nelson method is now widely used in the tropics (Goswami et al., 1971; Roufunnisa et al., 1971; Subramanian, 1971; Baynes and Walmsley, 1973; Palencia, 1974; Cano, 1973; Perur et al., 1974). In some laboratories, the interpretation is modified by splitting soils into the levels "low," "medium," and "high." The critical level usually falls between "medium" and "high" in such cases (Hauser, 1973). The Cate–Nelson approach has not been used in United States or Australian soil-testing programs. Soil test correlation methods in these countries are based on multiple regression computer programs with strong emphasis on nutrient interactions (Cope and Rouse, 1973; Hanway, 1973; Colwell, 1971).

Interpretation and Recommendations

Although the critical level separates soils with high probabilities of fertilizer response from soils with low probabilities, it tells nothing about the rate of fertilizer to be recommended. Likewise, correlations made according to continuous regressions do not indicate how much fertilizer to recommend. Fertilizer recommendations are obtained only through field trials. Since

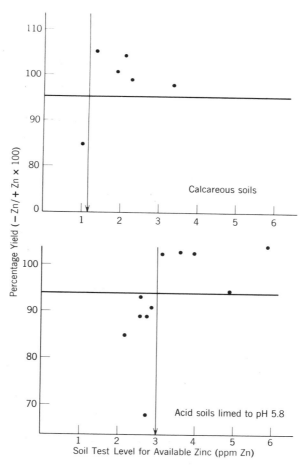

Fig. 9.6 Correlation between soil analysis for available zinc and yield response to applied zinc for sorghum in soils of the Coast, Sierra, and Selva of Peru. *Source:* ISFEIP (1968).

field trials are expensive to conduct, the soils should be carefully selected and the number of experiments kept relatively small in order to be able to manage them carefully.

The purpose of soil test interpretation is to establish how much of each nutrient must be applied to bring about a given yield response within a predictable crop–soil category (Waugh et al., 1973). A separate crop-soil category indicates that interpretation must be different for soils above or below the critical level and also for different crops. Figure 9.7 summarizes the results of several years of wheat trials in Bolivia, under both irrigated and rainfed conditions. This example suggests that trials should be run in soils testing *below* the critical level after this level has been determined, and that the absence of response in soils *above* the critical level should be confirmed in the field.

As in the case of correlations, two main methods of soil test interpretation are used in the tropics: the continuous (curvilinear) and discontinuous (linear) models.

Continuous models. These classic models are based on the law of diminishing returns, where appropriate curvilinear functions are fitted to yield response data. The most commonly used functions are the quadratic, square-root, logarithmic, and Mitscherlich. Statistical techniques determine which function fits the data best by providing the largest coefficient of determination (R^2). The optimum fertilizer rate occurs at that point in the

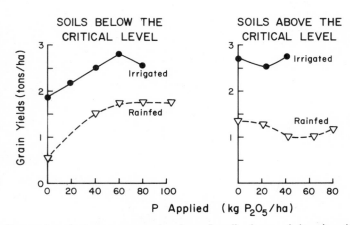

Fig. 9.7 Contrasting wheat responses to phosphorus in soils above or below the critical level (7 ppm P—Olsen) in Bolivia under irrigated and rainfed conditions. *Source:* Adapted from Waugh and Manzano (1971).

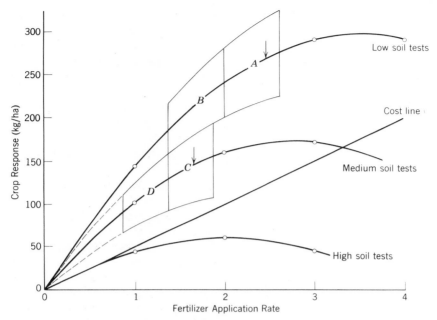

Fig. 9.8 Interpretation graph, using continuous curvilinear response functions for the cotton fertilization experiments of Fig. 9.1. Arrows indicate optimum economic rates. *Source:* Adapted from Hauser (1973).

curve where the marginal revenue equals the marginal cost (i.e., the point where the price of the last yield increment equals the cost of the last increment of fertilizer). The point can be determined mathematically or graphically by drawing a price:cost ratio line, expressed in the agronomic equivalents, in the yield response diagram. The optimum yield is then determined at the point where a tangent of the price:cost ratio line intersects the response curve.

Separate yield response equations are prepared according to the groupings made during the correlation process for each crop. If the Cate–Nelson technique is used, equations for soils testing above and below the critical level can be developed. If the continuum is split into "very low," "low," "medium," "high," and "very high" categories, there will be five response functions and an optimum fertilizer recommendation for each soil test level. An example of this second approach is shown in Fig. 9.8, based on the correlation data presented in Figs. 9.1 and 9.2. Figure 9.8 presents a helpful modification: an optimum range rather than an optimum level is indicated.

Ranges *A* and *C* reflect recommendations for the highest profit per hectare, while *B* and *D* are areas of lower fertilizer cost and higher return per unit of fertilizer (Hauser, 1973). Similar interpretations for other economic alternatives have been presented by Heady and Ray (1971). This modification is helpful in handling the economic uncertainties typical of developing countries.

When more than one element is deficient, regression models take this into account, as well as the interactions between these elements. The recommended rates are determined by solving simultaneous equations.

Continuous models are also used in a different manner to produce one function that takes into account soil test levels as variables, as well as other variables related to soil, climate, and management properties, in an attempt to account for the uncontrolled variables. This is the prevailing philosophy in the United States (Hanway, 1971; Walsh and Beaton, 1973) and in Australia (Colwell, 1971). A yield equation relating corn yields in Iowa with controlled and uncontrolled factors included 30 variables (Hanway, 1971). One for potato fertilization in Peru needed 27 variables (Ryan and Perrin, 1973). Optimum fertilizer levels are calculated on the basis of certain levels of crop prices, fertilizer costs, expected rainfall, and soil properties.

Such complex models are effective when there is adequate information about the variables involved, and when prices are stable. They usually fail in the tropics, however, because there are insufficient data to quantify all variables. Their use in tropical regions is limited to after-the-fact analysis in areas with detailed information; they do not serve successfully as predicting tools.

The quadratic models also have limitations. Anderson and Nelson (1975) found that quadratic models are biased when there is a marked response to the first fertilizer increment, followed by little or no response at higher rates. In such cases optimum fertilizer recommendations are unrealistically high.

Optimum fertilizer recommendations vary with price:cost ratios, which are characteristically more erratic in the tropics than in the developed countries. Recommendations are as good as the predicted prices.

Discontinuous linear model.   A series of studies conducted in England by Boyd (1970, 1974) and in the United States by Bartholomew (1972) summarized many fertilizer response functions over the world and concluded separately that in most instances fertilizer response curves can be characterized by a sharp linear increase followed by a flat horizontal line.

Waugh, Cate, and Nelson then developed the "linear response and plateau" model, which is also based on Leibig's law of the minimum and is a logical extension of the Cate–Nelson correlation model (Waugh et al., 1973,

1975). In this model fertilizer response from a field or a group of fields is represented by two straight lines for each individual nutrient. The first line represents the relatively steep response of an added nutrient until it ceases to be a limiting factor. This is followed by a flat plateau, where further additions no longer increase yields. The fertilizer response curve so constructed consists of three main points. The "threshold yield" is the yield at the zero level of the nutrient in question, but not of all nutrients. The "plateau yield" is the yield at the point where the nutrient ceases to be a limiting factor; it is not the maximum yield because other factors may still limit yields. The "relative yield" is simply the threshold yield divided by the plateau yield. The fertilizer rate needed to reach the plateau yield is the recommended rate for the particular nutrient.

As one nutrient ceases to be limiting, others may still be so. This concept, which is the classic interpretation of Leibig's law of the minimum, is illustrated in Fig. 9.9. The final plateau yield is the effect of genetic limitations, solar radiation, and other uncontrolled variables.

Regression studies indicate that there is no difference between this model

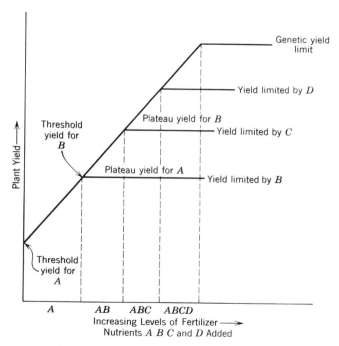

Fig. 9.9 Linear response and plateau (LRP) model, based on Leibig's law of the minimum. *Source:* Waugh et al. (1973).

and the curvilinear ones in terms of coefficients of determination (R^2). In fact, the linear response and plateau model provided the highest R^2 in 27 out of 37 response functions studied by Waugh et al. (1973). Although there is no question that the actual biological function is curvilinear, the interpretation of it as two straight lines is of considerable value in simplifying soil-testing interpretations. It also provides more realistic rate recommendations, since the bias toward unrealistically high fertilizer recommendations of the quadratic equation is eliminated.

The fitting of the two straight lines is described in detail by Waugh et al. (1973). It can be done by individual soil or soil–crop categories split by Cate–Nelson techniques for each crop. The fitting can be done graphically by drawing a straight line along the points that show an increase and another along the flat points. A statistical fit can be determined by least significant differences between the plateau points and other, more sophisticated statistical techniques. In most cases a visual fit is sufficient. This model ignores yield decreases with excessively high nutrient rates. This part of the curve, when it exists, is of no significance to soil test recommendations.

An example of fitting a linear response and plateau function is shown on the left-hand side of Fig. 9.10 for an individual nutrient (nitrogen) in a

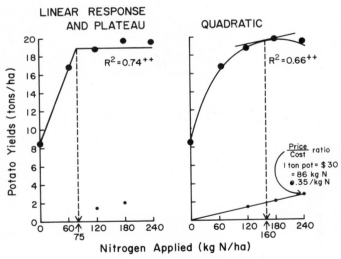

Fig. 9.10 Determination of nitrogen recommendations for potatoes in a set of field experiments from Bolivia according to the linear plateau response and the conventional curvilinear models. Each dot is the mean of several field experiments in a given crop–soil category. *Source:* Adapted from Waugh et al. (1973).

group of soils from Bolivia. The interpretation of the responses to separate nutrients is illustrated in Fig. 9.11, based on the same crop–soil category as Fig. 9.10. Figure 9.11 shows a strong response to nitrogen and phosphorous and a slight response to potassium. The recommendation choices depend on the yield level desired. A rate of 25-0-0 is required to raise the yield to the phosphorus threshold yield (no P) level; a rate of 65-60-0, to raise it to the potassium threshold yield (no K); and a rate of 75-70-20, to reach the maximum yield under these conditions. The decision between these four choices (including no fertilization) is based on a stepwise marginal analysis, shown in Table 9.4. If the rule of not recommending rates that return less than $1.5 per dollar invested in fertilizer is used, the recommendation is 60-55-0. Other factors such as risk can be considered in making the final decision.

The comparison between the two approaches is shown in Fig. 9.10 for the same data set. The dotted lines indicate how the fertilizer recommendations are arrived at. The linear response model has only one optimum point independent of cost and prices. The curvilinear model shows an optimum point based on the particular price:cost ratio at the time the experiments were conducted. It is very interesting to note the wide differences in optimum recommended rates between the two methods. The linear response and plateau model has a lower recommended rate because it attempts to reach a yield plateau of only about 19 tons/ha. This rate in effect provides nearly maximum yields, while preserving an efficient return per unit of fertilizer, because it is still along the increasing slope. The quadratic model in this case almost doubles the optimum recommendations in order to obtain a yield of 20 tons/ha. This is in the relatively flat part of the curve, where variability is quite high. Small changes in yields result in large changes in recommended rates. In other comparisons, however, the difference in recommended rates might not be as large as in this example.

Which approach is right? The answer can be derived from after-the-fact economic analysis. Table 9.5 shows data on nitrogen response in rice which this writer analyzed according to the quadratic model and published (Sanchez et al. 1973). By using the graphic technique, plateau yields and recommended rates were computed. The average nitrogen recommendation was 224 kg N/ha according to the quadratic model and 170 kg N/ha with the linear model. The average gross return for fertilization at the recommended rates was $486/ha for the quadratic and $457/ha for the linear response and plateau model. This difference is not statistically significant. The net return per dollar invested in fertilizer was superior in the linear plateau model, $8.8 versus $6.1 per dollar invested.

A comparison between the linear response and plateau model and a generalized quadratic response function with 20 variables on 26 corn experi-

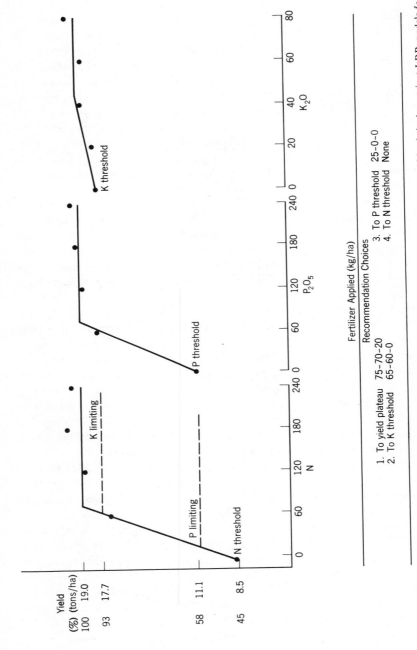

Fig. 9.11 Multinutrient response interpretation and development of fertilizer recommendations, using combined single-nutrient LRP models for potato fertilization in Bolivia. *Source:* Waugh et al. (1973).

Table 9.4 *Economic Analysis of Response Data of Fig. 9.11 Used to Arrive at the Fertilizer Recommendation (U.S. $/ha)*

Fertilizer Recommendation Choice (kg $N-P_2O_5-K_2O$/ha)	Average Potato Yield (tons/ha)	Crop Value at U.S. $30/ton	Marginal Increase in Crop Value	Fertilizer Cost	Marginal Increase in Fertilizer Cost	Marginal Value: Marginal Cost Ratio
0-0-0	8.5	255	—	0	—	—
25-0-0	11.1	333	78	10	10	7.8
60-55-0	17.7	531	198	40	30	6.6
70-65-40	19.0	570	39	60	20	1.9

Source: Adapted from Waugh et al. (1973).

Table 9.5   Nitrogen Recommendation Rates of a Series of Rice Experiments with Sulfur-Coated Urea in Peru, Based on the Quadratic Curve Marginal Analysis (as Originally Published) and According to the Graphic Linear Response and Plateau Model

Experiment No.	N Recommendation (kg N/ha)		Return on Fertilization (U.S. $/ha)		Net Return per Dollar of Fertilizer Invested (U.S. $/ha)	
	Quadr.	LRP	Quadr.	LRP	Quadr.	LRP
1	266	160	406	385	4.5	7.1
2	201	180	522	493	7.7	8.1
3	372	240	617	648	4.9	7.9
4	258	160	529	475	6.0	8.8
5	290	160	708	617	7.2	11.4
6	147	90	465	448	9.3	14.9
7	186	120	300	225	4.7	5.5
8	222	160	338	361	4.5	6.7
Mean	224	170	486	459	6.1	8.8
LSD$_{0.05}$.			52			

Source: Adapted from Sanchez et al. (1973).

ments from Minas Gerais, Brazil, is shown in Table 9.6. In this case there were no significant differences in recommended rates or profitability of the prediction, but both models overestimated the rates required for the actually most profitable combination.

The decision on which model to use may vary with individual inclinations and the availability of computers. The linear response and plateau model is

Table 9.6   Comparison of Recommended Rates and Profitability as Determined by the Graphic Linear Response and Plateau Model and by a Generalized Quadratic Equation with 20 Variables: Mean of 26 Corn Experiments from Minas Gerais, Brazil

Model Used	Recommended Rate (kg/ha)		Profitability of Recommended Rates (Cr $/ha)
	N	P$_2$O$_5$	
Actual best treatment	107	50	1460
Graphic LRP	120	85	1118
Computed quadratic	127	64	1109

Source: North Carolina State University (1973).

new and controversial. It has been successfully applied in several countries (Perur et al., 1974; Palencia, 1974) and to this writer's knowledge is at least as good in predicting yield response as the quadratic model. Moreover, the linear response and plateau model has a very practical advantage: it does not require a computer or complex calculations. The fact that prices enter into consideration only at the last stages of interpretation is also helpful. One significant aspect of the interpretation process is that fertilizer recommendations can be arrived at without the use of soil test data. This is how nitrogen recommendations usually are made because of the general lack of acceptable nitrogen soil tests, but it is not a valuable shortcut for other nutrients. Soil test values help greatly in separating the soil categories on which interpretation is made. The best interpretations are made for specific soil–crop categories, according to Waugh et al. (1973).

Using Soil Test Recommendations

Regardless of how fertilizer recommendations are determined, they are worthless unless farmers use them. Their effective application depends on the educational and extension aspects of the soil evaluation program, as well as on the validity of the recommendations themselves. The educational and promotional aspects are initially geared to capture the farmer's attention. Usually this effort involves field days, demonstration plots, radio programs, and audiovisual aids. After farmers are aware of the program, they will accept or reject the recommendations. In tropical areas where yields are low, the expected yield, to be appealing, must be at least more than twice the yield without fertilization (Fitts, 1971). As average yield levels increase, smaller increments may suffice, but at the beginning the soil fertility evaluation program must give rather spectacular results.

Failures in soil testing efforts are due to many factors. A principal one is the inability to obtain yield responses of such magnitude that yields can be doubled in areas with low average yields. E. W. Russell (1974) recognized this as one of the limiting factors preventing widespread fertilizer use in Africa. No rational farmer is going to adopt a practice that may increase his yield by 10 or 20 percent if his risks are higher than this. The lack of such yield responses may be due to incomplete research, such as limiting correlation studies to the greenhouse stage, or to extrapolating a few field results with one crop to widely divergent ones.

Successful soil fertility evaluation programs also summarize their information by regions or political subdivisions for each major crop. Examples of these country summaries are available for Central America (ISFEIP, 1967), Venezuela (Chirinos et al., 1971), Peru (Cano, 1973) and many other countries. They indicate the major fertility problems and are useful in determining the P:K ratios of mixed fertilizers that a region should use. Specific fertilizer recommendations must be clear and straightforward. They should include not only recommended rates for specific crops and soils but also the best fertilizer sources, placement methods, and timing of applications, obtained from other research projects. An excellent example is the fertilizer recommendations of Minas Gerais State in Brazil (PIPAEMG, 1972).

In areas where soil fertility evaluations of the type described have been put into practice, positive correlations between the number of soil tests and fertilizer consumption exist (Kanwar, 1971; ISFEIP, 1972). Although a correlation does not imply a cause–effect relationship, it is clear that successful soil testing and increased fertilizer consumption go hand in hand. At the beginning of a program it is probable that soil tests increase fertilizer use; as fertilization practices become widespread, increasing fertilizer demand augments the need for additional soil analysis.

Research

The last phase of a soil fertility evaluation is research to improve the quality of recommendations. This involves troubleshooting on problem samples through greenhouse research, plant analysis, refining soil–crop categories, and finding out why some soils are in the wrong quadrangle of the correlation graph. Examples of research projects are found in the list of references. New extraction methods and improved laboratory techniques must be evaluated under local conditions. Like any other service function, soil fertility evaluation requires research to keep it up-to-date. From the advances in the past few years, this appears to be a very productive area of research and innovation.

FERTILITY EVALUATION SYSTEMS BASED ON PLANT ANALYSIS

Other soil fertility evaluation systems are based on plant analysis. These are particularly widespread in areas without effective soil-testing systems and are especially popular for permanent crops. The division between fertility evaluation systems based on soil testing and those utilizing plant analysis is somewhat arbitrary because plant analysis is a component of systems based on soil testing and vice versa. Nevertheless, some evaluation systems are primarily based on plant analysis.

Plant analysis has one fundamental advantage: it integrates the effects of the soil, plant, climate, and management variables. In this way it is the ultimate measure of nutrient availability. Plant analysis has, however, one fundamental disadvantage: by the time the analysis indicates a nutritional problem, it is probably too late to correct it without considerable yield loss.

Plant analyses are used for three main purposes: (1) to identify nutritional problems and quantify their correction through the establishment of critical levels; (2) to compute nutrient uptake values as a key for fertilizer use; and (3) to monitor the nutrition of permanent crops, a practice called "crop logging."

Critical Levels

When plant samples are taken from the same anatomical part and at the same growth stage, certain critical levels can be established above which the plant is sufficiently supplied with the nutrient in question and below which it is not. A second set of critical levels is needed for nutrients that can be present in toxic amounts, such as boron, iron, and manganese, or in cases where an excessive amount of the nutrient can cause yield declines. An example of the latter is excessively high nitrogen contents in tall-statured rice varieties. The range between these two critical levels is called the "sufficiency range." The term "critical level" is normally used in conjunction with the deficiency–adequacy threshold.

Critical levels for plant analysis follow the law of the minimum, and essentially the same approach as the Cate–Nelson concept is used. In fact, the shape of nutrient concentration versus relative yield data (Ulrich and Hills, 1973) could be easily split by using the Cate–Nelson technique. The critical range for toxicity is an added factor not commonly considered in soil analysis.

Plant analysis critical levels are less site- and situation-specific than those obtained from soil tests, as long as they are standardized with respect to plant part, age, and in some cases varieties. Table 9.7 shows some typical examples from standardized plant parts and growth stages. The reader is referred to the references cited in this table for more detail, as well as to articles on this subject by Wallace (1956), INEAC (1958), Malavolta et al. (1962), Ishizuka (1971), Walsh and Beaton (1973), and Okajima et al. (1975). Some excellent pictures of nutrient deficiency symptoms appear in some of these publications, as well as in *Hunger Signs in Crops* by Sprague (1965).

A cursory mention of the visual identification of nutrient deficiency symptoms is not intended to deny the value of this technique. A description of the visual nutrient deficiency symptoms of major tropical crops, however, is beyond the scope of this book. The ability of agronomists to

Table 9.7   Examples of Critical Levels Separating Deficiency and Adequacy in Some Crops

Element	Sugarcane[a]	Rice[b]	Corn[c]	Soybeans[d]
N (%)	1.5	2.5	3.0	4.2
P (%)	0.05	0.10	0.25	0.26
K (%)	2.25	1.0	1.90	1.71
Ca (%)	0.15	0.15	0.40	0.36
Mg (%)	0.10	0.10	0.25	0.26
S (%)	0.01	0.10	—	—
B (ppm)	1	3.4	10	21
Cu (ppm)	5	6	5	10
Fe (ppm)	±10	70	15	51
Mn (ppm)	10–20	20	15	21
Mo (ppm)	—	—	0.1	1.0
Zn (ppm)	10	10	15	21
Si (%)	—	5	—	—

[a] From Humbert (1973). Values refer to leaf blades 2, 4, 5 and 6 except for P, K, Ca, Mg, and Zn, which represent leaf sheaths 2, 4, 5, and 6.
[b] From Tanaka and Yoshida (1970). N, P, K, and Fe values are for leaf blades at tillering stage; Ca, Mg, S, B, and Cu values, for straw at harvest; Mn and Zn values, for shoots at the tillering stage.
[c] From Jones and Eck (1973). All values are for the ear leaf at tasseling except for Mo, which represents the entire plant.
[d] From Small and Öhlrogge (1973). Lower limit of sufficiency range for upper fully developed trifoliate leaf blades sampled before pod set.

identify visual deficiency symptoms is a major asset in a soil fertility evaluation program. These visual observations should be verified with plant chemical analyses and related to critical levels. Unfortunately, both visual and chemical determinations are hampered when deficiencies of more than one nutrient are present. This situation may lead to an accumulation of other nutrients or just the opposite. Visual symptoms are further compounded when disease or insect attack symptoms appear.

Nutrient Uptake as a Predictive Tool

The general failure of soil tests for nitrogen has encouraged a completely different approach for estimating fertilizer nitrogen rates. In a review of the world literature Bartholomew (1972) observed a constant relationship between grain yields of cereal crops and their total nitrogen uptake (including roots). This relationship is shown for commercial grain yields of corn, wheat, and rice in Fig. 9.12. The slopes of the curves indicate that corn and rice averaged from 30 to 35 kg grain/kg N added, whereas wheat averaged only 15 to 20 kg. The lower efficiency of wheat is thought to be the result of generally drier climate and the lodging susceptibility of tall-statured varieties. If one knows the average yields without nitrogen fertilization (threshold yields) and the plateau yields for a specific area, the amount of nitrogen required for increasing yields from the threshold to the plateau can be obtained from such a graph. Bartholomew's uptake–grain yield rela-

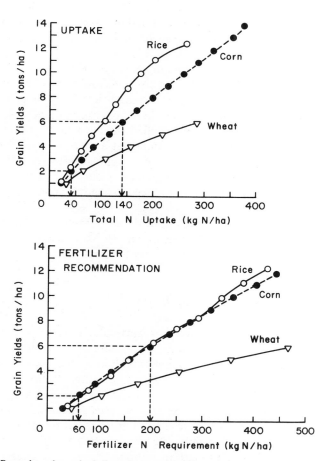

Fig. 9.12 Procedure for calculating nitrogen rates from crop uptake data based on average field trials and average fertilizer recovery. Rice expressed in unhulled (paddy) grain fields. *Source:* Adapted from data by Bartholomew (1972).

tionships are illustrated in the top half of Fig. 9.12. For example, if the threshold yield of corn is 2 tons/ha and it is known that with nitrogen fertilization and good management 6 tons/ha is probable, the crop will need to take up 100 kg N/ha extra (140–40) to attain 6 tons/ha, provided that nitrogen is the main limiting factor within this yield range. Figure 9.12 also shows the differences in nitrogen requirement between the three cereals. Rice required about 17 kg N to produce an extra ton of rough rice, corn about 25 kg N, and wheat about 42 kg N. These figures show that wheat is the least efficient user of nitrogen of these three crops.

Not all the fertilizer nitrogen will be taken up by the plant. The efficiency of fertilizer utilization varies with soil, varietys, climate, and management. On the basis of a worldwide survey, Bartholomew (1972) calculated the actual fertilizer nitrogen requirements at high, medium, and low levels of fertilizer efficiency. The lower half of Fig. 9.12 is based on the average nitrogen recovery rate. The percentage rates are 36 for rice, 72 for corn, and 62 for wheat. The amounts of fertilizer nitrogen required for a given yield increase are almost identical for corn and rice. They range from 30 to 35 kg grain/kg N. The high efficiency in nitrogen uptake of rice, however, is offset by the lower efficiency of applied nitrogen under flooded conditions, as compared to corn. Wheat responds at a rate of 15 to 20 kg grain/kg fertilizer, in spite of the high recovery of applied fertilizer nitrogen. The nitrogen fertilizer recommendation to increase yields from 2 to 6 tons/ha of corn would be 140 kg N/ha, the difference between the points marked with the arrows on Fig. 9.12.

These estimates, as well as the more detailed tables in Bartholomew's publication, are based on broad generalizations. More accurate predictions can be given with local uptake and fertilizer efficiency data.

Crop Logging

As mentioned before, plant analysis is particularly useful for permanent crops, where soil tests are less meaningful because of the high proportion of roots in the subsoil. An intensive plant-sampling system was developed by Clements (1960) to monitor the nutrient and water status of sugarcane plantations as a guide for fertilization and irrigation. This system, subsequently called "crop logging," is used in various sugarcane-growing areas but principally in Hawaii. It is described in great detail by Humbert (1963). Individual fields are sampled every 35 days during the first 6 months of growth, and the results are plotted on running graphs. The charts show the percentage of nitrogen in leaf blades and the percentages of phosphorus and potassium in elongated leaf sheaths. Rainfall, irrigation, temperature, and plant height records are also kept, and fertilizer and irrigation practices are recorded. When the plant analysis laboratory is capable of analyzing plant samples quickly, crop logging gives an excellent measure of crop growth and increases the efficiency of fertilizers and irrigation. Crop logging has also been practiced on other permanent crops, but on short-term crops such as corn because there is not enough time to make use of the analyses.

FERTILITY EVALUATION SYSTEMS BASED ON MISSING ELEMENT TECHNIQUES

A third major approach is based on the identification of nutrient deficiencies by the missing element technique. This involves growing indicator plants in the greenhouse or in the field on a soil to which a "complete" fertilization treatment is added, and a series of treatments in which one element is not added. The plants are normally harvested at an immature stage.

This technique has been used on a routine basis by IRAT researchers in Francophone Africa, where greenhouse pot tests with cereal plants as indicators constitute the first stage of the fertility evaluation process (Kilian and Velly, 1964; Chaminade, 1972; Charreau, 1974). This approach has also been used in greenhouse tests in Costa Rica (Martini, 1969b) and the Gambia (Webb, 1955), with small plastic cups by ISFEIP (1971, 1972, 1973) in Latin America, and in the field with "microplots" of corn of less than 1 m² area per treatment in Costa Rica and Brazil (Hardy and Bazán, 1966; Alvim, 1968; Martini, 1969a; Cordero and Salas, 1971).

According to Chaminade (1972), pot experiments with the missing element technique give three types of information: (1) which elements are deficient, (2) the relative importance of the deficiencies, and (3) the rate at which fertility is depleted with successive cuttings when a pasture indicator crop is used. The IRAT researchers consider deficiencies serious if the dry matter production is depleted to 40 percent or less of the complete treatment. Field trials with the appropriate missing elements follow the pot experiments and serve as the basis for fertilizer recommendations. The French approach places strong emphasis on soil surveys to select the soils to work with; the rate determination experiments are supplemented by research on rotation systems, the residual effects of fertilization, and the quantities needed to maintain the appropriate fertility level. Nutrient uptake serves as a basis for those maintenance applications.

Missing element techniques have the disadvantages that only early growth data are obtained, that nutrient deficiencies may be exaggerated because of the small volume of soil, and that the effort necessary precludes their use for large numbers of soils from farmers' fields. Their greatest value is for preliminary screening in research projects or for obtaining needed additional information on soils for which poor soil test–yield response correlations are obtained. The ISFEIP modifications provide a simpler methodology for such troubleshooting. Hunter (1975) found that a soil volume of 150 ml was as good as larger amounts for identifying missing nutrients when sorghum was grown for 6 to 8 weeks. Styrofoam coffee cups are used, and the cups are automatically watered by cigarette filter wicks connected to a water reservoir. This technique is also useful for diagnosing micronutrient deficiencies or toxicities, as well as nutrient imbalance problems.

In all cases the crucial step is the determination of the nutrient rates for the "complete" treatment. Serious errors can occur when arbitrary rates are used. Soil tests before initiating the greenhouse experiments should eliminate this problem.

SIMPLE FERTILIZER TRIALS ON FARMERS' FIELDS

A fourth major approach used in evaluating soil fertility in the tropics consists of the simple fertilizer trials conducted under the leadership of the Food and Agricultural Organization (FAO) of the United Nations. The purpose of this program is to introduce fertilizers as a means for increasing yields in large areas of the tropics. This program has been described at its different stages by Mukerjee (1963) and Hauser (1971, 1973, 1974).

The FAO program is based on H. N. Mukerjee's "method of dispersed experiments." Its basic assumption is that fertilizer needs are best estimated by conducting a large number of unreplicated fertilizer trials in farmers' fields chosen at random without prior analysis. Individual trials dispersed over a visually uniform area or soil type are considered replicates, and the average yield of such areas is determined. The individual experiments consist of a small number of treatments, usually a $2 \times 2 \times 2$ factorial NPK design. The rates used are rather small (20 and 40 kg/ha) because the purpose is to provide maximum efficiency of fertilizer investment, which is normally attained in the initial part of the response curve. Later designs include up to 16 treatments to establish the response curve of two nutrients. The underlying philosophy of this program is that fertility response must be evaluated in farmers' fields and not under controlled conditions at experiment stations.

Hundreds of thousands of such trials have been conducted with major crops in all or large parts of 40 countries, including the Philippines, Pakistan, Bangladesh, India, Indonesia, Malaysia, Vietnam, Ceylon, Thailand, Ghana, and Brazil. The results are published in individual country reports with limited distribution. Whenever soil maps are available, the results are given by soil groups; in the absence of maps, by geographical areas thought to be uniform. Whenever soil testing laboratories are available, correlation between yield response and soil test levels are attempted but usually with limited success. Mukerjee (1963) stated that the average estimate of nutrient needs for a visually uniform area is insufficient to provide a fertilizer recommendation to an individual farmer. He considered soil tests necessary to attain this objective.

The overall results of the program were recently summarized by Hauser (1974). The average yield response with the modest fertilizer rates used was 60 percent of the check yields. In grain crops this represented a yield increase from 1.0 to 1.6 tons/ha and an average return of $3.3 per dollar invested in fertilizer. The overall increase in fertilizer cost in all countries involved averaged $40 million per year during the last 10 years.

Examples of some country or regional summaries are shown in Table 9.8 (and in Table 10.11 of the next chapter). In general these trials show that moderate yield increases are obtained with moderate fertilizer rates. The response of individual crops by areas indicates whether or not such areas are deficient in nitrogen, phosphorus, or potassium. It also shows the different yield potential of different regions.

The benefits of this approach, considering the large numbers of experiments and the efforts invested in them, are essentially limited to the above conclusions. The predictability of the method is minimal because it ignores local soil variability. Consequently, no site-specific recommendations can be made.

Table 9.8   Results of Simple Fertilizer Trials Conducted on Corn in the Philippines, Arranged by Regions

Treatment (kg/ha)			Corn Yield (tons/ha)		
N	P_2O_5	K_2O	Isabela (20 trials)	Central Luzon (35 trials)	Bicol Peninsula (14 trials)
0	0	0	1.0	2.0	2.1
45	0	0	1.4	2.9	2.5
90	0	0	1.6	3.2	3.2
0	45	0	1.0	2.2	2.6
45	45	0	1.9	3.2	3.1
90	45	0	1.8	3.1	2.9
45	90	0	1.6	3.2	2.9
45	45	90	1.9	3.2	3.1

Source: FAO (1969).

The advent of high-yielding varieties of rice and wheat rendered the thousands of experiments conducted on tall-statured varieties obsolete when farmers adopted the short-statured varieties and applied nitrogen profitably at rates way beyond those used in these trials. Although the approach of conducting experiments under farmers' field conditions is certainly a valid one, the lack of on-site replication, the low fertilizer rates used, and the ignoring of local soil variability have limited its usefulness.

Efforts to select sites according to soil properties and soil test criteria, as well as the use of improved varieties and higher fertilizer rates to reach some sort of yield plateau, are improving the quality and predictability of the FAO approach. At later stages in the Philippines project (FAO, 1969) a small proportion of the experiments (about 7 percent) was conducted with both short- and tall-statured rice varieties at nitrogen rates reaching 130 kg N/ha. Grain yields reached 5 to 7 tons/ha. Such experiments conducted in farmers' fields provide extremely valuable information for determining nitrogen recommendations.

RELATIONSHIP BETWEEN SOIL FERTILITY AND SOIL CLASSIFICATION

The underlying assumption of the main soil fertility evaluation approaches is that fertilizer recommendations are site-specific. Differences in soil properties are one of the major reasons for site specificity. A logical conclusion is that soil fertility evaluation programs must be closely related to soil survey and classification programs. Unfortunately, this is seldom the case in either tropical or temperate regions. Buol et al. (1975) have summarized the problem in the following two paragraphs:

Within the field of soil science there is a clear difference between the subdisciplines of soil survey and soil fertility. Frequently these groups compete against each other for the dominant role in providing information about the agricultural potential of a country. The soil survey faction desires to produce maps which will inventory existing conditions, while the fertility group evaluates the potential of the soil for crop production through soil tests and field experiments. The roles of these two groups have tended to mutually exclude each other and have resulted in different approaches in providing information. The major difference is one of emphasis. Soil taxonomic systems stress subsoil features as major diagnostic criteria in the hierarchical grouping of soils and use the characteristics of the topsoil only at the lower categories. Fertility evaluation specialists confine their sampling to the plowed layer or the upper 20 centimeters of the soil. Thus the two groups really see two different soils while examining the same pedon.

The two groups are further alienated because of the divergent nature of their immediate, but not long-range, objectives. Soil surveyors attempt to provide information that will serve the needs of all potential land users over several decades, while the soil fertility specialist attempts to evaluate the fertility needs of a given crop for one or at most a few years, after which he will then reevaluate the soil. Thus the survey system attempts to be the storehouse of those physical and chemical soil properties with quasistable stature; the fertility group deals with the less stable components.

The amount of information needed by soil fertility specialists is only a fraction of the data gathered by soil survey groups. Consequently efforts have been made to interpret soil maps in terms of the productivity potential of the soil. The most widespread of these systems is the land use capability system developed by the U.S. Department of Agriculture several decades ago (Klingebiel, 1958). This system groups soils into classes of increasing land use limitations, designated as Classes I to VIII. The criteria used involve primarily soil physical properties such as permeability, slope, and erosiveness. No criteria related specifically to soil fertility are included. This system, with local modifications, is probably the most widely used technical system in the tropics (Steele et al., 1954; ONERN, 1971). In some cases soil fertility is included as an additional parameter and is classified as "high", "medium", or "low" (DaSilva and Carvalho, 1971; Equipe de Pedología e Fertilidade do Solo, 1972).

In the absence of a technical classification system for soil fertility, many interpretations are made empirically. In Africa, after soil surveys are made, interpretations of the capability of soil mapping units are based on personal experience and observations of general crop behavior without fertilization (Sys and Frankart, 1971; Bertrand, 1972). In some cases plant growth observations are quantified either by sampling farmers' fields or establishing field experiments (Waffelaert, 1963; Bouyer, 1963).

A more quantitative approach has been developed by Beek et al. (1965) as a land capability system for interpreting soil surveys of Brazil. Capability classes ("good", "fair", "restricted", "none") were developed for six management systems—ranging from the most primitive to the most sophisticated—according to five limiting factors: native soil fertility, water deficiency, water excess, erosion, and mechanization. Information is given as to the extent of limitations and the possibility for their improvement. For example, low soil fertility will severely restrict growth in a traditional unfertilized management system, whereas it will not limit a more intensive management system with heavy fertilization. The four native soil fertility limitation classes include in their definitions some qualitative limits. For example, the "best" class is defined as including soils with an oxic or argillic horizon, well drained, with more than 3 meq/100 g of exchangeable bases, less than 50 percent aluminum saturation, and less than 4 mmhos/cm of electrical conductivity. The other classes are also defined quantitatively. This system has been applied to the most recent soil surveys of Brazil, where suitability maps are drawn according to management systems (Divisão de Pesquisa Pedológica, 1973).

Special emphasis has been given to classifying soils under lowland rice production for productivity purposes. The reason for this is the obvious inapplicability of the USDA land use capability system to paddy soils. Capability systems for rice soils are based on limitations such as excessive drainage, salinity, and flood damage (Moormann and Dudal, 1964) and texture and drainage (Kilian, 1972; Kilian and Teissier, 1973), or on soil mapping units with rice yield potentials (Panabokke, 1969).

Another group of researchers attempted to correlate rice yields with soil properties. The most comprehensive effort of this type, undertaken by Kyuma and Kawaguchi (1973), was based on profile characteristics of 41 paddy fields of Malaysia. A total of 23 soil physical and chemical parameters was examined. Some of the parameters that were not mutually exclusive were grouped into four factors by principal component analysis and other mathematical means: organic matter, available phosphorus, "inherent potentiality," and available NH_4^+-nitrogen. Inherent potentiality is a function of cation exchange capacity, base status, clay content, and available silica. A numerical score was given to each factor. A quadratic regression between rice yields and these four factors plus their interactions produced an R^2 of 57 percent.

A technical classification system for grouping soils according to fertility limitations has been proposed by Buol et al. (1975). This fertility–capability soil classification system is the first one to group soils with similar fertility limitations, using quantitative limits. The second approximation of this system is shown in Table 9.9. Soils are grouped at the highest categorical level according to topsoil and subsoil texture. Thirteen modifiers are defined to delimit specific fertility-related parameters. These include identification of poor drainage (which is good for rice), strong dry season, low CEC, high and medium aluminum saturation levels, phosphorus fixation problems, severe physical limitations of Vertisols, potassium deficiency, and problems of calcareous soils, including salinity and alkalinity. Initial application of this system shows that some of the groupings have different fertilizer response patterns. The quantitative parameters act as critical levels, again following Leibig's concept. When fertility–capability groupings were used in conjunction with soil test critical levels, the profitability of fertilizer recommendations increased for potatoes in the Sierra of Peru (North Carolina State University, 1973). Although some of the parameters need further refining, this approach appears useful as a way of grouping soils with similar limitations for fertility evaluation purposes.

Table 9.9 Fertility–Capability Soil Classification System: 1974 Version

---

Type: Texture is average of plowed layer or 20 cm (8 in.) depth, whichever is shallower.

S = Sandy topsoils: loamy sands and sands (USDA).
L = Loamy topsoils: <35% clay but not loamy sand or sand.
C = Clayey topsoils: >35% clay.
O = Organic soil: >30% O.M. to a depth of 50 cm or more.

Substrata type: Used if textural change or hard root restricting layer is encountered within 50 cm (20 in.).

S = Sandy subsoil: texture as in type.
L = Loamy subsoil: texture as in type.
C = Clayey subsoil: texture as in type.
R = Rock or other hard root restricting layer.

Conditioner Modifiers: In plowed layer or 20 cm (8 in.), whichever is shallower, unless otherwise specified (*).

*g = (gley): Mottles $\leq$ 2 chroma within 60 cm for surface and below all A horizons or saturated with H_2O for > 60 days in most years.
*d = (dry): Ustic or xeric environment; dry > 60 consecutive days per year within 20–60 cm depth.
 e = (low CEC): <4 meq/100 g soil by Σ bases + unbuffered Al.
  <7 meq/100 g soil by Σ cations at pH 7.
  <10 meq/100 g soil by Σ cations + Al + H at pH 8.2.
*a = (Al toxic): >60% Al saturation of CEC by Σ bases and unbuffered Al within 50 cm.
  >67% Al saturation of CEC by Σ cations at pH 7 within 50 cm.
  >86% Al saturation of CEC by Σ cations at pH 8.2 within 50 cm *or* pH <5.0 in 1:1 H_2O except in organic soils.
*h = (acid): 10–60% Al saturation of CEC by Σ bases and unbuffered Al within 50 cm *or* pH in 1:1 H_2O between 5.0 and 6.0.
 i = (Fe-P fixation): % free Fe_2O_2-clay > .20 *or* hues redder than 5 YR and granular structure.
 x = (X-ray amorphous): pH >10 in 1 N NaF *or* positive to field NaF test *or* other indirect evidences of allophane dominance in clay fraction.
 v = (Vertisol): Very sticky plastic clay, > 35% clay and > 50% of 2:1 expanding clays. COLE > 0.09. Severe topsoil shrinking and swelling.
*k = (k deff): <10% weatherable minerals in silt and sand fraction within 50 cm *or* exch. K <0.20 meq/100 g *or* K <2% of Σ of bases, if Σ of bases <10 meq/100 g.
*b = (carbonate): Free $CaCO_3$ within 50 cm (fizzing with HCl) or pH >7.3.
*s = (salinity): >4 mmhos/cm of saturated extract at 25° within 1 m.
*n = (sodic): >15% Na saturation of CEC within 50 cm.
*c = (cat clay): pH in 1.1 H_2O <3.5 after drying; Jarosite mottles with hues 2.5Y or yellower and chromas 6 or more within 60 cm.

Source: Buol et al. (1975).

SUMMARY AND CONCLUSIONS

1. The process of diagnosing plant nutritional problems and making fertilizer recommendations in the tropics is based on different approaches at different stages of sophistication. Efficient fertility evaluation programs are well established in several tropical countries and have played a major role in increasing food production through increased fertilizer use.

2. Ongoing fertility evaluation programs can be grouped according to the principal technique used. The main ones are soil testing, plant analysis, missing element greenhouse techniques, and simple fertilizer trials. The first three emphasize the central concept that fertility evaluation is a site- and crop-specific undertaking.

3. The purpose of a soil fertility evaluation laboratory is to provide rapid diagnostic services to farmers. This involves. running large numbers of samples daily with simple but accurate techniques. A successful soil test procedure must extract all or a proportionate part of the available form of nutrients from widely different soils, through speedy and accurate techniques. The results must be correlated with growth response under field conditions. Few existing soil test procedures meet these criteria.

4. No successful nitrogen soil test procedures are presently available. A large number of phosphorus soil tests are effective for specific groups of soils because of differences in the forms of inorganic phosphorus present. Effective soil tests are available for potassium, secondary elements, and micronutrients and for diagnosing acidity, salinity, and alkalinity problems. With the use of specially designed equipment most nutrients can be determined with one or two extractants. The exceptions are sulfur, boron, molybdenum, and nitrogen.

5. Values obtained in soil analyses are empirical numbers that are meaningful only when correlated with yield responses. The concept of defining a critical level above which a nutrient element is no longer a limiting factor has greatly sharpened the focus of these techniques. A seemly scattered array of data points on a soil test–yield response graph is simply divided into soils with a high probability of yield response to the added nutrients, and those in which the probability is low.

6. Soil test correlations in the greenhouse are useful for comparing extractants and determining tentative critical levels. The definite critical levels for soil tests as well as plant analysis are established only through field trials.

7. The interpretation of field trials is the basis for formulating fertilizer rate recommendations, regardless of the type of soil fertility evaluation approach used. The optimum levels should be determined for specific soil–

crop categories separated on the basis of critical levels, soil properties, and crop or cropping system used. Two main methods of interpretation can be employed. The classic one involves the development of curvilinear regression equations and marginal analysis. A new linear response and plateau model requires only simple graphic techniques and the determination of an optimum level independent of price:cost ratios. The type of recommendations obtained through the latter method appears well suited to tropical conditions.

8. Plant analysis is especially useful for long-term crops and can be refined through crop logging methods developed by plantations. The use of plant uptake levels is a valuable way of determining nitrogen fertilizer recommendations.

9. Missing element techniques are used in many tropical countries. The amount of work and time involved, however, limits their use for routine purposes.

10. Simple fertilizer trials scattered at random through seemingly homogeneous areas are of value in determining the general nutrient deficiencies and average fertilizer requirements of a region. They cannot serve as a basis for fertilizer recommendations for a specific field, however, unless correlated with laboratory or greenhouse analytical techniques.

11. Efforts must be increased to bring soil survey data into soil fertility evaluation projects. The immediate goals of soil classification and soil fertility groups are different. With one exception, technical classification systems do not include fertility data. The development of such systems for grouping soils with similar fertility limitations is likely to improve the effectiveness of soil fertility evaluation programs and to bridge the present gap between the two subdisciplines.

REFERENCES

Agboola, A. A. 1972. The relationship between yields of eight varieties of Nigerian maize and content of nitrogen, phosphorus and potassium in the leaf at flowering stage. *J. Agr. Sci.* **79**:391–396.

Agboola, A. A. 1973. Correlations of soil tests for available phosphorus with maize in soils derived from metamorphic and igneous rocks in the Western State of Nigeria. *West Afr. J. Biol. Appl. Chem.* **16**:14–23.

Ahmad, N. 1967. Seasonal changes and availability of phosphorus in swamp rice soils of North Trinidad. *Trop. Agr. (Trinidad)* **44**:21–32.

Ahmad, N., I. S. Cornforth, and D. Walmsley. 1973. Methods of measuring available nutrients in West Indian soils. III. Potassium. *Plant and Soil* **39**:635–647.

Ahmed, B. and A. Islam. 1975. Extractable phosphate in relation to the forms of phosphate fractions in some humid tropical soils. *Trop. Agr. (Trinidad)* **52:**113–118.

Alvim, P. T. 1968. Avaliacião de fertilidade dos solos de Cerrado por testes de microparcelas. *Ciência è Cultura* **20:**613–619.

Anderson, R. L. and L. A. Nelson. 1975. A family of models involving intersecting straight lines and concomitant experimental designs useful in evaluating response to fertilizer nutrients. *Biometrics* **31:**303–318.

Balerdi, F., L. Muller, and H. W. Fassbender. 1968. A study of phosphorus in Central America. III. Comparison of five chemical analysis methods for available phosphorus. *Turrialba* **18:**348–360.

Bartholomew, W. V. 1972. Soil nitrogen: Supply processes and crop requirements. *Int. Soil Fert. Eval. Improvement Program Tech. Bull.* 6. North Carolina State University, Raleigh.

Baver, L. D. and A. S. Ayres. 1962. Soil analysis as basis for fertilizer recommendations in sugarcane. Pp. 835–841. In *Trans. Comm. IV and V, Int. Soc. Soil Sci. (New Zealand)*.

Baynes, R. A. and D. Walmsley. 1973. Fertility evaluation of some soils in the Eastern Caribbean. *Univ. West Indies Dept. Soil Sci. Rept.* 17. 80 pp.

Beek, K. J., J. Bennema, and M. N. Camargo. 1965. Primer intento de interpretación de un levantamiento de suelos en Brasil. *Centro Interamericano de Fotointerpretación Informe* 3677. Bogotá, Colombia.

Bertrand, R. 1972. Morphopédologie et orientations culturales des régions soudaniennes du Siné Saloum (Sénégal). *Agron. Tropicale (France)* **27:**1115–1190.

Beufils, E. R. 1971. Physiological diagnosis. Guides for improving maize production based on principles developed for rubber trees. *J. Fert. Soc. South Afr.* **1:**1–30.

Bhan, C. and H. Shanker. 1973. Correlation of available phosphorus values obtained by different methods to phosphorus uptake by paddy. *J. Indian Soc. Soil Sci.* **21:**177–180.

Bouldin, D. R., T. Greweling, and J. Lui. 1971. The effect of drying soils from the humid tropics on selected soil properties. *AID Rept.* 1. Agronomy Department, Cornell University, Ithaca, N.Y.

Bouyer, S. 1963. Considerations d'ordre practique sur l'etude de la fertilité des sols tropicaux. *Agron. Tropicale (France)* **18:**933–938.

Boyd, D. A. 1970. Some recent ideas on fertilizer response curves. Pp. 461–473. *9th Congr. Int. Potash Inst.* Berne, Switzerland.

Boyd, D. A. 1974. Developments in field experimentation with fertilizers. *Phosphorus in Agr.* **1974:**7–17.

Boyer, J. 1972. Soil potassium. Pp. 102–135. In *Soils of the Humid Tropics*. National Academy of Sciences, Washington.

Bray, R. H. 1948. Requirements for successful soil tests. *Soil Sci.* **66:**83–89.

Bruce, R. C. and I. J. Bruce. 1972. The correlation of soil phosphorus analysis with response of tropical pastures to superphosphates on some north Queensland soils. *Aust. J. Exptal. Agr. Anim. Husb.* **12:**188–194.

Buol, S. W. 1971. A soil-fertility capability classification system. Pp. 44–50. Contract AID/csd 2806, Annual Report. Soil Science Department, North Carolina State University, Raleigh.

Buol, S. W., P. A. Sanchez, R. B. Cate, Jr., and M. A. Granger. 1975. Soil fertility capability classification: a technical soil classification for fertility management. Pp. 126–141. In E.

Bornemisza and A. Alvarado (eds.), *Soil Management in Tropical America.* North Carolina State University, Raleigh.

Cabala-Rosand, F. P. 1972. *A Disponibilidade de Fósforo e o Uso de Extratores Químicos no Brasil.* Centro de Pesquisas do Cacao, Itabuna, Bahia. 27 pp.

Cabala-Rosand, F. P. and M. B. M. Santana. 1972. Comparacão de extratores químicos de fósforo em solos do sul da Bahia. *Turrialba* **22**:19–26.

Cajuste, L. J. and W. R. Kussow. 1974. Use and limitations of the North Carolina method to predict available phosphorus in some Oxisols. *Trop. Agr. (Trinidad)* **51**:246–252.

Cano, M. 1973. Evaluación de la fertilidad de los suelos en el Perú. Min. de Agr. *Dirección de Investigación Agraria Bol. Tec.* 78. La Molina, Peru.

Carandang, D. A. 1973. The fertility status of soils of the Philippines. *ASPAC Tech. Bull.* 12 pp., 29–49. Taipei, Taiwan.

Cate, R. B. Jr., and L. A. Nelson. 1965. A rapid method for correlation of soil test analysis with plant response data. *Int. Soil Testing Ser. Tech. Bull.* 1. North Carolina State University, Raleigh.

Cate, R. B. Jr., and L. A. Nelson. 1971. A single statistical procedure for partitioning soil test correlation data into two classes. *Soil Sci. Soc. Amer. Proc.* **35**:658–659.

Cate, R. B., Jr., A. H. Hunter, and J. W. Fitts. 1971. Economically sound fertilizer recommendations based on soil analysis. *Proc. Int. Symp. Soil Fert. Eval. (New Delhi)* **1**:1083–1091.

Chakravarti, S. N. 1956. Nutrient status of some West Bengal soils as determined by rapid chemical methods. *J. Indian Soc. Soil Sci.* **3**(2):83–86.

Chaminade, R. 1972. Recherches sur la fertilité et la fertilisation des sols en régions tropicales. *Agron. Tropicale (France)* **27**:891–904.

Chang, S. C. 1964. Phosphorus and potassium tests on rice soils. Pp. 373–381. In *The Mineral Nutrition of the Rice Plant.* Johns Hopkins Press, Baltimore.

Chang, S. C. and S. R. Juo. 1963. Available phosphorus in relation to forms of phosphorus in soils. *Soil Sci.* **95**:91–96.

Charreau, C. 1974. *Soils of Tropical Dry and Dry-Wet Climatic Areas of West Africa and Their Use and Management* (Preliminary draft). Agronomy Department, Cornell University, Ithaca, N.Y.

Chau, C. C. 1970. Study on the correlations of soil tests and response of rice to added fertilizers in regions north of Maiolie in Taiwan. *Soils Fert. Taiwan* **170**:1–20.

Chirinos, A. V., J. de Brito, and I. de Rojas. 1971. Características de fertilidad de algunos suelos venezolanos vistos a traves de los resúmenes de análisis rutinarios. *Agron. Trop. (Venezuela)* **21**:397–409.

Cholitkul, W. and E. H. Tyner. 1971. Inorganic phosphorus fractions and their relation to some chemical indices of phosphorus availability for some lowland rice soils of Thailand. *Proc. Symp. Soil Fert. Eval. (New Delhi)* **1**:7–20.

Clements, H. F. 1960. Crop logging of sugar cane in Hawaii. Pp. 131–147. In H. E. Reuther (ed.), *Plant Analysis and Fertilizer Problems.* American Institute of Biological Science, Washington.

Colwell, J. D. 1971. Effects of variation in soil composition on soil test values for phosphorus fertilizer requirements. *Proc. Int. Symp. Soil Fert. Eval. (New Delhi)* **1**:327–336.

Committee on Tropical Soils. 1972. Soil testing and soil fertility evaluation services. Pp. 198–202. In *Soils of the Humid Tropics.* National Academy of Sciences, Washington.

Cope, J. T., Jr., and R. D. Rouse. 1973. Interpretation of soil test results. Pp. 35–54. In L. M. Walsh and J. D. Beaton (eds.), *Soil Testing and Plant Analysis,* rev. ed. Soil Science Society of America, Madison, Wisc.

Cordero, A. and J. Salas. 1971. Evaluación de la fertilidad de tres suelos aluviales de Costa Rica mediante el método de las microparcelas de maíz. *Min. Agr. Ganad.* (*Costa Rica*) *Bol. Tec.* 58.

Cordero, A. and G. Miner. 1975. A field research program for obtaining interpretation data. Pp. 518–532. In E. Bornemisza and A. Alvarado (eds.), *Soil Management in Tropical America.* North Carolina State University, Raleigh.

Cornforth, I. S. and D. Walmsley. 1971. Methods for measuring available nutrients in West Indian soils. 1. Nitrogen. *Plant and Soil* **35**:389–399.

Cox, F. R. and E. J. Kamprath. 1971. Micronutrient soil tests. Pp. 289–317. In J. J. Mordvedt, (ed.), *Micronutrients in Agriculture.* Soil Science Society of America, Madison, Wisc.

DaSilva, L. F. and R. Carvalho. 1971. Suitability classes for cacao soils of Bahia, Brazil. *Theobroma* **1**:39–54.

Datta, N. P. and A. R. Kalbande. 1967. Correlation of response in paddy with soil test potassium in different Indian soils. *J. Indian Soc. Soil Sci.* **15**:1–6.

Divisão de Pesquisa Pedológica. 1973. Levantamento exploratorio—reconhecimento de solos do Estado de Ceará, Vols. I, II. *Boletim. Tec.* 28. Río de Janeiro, Brasil.

Doll, E. C. and R. E. Lucas. 1973. Testing soils for potassium, calcium and magnesium. Pp. 133–154. In L. M. Walsh and J. D. Beaton, *Soil Testing and Plant Analysis,* rev. ed. Soil Science Society of America, Madison, Wisc.

Ekpete, D. M. 1972. Predicting response to potassium for soils of eastern Nigeria. *Geoderma* **1**:177–189.

Equipe de Pedología e Fertilidade do Solo. 1972. Levantamento exploratorio—reconhecimento de solos do Estado de Paraíba. *Boletim Tec.* 15 pp., 664–670.

Estrada, J. A. 1966. Reconocimiento del estado de fertilidad de los suelos del Huallaga Central. *Anal. Cient.* (*Peru*) **4**:127–146.

Estrada, J. A. 1967. Interrelaciones suelo-planta-nutrición. III. Reconocimiento del estado de fertilidad de algunos suelos del Departamento de Arequipa. *Anal. Cient.* (*Peru*) **5**:169–182.

FAO. 1963. Report of the first meeting on soil fertility and fertilizer use in West Africa. Food and Agricultural Organization of the United Nations, Rome.

FAO. 1969. Soil fertility survey and research. The Philippines. Field experiments, plant nutrition, and soil classification. LA:SF/PH1 10. Food and Agricultural Organization of the United Nations, Rome.

FAO. 1971. Improving soil fertility in Africa. *FAO Soils Bull.* 14. Rome.

Fitts, J. W. 1955. Using soil tests to predict a probable response from fertilizer application. *Better Crops with Plant Food* **39**(3):17–28.

Fitts, J. W. 1971. Using soil fertility evaluation and improvement information. *Proc. Int. Symp. Soil Fert. Eval.* (*New Delhi*) **1**:1065–1072.

Fitts, J. W. 1972. Nécessité pour l'agriculture de disposer d'informations précises sur les besoins en matiere de fertilité du sol. *Phosphore et Agr.* **26**(59):1–7.

Fitts, J. W. 1974. Proper soil fertility evaluation as an important key to increased crop yields. Pp. 5–44. In V. Hernando (ed.), *Fertilizers, Crop Quality and Economy.* Elsevier, Amsterdam.

Fitts, J. W., R. B. Cate, Jr., A. H. Hunter, J. L. Walker, and D. L. Waugh. 1965. Evaluation of soil fertility in Latin America: soil testing plant analysis. *Int. Soil Testing Ser. Tech. Bull.* 2. North Carolina State University, Raleigh.

Foster, H. L. 1973. Fertilizer recommendations for cereals grown on soils derived from volcanic rocks in Uganda. *East Afr. Agr. For. J.* **38**:303–313.

Galiano, F. 1968. Diagnóstico foliar en maíz: Ensayo de campo con Diacol V-103. *Rev. Inst. Inv. Tecnolog. (Colombia)* **J2**:14–24.

Gallo, J. R. and F. A. S. Coelho. 1963. Diagnose da nutricão nitrogenada do milho pela análise química das folhas. *Bragantia* **23**:537–548.

Gallo, J. R., R. Hiroce, and L. T. de Miranda. 1965. Análise foliar na nutrição de plantas de milho. II. Relatório em progresso sobre estudos em N, P, K e elementos menores. *Bragantia* **24**:41–47.

Gallo, J. R., R. Hiroce and R. Alvarez. 1968a. Levantamento do estado nutricional de canaviais de São Paulo pela análise foliar. *Bragantia* **27**:365–382.

Gallo, J. R., R. Hiroce, and L. T. de Miranda. 1968b. Análise foliar na nutrição das plantas de milho. III. Correlações de análise das folhas com produção. *Bragantia* **27**:177–186.

Goswami, N. N., S. R. Bapat, and V. N. Pathak. 1971. Studies on the relationship between soil tests and crop responses to phosphorus under field conditions. *Proc. Int. Symp. Soil. Fert. Eval. (New Delhi)* **1**:351–359.

Govinda Iyer, T. A., V. Renganathau, L. M. Ghouse, and G. Teekaramaran. 1970. Studies on the productivity rating of alluvial soils. *Madras Agr. J.* **57**:221–226.

Guajardo, R. and E. Ortega. 1968. Estudio de calibración y correlación de un método químico para el análisis de fósforo en suelos del valle del Yaqui. *Agr. Técnica (Mexico)* **2**:396–399.

Hanway, J. J. 1971. Relating laboratory test results and field crop response: principles and practices. *Int. Symp. Soil-Fert. Eval. Proc. (New Delhi)* **1**:337–343.

Hanway, J. J. 1972. Future direction of research on soil testing in agronomy. *Indian J. Agr. Sci.* **42**:357–359.

Hanway, J. J. 1973. Experimental methods for correlating and calibrating soil tests. Pp. 55–66. In L. M. Walsh and J. D. Beaton (eds.); *Soil Testing and Plant Analysis*, rev. ed. Soil Science Society of America, Madison, Wisc.

Hardy, F. and R. Bazán. 1966. The maize microplot method of soil testing. *Turrialba* **16**:267–270.

Hauck, F. W. 1963. Die dedeutung der mineraldüngung für die landwirtscaft Westafrikas. Pp. 250–266. In *Afrika Heute, Jahrb.*

Hauser, G. F. 1971. Specialized techniques and experimental designs used by FAO in large scale soil fertility investigations in developing countries. *Proc. Int. Symp. Soil Fert. Eval. (New Delhi)* **1**:549–561.

Hauser, G. F. 1973. Guide to claibration of soil tests for fertilizer recommendation. *FAO Soils Bull.* 18. 71 pp.

Hauser, G. F. 1974. FAO's efforts to increase food production by promoting fertilizer use. Pp. 343–361. In V. Hernando (ed.), *Fertilizers, Crop Quality and Economy*. Elsevier, Amsterdam.

Heady, E. O. and H. E. Ray. 1971. Application of soil test data, fertilizer response research and economic models in improving fertilizer use. *Int. Symp. Soil Fert. Eval. (New Delhi). Proc.* **1**:1073–1082.

Hiroce, R., J. R. Gallo, and H. A. A. Mascarenhas. 1970. Análise foliar do feijão (*Phaseolus vulgaris*): Diagnose de necessidades de fósforo. *Bragantia* **29**(2):7–9.

Humbert, R. P. 1963. *The Growing of Sugarcane.* Elsevier, Amsterdam.

Humbert, R. P. 1973. Plant analysis as an aid in fertilizing sugar crops. II. Sugarcane. Pp. 289–297. In L. M. Walsh and J. D. Beaton (eds.), *Soil Testing and Plant Analysis,* rev. ed. Soil Science Society of America, Madison, Wisc.

Hunter, A. H. 1975. New techniques and equipment for routine plant analytical procedures. Pp. 467–483. In E. Bornemisza and A. Alvarado (eds.), *Soil Management in Tropical America.* North Carolina State University, Raleigh.

Hunter, A. H. and J. W. Fitts. 1969. Soil test interpretation studies: field trials. *Int. Soil Testing Program Tech. Bull.* 5. North Carolina State University, Raleigh.

INEAC. 1958. Techniques de prélèvements en vue du diagnostic chimique du besoin en engrais. *Inst. Nat. Etud. Agron. Congo Belgue Bull. Infom.* 7(5):273/302.

ISFEIP. 1967–1974. Annual Reports. International Soil Fertility Evaluation and Improvement Program, Soil Science Department, North Carolina State University, Raleigh.

Ishaque, M. and A. F. M. Hafizar Rahman. 1971. *Some Studies on Fertilizers and Soils of Bangladesh.* Bangladesh Soil Fertility and Soil Testing Institute, Tejgaon, Dacca. 22 pp.

Ishizuka, Y. 1971a. *Nutrient Deficiencies of Crops.* ASPAC Food and Fertilizer Technology Center, Taipei, Taiwan. 112 pp.

Ishizuka, Y. 1971b. Physiology of the rice plant. *Adv. Agron.* 23:241–315.

John, M. K. 1972. Extractable phosphorus related to forms of P and other soil properties. *J. Sci. Food Agr.* 23:1425–1433.

Jones, J. B., Jr., and H. V. Eck. 1973. Plant analysis as an aid in fertilizing corn and grain sorghum. Pp. 349–364. In L. M. Walsh and J. D. Beaton (eds.), *Soil Testing and Plant Analysis,* rev. ed. Soil Science Society of America, Madison, Wisc.

Joseph, K. T. 1971. Nutrient content and nutrient removal in bananas as an initial guide for assessing fertilizer needs. *Planter* 47(538):7–10.

Kamprath, E. J. 1967. Soil acidity and response to liming. *Int. Soil Testing Ser. Tech. Bull.* 4. North Carolina State University, Raleigh.

Kanwar, J. S. 1971. Soil testing service in India—Retrospect and prospect. *Proc. Int. Symp. Soil Fert. Eval. (New Delhi)* 1:1103–1113.

Kilian, J. 1972. Contribution a l'etude d'aptitude des terres a la riziculture en Dahomey septentrionale. *Agron. Tropicale (France)* 27:321–357.

Kilian, J. and J. Velly. 1964. Diagnostic des carences minérales en vases de végétation sur quelques sols de Madagascar. *Agron. Tropicale (France)* 19:413–443.

Kilian, J. and J. Teissier. 1973. Methodes d'investigation pour l'analyse et le classement des bas-fonds dans quelques regions de l'Afrique de l'Ouest. Proposition de classification d'aptitudes des terres a la riziculture. *Agron. Tropicale (France)* 28:156–172.

Kim, Y. K., C. W. Hong, C. S. Park, and Y. S. Kim. 1971. Rapid examination of available silica for paddy soil. *Res. Repts. Office Rural Dev. (Korea)* 14:39–44.

Klingebiel, A. A. 1958. Soil survey interpretations—capability groupings. *Soil Sci. Soc. Amer. Proc.* 22:160–163.

Kyuma, K. and K. Kawaguchi. 1973. A method of fertility evaluation for paddy soils. I. First approximation: chemical potentiality grading. II. Second approximation: evaluation of four independent constituents of soil fertility III. Third approximation: synthesis of fertility constituents for soil fertility evaluation. *Soil Sci. Plant Nutr.* 19:1–27.

Lathwell, D. J., H. D. Dubey, and R. H. Fox. 1972. Nitrogen supplying power of some tropical soils of Puerto Rico and methods for its evaluation. *Agron. J.* 64:763–766.

Layese, M. F. and S. N. Tilo. 1970. Evaluation of three phosphorus test methods for lowland rice soils. *Philipp. Agr.* **54**:302–311.

Lizaraso, B. and R. Tinnermeier. 1961. Normas para el uso óptimo de los fertilizantes en algunos valles de la Costa. *Min. Agr. y Pesquería Bol.* 3. Lima, Perú.

Lopes, A. S. 1975. Comparative effects of methyl bromide, propylene oxide and autoclave sterilization on specific soil chemical characteristics. Unpublished paper, Soil Science Department, North Carolina State University, Raleigh. 20 pp.

Lott, W. L., A. C. McClung, R. de Vita, and J. R. Gallo. 1961. Survey of coffee fields in São Paulo and Paraná by foliar analysis. *IBEC. Res. Inst. Bull.* 26.

Malavolta, E., H. P. Haag, F. A. F. Mello, and M. O. C. Brasil. 1962. *On the Mineral Nutrition of Some Tropical Crops.* International Potash Institute, Berne, Switzerland. 155 pp.

Manzano, A., W. Carrera, and D. L. Waugh. 1969. La fertilización del trigo y su relación con el analisis de suelo. *ISFEIP Bol. Esp.* 1. North Carolina State University, Raleigh. 46 pp.

Marín, G. 1968. Recomendaciones tentativas de fertilizantes y cal para diversos cultivos de acuerdo con los resultados de analisis de suelos. *Rev. ICA (Colombia)* **3**:91–102.

Marín, G. and L. A. León. 1970. El analisis de suelos como guía para hacer recomendaciones de fertilizantes y enmiendas. *Agr. Tropical (Colombia)* **26**:24–33.

Martini, J. A. 1969a. La microparcela de campo como un método biológico rápido para evaluar la fertilidad del suelo. *Turrialba* **19**:161–266.

Martini, J. A. 1969b. Caracterización del estado nutricional de los "Latosoles" principales de Costa Rica utilizando la técnica del elemento faltante en en experimento de invernadero. *Turrialba* **19**:394–408.

McCollum, R. E. and C. Valverde. 1958. The fertilization of potatoes in Peru. *North Carolina Agr. Exp. Sta. Tech. Bull.* 185.

McLean, E. O. 1973. Testing soils for pH and lime requirement. Pp. 77–96. In L. M. Walsh and J. D. Beaton (eds.), *Soil Testing and Plant Analysis,* rev. ed. Soil Science Society of America, Madison, Wisc.

Miranda, L. T. 1960. Relação entre teores de nitrogênio e fósforo e pH do solo, e a resposta a adubação fosfatada em milho. *Bragantia* **19**:503–513.

Moormann, F. R. and R. Dudal. 1964. Characteristics of soils on which paddy is grown in relation to their capability classification. *Proc. 9th Meeting Working Party on Rice, Soils, Water, and Fertilizer Practices.* Mimeo IRC/SF-64/4. Food and Agriculture Organization of the United Nations, Rome.

Morillo, M. R. and H. W. Fassbender. 1968. Formas y disponibilidad de fosfatos de los suelos de la cuenca baja del río Choluteca, Honduras. *Turrialba* **18**:26–33.

Mukerjee, H. N. 1963. Determination of nutrient needs of tropical soils. *Soil Sci.* **95**:276–280.

Muller, A. and F. A. Van Baren. 1971. Standard method of soil characterization as a first approach to the evaluation of soil fertility. *Proc. Int. Symp. Soil Fert. Eval. (New Delhi)* **1**:1–6.

Muzilli, O. and R. E. Kalckmann. 1971a. Análise de assistência - interpretação de resultados e determinação de níveis críticos. I. Determinação de níveis críticos de acidez. *Bol. Univ. Fed. Paraná* **1**:1–18.

Muzilli, O. and R. E. Kalckmann. 1971b. Suggestoes de calagem e adubação para recuperação dos solos da região nordeste do Estado do Paraná. *Bol. Univ. Fed. Paraná* **2**:1–12.

Navas, J., H. Manzano, and A. C. McClung. 1966. Calibraciones del analisis de fósforo. *Agr. Tropical (Colombia)* **22**:23–32.

Nelson, D. W. and J. M. Bremner. 1972. Preservation of soil samples for inorganic nitrogen analyses. *Agron. J.* **64**:196–199.

Ng, S. K., G. C. Iyer, and K. Ratnasinigam. 1974. Laboratory errors in soil analysis. *J. Rubber Res. Inst. Malaysia* **24**:39–44.

North Carolina State University. 1973. Research on tropical soils. Annual Report. Soil Science Department, North Carolina State University, Raleigh.

Okajima, H., I. Uritani, and K. H. Houng, 1975. *The Significance of Minor Elements in Plant Physiology.* ASPAC Food and Fertilizer Technology Center, Taipei, Taiwan. 76 pp.

Oko, B. F. D. and A. A. Agboola. 1974. Comparison of different phosphorus extractants in soils of the Western State of Nigeria. *Agron. J.* **66**:639–642.

ONERN. 1971. *Capacidad de Uso de Los Suelos de Perú (Tercera Aproximación).* Oficina Nacional de Evaluación de Recursos Naturales, Lima, Perú. 57 pp.

Ortega, E. 1971. Correlation and calibration studies of chemical analysis in soils and plant tissues for nitrogen and available phosphorus. *J. Indian Soc. Soil Sci.* **19**:147–153.

Oteng, J. W. and D. K. Acquaye. 1971. Studies on the availability of phosphorus in representative soils of Ghana. I. Availability tests by conventional methods. *Ghana J. Agr. Sci.* **4**:171–183.

Pagel, H. 1972. Comparison of various extraction methods for determining the available phosphates in important soils of the arid and humid tropics (in German). *Beitr. Trop. Subtrop. Landwirt. Tropenveterinaermed.* **10**:123–138.

Pal, A. R. and D. P. Motiramani. 1971. Evaluation of some soil test methods for measuring available sulfur in medium black soils. *Proc. Int. Symp. Soil Fert. Eval. (New Delhi)* **1**:297–307.

Palencia, J. A. 1974. Programa de nutrición vegetal. *Memoria Anual de 1973.* Instituto de Ciencia y Tecnología Agrícola, Guatemala. 73 pp.

Palencia, J. A., E. Estrada, and J. L. Walker. 1975. A soil fertility evaluation program. Pp. 455–466. In E. Bornemisza and A. Alvarado (eds.), *Soil Management in Tropical America.* North Carolina State University, Raleigh.

Panabokke, C. R. 1969. Soil science and agricultural development in Ceylon. *Proc. Ceylon Assoc. Adv. Sci.* **2**:124–144.

Parra, J. and G. Quiceno. 1958. Los sistemas biológicos en la evaluación de la fertilidad de los suelos. *Cenicafé* **9** (1-2):5/22.

Perur, N. C., C. K. Subramanian, G. R. Muhr, and H. E. Ray. 1974. *Soil Fertility Evaluation to Serve Indian Farmers.* Mysore Department of Agriculture, Bangalore, India. 123 pp.

Peterson, F. J., R. H. Brupbacher, H. L. Clark, and J. E. Sedberry, Jr. 1971. Rice fertilization as related to soil type and soil test. *Proc. Int. Symp. Soil Fert. Eval. (New Delhi)* **1**:445–454.

PIPAEMG. 1972. Recomendações do uso de fertilizantes para o Estado de Minas Gerais. 2ª Tentativa. Programa Integrado de Pesquisas Agropecuárias do Estado de Minas Gerais. Secretaría da Agricultura, Belo Horizonte, M.G., Brazil.

Primavesi, A. and A. M. Primavesi. 1971a. Constant relationships between available nutrients in different soil-mapping units in Río Grande do Sul, Brazil (in German). *Agrochimica* **15**:212–227.

Primavesi, A. M. and A. Primavesi. 1971b. Various analytical results with dry and wet soils (in German). *Agrochimica* **15**:454–460.

Reeve, N. G. and M. E. Sumner. 1970. Lime requirements of Natal Oxisols based on exchangeable aluminum. *Soil Sci. Soc. Amer. Proc.* **34**:595–598.

Reisenauer, H. M., L. McWalsh, and R. G. Hoeft. 1973. Testing soils for sulfur, boron, molybdenum and chlorine. Pp. 173–200. In L. M. Walsh and J. D. Beaton (eds.), *Soil Testing and Plant Analysis,* rev. ed. Soil Science Society of America, Madison, Wisc.

Rodriguez, N. 1974. Potassium supplying capacity of some Venezuelan soils. *Trop. Agr. (Trinidad)* **51:**189–199.

Roufunnisa, K. V. Seshagiri Rao, T. Seshagiri Rao, and K. Venkateswarlu. 1971. Crop response to available phosphorus and potassium in the soils of Nagarjunagasar Project. *Proc. Int. Symp. Soil Fert. Eval. (New Delhi)* **1:**407–414.

Russell, E. W. 1974. The role of fertilizer in African agriculture. Pp. 213–250. In V. Hernando (ed.), *Fertilizers, Crop Quality and Economy.* Elsevier, Amsterdam.

Ryan, J. C. and R. K. Perrin. 1973. The estimation and use of a generalized response function for potatoes in the Sierra of Peru. *North Carolina Exp. Sta. Tech. Bull.* 214.

Ryu, I. S., Y. S. Kim, and C. S. Park. 1971. Relationship between the productivity of paddy soils and their physical and chemical properties. *Res. Rept. Office Rural Dev. and Plant Environment (Korea)* **14:**1–6.

Salazar, J. R. 1968. Clasificación de las zonas de respuesta a la aplicación de fósforo en los suelos de El Salvador. *Agr. El Salvador* 8 (Num. esp.):27–29.

Sanchez, P. A., A. Gavidia, G. E. Ramirez, R. Vergara, and F. Minguillo. 1973. Performance of sulfur-coated urea under intermittently flooded rice culture in Peru. *Soil Sci. Soc. Amer. Proc.* **37:**789–792.

Sarkar, M. C. 1971. Equilibrium-phosphate potential as a measure of available soil phosphorus as judged from the crop-response correlation studies. *Punjab Agr. Univ. J. Res.* **8:**206–210.

Saunder, D. H., B. S. Ellis, and A. Hall. 1957. Estimation of available nitrogen for advisory purposes in Southern Rhodesia. *J. Soil Sci.* **8:**301–312.

Saxena, M. C. and O. P. Gautam. 1971. Foliar diagnosis of nitrogen and phosphate status of hybrid maize. *Indian J. Agr. Sci.* **41:**803–807.

Sims, J. L. and B. G. Blackman. 1967. Predicting nitrogen availability to rice. II. Assessing available nitrogen in silt loam with different previous year crop history. *Soil Sci. Soc. Amer. Proc.* **31:**676–680.

Sims, J. L., J. B. Wells, and D. L. Tackett. 1967. Predicting nitrogen availability to rice. I. Comparison of methods for determining available nitrogen to rice from field and reservoir soils. *Soil Sci. Soc. Amer. Proc.* **31:**672–675.

Singh, B. 1973. Soil test-crop response correlation studies on wheat crop in alluvial soil associations of Uttar Pradesh. *J. Indian Soc. Soil Sci.* **21:**467–474.

Singh, R. M. and B. R. Tripathi. 1970. Correlation studies of soil-tests for available nitrogen and response of paddy on certain soils of Uttar Pradesh. *J. Indian Soc. Soil Sci.* **18:**313–318.

Small, H. G., Jr., and A. J. Ohlrogge. 1973. Plant analysis as an aid in fertilizing soybeans and peanuts. Pp. 315–328. In L. M. Walsh and J. D. Beaton (eds.), *Soil testing and Plant Analysis,* rev. ed. Soil Science Society of America, Madison, Wisc.

Sopher, C. D. and R. J. McCracken. 1973. Relationships between soil properties, management practices and corn yields in South Atlantic Coastal Plain Soils. *Agron. J.* **65:**595–600.

Spence, J. A. and N. Ahmad. 1967. Plant nutrient deficiencies and related tissue composition of sweet potatoes. *Agron. J.* **59:**59–62.

Sprague, H. 1965. *Hunger Signs in Crops.* McVay, New York.

Srivatsava, O. P., and A. N. Pathak. 1971. Available phosphorus in relation to the forms of phosphorus fractions in Uttar Pradesh soils. *Geoderma* **5**:287–296.

Stanford, G., J. O. Legg, and S. J. Smith. 1973. Soil nitrogen availability evaluations based on nitrogen mineralization potentials of soils and uptake of labelled and unlabelled nitrogen by plants. *Plant and Soil* **39**:113–124.

Steele, J. C., K. C. Vernon, and C. W. Hewitt. 1954. A capability grouping of the soils of Jamaica, B.W.I. *Trans. 5th Int. Congr. Soil Sci.* **3**:402–406.

Stefanson, R. C. and N. Collis-George. 1974. The importance of environmental factors in soil fertility assessment. I. Dry matter production. II. Nutrient concentration and uptake. *Aust. J. Agr. Res.* **25**:299–316.

Su, N. R. 1972. The fertility status of Taiwan soils. *ASPAC Tech. Bull.* **18**:15–100.

Subramanian, C. K. 1971. A quick method for correlation of soil test values with plant response. *Proc. Int. Symp. Soil Fert. Eval. (New Delhi)* **1**:371–375.

Sudjadi, M. 1971. Soil fertility studies to support agricultural intensification program in Indonesia. *FAO World Soil Resources Rept.* 41, pp. 103–105.

Sys, C. and R. Frankart. 1971. Land capability classification in the humid tropics. *Sols Afr.* **16**:153–200.

Tanaka, A. and S. Yoshida. 1970. Nutritional disorders of the rice plant in Asia. *IRRI Tech. Bull.* 10.

Thenebadu, M. W. 1972. Evaluation of the nitrogen status of rice by plant analysis. *Plant and Soil* **37**:41–48.

Tisdale, S. L. 1971. Soil and plant tests for the evaluation of the sulfur status of soils. *Proc. Int. Symp. Soil Fert. Eval. (New Delhi)* **1**:119–133.

Tiwari, K. N. and M. P. Singh. 1972. Effect of soil types on nutrient responses of high yielding varieties of kharif crops. *J. Indian Soc. Soil Sci.* **20**:211–217.

Tobón, J. M. and L. A. León. 1970. Comparación de varios métodos para determinar requerimientos de cal en algunos suelos colombianos. *Rev. ICA* **5**:307–326.

Twyford, I. T. and J. K. Coulter. 1962. Soil capability assessment in the West Indies. Pp. 761–769. In *Trans. Comm. IV and V, Int. Soc. Soil Sci. (New Zealand)*.

Tyner, E. H. and J. G. Davide. 1962. Some criteria for evaluating soil phosphorus tests for lowland rice soils. Pp. 625–634. In *Trans. Comm. IV and V, Int. Soc. Soil Sci. (New Zealand)*.

Ulrich, A. and F. J. Hills. 1973. Plant analysis as an aid in fertilizing sugar crops. Part I. Sugar beets. Pp. 271–288. In L. M. Walsh and J. D. Beaton (eds.), *Soil Testing and Plant Analysis*. Soil Science Society of America, Madison, Wisc.

United States Soil Salinity Laboratory. 1954. Diagnosis and improvement of saline and alkaline soils. *Handbook* 60. U.S. Department of Agriculture, Washington.

Vajragupta, Y., L. E. Haley, and S. W. Melsted. Correlation of phosphorus soil test values with rice yields in Thailand. *Soil Sci. Soc. Amer. Proc.* **27**:395–397.

Velayotham, M. and J. M. Jain. 1971. Preliminary studies on the development of an ideal soil test for available phosphorus for rice under field conditions. *Proc. Int. Symp. Soil Fert. Eval. (New Delhi)* **1**:135–143.

Vettori, L. 1969. Métodos de análise de solo. *Equipe de pedologia e fertilidade de solo (Brazil) Bol. Tec.* 7. Rio de Janeiro.

Virmani, S. M. 1971. Comparative efficacy of different methods for evaluating available sulphur in soils. *Indian J. Agr. Sci.* **41**:119–125.

Waffelaert, T. 1963. Essai d'estimation de la valeur agricole de families de sols au Congo. *Ann. Gembloux* **69**:688–699.

Walker, J. L. 1971. *El Analisis de Suelo y Recomendaciones para Fertilizantes*. Min. de Agr. DIGESA, Guatemala. 16 pp.

Wallace, T. 1956. *Plant Analysis and Fertilizer Problems*. Institut de Recherches pour les Huiles et Oléagineux, Paris. 410 pp.

Walmsley, D. et al. 1971. An evaluation of soil analysis methods for nitrogen, phosphorus and potassium using banana. *Trop. Agr. (Trinidad)* **48**:141–155.

Walmsley, D. and I. S. Cornforth. 1973. Methods of measuring available nutrients in West Indian soils. II. Phosphorus. *Plant and Soil* **39**:93–101.

Walsh, L. M. and J. D. Beaton. (eds.). 1973. *Soil Testing and Plant Analysis*, rev. ed. Soil Science Society of America, Madison, Wisc.

Waugh, D. L. and A. Manzano. 1971. The correlation of phosphorus response with soil analysis in tall and dwarf wheat varieties in Bolivia. *Proc. Int. Symp. Soil Fert. Eval. (New Delhi)* **1**:377–382.

Waugh, D. L., R. B. Cate, Jr., and L. A. Nelson. 1973. Discontinuous models for rapid correlation, interpretation and utilization of soil analysis and fertilizer response data. *Int. Soil Fert. Eval. Improvement Program. Tech. Bull.* 7, North Carolina State University, Raleigh.

Waugh, D. L., R. B. Cate, Jr., L. A. Nelson, and A. Manzano. 1975. New concepts in biological and economical interpretation and fertilizer response. Pp. 484–501. In E. Bornemisza and A. Alvarado (eds.), *Soil Management in Tropical America*. North Carolina State University, Raleigh.

Webb, R. A. 1955. A new approach to the study of the fertility of tropical soils. *Sols Afr.* **3**:379/391.

Weir, C. C. 1962. Evaluation of chemical soil tests for measuring available phosphorus on some Jamaican soils. *Trop. Agr. (Trinidad)* **39**:67–72.

Weir, C. C. 1966. Phosphorus and potassium states of some Trinidad soils. *Trop. Agr. (Trinidad)* **43**:315–321.

Wong, C. M. 1971. Soil testing and its application with special reference to Taiwan. *ASPAC Ext. Bull.* 10.

10

SOIL MANAGEMENT IN SHIFTING CULTIVATION AREAS

The rest of the chapters in this book attempt to consolidate the information previously presented in terms of the four most common management systems found in the tropics: shifting cultivation, flooded rice culture, intercropping, and tropical pastures.

Of these, shifting cultivation covers the largest proportion of the tropical land mass and is by far the most widespread tropical soil management system.

SHIFTING CULTIVATION SYSTEMS

Shifting cultivation can be defined as an agricultural system in which temporary clearings are cropped for fewer years than they are allowed to remain fallow. It is the predominant practice in approximately 30 percent of the exploitable soils of the world, 360 million ha, as well as the means of subsistence of over 250 million people or 8 percent of the world's population (Hauck, 1974). Even in the most densely populated tropical area, Southeast Asia, about one-third of the total farmland is under shifting cultivation (Dobby, 1950). In terms of total area, shifting cultivation is also the pre-

Plate 11   Pattern of shifting cultivation near Yurimaguas, in the Amazon Jungle of Peru.

Plate 12   Burning after felled material has dried in Yurimaguas, Peru.

dominant agricultural system in tropical America and Africa. It is found primarily in sparsely populated areas where power implements and fertilizers are not available.

This practice is known by a variety of names, such as "milpa" in Mexico and Central America, "conuco," "roza," "monte," and "chaco" in South America, "kaingin" in the Philippines, "chena" in Sri Lanka, "lua" in Vietnam, "ray" in Laos, and "ladang" in Indonesia. In English it is also called "slash and burn agriculture," "bush fallowing," and "swidden farming." Recently the term "land rotation" was coined for shifting cultivation systems in which the farmers live in permanent settlements, as opposed to "true" shifting cultivation, in which both fields and settlements move. The term "shifting cultivation" is used in this book to encompass both situations.

Throughout the world, however, a remarkable similarity in shifting cultivation practices exists. Two main forms can be easily distinguished: shifting cultivation in the forests and in the savannas. In forested areas small fields are cleared by ax and machete during periods of least rainfall and are burned shortly before the first rains. Without further removal of debris, crops such as corn, rice, beans, cassava, yams, and plantains are planted in holes dug with a planting stick. Building mounds with a hoe is often practiced in Africa for root crops. Intercropping is very common, and some degree of manual weeding is practiced. After the first or second harvest the fields are abandoned to rapid forest regrowth. The secondary fallow may grow for 4 to 20 years before it is cut again. This is the most common form of shifting cultivation throughout the tropics.

Shifting cultivation is also practiced in the African savannas located between forest and desert regions. In these ustic environments the vegetation is removed during the dry season by cutting the scattered trees, uprooting the grasses, and burning. Nye and Greenland (1960) report a typical example. After clearing the burning, the soil is then scraped into mounds about 50 cm high with a hoe. Yams are planted in the mounds, intercropped with corn, squashes, or beans. After the first year the yams mounds are destroyed, and corn and beans are planted on narrow ridges, followed by peanuts and millets. The land is then abandoned to the regrowth of several coarse grasses, particularly *Imperata cylindrica*. After 10 years the cycle is repeated.

Shifting cultivation in the savannas, therefore, differs from that in forested regions in four main aspects: (1) the topsoil is thoroughly disturbed in the processes of removing the grass roots, mounding, and ridging; (2) the cropping period is longer; (3) the land lies bare during the dry season, creating considerable erosion hazards; and (4) weeding is a much more serious problem.

Most soil types are used, regardless of fertility status. Shifting cultivation is practiced in areas with annual rainfall ranging from 750 to 7500 mm. In the highest-rainfall areas, where burning is not possible, as on the Pacific coast of Colombia, the cleared vegetation is used as a mulch instead of burning (Snedaker and Gamble, 1969). Other variations in forest and savanna systems are discussed by Ruthenberg (1971) and Greenland (1974).

The literature on shifting cultivation is extensive. Conklin (1963) compiled a bibliography with over 1300 references, dealing primarily with geographical and anthropological matters. Extensive soil research was conducted in Africa by the colonial powers before independence. These studies are summarized in three excellent publications: Nye and Greenland (1960), Newton (1960), and Jurion and Henry (1969). The Latin American literature has been summarized by Sanchez (1973). The evolution of shifting cultivation in relation to other agricultural systems of the world is well covered by Grigg (1974). Valuable up-to-date information has also been published from a seminar on shifting cultivation and soil conservation in Africa (FAO–SIDA, 1974).

In spite of the extensive literature on this subject and the large amounts of land and numbers of people involved, there is considerable controversy in academic and popular circles as to the value of shifting cultivation. In 1957 the Food and Agricultural Organization of the United Nations officially condemned shifting cultivation as a waste of land and human resources and a major cause of soil erosion and deterioration (FAO, 1957). This agency has urged member governments to eliminate shifting cultivation as quickly as possible. Subsequent FAO studies support this view but with less vehemence (Watters, 1971; FAO–SIDA, 1974). This position is often reinforced by the common belief that tropical soils deteriorate rapidly both physically and chemically upon exposure and often turn into laterite.

In the past few years the worldwide ecological movement has provided arguments at the other extreme. Shifting cultivation is considered an ecologically sound system for underpopulated, low capital-input areas.

The amplitude of opinion is paradoxical, considering the large amounts of information available. To reach some reasonable balance, this chapter describes the different phases of shifting cultivation and the attempts made to either improve or replace this practice.

FOREST–SOIL NUTRIENT CYCLES

The central concept of shifting cultivation is the dependence on forest or savanna fallow periods as the source of nutrients to crops. These nutrients are gradually accumulated during the fallow period and provide an alterna-

tive to fertilization. Consequently, an examination of nutrient cycling during the fallow period is in order.

The existence of a nearly closed nutrient cycle between a mature tropical
forest and the soil was first recognized by Hardy (1936) in Trinidad. Since
then, many studies have attempted to quantify this phenomenon as a means
of understanding the mechanism responsible for lush forest growth in
otherwise infertile soils. The nutrient cycle has two main storage areas: the
biomass and the topsoil. These are connected by several nutrient pathways,
illustrated in Fig. 10.1.

Vegetation Storage

The total biomass of mature tropical forests ranges between 200 and 400
tons/ha of dry matter. Studies conducted in Zaïre (Bartholomew et al.,

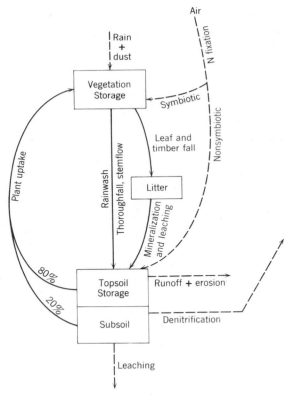

Fig. 10.1 A simplified forest–soil nutrient cycle. Dotted lines represent additions or removals
from the system.

Table 10.1   *Distribution of the Biomass of Some Mature Tropical Forests*

	Dry Matter (tons/ha)				
				Darién, Panamá	
Biomass Component	Kade, Ghana (40 years old)	Yangambi, Zaïre (18 years old)	El Verde, Puerto Rico (virgin)	Tropical Moist (virgin)	Premontane (virgin)
Leaves	25[a]	6	8	12	11
Wood	278	132	190	377	259
Litter	2	6	6	3	5
Roots	54	31	65	10[b]	13[b]
Total	361	175	269	402	288

Source: Calculated from data by Bartholomew et al. (1953), Golley et al. (1969), Greenland and Kowal (1960), and Ovington and Olson (1970).
[a] Includes twigs and other materials.
[b] Superficial roots only.

1953), Ghana (Greenland and Kowal, 1960), Panama (Golley et al., 1969) and Puerto Rico (Ovington and Olson, 1970) indicate that the proportion of the main forest parts is fairly constant. Approximately 75 percent of the biomass consists of branches and trunks, 15 to 20 percent of roots, 4 to 6 percent of leaves, and 1 to 2 percent of litter (Table 10.1).

Studies in Guatemala show that the rate of secondary forest regrowth is about 10 tons/ha a year during the first 9 years (Snedaker, 1970). Studies in the Zaïre, shown in Fig. 10.2, indicate that about 90 percent of the maximum biomass is attained during the first 8 years of regrowth. This dry matter curve is similar to the curves for many crops except for different orders of magnitude in both the dry matter and the time axes.

The first year of forest regrowth is similar to crop growth in terms of dry matter production. Tergas and Popenoe (1971) found the dry matter production of a 10 month old forest to be 9.7 tons/ha, while unfertilized corn produced 9.9 tons/ha near Izabal, Guatemala. When corn was adequately fertilized, it produced 30 tons/ha of dry matter during the same period. Results of longer-term comparisons in Nigeria showed that the secondary forest produced 40 tons/ha after 6 years of regrowth, while star grass pasture (*Cynodon dactylon*) established at the same time produced 17 tons/ha (Jaiyebo and Moore, 1964).

The amounts of nutrients accumulated by forest fallows is remarkable. Figure 10.3 shows that within an 8 year period the biomass accumulates over 500 kg/ha of N, K, Ca, and Mg, as well as considerable quantities of S

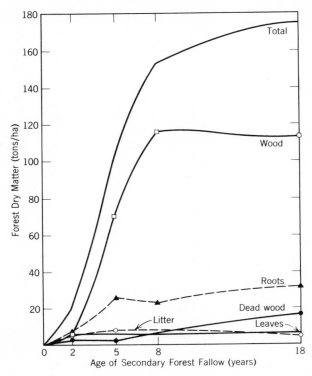

Fig. 10.2 Increases in biomass of forest components with age of a secondary forest in Yangambi, Zaïre. *Source:* Adapted from Bartholomew et al. (1953).

and P. The range in nutrient content of forest biomass is shown in Table 10.2. Nye and Greenland (1960) explain the mechanism as follows:

When a cultivated plot is abandoned, presumably due to fertility depletion, the seedlings and regrowth from the previous forest quickly form a canopy which reduces soil temperatures and stops erosion. The litter additions are rapidly decomposed, adding nutrients to the soils which are not leached away because of the quickly established forest roots. A nearly closed nutrient cycle is formed.

The amounts of nutrients in the cycle increase with added litter fall and eventually reach a plateau. According to Figs. 10.2 and 10.3, this plateau is achieved in about 8 years in a udic tropical forest grown on an Ultisol in Zaïre. No comparable data are available from other areas. Presumably, growth rates may be faster in soils with higher base status.

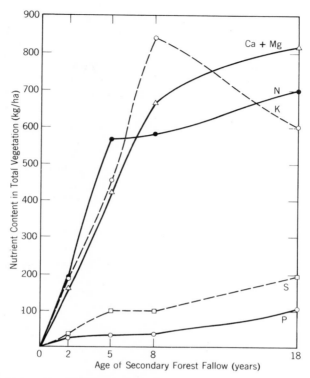

Fig. 10.3 Nutrient accumulation pattern in a secondary forest fallow in Yangambi, Zaïre. *Source:* Adapted from Bartholomew et al. (1953).

Table 10.2 Range in Nutrient Content in Total Biomass of Mature Forests in Zaïre, Ghana, Panama, and Puerto Rico

Element	Range (kg/ha)	Element	One Observation (kg/ha)
Nitrogen	701–2044	Sulfur	196
Phosphorus	33–137	Iron	43
Potassium	600–1017	Zinc	13
Calcium	653–2760	Manganese	5
Magnesium	381–3890	Copper	3

Source: Sanchez (1973).

353

Absent from this explanation is the answer as to why a secondary forest regrows at such a fast rate in a soil where crops cannot. An adequate supply of nutrients must be there for tree species which is not available to crops. There is no adequate answer to this question, but obviously the nutrient requirements are different for forest regrowth than for crops. Also, many of the felled trees sprout again and take advantage of a large existing root system. It is well established, however, that if the cropping period is extended for too long, the forest will not regrow. Coarse grasses take its place.

Tropical forests accumulate nutrients at a faster rate than temperate forests. According to Greenland and Kowal (1960), 18 year old secondary tropical forests accumulate in their biomass more nutrients than 50 to 100 year old temperate coniferous or hardwood forests.

Only about 40 or 50 percent of the total biomass is added to the soil, mostly as leaves, small branches, and roots. Fortunately, the nutrient accumulation pattern in these parts is faster than in the rest of the vegetation. Leaves accumulate about 94 percent of their maximum uptake within 2 years of forest regrowth. According to calculations based on the Zaïre study, about 80 percent of the maximum litter production and 64 percent of the maximum root production are achieved within the first 5 years (Fig. 10.4).

Soil Storage

The magnitude of the nutrient storage capacity of West African Alfisols in equilibrium with a mature forest was established by Greenland and Kowal (1960). The top 30 cm layer of an Alfisol contained 2.6 times as much total nitrogen as the biomass, and about the same amount of exchangeable calcium and magnesium as the total plant calcium and magnesium. The topsoil contained 75 percent of the biomass potassium as exchangeable potassium but only 9 percent of the biomass phosphorus as available phosphorus.

The phosphorus and base status of this soil is considerably higher than that of Ultisols, Oxisols, and some Inceptisols. The relationships in such soils may be considerably different. The Ghana study also shows that more than two-thirds of the root system of mature forests is found within the top 20 to 30 cm (Greenland and Kowal, 1960), so that the subsoil plays a secondary role in the nutrient cycle. The shallow nature of tropical forest root systems has been observed in other areas as well.

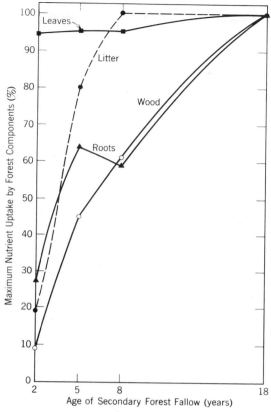

Fig. 10.4 Nutrient accumulation patterns in major components of secondary forests in Yangambi, Zaïre. *Source:* Adapted from Bartholomew et al. (1953).

Nutrient Transfer from the Vegetation to the Soil

The recognized mechanisms for nutrient transfer are rainwash (stemflow and thoroughfall); litter fall, timber fall, and root decomposition. Although rainwash varies considerably with seasons and species, its contributions are considerable. Nye (1961) estimated that rainwash added to the soil about 12 kg N, 4 kg P, 220 kg K, 311 kg Ca, and 70 kg Mg/ha a year. Sollins and Drewry (1970) observed that the concentrations of N, K, Ca, Mg, and S in rainwater increased from two to six times as it passed through a rainforest in Puerto Rico. Their estimates are shown in Table 10.3. Although the

Table 10.3   Increases in Nutrient Composition of Rainwater as It Passes through a Tropical Rainforest

Rainwater	Nutrient (ppm)				
as	NO_3	K + Na	Ca	Mg	SO_4
Rainfall[a]	0.1	3.6	0.9	0.7	1.1
Thoroughfall	0.6	14.1	2.1	1.2	3.0
Stemflow	1.7	4.7	1.7	0.9	0.1

Source: Adapted from data by Sollins and Drewry (1970).
[a] Before it reached forest canopy.

nutrient content of rains is low, considerable quantities of nutrients reach the soil as the rain washes through the vegetation.

The annual rates of litter fall range from 5.5 to 15.3 tons/ha in the tropics, as compared with 1.0 to 8.1 tons/ha in temperate forests (Ewell, 1968). The nutrient composition of the litter is similar in tropical and temperate forests except for a substantially higher nitrogen content in the tropics. The nutrient compositions of litter layers in Guatemala, Puerto Rico, Ghana, and Zaïre are shown in Table 10.4.

The relationships between litter production, accumulation, and decomposition have been studied in Colombia, Guatemala, and Puerto Rico, as well as by Nye's and Bartholomew's groups in Africa. Although considerable argument exists about which parameters are the proper ones, litter decomposition rates vary from 50 to 500 percent per year (McGinnis and Golley, 1967). Decomposition studies with time in Zaïre (Bartholomew et

Table 10.4   Ranges in Nutrient Composition of Litter Layers from Zaïre, Ghana, Guatemala, and Puerto Rico

Nutrient	Range (kg/ha)
Nitrogen	74–200
Phosphorus	1–7
Potassium	8–81
Calcium	45–220
Magnesium	10–94
Sulfur	9–10

Source: Sanchez (1973).

al., 1953) and Guatemala (Ewell, 1968) show very similar trends. Approximately half of the dry matter in the litter is mineralized within the first 8 to 10 weeks, after which the rate decreases. About 80 percent of the potassium is mineralized within the first month. Phosphorus, calcium, magnesium, and sulfur are mineralized at a faster rate than dry matter, but nitrogen is mineralized more slowly. Figure 10.5 represents in graph form Bartholomew's data from Zaïre. Ewell's results in Guatemala, produced 15 years later, are essentially identical.

The only estimate in the available literature that includes all four sources of nutrient transfer is that of Nye (1961) for an Alfisol from Ghana. On the assumption that 10 percent of the roots is decomposed per year, the annual rate of nutrient transfer from the vegetation to the soil is shown in Table

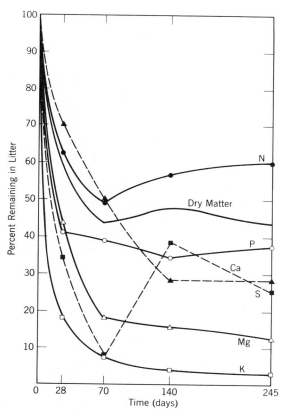

Fig. 10.5 Litter decomposition rates in Yangambi, Zaïre, *Source:* Adapted from Bartholomew et al. (1953).

Table 10.5   Annual Nutrient Additions from a Mature Forest to an Alfisol in Ghana

Transfer Pathway	Dry Matter (kg/ha)	Nutrient (kg/ha)				
		N	P	K	Ca	Mg
Rainwash	—	12	3.7	220	29	18
Litter fall	10,528	199	7.3	68	206	45
Timber fall	11,200	36	2.9	6	82	8
Root decomposition	2,576	21	1.1	9	15	4
Total	24,304	268	15.0	303	332	75
Annual turnover (%)	7	13	11	33	12	19

Source: Calculated from data by Nye (1961).

10.5. This extraordinary natural fertilization program is what keeps tropical forests so well supplied with nutrients!

Nutrient Losses from the System

The nutrient additions from the vegetation to the soil are nicely balanced by the nutrient uptake by the vegetation from the topsoil. Apparently, the shallow nature of tropical forest roots provides a very effective means of maintaining a nearly closed nutrient cycle (Nye and Greenland, 1960). Nevertheless, nutrient uptake from the subsoil may account for 20 percent of the total, thus undoubtedly contributing to the efficiency of the system. Data on leaching losses are almost nonexistent. An analysis of Amazon River waters provided the following leaching estimates: 0.5 kg NO_3, 4.4 kg K, 37 kg Mg, and 5 kg SO_4/ha a year (Russell, 1950). These small amounts support the thesis that nutrient losses from the closed cycle are indeed small.

The Cycle as a Whole

A quantitative study of nutrient cycling was attempted in Puerto Rico by Odum and his coworkers. Unfortunately the quality of the data is questionable in many of their reports. The following generalizations, however, may be of value. The nitrogen cycle consists of an annual flow of 102 kg N/ha a year. This element is tightly held in the cycle. Additions from rainwash and biological fixation plus possible denitrification losses are considered relevant (Edminsten, 1970). The phosphorus cycle is restricted to the superficial soil layer. The ^{32}P released from decaying leaves was quantitatively captured by the surface roots in the top 5 cm of soil and subsequently taken up by plants (Luse, 1970). Potassium is cycled very rapidly but is loosely held. On the other hand, Ca, Mg, Mn, Fe, and Cu are cycled slowly and bound very tightly (Jordan, 1970; Odum, 1970).

CHANGES IN SOIL PHYSICAL PROPERTIES UNDER SHIFTING CULTIVATION

When the equilibrium described in the preceding section is destroyed, the soil undergoes a series of changes caused by clearing, burning, cropping, and forest regrowth. This section describes what is known about this process. The reader should be aware that data from a wide variety of soils are presented here and that the magnitude of changes may be soil-dependent.

Soil Temperature

Air temperatures over a burning tropical forest may reach 450 to 650°C at 2 cm above the soil surface. Temperatures decrease at a rate of 100°C/cm below the soil surface for the first 5 cm, according to Zinke et al. (1970). Below this depth Suarez de Castro (1957) found no changes in soil temperature during burning in Colombia.

The variability in soil temperatures during burning depends on the intensity and duration of the burn itself. A study in Thailand provided some interesting information. Figure 10.6 shows the temperature profile of a "moderate" burn, of a "heavy" burn, and around a reburn pile, where the unburned material is grouped together and burned for several days. This figure shows that the increase in soil temperature is moderate in the normal and heavy burns, but considerable around the reburn piles.

After the burning process is over, cleared areas have higher mean air and soil temperatures than before (Budowski, 1956; Ahn, 1974). The maximum air temperature increased from 25 to 32°C after a forest was cleared in Thailand, but no change was registered in minimum air temperature (ASRCT, 1968). The maximum soil temperature at 7.5 cm depth increased from 27 to 38°C when a forest was cleared in Ghana, whereas the minimum temperature remained at 24°C (Cunningham, 1963).

The presence of many charcoal particles of varying size can change the

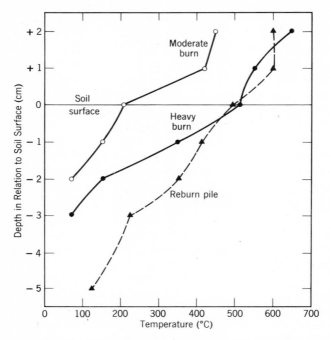

Fig. 10.6 Air and soil temperatures at time of burning on an Alfisol in Ban Pa Paes, Thailand. *Source:* Adapted from data by Zinke et al. (1970).

topsoil color, particularly in sandy soils, and increase the heat capacity of such soils. Many of these particles do not decompose rapidly. This author has observed charcoal bits on the soil surface of 18 year old forest fallows in the Amazon of Peru.

Studies by Lal and his coworkers (1975) on Nigerian Alfisols show that vegetation removal also increases diurnal soil temperature fluctuations to levels as high as 20 to 30°C. Germination, emergence, seedling growth, and nutrient and water uptake of crops such as corn, soybeans, and yams are severely affected by high soil temperatures on the order of 40 to 50°C. These extreme values are due to the low heat-conducting capacity of these Alfisols, which have a sandy topsoil. The low clay content and moisture-holding capacity make these soils hotter than heavier-textured soils.

In a virgin forest of Guatemala only 4 percent of the total solar radiation reaches the soil surface (Snedaker, 1970). Consequently, the incoming solar energy reaching a cleared soil is 25 times greater than in the forest. There is little doubt that biological processes, particularly organic matter decomposition, will proceed at a faster rate in a cleared soil. In traditional shifting cultivation, the soil is exposed only for 1 or 2 months after clearing, until the first crop canopy forms. The effect is more important when the change is made from shifting to continuous cultivation.

When the fields are abandoned to forest regrowth, soil temperatures decrease. Snedaker reported that during the first year of forest regrowth the solar radiation reaching the soil surface decreased to 35 percent of the amount when the land was cleared. In older secondary forests only 5 to 10 percent of the incoming solar radiation reaches the soil surface, as is readily apparent when one enters a tropical forest after walking through a cleared patch. It is always cooler (but more humid) inside the forest.

No equivalent information is available from savanna areas. The differences in soil and air temperatures with clearing are probably minor in relation to the changes just described in forested areas.

Soil Moisture

The soil moisture regime also changes when a forest is cleared. The first effect is equivalent to a sharp increase in rainfall. Greenland and Kowal (1960) found that a mature 40 year old forest canopy in Ghana intercepted and evaporated about 16 percent of the total annual rainfall. After clearing, therefore, the soil received 16 percent more rain. After a forest in Thailand was cleared, daily evapotranspiration increased eight times (ASRCT, 1968). Most of the evapotranspiration losses came from the top horizons, which resulted in a disuniform moisture supply. Fluctuating water tables may be formed after clearing (Budowski, 1956), and soil moisture contents may increase in the B horizons (ASRCT, 1968).

Soil Structure

It is widely believed that clearing and burning cause a deterioration of soil physical properties. The evidence shows, however, that this effect is dependent on soil properties. Suarez de Castro (1957) observed that burning increased infiltration rates and the soil aggregate fraction larger than 0.25 mm in an Andept of Colombia. Popenoe (1957) found that the bulk density of the 5 to 10 cm layer of Guatemalan volcanic soils increased from 0.56 to 0.66 g/cc after clearing and decreased from 0.74 to 0.70 g/cc after 3 or 5 years of forest regrowth in another observation. These differences are probably not significant enough to affect plant growth. Popenoe (unpublished) reports cases of irreversible dehydration in the soil surrounding long-burning logs or in reburn piles.

Observations by Moura and Buol (1972) on an Oxisol from Brazil indicate a rather sharp decrease in infiltration rate from 82 to 12 cm/hour when the original forest was cleared and the land tilled for 15 years. This decrease is actually beneficial because the original infiltration rate was excessive. Consequently, in soils high in oxides, such as Andepts and Oxisols, the structural changes occurring with clearing are by no means detrimental.

In soils with less desirable physical properties, however, the situation is quite different. Lal et al. (1975) reported that, when certain Nigerian Alfisols with sandy topsoils were cleared, severe surface crustation took place, resulting in serious erosion losses. Infiltration rates declined rapidly after forest clearing and cultivation. Cunningham (1963) observed a decrease in total porosity from 52 to 43 percent and a similar decrease in water-stable aggregates over 3 years of clearing an Alfisol from Ghana, which was kept bare for that period of time. Seubert (1975) observed the infiltration rate of a sandy Ultisol of the Peruvian Amazon to be 26 cm/hour under a mature secondary forest. After clearing, the infiltration rate decreased to 14 cm/hour and remained so during the first year. Surface crust problems similar to those reported by Lal were observed in this Amazon site, particularly after the second year of continuous cropping. Friese (1939) reported that the total porosity in some soils in southern Brazil decreased from 51 percent in a virgin forest to 12 percent in a derived savanna. He attributed the decrease to clay shrinkage and cracking, suggesting the presence of montmorillonite. These observations suggest that detrimental changes in soil physical properties due to clearing do occur in soils with initially poor properties, such as the sandy Alfisols or Ultisols, and also in montmorillonitic families.

Runoff and Erosion

Runoff and erosion are negligible in most soils that are protected by a forest canopy. Upon clearing, the magnitude of these processes will depend on soil properties and management.

In Guatemala, Popenoe (1957) found little surface erosion in cleared fields with steep slopes and attributed this to the low bulk density of the soils. He observed that most of the erosion takes place as landslides during heavy rainfall. Suarez de Castro (1957) showed that runoff decreases after clearing in Colombia and attributed the decline to increases in permeability due to burning. Both sites were affected by volcanic ash and are probably classified as Andepts.

The effects of clearing on runoff and erosion losses may be severe in other soils. Studies in West Africa conducted by Le Buanec (1972) in the

Ivory Coast and by Lal (1974) and Lal et al. (1975) in Nigeria show the tremendous amounts of soil are lost via runoff and erosion when Alfisols with sandy topsoils are cleared. Lal reports that, whereas erosion is negligible under the forest, it can be as high as 115 tons/ha a year of soil in unprotected plowed land. When the first inch of soil was lost, crop yields were reduced by as much as 50 percent. Fertilizer additions are no substitute for loss of surface soil because of the extremely poor root penetrability of the clayey, gravelly subsoil in Lal's sites. Consequently, the magnitude of the threat of erosion and runoff losses depends on soil properties.

Management is another important parameter. Erosion proceeds only when there is no crop canopy to protect the soil. In traditional shifting cultivation, the soil is devoid of a canopy for just a few weeks. The abundant debris in the form of trunks, branches, pieces of charcoal, and ash protects most soils during this critical period. This writer has observed virtually no erosion in cleared fields located on steep Ultisols in the Amazon Jungle under traditional management. When large tracts of land are mechanically cleared, however, erosion can and does occur. Also, when shifting cultivation is attempted by displaced urban dwellers, tremendous amounts of erosion result. Watters (1971) provides an excellent description of this process as observed in Venezuela. On the same land types, traditional shifting cultivators produced no erosion, whereas newcomers to this practice destroyed large amounts of land. Consequently, erosion problems are closely related to the social makeup of the farmers.

It is precisely in this kind of situation that the dangers of exposing plinthite layers occur. This problem is particularly serious in West African savannas, where population pressures have decreased the fallow period, erosion occurs when the land is exposed because of poor crop stands, and plinthite in the subsoil is common.

CHANGES IN SOIL NUTRIENT STATUS UNDER SHIFTING CULTIVATION

Soils undergo a series of changes in chemical properties when cleared, burned, cropped, and abandoned to forest regrowth. As described before, clearing increases soil temperature and, therefore, the rate of biological activity. Burning increases soil temperatures temporarily and produces large quantities of ash, which are equivalent to a good dose of fertilizer.

Two types of studies have been made to follow these changes. The first and most common is to sample soils simultaneously in areas known to have specific ages after clearing. Such studies confound soil variability with time effects, however, and do not give a clear trend. A very limited number of studies have followed the changes taking place in a soil as a function of time. The following discussion is based on the latter studies.

Ash Composition

Very little is known about the quantity and chemical composition of the ashes after a tropical forest is burned. Seubert (1975) obtained ash samples immediately after burning a 17 year old secondary forest growing in an Ultisol of Yurimaguas, Peru. The mean ash weight of 64 plots was 4 tons/ ha on a dry weight basis. This included partially burned or charred plant material as well as true ash. The fact that this material added the equivalent of approximately 70-14-45 kg of N, P_2O_5, and K_2O/ha, 240 kg/ha of dolomitic lime, and substantial quantities of Mn, Fe, Cn, and Zn (Table 10.6) indicates that, contrary to commonly held views, not all the nitrogen is lost in the process of burning because not all the vegetation is completely burned. Most of the burned or partially burned material consisted of leaves, litter, small twigs, and the bark of tree trunks and branches. The bulk of the

Table 10.6   Elemental Composition of Ash and Partially Burned Material, and Total Nutrients Added to an Ultisol, after Burning a 17 Year Old Secondary Forest in Yurimaguas, Peru

Element	Composition	Total Additions (kg/ha)
N	1.72%	67
P	0.14%	6
K	0.97%	38
Ca	1.92%	75
Mg	0.41%	16
Fe	0.19%	7.6
Mn	0.19%	7.3
Na	180 ppm	0.7
Zn	137 ppm	0.7
Cu	79 ppm	0.3

Source: North Carolina State University (1973) and Seubert (1975).

forest dry matter (trunks, branches, stumps, and roots) was not burned in this udic environment.

No similar data are available from the tropical literature. Boyle (1973) reported that the ash produced after burning a 32 year old jack pine forest on a sandy soil in Wisconsin averaged about 16 tons/ha of dry matter. His variability among samples ranged from 5 to 54 tons/ha of ash, which added to the soil from 46 to 595 kg K/ha and from 37 to 1128 kg Ca/ha.

Indirect evidence of the magnitude of nutrient additions to the soil by the process of clearing and burning is available from soil analysis before and after burning. In Alfisols of Ghana, ashes contributed from 1.5 to 3 tons Ca/ha, about 180 kg Mg/ha, and from 600 to 800 kg K/ha, according to calculations from data by Nye and Greenland (1960, 1964). For Oxisols or Ultisols, however, the estimates are much lower. Calculations with these soil orders have been made by Nye and Greenland (1960) on Ultisols from Yangambi and Zaïre and a large number of Liberian soils, presumably Ultisols. This author has also made similar calculations on six yellow Latosols (Oxisols or Ultisols) from data published by Brinkmann and Nascemento (1973) from Manaus, Brazil. The amounts of calcium added ranged from 275 to over 600 kg/ha, magnesium from 30 to 80 kg/ha, and potassium from 90 to 240 kg/ha. It is obvious that the lower base status of the Oxisols and Ultisols is dramatically reflected in the nutrient value of ash, as compared with the Alfisols.

The intensity of the burn is no doubt a major variable. Unfortunately no data are available on this aspect. However, farmers in the Amazon assess the quality of the burn and predict how good the crop will be on that basis. The burning process is more thorough in ustic than in udic moisture regimes because of the lower moisture content of the felled vegetation in the former regime during the dry season.

Changes in Soil Acidity

Soil pH increases after burning and decreases gradually with time because of the leaching of bases. The magnitude and the speed of these changes vary with the soil properties and the quantities of ash. In an Alfisol from Ghana, Nye and Greenland (1964) found that the pH increased from 5.2 to 8.1 in the top 5 cm layer right after burning and decreased to 7.0 after 2 years (Fig. 10.7). The pH in the 5 to 15 and 15 to 30 cm layers increased from 4.9 to 6.2 with burning and decreased to 5 after 2 years. Popenoe (unpublished data) also observed pH increases down to 40 cm in a soil affected by volcanic ash in Guatemala.

Studies in the Amazon Jungle indicate that pH increases with burning are

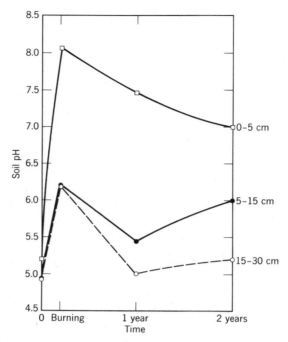

Fig. 10.7   Changes in pH at different soil depths in Kade, Ghana. *Source:* Adapted from Nye and Greenland (1964).

of a different order of magnitude in more acid soils. Brinkmann and Nascimento (1973) observed that the pH of Yellow Latosol (Oxisol) topsoils increased from 3.8 to 4.5 with burning and decreased quickly to the original level within about 4 months. Seubert (1975) found that the pH increased from 4.0 to 4.5 in the topsoil of an Ultisol and that this level remained stable during the first year. The lower quantity of bases in the ashes of these acid soils minimized the increases in pH.

The effect of these changes on plant growth will depend on their magnitude. The very high pH level observed after burning West African Alfisols has caused iron deficiency in upland rice, according to Lal et al. (1975). Burning is perhaps immediately detrimental in such soils. The more moderate increases in the Amazon Oxisols and Ultisols, on the other hand, are definitely beneficial to plant growth. Nye and Greenland's data show that the pH changes take place at considerable depths, suggesting a rapid downward movement of ash. These West African Alfisols have a sandy and gravelly topsoil that facilitates such movement. In an Ultisol from the Amazon, no evidence of such downward movement of calcium and magnesium was observed, but potassium moved rapidly during the first year after clearing (North Carolina State University, 1974).

Exchangeable Bases

The basic cations in the ash cause dramatic increases in exchangeable calcium, magnesium, and potassium levels after burning. These are followed by a gradual decrease during the cropping period due to leaching and crop uptake. The magnitude of these changes varies with soil properties and ash composition. Figure 10.8 compares the dynamics of exchangeable bases and pH in an Ultisol from Peru and an Alfisol from Ghana. In both soils the exchangeable calcium content tripled with burning. Exchangeable magne-

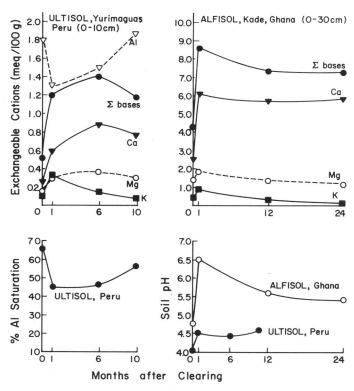

Fig. 10.8 Dynamics of exchangeable bases and soil acidity upon clearing and burning an Ultisol from Yurimaguas, Peru, and an Alfisol from Kade, Ghana. *Source:* Adapted from Nye and Greenland (1964), North Carolina State University (1974), and Seubert (1975).

sium tripled in the Ultisol but increased only slightly in the Alfisol. In both cases exchangeable potassium showed a sharp increase after burning, followed by a sharp decrease. This suggests that potassium was leached at a faster rate than calcium or magnesium. It is interesting to note the differences between the two soils after burning. In the Ultisol there was evidence of base depletion after the sixth month after burning, whereas in the Alfisol no significant change occurred in the base status in 2 years after burning.

In acid soils aluminum is the dominant cation. The effect of burning on decreasing exchangeable aluminum and aluminum saturation was quite dramatic in the Ultisol. After 6 months, base depletion and perhaps a high rate of organic matter decomposition changed the direction of the curve (Fig. 10.8). These data show how delicate the situation is in these highly weathered soils. Burning changed the soil–test status from potassium deficiency to potassium sufficiency and from aluminum toxicity to an aluminum-problem-free state. Within the first year, however, these soil properties reverted to the preburn levels. Similar observations in Oxisols from Manaus, Brazil, have been made by Brinkmann and Nascimento (1973).

No comparable data are available for the Ghana Alfisol. The pH curve suggests essentially 100 percent base saturation within 2 years after burning. It seems reasonable to assume that the reasons for shifting cultivation in the Alfisols after 2 years are other than base status. The beneficial effects of the ash, therefore, can extend to the initiation of fallow regrowth. The levels of exchangeable calcium, magnesium, and potassium have increased with time in secondary vegetation in Guatemala (Popenoe, 1957; Urrutia, 1967).

Soil Organic Matter and Organic Nitrogen

Although burning volatilizes most of the carbon, sulfur, and nitrogen present in the vegetation, it has little effect on soil organic matter. Contrary to popular belief, burning in the process of shifting cultivation does not destroy soil organic matter. Soil temperatures during burning are not high enough for a sufficiently long period of time for combustion, except in reburn pile situations. Studies by Nye and Greenland (1964) in Ghana, Popenoe in Guatemala, and Seubert (1975) in Peru actually show small increases in soil organic carbon and total nitrogen after burning. These have been attributed to incomplete combustion of the vegetation and the measurement of charcoal particles as organic carbon. In instances where sharp organic matter decreases are reported, they are probably associated with erosion losses from the topsoil.

Upon cultivation and exposure, soil organic carbon contents decrease in the top few centimeters because of increased soil temperatures and disturbance by tillage implements. Also, a "flush" of organic matter mineralization after burning has been reported by Laudelot (1961) in Zaïre. He attributed it to a sharp rise in microbial population after burning and, presumably, some rain.

The magnitude of the subsequent decrease seems to be related to the topsoil organic carbon content before burning. Figure 10.9 shows the changes observed by Popenoe in an Andept from Guatemala. A slight increase in soil organic matter right after burning took place; it was followed by a marked decrease and a trend toward equilibrium after about 4 months. This soil is affected by volcanic ash, which explains the high organic matter content. Figure 10.10 shows the changes in topsoil organic matter of two Oxisols during 5 years of continuous cropping after forest clearing in Sierra Leone. In spite of the difference in the original organic matter contents (3.5 and 10 percent) each Oxisol lost about half of its organic matter within 5 years. The sharpest drops occurred during the first year.

The situation in soils with initially lower organic matter contents is different. In an Alfisol from Ghana, Nye and Greenland (1964) found that the topsoil organic carbon content increased from 0.94 to 1.25 percent with burning, followed by a decrease to 0.94 percent after 2 years of clearing. In other words, there was no marked organic carbon depletion in relation to

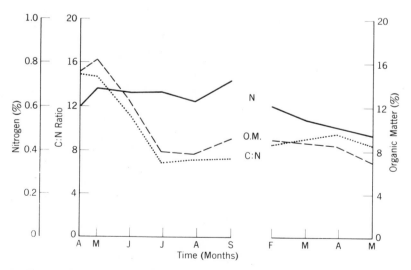

Fig. 10.9 Changes in total soil nitrogen organic matter and C:N ratio after clearing and burning a soil from Murciélago, Izabal, Guatemala *Source:* Popenoe, unpublished data.

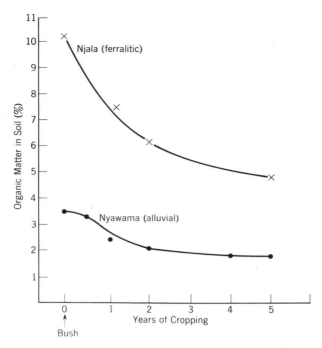

Fig. 10.10 Organic matter diminution in two Oxisols from Sierra Leone (2.5–18 cm depths) as affected by 5 years of continuous cropping after clearing. *Source:* Brams (1971).

the levels before clearing. In a 10 year study by Le Buanec (1972) on Alfisols of the Ivory Coast, no significant differences were observed in organic matter contents in three areas with adequate erosion control and fertilization practices. In Le Buanec's study topsoil organic matter contents ranged from 1.3 to 2.9 percent.

These very limited examples suggest that the magnitude of organic carbon depletion after clearing is greater in soils with higher organic carbon contents.

The changes in topsoil organic nitrogen with clearing show a different trend. Reports from soils affected by volcanic ash from Colombia and Guatemala indicate very little change in nitrogen and a decrease at a much slower rate than that for organic carbon. This is illustrated by Popenoe's data in Fig. 10.9 and has also been observed by Suarez de Castro (1957). In such cases the $C:N$ ratios drop sharply, suggesting that the increased biological activity produced large quantities of CO_2 but that part of the organic nitrogen either was in a fairly resistant form or was retained in the soil as microbial tissue. In soils with lower organic nitrogen contents, the $C:N$

ratios remain essentially the same after clearing and cultivation, as observed in West African Alfisols by Cunningham (1963), Nye and Greenland (1964), and Le Buanec (1972) and in Peru by North Carolina State University (1974).

The available data also indicate that changes in organic carbon, nitrogen, or C:N ratios are limited to the topsoil layer, except when the effect of crop root decomposition is evident in the subsoil.

When fields are abandoned to forest regrowth, however, the trend described above is reversed. Both organic carbon and organic nitrogen contents increase gradually without changes in C:N ratios. Reed (1951) in Liberia, Greenland and Nye (1959) in Ghana, and Urrutia (1967) in Guatemala found positive correlations between the age of secondary forest fallows and the soil organic carbon and nitrogen contents. The overall effect of traditional forest shifting cultivation is one, not of organic matter depletion, but the reaching of new equilibrium levels slightly below those of virgin forests. After analyzing 100 shifting cultivation sites in Liberia, Reed (1951) found that organic carbon reached equilibrium at a level equivalent to 75 percent of the virgin forest values. This is illustrated in the top line of Fig. 10.11, which represents a crop:fallow ratio of 2:12 years. According to Nye and Greenland (1960), when the crop:fallow ratio becomes narrower because of population pressures, a new equilibrium level is attained at about 50 percent of the virgin forest values. This is illustrated in the lower line of

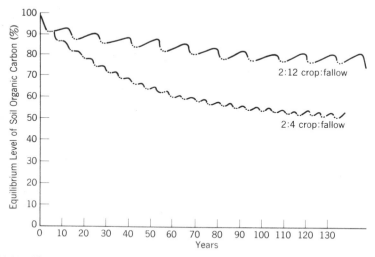

Fig. 10.11 Theoretical representation of the changes in soil organic carbon in the top 30 cm as affected by crop:fallow ratio. Broken line represents a 2 year lag period for the fallow to reestablish itself. *Source:* Nye and Greenland (1960).

Fig. 10.11. In the case of savanna areas, a wider crop:fallow ratio of 1:12 years is needed to reach an equilibrium organic carbon content of 75 percent, because organic matter additions are much lower than in the forests, according to calculations by Greenland and Nye (1959). They estimated the average annual decomposition rate of organic carbon to be on the order of 3 percent in shifting cultivation areas.

Cation Exchange Capacity

The changes in pH and organic matter contents of soils under shifting cultivation affect their effective cation exchange capacity because of the predominant pH-dependent status in many shifting cultivation areas. During the first few months after burning the effective CEC of a Peruvian Ultisol increased from 2.9 to 3.4 meq/100 g, probably as a result of the pH increase due to burning. At later stages the CEC decreases because of decreases in pH and soil organic matter contents. In two Oxisols from Sierra Leone, Brams (1971) observed that a 50 percent reduction in soil organic matter within 5 years after clearing resulted in a 30 percent reduction in CEC. This observation emphasizes the close dependency of organic matter negative radicals on the CECs of highly weathered soils. Although no comparable data are available, it seems reasonable to assume that the CEC changes in high-base-status soils moderately supplied with organic matter will not be as large as those described.

Available Phosphorus

The available phosphorus level of a soil increases upon clearing and burning because of the phosphorus contents in the ash. The magnitude of these additions is on the order of 7 to 25 kg P/ha, according to Nye and Greenland (1960) and North Carolina State University (1974).

These changes are reflected in soil test values. Upon burning, the Bray-extractable phosphorus in the top 5 cm layer of a Guatemalan Inceptisol increased about four times and remained at this level for about 6 months. At the end of 1 year it was still twice the original value (Popenoe, unpublished). No changes in available phosphorus were observed below that layer. These relationships are illustrated in Fig. 10.12. Suarez de Castro (1957) in Colombia, Watters and Bascones (1971) in Venezuela, and North Carolina State University (1974) in Peru obtained similar results.

The decrease in available phosphorus with cropping is not well understood. It may be due to fixation and/or crop removal. The decline in avail-

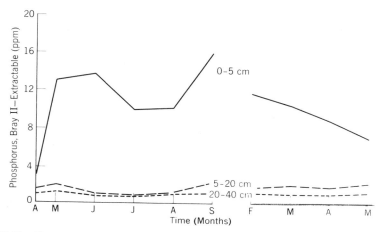

Fig. 10.12 Changes in available phosphorus with clearing and burning in a soil from Murciélagos, Izabal, Guatemala. *Source:* Popenoe (unpublished data).

able phosphorus may be one of the most important reasons for abandoning the field to the forest regrowth.

Urrutia (1967) found that secondary forest regrowth increases the Bray-extractable phosphorus in Rendolls of Petén, Guatemala. Tergas and Popenoe (1971) observed drastic species differences in phosphorus uptake after 10 months of secondary forest regrowth in Guatemala. At similar levels of dry matter production, *Heliconia* spp. (a relative of the banana) and *Gynerium* spp. (a grass) accumulated about three times as much as the average vegetation.

Many of the soils under shifting cultivation are deficient in phosphorus. Nye and Bertheux (1957) indicate that the C:P and N:P ratios of the organic matter of forested Alfisols of Ghana are considerably higher than those of the temperate region. They assert that this is an indication of phosphorus deficiency. There are, however, no phosphorus deficiency symptoms in mature tropical forests. The small amounts of phosphorus flowing through the closed nutrient cycle are apparently sufficient to prevent phosphorus deficiency under natural conditions.

Microbial Population

Clearing and burning also cause significant changes in the soil microflora. A review of the literature by Laudelot (1961) indicates that burning causes partial sterilization of the soil and that this is followed by a "flush" of microbial population and eventually by a decline approaching new equilibrium levels. A larger proportion of bacteria is also found after clearing. The total microbial population decreases during the dry season and increases during the rainy season and also with mulching and fertilization. The only evidence to the contrary is a study by Mieklejohn (1955) in Kenya, where she observed that burning decreased the soil microflora and noted that anaerobic nitrogen fixers survived whereas aerobic fixers were killed. Studies by Suarez de Castro (1957) and many others, however, indicate that such detrimental effects are short lived. A review of this subject in the temperate region by Ahlgren (1974) agrees with Laudelot's and Suarez de Castro's observations.

Little is known about the effect of burning on symbiotic and asymbiotic nitrogen-fixing bacteria, but the changes in pH no doubt affect their population. Laudelot points out that *Rhizobium* of the cowpea innoculation group is effective in acid soils and innoculates a number of aluminum-tolerant species such as *Stylosanthes guyanensis*.

CROP YIELD DECLINES UNDER SHIFTING CULTIVATION

Ash additions plus the rapid mineralization of organic matter after clearing and burning provide a sharp increase of available nutrients to the first crop planted. Afterwards, crop yields gradually decline, but the rate at which this process takes place varies with soil properties, cropping systems, and management. Yield declines with successive cropping constitute, of course, the fundamental reason why the cultivators shift.

Evidence

Farm surveys in Latin America indicate that farmers abandon their fields when they cannot expect the yield of a subsequent crop to be more than half that of the first crop. Cowgill (1962) and Watters (1971) report that a 50 percent yield decline is usually reached with the second consecutive cereal crop in Central and South America. In very sandy soils of Trinidad a second crop is impossible, according to Pendleton (1954). On the other hand, three or more consecutive crops are normally grown in shifting cultivation systems in West Africa.

These apparent discrepancies can be related to soil and climatic properties by comparing studies on the continuous cultivation of unfertilized monocultures in shifting cultivation regions. Some representative examples are illustrated in Fig. 10.13. At a glance it can be clearly seen that the rate of yield decline increases as the soil pH decreases.

Experimental data on high-base-status soils with pH levels above 6 are available from Entisols of Bolivia (Cordero, 1964) and New Britain (Newton and Jamieson, 1968); Alfisols of Africa by Nye and Greenland (1960), Nye and Stephens (1962), and IITA (1974); and eutric Oxisols of Africa by Anthony and Willimott (1956), Abu-Zeid (1973), and Cutting et al. (1959). These were all unfertilized experiments with adequate weed control. In general, 50 percent yield declines were reached by the fourth to sixth consecutive planting. In some cases, however, yields actually increased because of better weather conditions or fewer pest problems. In some cases a 50 percent yield decline was not reached in 8 to 15 years (Nye and Greenland, 1960; Nye and Stephens, 1962). Yields declined to 30 percent of the original values in the Guatemalan Mollisols. These are the only data gathered from farmers' fields, where weeds were an uncontrolled variable.

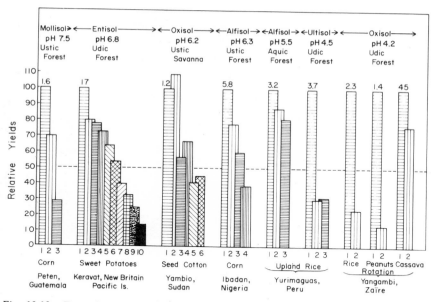

Fig. 10.13. Examples of yield declines under continuous cropping without fertilization in shifting cultivation areas as a function of soil, climate, and vegetation. Numbers on top of histograms refer to economic crop yields (tons/ha). Numbers on *x*-axis refer to consecutive crops. *Source:* Adapted from Cowgill (1962), Newton and Jamieson (1968), Anthony and Willimott (1956), IITA (1974), Sanchez and Nureña (1972), North Carolina State University (1974), and Nye and Greenland (1960).

In sharp contrast, experimental data on low-base-status soils show that yields of the second crop decrease to about 30 percent of the first-crop yields. Local soil variation is extremely important. The two examples shown in Fig. 10.13 from the same experiment station in Yurimaguas, Peru, depict a slow yield decrease in the poorly drained Alfisol with a pH of 5.5 and a precipitous decline in the well-drained Ultisol with a pH of 4.5. The three examples from Yangambi, Zaïre, underscore the differences in crops. Upland rice and peanuts suffered a sharp decline, whereas cassava did not.

In addition to soils, the type of fallow influences the rate of yield decline. Several long-term experiments conducted in Ghana and Nigeria on Ustalfs indicate that yield declines were faster in forested than in savanna areas (Nye and Greenland, 1960; Nye and Stephens, 1962). Figures 10.14 and 10.15 illustrate the influence of fallow vegetation. Figure 10.14 shows a rather rapid yield decline in a ustic forested region. The zigzag pattern is typical of double cropping in ustic regimes, where soil moisture is marginal for corn during the dry season. Figure 10.15 shows long-term experiments in an ustic savanna region. From 10 to 15 years was required to reach a 50 percent yield decline in this area, where moisture permits only one crop a year. The faster yield decline in the forest areas is probably due to the larger quantity of nutrients released by clearing and burning. Since the nutrient release in the savannas is smaller, there is less fertility to lose.

As mentioned before, shifting cultivators seldom attempt to grow the same crop consecutively. Experience has shown that certain cropping sequences, usually intercropped, produce the best results. The choice of crops seems to obey two principles: (1) plant first the crops that are most demanding for nutrients, such as grain crops, and follow them by less

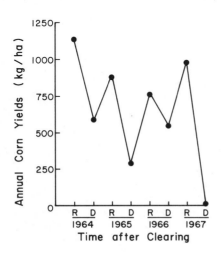

Fig. 10.14 Corn yield declines in 4 consecutive years after clearing a high forest on an Alfisol from Kade, Ghana. R = rainy season crop, D = dry season crop. *Source:* Adapted from Ahn (1974).

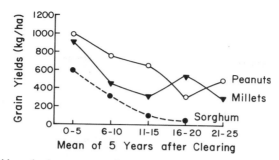

Fig. 10.15 Yields under long-term continuous cropping in savanna areas near Kano, Nigeria. *Source:* Adapted from Nye and Greenland (1960).

demanding crops such as cassava; and (2) plant successively taller crops to simulate actual forest regrowth. A common case is rice followed by cassava followed by plantains. The taller crops are also of longer growth duration and may develop some degree of nutrient cycling. The last example in Fig. 10.13 illustrates this principle in a rice–peanut–cassava succession. The last crop, cassava, is certainly more tolerant than the preceding grain crop. Nevertheless yield declines are evident when the second set of successions is considered. Farmers seldom if ever grow a second cycle of such successions.

Farmers and scientists alike identify five major reasons for yield declines and thus for shifting cultivation: soil fertility depletion, increased weed infestation, deterioration of soil physical properties, increased insect and disease attacks, and social customs. An examination of each of these reasons follows.

Soil Fertility Depletion

The preceding section of this chapter suggests that the decrease in available nutrients during the cropping period could be the main reason why yields decline. Indeed the experimental data presented in Fig. 10.13 indicate that this is the case. Unfortunately very little correlation has been found between yield declines and measured soil changes before and after cropping. This is particularly true of the bulk of the African data, where the changes in soil properties are not of sufficient magnitude to assert that fertility depletion is the primary cause of yield declines in Alfisols. The same applies to studies on Guatemalan Andepts and Mollisols by Popenoe (1957) and Cowgill (1962). For example, the changes in base status shown for the Alfisol in Fig. 10.8 and those in other parameters are too slow to account for yield declines (Nye and Greenland, 1964). In all these high-base-status soils, nitrogen deficiency seems to be a major limiting factor that cannot be adequately quantified by soil analyses.

Very clear examples of the effect of fertility depletion on yield declines are available from low-base-status Ultisols and Oxisols. The decline in continuously grown upland rice in Yurimaguas, Peru, is clearly correlated with decreases in exchangeable potassium and increases in aluminum saturation (North Carolina State University, 1974).

Increased Weed Infestation

The sharper yield declines observed in farmers' fields in comparison to unfertilized experimental plots suggest that increased weed infestation is an extremely important factor. Experimental plots are usually weed free; this may account for the generally observed slower yield decline rates in a wide variety of soils. The labor input for weed control increases dramatically with cropping. The amount of labor needed to control weeds during the second crop is often twice that required for this purpose during the first crop after clearing, according to Nye and Greenland (1960). When the labor input for weeding exceeds that for clearing a new patch of forest, the cultivators shift. In the two major African books on shifting cultivation (Nye and Greenland, 1960; Jurion and Henry, 1969) the authors clearly assert that weed infestation is the primary reason why cultivators shift.

The traditional shifting cultivation system is well adapted to temporary weed control. Crops with successively higher canopies are planted (cereals, cassava, plantains). Thus the least competitive crop, the cereal, is grown when weed infestations are the least intensive right after burning. As the taller crops grow, they permit the regrowth of woody, shade-tolerant species that will start the secondary forest. The traditional system, therefore, is well in tune with minimum weed control and permits a good forest regrowth.

Excessive weeding can destroy the traditional system. In acid soils of Sierra Leone, 2 consecutive years of rice–peanut rotations induced a grass savanna when weeding was done (Brams, 1971). Weeding methods are also important. In the Amazon Jungle minimal weeding is done with machetes, without killing the roots of the forest regrowth. It is this writer's opinion that the need for weed control may be the primary reason why fields are abandoned in high-base-status soils, whereas fertility depletion may be the primary cause in lower-base-status soils.

Deterioration of Soil Physical Properties

The preceding section showed rather dramatic instances of deterioration of physical properties related to water movement in soils with inherently poor physical properties (such as sandy Alfisols, Vertisols) when unprotected by a crop canopy for considerable periods of time. The changes previously described for such soils are clearly responsible for yield declines.

There are also abundant examples in which soil erosion can be a primary cause of yield declines (Watters, 1971; Nye and Greenland, 1960; IITA, 1973). Again, this is most serious in soils susceptible to erosion. In the extreme case uncontrolled erosion can expose soft plinthite layers that indurate upon exposure. Occurrence of this phenomenon, as mentioned previously, is of limited areal extent. It should be remembered that the traditional shifting cultivator is a good manager of soil physical properties. His minimum tillage techniques cause minimal soil disruption. His intercropped sequence of gradually taller crops provides a good canopy protection. He seldom causes soil erosion. The dangers occur when population pressures force him to narrow the crop:fallow ratio or when displaced urban dwellers unfamiliar with the system attempt to make a living out of it. Such situations produce tremendous soil erosion losses (Watters, 1971; Lal et al., 1975).

Increasing Incidence of Pests and Diseases

Continuous monoculture normally intensifies pest and disease attacks, particularly in udic tropical environments. Experiments on continuous rice cropping in Yurimaguas, Peru, have shown a greater incidence of mole cricket attacks and several diseases caused by *Helminthosporium oryzae* and *Rinchosporium orzyae*. Damages by birds, rodents, and larger animals increase as the area planted to one crop increases. The traditional shifting cultivator has learned to avoid or minimize these problems by using intercropped sequences and often by planting as a mixture several crop varieties that have different disease tolerances. It is probable that some reasonable degree of equilibrium exists between these factors. Therefore most pest and disease problems are bound to increase drastically when crop:fallow ratios decrease or continuous cropping is attempted.

Social Customs

Social and cultural factors can cause farmers to abandon fields. Farmers are known to give less care to crops they expect to yield less. Consequently, weeding and other good management practices are less intense after the first crop because the expectations for subsequent ones are lower (Ahn, 1974).

The seasonality in labor requirements is an additional factor. Weeding has to be done during the rainy season, when people are very busy. Cutting and burning, on the other hand, is carried out during the drier months, when there is less activity. Coulter (1972) points out that in some areas of Asia weeding is done by women whereas men do the cutting. Since many of these societies are matriarchial, the decision may be in favor of shifting to a new plot. Jurion and Henry (1969) reported that other social factors such as death in the family, superstition, and arbitrary decisions by tribal chiefs also cause shifting for no apparent agronomic reason.

IMPROVING CROP PRODUCTION UNDER SHIFTING CULTIVATION

Crop yields under shifting cultivation are extremely low. They range from 0.5 to 1.5 tons/ha of cereal grains and about 8 tons/ha of cassava and yams. They are, however, relatively stable and usually sufficient for subsistence agriculture. Because of the large proportion of the arable tropics persistently under shifting cultivation (44 percent), there is little doubt that efforts should be made to increase the productivity of these areas, considering the world's food needs as well as the well-being of the estimated 250 million people who depend on this method of farming.

The approaches for increasing production vary widely. They range from the "hands off" policy presently favored by many ecologists to advocacy of the total elimination of such farming—the stated policy of the Food and Agricultural Organization of the United Nations (FAO, 1957). Shifting cultivation was actually outlawed in Indonesia (Van Beukering, 1947). All sides, however, are united in requesting basic soil management data, and a substantial quantity of work on this subject has been done, primarily in Africa. The various solutions can be grouped into those that aim at improving the present system and those that seek to replace it by other types of agriculture.

Do Nothing

The first alternative is to do nothing. This is probably a reasonable one for remote, thinly populated areas where transportation costs are extremely high, markets are distant, and no major changes in population growth rates are contemplated. Such areas are actually few in the tropics, limited mainly to scattered aboriginal tribes in the Amazon Jungle, the hills of Southeast Asia, and the Congo Basin, located far from existing roads or river systems. A "hands off" policy implies a realization that traditional shifting cultivation is indeed a sound ecological system that permits subsistence but little else. This policy also forces those who practice it to keep their present standard of living. These persons are usually not politically visible and thus exert little or no pressure on development authorities.

Rationalize It: The Corridor System

After years of attempting to eliminate shifting cultivation in what is now Zaïre, Belgian agronomists developed a scheme for rationalizing shifting cultivation in thinly populated areas. This story is told in great detail in a book by Jurion and Henry (1969). Large forested areas were divided into strips or corridors approximately 100 m wide and oriented in the east–west direction to maximize sunlight penetration. Every other corridor was cleared every year. In this fashion every cultivated corridor was flanked by a forest fallow on each side. The sum of the number of years in crops plus the number of years in fallow determined the number of corridors in a management unit.

The corridors were as long as the topography permitted. Individual or communal boundaries were drawn in right angles. In Yamgambi, Zaïre, the cycle consisted of 3 years of cropping followed by 12 years of fallow. Thus the available land was divided into 15 corridors, laid out in contours along the slopes in rolling terrain. The system is illustrated in Fig. 10.16.

The principal advantage of the corridor system is control of the crop:fallow ratio. Also, the forested strips provide a good source of tree seeds for regrowth, and in some cases prevent erosion. The principal disadvantage is that they require complete control of the population—a situation not likely to be encountered in many shifting cultivation areas. In fact, when the Congo became independent the corridor system was abandoned (Ruthenberg, 1971).

Increasing Yields within the Cropping Period

The strategy of increasing crop yields without altering the shifting cultivation cycle has been successful in instances where the recommended practices are realistic in terms of the farmer's economic resources. An example is the introduction of new varieties and cultural practices in an upland rice-growing area in Yurimaguas, Peru. A series of experiments reported by Sanchez and Nureña (1972) showed that rice yields could be doubled or tripled by using short-statured, disease-resistant varieties and by reducing plant spacing to 25 × 25 cm. The labor involved in harvesting was also reduced because the new varieties matured uniformly. Harvesting could

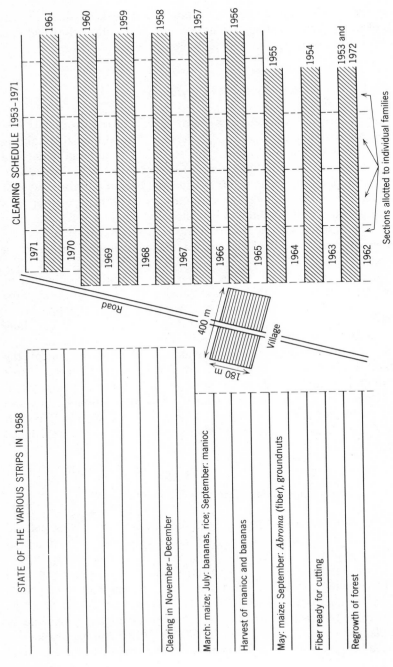

Fig. 10.16 Controlled shifting cultivation in the corridor system. *Source:* **Ruthenberg** (1971).

382

be done in one operation utilizing sickles (which were introduced into the region), and this replaced the conventional process of hand-harvesting panicle by panicle. Since fertilizers were not available, these recommendations were limited to the high-base-status soils along the river margins and near inland swamps. Research showed that continuous cropping was feasible with fertilizer applications, but it was not realistic to recommend these. On-farm field comparisons throughout the region showed that the new variety and closer spacing averaged 1.67 tons/ha and the traditional variety and spacing 0.95 ton/ha, a 76 percent yield increase without substantial investment (Donovan, 1973).

Fertilizer applications to improve crop yields and extend the duration of the cropping period should be economical under certain conditions. The effect of fertilizers will be discussed in conjunction with attempts to replace shifting with continuous cultivation.

Shortening the Fallow Period

The limited studies conducted on nutrient accumulation by forest fallows, discussed in the first section of this chapter, suggest that no more than 8 to 10 years is needed for nutrient uptake to reach maximum levels. In many shifting cultivation areas cutting 20 year old fallows is not uncommon. One possible improvement is to reduce the age of fallows to that necessary for maximum nutrient uptake. The problem is that usually no local information is available. Figure 10.17 illustrates this concept and the dangers of underestimating the optimum duration of the fallow period.

Figure 10.17 ignores, however, the crucial issue of weed control. Clearing older fallows not only is less time consuming but also reduces the degree of weed infestation, particularly by grassy weeds, encountered in clearing younger fallows. Extension demonstration trials conducted by Donovan (1973) in Yurimaguas showed that the average upland rice yield was 2.1 tons/ha in old fallow clearings and about half of this value, 1.1 tons/ha, in young fallow clearings. Farmers consider weed control to be the primary reason for these differences.

Improving the Fallows

Secondary forest or savanna regrowth is not the only type of vegetation capable of building up available nutrients through the development of a closed nutrient cycle. Several attempts have been made to increase the effi-

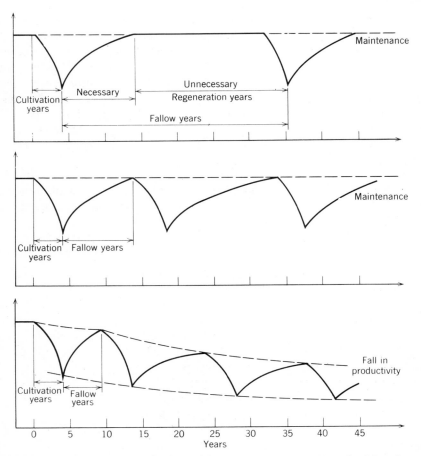

Fig. 10.17 Theoretical relationship between length of fallow and soil productivity. *Source:* Guillemin (1956).

ciency of the fallow period either by accelerating the nutrient buildup process or by substituting for the natural fallow crops with economic value.

In forested areas a few native species have been selected for their ability to accumulate certain nutrients at a faster rate than occurs in the mixed secondary forest fallows. Tergas and Popenoe (1971) reported that 10 month old pure stands of *Heliconia* spp. and *Gynerium* spp. accumulated three to four times more phosphorus than mixed fallows of the same age. In phosphorus-deficient soils such as those in Guatemala, these differences can be very important. Unfortunately, because of differences in shade tolerance, such species are replaced by others as the forest fallow grows.

The fast-growing tree species *Acioa barteri* has been used as an artificial fallow in southern Nigeria. Nye and Stephens (1962) reported that tree seedlings were planted in rows right after clearing, and grew while food crops were produced for 1 or 2 years. These fast-growing trees accumulated more calcium and magnesium but only about half the phosphorus and potassium of the natural secondary forest. Unfortunately, phosphorus and potassium were the limiting nutrients in these Alfisols. After 4 years of regrowth, crop yields were no different on land cleared than in natural fallow and *Acioa* (Newton, 1960).

Casuarina pines have effectively substituted for natural fallows in the highlands of New Guinea. Their main advantage is that this species fixes nitrogen symbiotically. Newton (1960) reports that *Casuarina* was planted with the last crops and allowed to grow for 10 to 20 years before clearing. The trees are also used as building materials and for firewood.

The use of a legume cover crop as a fallow was considered as a logical possibility by European colonial research services. Short-term experiments conducted in Africa all showed good possibilities, but recommendations were not adopted by farmers (Vine, 1968; Nye and Stephens, 1962). A long-time experiment by Jaiyebo and Moore (1964) provided the reasons why. They cleared an Alfisol forest in Ibadan, Nigeria, and planted several kinds of fallows which they monitored for 7 years. Afterwards, they cleared and burned the fallows and planted them to unfertilized corn. The results shown in Table 10.7 indicate that the native secondary forest fallow accumulated two to three times more nitrogen, phosphorus, potassium, and calcium than a legume (kudzu) or a grass (star grass) fallow. This resulted in a somewhat

Table 10.7   Effects of 7 Year Fallows on Soil and Biomass Storage on an Alfisol in Ibadan, Nigeria

| Type of Fallow | Biomass before Clearing | | | | | | Soil Properties (0–10 cm) | | | |
	Dry Matter (tons/ha)	N	P	K	Ca	Mg	pH	O.M. (%)	Bray P (ppm)	Exch. K (meq/100 g)
				(kg/ha)						
Forest	39.6	433	27	304	407	52	6.6	4.4	5	0.4
Kudzu	8.7	187	11	103	113	38	6.4	3.4	4	0.3
Stargrass	16.9	120	10	146	64	23	7.1	2.8	4	0.3
Grass mulch	47.7[a]	—	—	—	—	—	6.5	2.3	8	0.4
Bare soil	—	—	—	—	—	—	6.8	1.4	4	0.1

Source: Adapted from Jaiyebo and Moore (1964).
[a] Total added in 7 years as *Imperata cyclindrica.*

lower soil organic matter content in the artificial fallows but no significant change in other soil properties. Corn yields and nutrient uptake were essentially the same when preceded by the natural fallow or the kudzu, but were reduced to about half when preceded by the grass fallow or grass mulch (Table 10.8).

The artificial fallows, therefore, were no better than the natural forest fallow as far as nutrient immobilization or subsequent crop yields were concerned. Only when the fallow itself can be used economically is such a change feasible. For example, grass fallows or grass–legume fallows should be recommended when cattle is raised. In most of the African forested areas the tsetse fly prevents widespread beef or milk production. The data in Tables 10.7 and 10.8 show that effective nutrient buildup can be obtained with pasture species such as kudzu, which create an adequate nutrient cycle. Research is needed on their use for grazing, including the effects of animal feeding, excreta, and trampling. For the jungle areas of tropical America, where beef is likely to be the most important commodity, this possibility needs examination.

Another economic alternative is to grow marketable trees instead of fallows. In most cases, however, this represents a change to continuous tree cultivation, and it will be discussed in the next section.

In savanna areas experience shows little hope for improved fallows. Attempts to replace the coarse native *Andropogon* or *Imperata* species with improved grass have failed. Experiments in eastern Zaïre indicate that planted *Setaria sphacelata* grass pastures accumulated only two-thirds as much nutrients as native grass fallows (Van Parijs, 1959). A significant amount of information has been gathered in the East African highlands on the use of elephant grass (*Pennisetum purpureum*) fallows planted along contour lines. This has been described by Kerr (1942) and Webster (1954) as

Table 10.8 *Effects of the Above Fallow Treatments on Unfertilized Corn Yield and Nutrient Uptake*

Type of Fallow	Grain Yield (tons/ha)	Nutrient Uptake (kg/ha)				
		N	P	K	Ca	Mg
Forest	3.76	99	8	68	23	20
Kudzu	3.71	97	10	74	24	19
Stargrass	1.95	57	4	35	12	13
Grass mulch	1.97	43	6	53	12	9
Bare soil	1.09	22	3	17	9	6

Source: Adapted from Jaiyebo and Moore (1964).

analogous to the "ley" grass fallow system used in England. Extensive research conducted by Jameson and Kerkham (1960), Pereira et al. (1954), and Stephens (1960, 1967) indicate no significant yield differences between native and improved species under both grazed and ungrazed conditions and with several crop:fallow ratios. Most experiments produced such extremely low crop yields (< 1.5 tons/ha of grains) that other factors undoubtedly were limiting. In several cases a 3 year elephant grass fallow improved soil structure, infiltration rates, and soil moisture stored. These beneficial effects, however, lasted only for 2 crop years. The problem with this system is that the absence of legumes causes nitrate depletion, and that the grasses can exhaust subsoil moisture that would otherwise be available to crops. Webster (1954) emphasized that the maintenance of soil fertility through grass leys in England involves the use of legumes, manures, fertilizer applications, and grazing animals.

The only positive report from savanna regions is that of Nye (1958) in northern Ghana. He found that a crop of pigeon peas (*Cajanus cajan*), if planted closely, establishes itself quickly, withstands the dry season, and accommodates three times as much phosphorus, potassium, calcium, and magnesium, as well-established *Andropogon* savannas. Pigeon peas can, of course, be harvested as a grain legume.

In summary, it seems difficult to improve fallows simply by changing species. In conjunction with income-producing practices such as planting grasses for grazing or trees for firewood, improved fallows may have a role in shifting cultivation systems.

Semipermanent Cultivation

An intermediate stage between shifting and continuous cultivation usually coexists with shifting cultivation. The concept of semipermanent cultivation is discussed in detail by Ruthenberg (1971), who describes several forms that are very important in Africa. They consist essentially of very narrow crop:fallow ratios with some use of the short-term grass fallows for cattle grazing. They are found in relatively fertile soils or in fields that are continuously fertilized with manures and debris from houses. Figure 10.18 illustrates a commonly found gradient around an African village, which consists of permanent cropping or orchards near the houses, usually a result of heavy manuring and intensive weed control, followed by semipermanent cultivation of staple food crops, by grass fallows used as pastures, and finally by true shifting cultivation. Although, agronomically, semipermanent cultivation is a separate system, it represents an intermediate step in soil fertility and weed control.

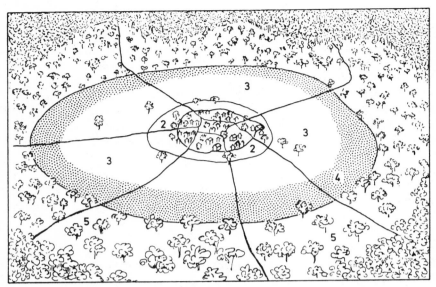

Fig. 10.18 Spatial arrangement of continuous, semipermanent, and shifting cultivation systems around a village in Senegal. (1) Houses and gardens, (2) continuous cultivation, (3) semipermanent cultivation, (4) intensive shifting cultivation, and (5) extensive shifting cultivation. *Source:* Ruthenberg (1971).

CHANGING FROM SHIFTING TO CONTINUOUS CULTIVATION

In most shifting cultivation areas, rapidly increasing populations require a change toward continuous cultivation. Population pressures have decreased crop:fallow ratios and plunged soil productivity into a downward spiral. This is particularly evident in highly populated parts of West Africa and in areas of South America that are receiving heavy migration from the congested Andean highlands and northeast Brazil. The construction of penetration highways and the discovery of oil in parts of the Amazon Jungle and southeast Mexico are aggravating these conditions. This section discusses soil management practices for establishing continuous cultivation under population pressure.

Land Clearing

Continuous cultivation schemes usually begin with the land-clearing process, which varies from the traditional manual slash and burn to a completely mechanized operation with specially equipped bulldozers and tree crushers. Tree poisoning, barking, and ringing have also been used. The effectiveness of a particular land-clearing method depends on soil properties, management practices, and the scale of operations.

Cordero (1964) compared the conventional slash-and-burn method with bulldozed clearing followed by plowing and land leveling, with and without the use of a root rake, in an Entisol near Santa Cruz, Bolivia. Yield records of sugarcane and upland rice were kept for 5 years, after which the soil was sampled. Table 10.9 shows no significant yield differences or changes in soil properties with the various land-clearing methods. The costs of land clearing about doubled, however, with the mechanized treatments. Cordero concluded that the traditional system was better because of its lower cost.

In sharp contrast with Cordero's results, several studies have shown that mechanized land clearing can have agronomic disadvantages as well. Removal of the felled vegetation by a D-6 bulldozer equipped with a conventional blade produced consistently lower crop yields than the traditional slash-and-burn system as practiced in the Amazon Jungle on Ultisols. Table 10.10 shows that crop yields under mechanized clearing were consistently inferior both with and without fertilization.

Three reasons have been advanced by Seubert (1975) to explain these dif-

Plate 13 Mechanized land-clearing operations such as this one on coarse-textured Ultisols produce severe soil compaction and lower yields than the traditional slash and burn system.

Table 10.9   Effects of Land-Clearing Treatments on Crop Production and Changes in the Top 50 cm of an Entisol from Santa Cruz, Bolivia, 5 years after Clearing

Land Clearing Method	Clearing Cost (1957) (U.S. $/ha)	Upland Rice Production (3 years) (tons/ha)	Sugarcane Production (5 years) (tons/ha)	Soil pH	Organic Matter (%)	Available P (Bray) (ppm)	Exch. K (meq/100 g)
Slash and burn, no plowing	41	5.0	338	7.1	4.5	37	0.60
Bulldozer clearing, plowing, harrowing, and leveling	92	5.5	343	7.1	3.2	17	0.32
Bulldozer clearing with root rake, plowing, harrowing, and leveling	141	5.8	365	6.7	3.9	21	0.53

Source: Adapted from Cordero (1964).

Table 10.10 Effects of Land-Clearing Methods and Fertilization on Crop Production on an Ultisol from Yurimaguas, Peru (tons/ha)

Land Clearing Method	Fertility Treatment	Continuous Upland Rice				Cassava	Soybeans	Guinea Grass (annual production)
		2nd	3rd	4th				
Slash and burn	None	1.93	1.36	0.77		22.5	0.72	9.9
	Complete[a]	3.20	3.53	2.00		34.2	2.34	24.1
Bulldozer clearing	None	1.09	0.92	0.20		10.1	0.12	8.3
	Complete	2.52	3.19	1.42		32.0	1.31	18.4

Source: North Carolina State University (1973, 1974).

[a] Initial application of lime to pH 6.2, 50-172-40 kg NPK/ha, and same N + K rates after every crop or pasture cutting.

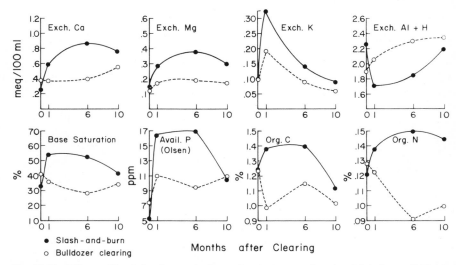

Fig. 10.19 Effects of land clearing methods on chemical properties of an Ultisol topsoil (0–10 cm layer) in Yurimaguas, Peru. *Source:* Adapted from North Carolina State University (1974).

ferences: the benefits of the ash in the native system, the compaction caused by the bulldozer, and topsoil disturbance by the bulldozer blade. The chemical benefits of the ash, whose composition is shown in Table 10.6, are reflected in the changes in the topsoil illustrated in Fig. 10.19. The supply of exchangeable calcium, magnesium, and potassium increased dramatically, exchangeable acidity decreased, available phosphorus tripled, and the decrease in organic matter and nitrogen was slower with the slash-and-burn than with the bulldozed system.

Infiltration rates at 1 and 11 months after clearing averaged 10.5 cm/hour for the slash-and-burn system in contrast to about 0.5 cm/hour for the bulldozed system. This difference is believed to be the result of bulldozing itself in sandy topsoils very susceptible to compaction. Similar data were obtained in the interior of Surinam by Van der Weert (1974) with a more sophisticated land-clearing system involving the use of a K.G. blade. The dramatic increase in bulk density throughout the profile resulted in decreased aeration, infiltration rates, and root development. Van der Weert's results are shown in Fig. 10.20.

The third undesirable factor in mechanized clearing is the partial removal of topsoil and its deposition elsewhere during the process. Although the bulldozer blades can be kept at a height that prevents contact with the soil, the process of stump and log removal usually involves carrying topsoil from high spots and depositing it in low spots, or close to the windrows of cleared debris. In most soils in equilibrium with a tropical forest, the bulk of the organic matter and available nutrients is in the topsoil, and therefore topsoil removal can be catastrophic. Lal and his coworkers (1975) observed a 50 percent corn yield reduction when the top 2.5 cm of a Nigerian Alfisol was removed, and a 90 percent yield reduction when removal reached 7.5 cm. Van der Weert (1974) in Surinam observed better crop growth near the windrows, where the vegetation was pushed. This was probably a direct result of topsoil accumulation.

Economic and social considerations must also be evaluated when comparing mechanized clearing with the traditional slash-and-burn system. In addition to the agronomic disadvantages of the example from Peru (Table 10.10), the actual costs of mechanized land clearing per hectare were three times as high as with slash and burn. Additional difficulties are encountered in the transportation and maintenance of heavy equipment in primitive areas. Perhaps more importantly, the immediate clearing of large areas poses serious management problems.

Farmers must cope with fertilizer shipment and the need to control

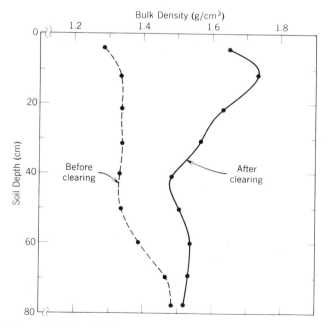

Fig. 10.20 Effects of bulldozed clearing with a K.G. blade on bulk density of a sandy kaolinitic soil at Coebiti, Surinam. *Source:* Van der Weert (1974).

rapidly growing weeds. Extensive mechanized land clearings were made with heavy machinery in Pucallpa, Peru (Abastos, 1971). Many of these fields have since been abandoned because there was not enough labor to control weeds or enough fertilizer available to counteract the fertility decline. It is this writer's belief that the transition from shifting to continuous cropping by small farmers is best accomplished gradually and that slash-and-burn clearings of small size permit the farmers to concentrate on the new practices, fertilization and weeding, on a manageable scale.

On the other hand, mechanical land clearing may be economically sound for the establishment of commercial plantations or large cooperatives with sufficient capital. The agronomic disadvantages encountered in the studies from Peru, Surinam, and Nigeria can be counteracted by improvements in the land-clearing techniques. Burning should be an integral part of the mechanized land-clearing system in order to take full advantage of the fertilizer value of the ash. This can be accomplished by felling the trees in one operation, burning when dry, and in a second operation piling the debris in windrows. The compaction effects can be minimized by the use of improved equipment, such as chain drags between two bulldozers, but more importantly by clearing at a soil moisture content least subject to puddling or compaction. In many soils with excellent structural properties compaction is not likely to be as serious as in the examples mentioned. Topsoil carryover can be minimized by the use of a root rake, floating blades, and careful operators.

Jurion and Henry (1964) summarized the Belgian experience in Zaïre by recommending that mechanized land clearing be used only in conjunction with plowing and the growth of monocultures for large-scale operations. This represents a drastic transition from shifting to a temperate-region type of cultivation.

Another land-clearing method that can be used in forested areas is the poisoning of trees with low-cost chemicals like 2,4-D. Such practice is not recommended, however, because of the detrimental environmental effects. In addition, a study conducted by Ahn (1970) in Ghana showed that higher yields of plantains and cocoyams were obtained with the traditional slash-and-burn method than with tree poisoning, the reason being the fertilizer value of the ash.

The land-clearing problems in savanna areas are simpler. The woody vegetation is easily removed either manually or mechanically. Serious compaction effects have not been reported, because tree removal does not involve the effort required to clear a forest. Topsoil carryover is less of a problem, since both organic matter and nutrients are better distributed in the soil profile because of the action of the grass roots. Burning is usually practiced because it is the only effective way to clear the grass. The main management problem is one of timing, if shifting cultivation is to be maintained. Ramsay and Rose-Innes (1963) found that burning early in the dry season minimizes damage to trees, whereas burning late in the dry season, when the vegetation is much drier, can kill many tree species and promote the development of undesirable grasses.

An additional problem always encountered is what to do with termite mounds in both forest and savanna areas in the process of land clearing. The number of termite mounds and the area covered by them can be very large. Lal et al. (1975) estimate an average of 60 active and abandoned mounds per hectare in Nigeria. These mounds consist primarily of subsoil material. Where the subsoil is relatively low in available nutrients, growth will be limited. In cases where the subsoil is higher in bases, there may be some fertility benefits. Termite mounds are not disturbed by the traditional slash-and-burn method. They are destroyed in mechanized land clearing, however, particularly if this is followed by plowing. The best policy depends on their chemical composition and local management considerations.

Fertilization

The transition from shifting to continuous cultivation invariably involves fertilization. The question is not whether to fertilize or not; rather, what must be determined is what sources (manures vs. inorganic) to use, how much fertilizer to apply, and what residual effects to expect.

A massive amount of research on fertilization has been conducted in shifting cultivation areas of Africa. These studies have been summarized periodically by Vine (1954), Nye and Stephens (1962), Touré (1964), Nye (1966), Richardson (1968), and FAO (1973). Their impact on food crop production, unfortunately, has been minimal. The bulk of the research has been limited to extremely low fertilizer rates on the order of 10 to 40 kg N, P_2O_5, and K_2O/ha. Yield responses have been understandably very modest. A summary of FAO "Freedom from Hunger" trials conducted in West Africa is shown in Table 10.11. The responses are typical of much of the research data produced in the region. It is no wonder that farmers are not using fertilizers with this type of response! Much of the research has been conducted without attention to soil characteristics or soil test data. It is disappointing to see reports on hundreds of lime trials on soils with pH levels between 6 and 7.

Some valid generalizations can be made from these data. Nitrogen responses almost invariably occur in the first crop after clearing a savanna but seldom after clearing a forest. Nye (1966) reports that nitrogen

Table 10.11   Summary of FAO Fertilizer Program Simple Trials in Ghana, Nigeria, and Senegal for the Period 1961–1964 (yields in tons/ha)

Crop	Zone	Number of Trials	Check Yields	Main Effects from 20 or 40 kg/ha		
				N	P_2O_5	K_2O
Corn	Forest	343	1.22	1.40	1.38	1.38
	Savanna	159	1.19	1.37	1.40	1.32
Rice	Forest	153	1.31	1.53	1.52	1.39
	Savanna	199	1.27	1.48	1.47	1.44
Yams	Forest	60	9.0	10.0	10.1	9.6
	Savanna	160	8.4	9.4	9.5	9.2
Sorghum	Savanna	21	0.40	0.49	0.43	0.43
Millet	Savanna	91	0.57	0.70	0.70	0.63

Source: Nye (1966).

responses in cereals are found after 6 years of continuous cropping in forest areas. The differences in nutrient accumulation patterns between forest and savanna fallows explains these differences. Savanna fallows are depleted of inorganic nitrogen by the grasses. Most of the nitrogen accumulated is lost to the atmosphere with annual burning. Thus sharp nitrogen responses can be expected. In the forest areas, on the other hand, total soil nitrogen contents are higher and are mineralized quickly after burning. Farmers usually plant nitrogen-demanding crops such as cereals after clearing the forest, whereas in the savannas root crops are the first ones planted because of their low nitrogen requirements and the abundance of available potassium.

Most of the West African soils under shifting cultivation are deficient in phosphorus but are not high phosphorus fixers because of their coarse top-soil texture. In both forest and savanna areas phosphorus responses are common after the first crop. The modest addition of phosphorus in the ash apparently prevents deficiencies in the first crop. Potassium deficiencies are common in forest areas after the second crop, but seldom occur in savannas until the eighth consecutive crop. Lime responses are positive in soils with pHs below 6 and are often negative at high pH values. Sulfur deficiencies are common in grain legumes.

With the exception of nitrogen and sulfur, simple soil test data should identify the possibilities of nutrient response. Much of the confusion about liming in the tropics stems from empirical lime applications made regardless of soil pH. At the same time it is not feasible to develop good soil test correlations with the low fertilizer responses obtained at the low rates applied.

A series of long-term experiments conducted throughout Africa proves that continuous cultivation is possible with the use of fertilizers and manures (Vine, 1953; Culot and Meyer, 1959; Djokoto and Stephens, 1961; Grimes and Clarke, 1962; Heathcote, 1969; Stephens, 1969; Heathcote and Stockinger, 1970; Dabin, 1971; Abu-Zeid, 1973; Ofori, 1973). In most cases only one fertilizer combination was used, in comparison with mulches or manures. A typical example is the results of Abu-Zeid's (1973) 10 year experiment with a cotton–peanut–eleusine rotation in Yambio, Sudan. The results shown in Table 10.12 illustrate that the rotation yields increased by about 60 percent when the soil was ridged, fertilized, manured, or mulched. The yield levels, however, are extremely low. A series of continuous cropping experiments conducted in Ibadan, Nigeria, from 1922 to 1951 and summarized by Vine (1953, 1954, 1968) shows that yields can be maintained at levels between 0.4 and 1.7 tons/ha of corn, about 7 tons/ha of yams, and 12 tons/ha of cassava almost indefinitely with sporadic applications of 60 kg N/ha and 20 kg P/ha as ammonium sulfate and ordinary superphosphate. In no case, however, was the economic validity of such treatments determined.

More recent work in which higher fertility rates were used shows that crop yields can be maintained at high levels. Jones (1972) obtained corn yields of 3 to 7 tons/ha and high yields of dry beans and seed cotton in Ulti-

Table 10.12   Effects of Fertilization, Manures, Mulches, and Ridging on a 10 Year Cotton–Peanut–Eleusine Continuous Rotation on an Oxisol from Yambio, Sudan (yields are means of 10 years)

Treatment, Annual Application	Seed Cotton (tons/ha)	Peanuts (unshelled) (tons/ha)	Eleusine (grains) (tons/ha)	Relative Rotation Yield
Traditional practice	0.29	0.73	0.61	100
Ridged	0.49	0.87	0.78	131
Ridged + 52-86-105	0.72	1.00	0.94	165
Ridged + manure[a]	0.74	0.98	0.84	157
Mulched	0.70	0.98	0.88	157

Source: Abu-Zeid (1973).
[a] Estimated nutrient contents of 108-80-18 kg NPK/ha.

sols of Namungolo, Uganda. The latter two, in which an annual fertilizer application of 51 kg N/ha, 22 kg P/ha, 63 kg K/ha, 32 kg S/ha, and 78 kg Ca/ha was used, are illustrated in Fig. 10.21.

The interaction between fertilization and weeding in continuous cropping is dramatically shown in Fig. 10.22. Without adequate fertilization and weeding, yields declined to less than 30 percent of the original in the second year. With weeding only, the decline was less pronounced. With fertilization only, yields began to decline during the third year, but with both fertilization and weeding yields actually increased and averaged about 6.5 tons/ha. Although no economic interpretation is available, the responses to fertilization and weeding are of such magnitude that they probably compensate for the costs involved.

To obtain a realistic economic appraisal of fertilization as a key element in continuous cultivation, the residual effects, particularly of phosphorus, lime, and micronutrients, must be evaluated over a period of several years. Some examples in the literature indicate the importance of residual effects, but no research has yet been conducted for a long enough period of time in shifting cultivation areas. With an initial lime application of 4 tons/ha, which raised the pH from 4.5 to 6.2 in an Ultisol from Peru, production of three successive crops of upland rice, corn, and soybeans per year has maintained a total annual grain yield of 8 to 10 tons/ha during the first 2 years (North Carolina State University, 1974). The residual effects of low rates of

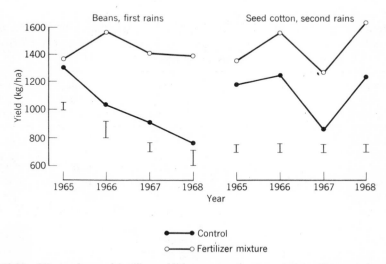

Fig. 10.21   Effects of annual fertilizer additions on continuous rotation of beans and cotton in Namungole, Uganda. *Source:* Jones (1972).

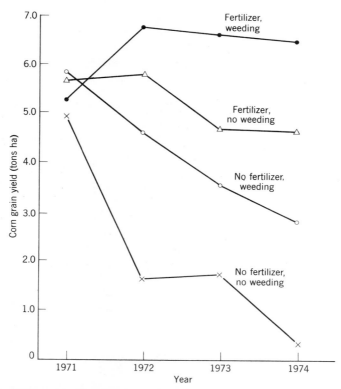

Fig. 10.22 Effects of fertilization (120 kg N/ha, 40 kg P/ha, and 50 kg K/ha per crop) and weeding on rainy season corn yields in a continuously cropped experiment on an Alfisol near Ibadan, Nigeria. Forest cleared in 1970 and subjected to a uniformity trial with corn. *Source:* IITA (1974).

simple superphosphate (10 kg P/ha) were sufficient to prevent yield declines in Ghana for 2 subsequent years, according to Stephens (1960). Continuous applications of ammonium sulfate and other fertilizers with acidifying effects have decreased soil pH to a point where liming is needed in East Africa (Grimes and Clark, 1962).

Manures

The high cost of fertilizers and in many cases the high cost of transporting them to shifting cultivation areas render their use uneconomical in spite of the sharp agronomic responses often obtained when the appropriate rates are used. Animal manures and crop rotation involving legumes were the means for transforming shifting into continuous cultivation in the temperate region, according to Grigg (1974). The use of animal manures has been primarily responsible for permanent cultivation patches near the homes of shifting cultivators, as illustrated in Fig. 10.18. Research in Africa has shown that manures can maintain continuous cropping at low yield levels similar to those produced by modest applications of inorganic fertilizers. Table 10.12 shows typical results. Culot and Meyer (1959), Cutting et al. (1959), Grimes and Clarke (1962), and Heathcote (1969) have shown that the effectiveness of manure applications is related to their nutrient contents. The problem in most areas is the availability of cattle manure in sufficient quantity. In many shifting cultivation areas of Africa cattle production is severely hampered by tsetse fly disease; in other areas extensive grazing prevents the accumulation of manures. In the highlands of eastern Zaïre Van Parijs (1959) developed a system for the effective use of manure. He found that small, 3 ha pastures produced enough manure to fertilize about 1 ha of crops for an indefinite period. Perhaps on this scale, manuring can be practical.

The role of green manures has also been explored and in many cases found agronomically sound. The problem has been to convince the shifting cultivator to grow a crop that will produce no direct economic advantage, instead of growing a cash crop. Also the need to incorporate the green manure into the soil requires power beyond that available to many farmers. Vine (1953, 1954) found that corn yields were maintained between 1.3 and 1.6 tons/ha for 17 years when a velvet bean crop was grown in Ibadan, Nigeria, once a year. Vine calculated that this practice added 20 kg N/ha to the soil. In most cases the same amount of nitrogen can be added as fertilizer or manure and an extra crop grown. Some green manure crops are producing interesting side effects. When kudzu was incorporated into an Ultisol in Peru, soybeans grew better, nodulated more, and yielded more than with comparable treatments at similar fertility levels and innoculation treatments (North Carolina State University, 1974). Intercropping green manure crops with cereals has increased crop yields without the loss of time that would be incurred by growing a green manure crop separately (Agboola and Fayemi, 1972). The effects of green manures in intercropping systems will be discussed in Chapter 12.

Mulching

One practice that has consistently been successful in shifting cultivation areas is mulching. When crop residues or straw from other areas is used,

Table 10.13 Effect of Mulching on Runoff Losses in Ibadan, Nigeria (% of rainfall)

Slope (%)	Corn Crop		Forest Cover
	Unmulched	Mulched	
1	6.4	2.0	1.7
5	40.3	7.7	1.3
10	42.7	5.7	1.7
15	17.6	1.9	2.0

Source: Lal et al. (1975).

yield responses have been as high as those obtained with the application of fertilizers or manures, or higher. This is shown in Abu-Zeid's results in Table 10.12 and also in several other reports (Djokoto and Stephens, 1961; Nye and Stephens, 1962; Ofori, 1972). Lal et al. (1975) have shown that mulching decreases soil temperatures, conserves moisture, prevents erosion, and adds nutrients to the soil.

Minimum tillage practices consisting of tilling the rows only, with the rest of the land mulched, have produced excellent results. Tables 10.13 and 10.14 show some of Lal's results. There is no question in this writer's mind that mulching and minimum tillage should be integral parts of continuous cropping systems in shifting cultivation areas. A model that integrates these management practices for West Africa has been developed by Greenland (1975).

Table 10.14 Comparative Crop Yields with Minimum and Conventional Tillage in Ibadan, Nigeria (tons/ha)

Crop	Minimum Tillage	Conventional Tillage
Corn	4.50	3.00
Cowpeas	0.78	0.63
Soybeans	0.84	1.06
Sweet potatoes	22.00	17.75
Pigeon peas (dry matter)	26.25	25.00

Source: Lal et al. (1975).

Crop Rotations

The well-known advantages of crop rotations in the temperate region are also applicable to tropical shifting cultivation areas where outside inputs must be minimized because of their high cost. The choice of crops depends on local conditions and market demand. One obvious advantage is the avoidance of insect and pest buildups common in continuous cropping. An annual rotation of upland rice, corn, and soybeans in Yurimaguas, Peru, has produced excellent results, whereas continuous rice cultivation has resulted in sharp yield declines at adequate fertility levels. A complex combination of insect and pathogen attacks seems to be responsible for the poor performance of continuously grown rice.

The literature on this subject has been critically reviewed by Newton (1960). He concludes that no rotation scheme involving legumes and/or green manures has been successful in maintaining continuous cropping in shifting cultivation areas. This is an unfortunate result of single-factor experiments. What is badly needed is to develop complete soil management systems that include rotation and intercropping schemes plus fertilization and mulching. In this way suitable systems could be developed for continuous food production.

Permanent Crops

In many shifting cultivation areas a more feasible alternative is to grow trees which produce a crop that can be harvested. Large plantations have been developed in the tropics in this fashion. The advantage of planting

Table 10.15   Total Oil Palm Production as a Function of Establishment Treatments on an Oxisol of Benin, Nigeria, Planted in 1940

Establishment Treatment	Total Fruit Production, 1945–1956 (tons bunches/ha)
Pure stand, normal weed cover	85.0
2 years of intercropping[a]	87.8
12 years of intercropping[a]	93.4
Pure stand, controlled kudzu cover	89.0
Pure stand, no weeding	75.6

Source: Adapted from Kowal and Tinker (1959).
[a] With yams, cassava, and corn.

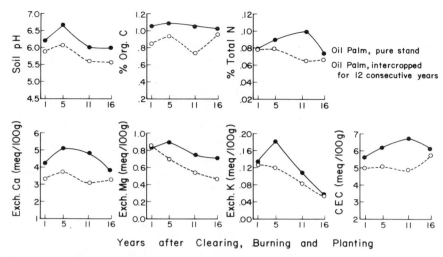

Years   after   Clearing,   Burning   and   Planting

Fig. 10.23 Changes in the soil of an oil palm plantation established from a high secondary forest on an Alfisol in Benin, Nigeria (0–15 cm depth). *Source:* Adapted from Kowal and Tinker (1959).

permanent tree crops is that they establish their own closed nutrient cycle. The amounts harvested yearly are usually a small fraction of the biomass in crops such as cacao, oil palm, rubber, and many tropical fruits. The soil dynamics after the establishment of an oil palm plantation in southern Nigeria are illustrated in Fig. 10.23. Most parameters remained static except for potassium, which is usually rapidly lost after burning. An additional advantage of establishing tree crops is that the land can be used for producing food crops or pastures while the trees grow. Figure 10.23 shows that certain nutrients were depleted when the oil palm was intercropped for 12 years with food crops (yams, cassava, and corn), but oil palm production was not affected (Table 10.15). This table also shows the detrimental effects of not weeding, even under these conditions.

Another alternative is the planting of commercial timber. Fast-growing species such as teak (*Tectona grandis*) have been planted in huge areas in ustic climates of Africa and Southeast Asia. Normally called the "taungya" system, this involves harvesting after a considerable period of time, and reverting to growing crops while new tree seedlings are being established. In Jarilandia, Brazil, several hundred thousand hectares of Amazon Jungle have been cleared and planted to a fast-growing tree species, *Gmelina arborea,* which can be harvested as pulpwood after 6 to 8 years, or as wood after 10 years (Lamb, 1973). There is no doubt that this form of land use can be very profitable if coupled with processing facilities.

SUMMARY AND CONCLUSIONS

1. Shifting cultivation is the most widespread form of soil management in the tropics, covering about 44 percent of the potentially arable and grazing lands in the region. Although practiced by widely divergent peoples, the main components are very similar, the principal difference being whether it is conducted in forest or savanna vegetation.

2. Before clearing, the soil and forest have a remarkably closed nutrient cycle in which most nutrients are stored in the biomass and topsoil, and transferred from one to the other via rainwash, litter fall, timber fall, root decomposition, and plant uptake. Losses from this system are usually negligible. Therefore lush tropical vegetation grows without nutrient deficiency symptoms in soils of very low native fertility.

3. When this nutrient cycle is broken by clearing and burning the vegetation, significant changes in soil physical properties take place. Soil temperatures increase during burning but seldom affect more than the top 2 to 5 cm of the soil. After burning, soil and air temperatures increase because more solar radiation reaches the soil surface. Soil moisture regimes are also altered, with less moisture removal from the subsoil than when forest roots were active. Soil structure deterioration which leads to runoff and erosion losses occurs in poorly aggregated topsoils subjected to inappropriate management practices, often caused by increasing population density. Changes in soil structure are minor in well-aggregated Oxisols and Andepts and in most soils when protected by ash or mulch cover and a continuous crop canopy, as in traditional shifting cultivation systems.

4. The chemical composition of the ash produces important changes in soil chemical properties, but the magnitude and duration of these changes depend on type of vegetation felled, climate, and soil properties. Soil pH values increase after burning because of the incorporation of basic cations, and gradually decrease with cultivation. In acid soils these changes are beneficial because they increase the exchangeable calcium and magnesium contents and neutralize part of the exchangeable aluminum. In high-base-status soils the ash may raise the pH to 7 or 8, possibly causing detrimental effects such as iron deficiency. In all cases, however, the soil test available phosphorus and potassium levels increase after burning because of the contribution of the ash. The organic matter and soil nitrogen contents increase slightly after burning because of the addition of partially burned vegetation, but decrease gradually with cultivation. The magnitude of this decrease is greater in soils with high organic matter contents and in cases where the soil surface is exposed for considerable periods of time.

5. The magnitude of yield declines without fertilization depends on soil properties, climate, and vegetation. Yield declines are sharper in acid soils,

in udic soil moisture regimes, and in forested areas. The reasons for abandoning cultivated fields one or a few years after clearing are soil fertility depletion, increased difficulties in weed control, soil erosion, higher incidence of pests and diseases, and certain social customs. Soil fertility depletion is believed to be the most important cause for abandonment of low-base-status soils; increased weed control problems, of high-base-status soils.

6. Several approaches have been proposed to increase crop production without changing the shifting cultivation system: the corridor system, improving varieties and plant spacing, shortening the length of fallows, and improving the quality of fallows. The most promising improvements are better or different varieties, closer plant spacing, and modest fertilization— changes that are not beyond the economic means of the farmers.

7. In some areas population pressures require a complete change to continuous cultivation. Land-clearing methods are crucial because certain mechanical operations may result in serious damage to soil physical properties, leading to compaction, topsoil removal, and erosion. The traditional slash-and-burn method does not cause such problems and adds valuable fertilizer in the form of ash. The choice of land-clearing practices also depends on soil properties, management, and capital available for management after clearing.

8. The fertility decline incurred in changing to continuous cultivation can be corrected by fertilizing or manuring in a straightforward manner. In many areas the cost of fertilizers and transportation makes heavy fertilization uneconomical. A combination of minimum tillage, mulching, and multiple cropping with modest fertilization practices has proved successful in maintaining good yields on a continuous basis in certain areas previously under shifting cultivation.

REFERENCES

Abastos, M. 1971. *Inventario y Evaluación de la Concesión Tournavista. Pucallpa.* LeTourneau del Peru, Inc., Lima. 76 pp.

Abu-Zeid, M. O. 1973. Continuous cropping in areas of shifting cultivation in Southern Sudan. *Trop. Agr. (Trinidad)* **50**:285–290.

Agboola, A. A. and A. A. Fayemi. 1972. Effect of soil management on corn yield and soil nutrients in the rainforest zone of western Nigeria. *Agron. J.* **64**:641–644.

Ahlgren, I. F. 1974. Effect of fire on soil organisms. Pp. 47–72. In T. T. Kozlowski (ed.), *Fire and Ecosystems.* Academic Press, New York.

Ahn, P. M. 1968. The effects of large scale mechanized agriculture on the physical properties of West African soils. *Ghana J. Agr. Sci.* **1**:35–40.

Ahn, P. M. 1970a. Shifting cultivation and mechanized agriculture. Pp. 232–253. In *West African Soils*. Oxford University Press, London.

Ahn, P. M. 1970b. The effects of clearing by poisoning as against felling and burning on plantations and cocoyam yields on acid Ghana forest soils. *Ghana J. Agr. Sci.* 3:93–97.

Ahn, P. M. 1974. Some observations on basic and applied research in shifting cultivation. *FAO Soils Bull.* 24, pp. 123–154.

Anthony, K. R. M. and S. G. Willimott. 1956. A study of soil fertility in Zandeland. *Emp. J. Exptal. Agr.* 24:75–88.

ASRCT. 1968. Semiannual Report 2. Cooperative Research Programme 27, Applied Scientific Research Corporation of Thailand, Bangkok.

Bartholomew, W. V., I. Meyer, and H. Laudelot. 1953. Mineral nutrient immobilization under forest and grass fallow in the Yangambi (Belgian Congo) region. *INEAC, Ser. Sci.* 57, pp. 1–27. Institut National pour l'Etude Agronomique du Congo, Brussels.

Bouyer, S., R. Tourte, and L. Collot. 1951. Deuxieme contribution a l'etude de la fumure des terres a arachide du Sénégal; effet résiduel des formules NPK sur la deuxieme année de culture. *Agron. Tropicale (France)* 6:287–293.

Boyle, J. R. 1973. Forest soil chemical changes following fire. *Commun. Soil Sci. Plant Anal.* 4:369–374.

Brams, E. A. 1971. Continuous cultivation of West African soils: organic matter diminution and effects of applied lime and phosphorus. *Plant and Soil* 35:401–414.

Brams, E. A. 1973. Residual soil phosphorus under sustained cropping in the humid tropics. *Soil Sci. Soc. Amer. Proc.* 37:579–583.

Briggs, G. W. G. 1938. Maintenance of soil fertility. *Third West Afr. Agr. Conf.* 1:1–8.

Brinkmann, W. L. F. and J. C. de Nascimento. 1973. The effect of slash and burn agriculture on plant nutrients in the tertiary region of Central Amazonia. *Turrialba* 23:284–290.

Budowski, G. 1956. Tropical savannas, a sequence of forest felling and repeated burnings. *Turrialba* 6:23–33.

Charreau, C. and R. Nicou. 1971. L'amèloration du profil cultural dans les sols sableux et sablo-argileux de la zone tropicale sèche ouest africaine et ses incidences agronomiques. *Agron. Tropicale (France)* 26:903–978, 1184–1247.

Clarke, R. T. 1962. The effect of some resting treatments on a tropical soil. *Emp. J. Exptal. Agr.* 30:57–62.

Conklin, H. C. 1963. *The Study of Shifting Cultivation.* Panamerican Union, Studies and Monographs, No. 6. Washington. 165 pp.

Cooke, C. W. 1931. Why the Mayan cities of the Petén district, Guatemala were abandoned. *J. Wash. Acad. Sci.* 21:283–287.

Cordero, A. 1964. The effect of land clearing on soil fertility in the tropical region of Santa Cruz, Bolivia. M. S. Thesis, University of Florida, Gainesville. 102 pp.

Coulter, J. K. 1972. Soil management systems. Pp. 189–197. In *Soils of the Humid Tropics*. National Academy of Sciences, Washington.

Cowgill, U. M. 1960. Soil fertility, population, and the ancient Maya. *Proc. Nat. Acad. Sci.* 46:1009–1011.

Cowgill, U. M. 1962. An anthropological study of the southern Maya lowlands. *Amer. Anthropologist* 64:273–286.

Culot, J. P. and J. Meyer. 1959. Possibilities des cultures vivrieres continues en conditions equatoriales. *Proc. Third Inter-Afr. Soils Conf. (Dalaba)* 2:831–841.

Cunningham, R. K. 1963. The effect of clearing a tropical forest soil. *J. Soil Sci.* **14**:334–345.

Cutting, C. V., R. A. Wood, P. Brown, et al. 1959. Assessment of fertility status and the maintenance of productivity of soils in Nyasaland. *Proc. Third Inter-Afr. Soils Conf.* (*Dalaba*) **2**:817–824.

Dabin, B. 1971. The use of phosphate fertilizer in a long-term experiment on ferralitic soil at Bambari, Central African Republic. *Phosphorus in Agr.* **25**:1–11.

Davide, J. G., L. M. Villegas, and E. H. Tyner. 1962. Rotation and fertilizer studies on abandoned *Kainging* lands in the Island of Palawan, Philippines. Pp. 556–573. In *Trans. Comm. IV and V, Int. Soc. Soil Sci.* (*New Zealand*).

Djokoto, R. K. and D. Stephens. 1961. Thirty long term fertilizer experiments under continuous cropping in Ghana I & II. *Emp. J. Exptal. Agr.* **29**:181–196, 245–258.

Dobby, E. G. H. 1950. *Southeast Asia*. Wiley, New York.

Donovan, K. 1973. *Informe de Ensayos con Arroces Chancay y Huallaga* 1972–73. Ministerio de Agricultura, Zona Agraria IX, Agencia de Extensión, Yurimaguas, Peru. 9 pp.

Doyne, H. C., K. T. Hartley, and W. A. Watson. 1938. Soil types and manurial experiments in Nigeria. Pp. 227–298. In *Third West Afr. Agr. Conf.*

Dumont, R. 1966. *Types of Rural Economy*. Methuen, London.

Edminsten, J. 1970. Preliminary studies of the nitrogen budget of a tropical rain forest. Pp. H211–H216. In H. T. Odum (ed.), *A Tropical Rain Forest*. U.S. Atomic Energy Commission, Washington.

Ewell, J. J. 1968. Dynamics of litter accumulation under forest succession in eastern Guatemala lowlands. M.S. Thesis, University of Florida, Gainesville.

FAO Staff. 1957. Shifting cultivation. *Trop. Agr.* (*Trinidad*) **34**:159–164.

FAO. 1973. Improving soil fertility in Africa. *FAO Soils Bull.* 14. 145 pp.

FAO–SIDA. 1974. Shifting cultivation and soil conservation in Africa. *FAO Soils Bull.* 24. 248 pp.

Ferwerda, J. D. 1970. Soil fertility in the tropics as affected by land use. Pp. 317–329. In *Int. Potash Inst. Proc. 9th Congr.* (*Berne*).

Friese, F. W. 1939. Untersuchungen uber die Falgen der Brandwirtschaft ans tropischen Boden. *Tropenpflanzer* **42**:1–22.

Gillier, P. 1960. La reconstitution et le maintein de la fertilité des sols du Sénégal et le probleme des jácheres. *Oleajenaux* **15**:699–704.

Golley, F. B., J. T. McGinnins, R. J. Clements, et al. 1969. The structure of tropical forests of Panama and Colombia. *Biol. Sci.* **19**:693–696.

Goswami, P. C. 1971. Shifting cultivation in the hills of northeastern India. *Indian Farming* **21**:10–13.

Grant, P. M. 1970. Restoration of productivity of depleted sands. *Rhodesia Agr. J.* **67**:131–137.

Greenland, D. J. 1974. Evolution and development of different types of shifting cultivation. *FAO Soils Bull.* 24, pp. 5–13.

Greenland, D. J. 1975. Bringing the green revolution to the shifting cultivator. *Science* **190**:841–844.

Greenland, D. J. and J. M. L. Kowal. 1960. Nutrient content of a moist tropical forest of Ghana. *Plant and Soil* **12**:154–174.

Greenland, D. J. and P. H. Nye. 1969. Increases in the carbon and nitrogen contents of tropical soils under natural fallows. *J. Soil Sci.* **9**:284–299.

Grigg, D. B. 1974. *The Agricultural Systems of the World*. Pp. 57–74. Cambridge University Press, London.

Grimes, R. C. and R. T. Clarke, 1962. Continuous arable cropping with the use of manures and fertilizers. *East Afr. Agr. For. J.* **28**:74–80.

Guillemin, R. 1956. Evolution de l'agriculture autochtone dans les savannes de l'Oubangui. *Agron. Tropicale (France)* **11**:143–176.

Hardy, F. 1936. Some aspects of tropical soils. *Trans. 3rd Int. Congr. Soil Sci.* **2**:150–163.

Hauck, F. W. 1974. Introduction. Shifting cultivation and soil conservation in Africa. *FAO Soils Bull.* 24, pp. 1–4.

Heathcote, R. G. 1959. Soil fertility under continuous cultivation in northern Nigeria. I. The role of organic manures. *Exptal. Agric.* **6**:229–237.

Heathcote, R. G. and K. R. Stockinger. 1970. Soil fertility under continuous cultivation in northern Nigeria. II. Responses to fertilizers in the absence of organic manures. *Exptal. Agric.* **6**:345–350.

Huttel, C. H. 1959. Vertical distribution of roots in a tropical rainforest in Southern Ivory Coast. *J. West Afr. Sci. Assoc.* **14**:65–72.

IITA. 1971–1974. Annual Reports. Farming Systems Program, International Institute for Tropical Agriculture, Ibadan, Nigeria.

Jaiyebo, E. O. and A. W. Moore. 1964. Soil fertility and nutrient storage in different soil-vegetation systems in a tropical rain forest environment. *Trop. Agr. (Trinidad)* **41**:129–139.

Jameson, J. D. and R. K. Kerkham. 1960. The maintenance of soil fertility in Uganda. *Emp. J. Exptal. Agr.* **28**:179–192.

Jewitt, T. N. 1950. Shifting cultivation in the clay plains of the central Sudan. Pp. 331–333. *Trans. Fourth Int. Congr. Soil Sci. (Amsterdam)*.

Jones, E. 1972. Principles for using fertilizers to improve red ferralitic soils in Uganda. *Exptal. Agr.* **8**:315–320.

Jordan, C. F. 1970. A progress report on studies of mineral cycles at El Verde. Pp. H217–H219. In H. T. Odum (ed.), *A Tropical Rain Forest*. U.S. Atomic Energy Commission Washington.

Jurion, F. and J. Henry. 1969. Can primitive farming be modernized? *INEAC, Ser. HORS* 1969. Institut National pour L'Etude Agronomique du Congo, Brussels. 445 pp.

Kawano, K., P. A. Sanchez, M. A. Nureña, and J. Velez. 1972. Upland rice in the Peruvian Jungle Pp. 637–644. *Rice Breeding*. International Rice Research Institute, Los Baños, Philippines.

Keen, B. A. and D. W. Duthie. 1953. Crop responses to fertilizers and manures in East Africa. *East Afr. Agr. For. J.* **19**:19–28.

Kellogg, C. E. 1963. Shifting cultivation. *Soil Sci.* **95**:221–230.

Kerr, A. I. 1942. A new system of grass fallow strip cropping for the maintenance of soil fertility. *Emp. J. Exptal. Agr.* **16**:125–132.

Kowal, J. M. L. and P. B. H. Tinker. 1959. Soil changes under a plantation established from high secondary forest. *J. West Afr. Inst. Oil Palm Res.* **2**:376–389.

Lafont, P. O. 1959. Slash and burn (ray) agriculture systems in mountain populations of Central Vietnam. *Proc. 9th Pacific Sci. Cong.* **7**:56–59.

Lal, R. 1974. Soil erosion and shifting agriculture. *FAO Soils Bull.* 24, pp. 48–71.

Lal, R. 1975. Role of mulching techniques in tropical soil and water management. *IITA Tech. Bull.* 1.

Lal, R., B. T. Kang, F. R. Moorman, A. S. R. Juo, and J. C. Moomaw. 1975. Soil management problems and possible solutions in Western Nigeria. Pp. 372–408. In E. Bornemisza and A. Alvarado (eds.), *Soil Management in Tropical America.* North Carolina State University, Raleigh.

Lamb, A. F. A. 1973. *Gmelina arborea.* Commonwealth Forestry Institute, Oxford University, 31 pp.

Laudelot, H. 1954. Investigation of the mineral-element supply by forest fallow burning. *Proc. Second Inter-Afr. Soil Congr.* 1:383–388.

Laudelot, H. 1959. Principles of the utilization of mineral fertilizers in the Belgian Congo. *Agr. Louvain* 7:451–470.

Laudelot, H. 1961. *Dynamics of Tropical Soils in Relation to Their Fallowing Techniques.* Paper 11266/E. Food and Agricultural Organization of the United Nations. Rome. 111 pp.

Lawes, D. A. 1961. Rainfall conservation and yield of cotton in northern Nigeria. *Emp. J. Exptal. Agr.* **29**:307–318.

Le Buanec, B. 1972. Dix ans de culture motorisee sur un bassin versant du centre Cote d'Ivoire. Evolution de la fertitité et de la production. *Agron. Tropicale (France)* **27**:1191–1211.

Lewin, C. J. 1931. The maintenance of soil fertility in southern Nigeria. *Third Spec. Bull. Agr. Dept.* Lagos, Nigeria. 43 pp.

Luse, R. A. 1970. The phosphorus cycle in a tropical rainforest. Pp. H161–H166. In H. T. Odum (ed.), *A Tropical Rain Forest.* U.S. Atomic Energy Commission, Washington.

Martin, W. W. and C. E. J. Biggs, 1937. Experiments on the maintenance of soil fertility in Uganda. *East Afr. Agr. For. J.* **2**:371–378.

McGinnis, J. T. and F. B. Golley. 1967. *Atlantic–Pacific Interocean Canal: Phase I.* Final Report. Bioenvironmental and Radiological Safety Feasibility Studies, Batelle Memorial Institute, Columbus, Ohio.

Mieklejohn, J. 1955. Effect of bush burning on the microflora of a Kenya upland soil. *J. Soil Sci.* **6**:111–118.

Moberg, J. P. 1972. Soil fertility problems in the West Lake Region of Tanzania. Effects of different forms of cultivation on the fertility of some Ferralsols. *East Afr. Agr. For. J.* **38**:35–46.

Moura, W. and S. W. Buol. 1972. Studies of a Latosol Roxo (Eutrustox) in Brazil. *Experientiae* **13**:201–217.

Newton, K. 1960. Shifting cultivation and crop rotation in the tropics. *Papua New Guinea Agr. J.* **13**:81–118.

Newton, K. and G. I. Jamieson. 1968. Cropping and soil fertility studies at Keravat, New Britain, 1954–1962. *Papua New Guinea Agr. J.* **20**:25–51.

North Carolina State University. 1973, 1974. Research on tropical soils. Annual Report. Soil Science Department, North Carolina State University, Raleigh.

Nye, P. H. 1958. The relative importance of fallows and soils in storing plant nutrients in Ghana. *J. West Afr. Sci. Assoc.* **4**:31–49.

Nye, P. H. 1961. Organic and nutrient cycles under a moist tropical forest. *Plant and Soil* **13**:333–346.

Nye, P. H. 1966. African experience of the use of fertilizers in the production of basic food crops. *Proc. Soil Crop. Sci. Soc. Fla.* **26**:306–313.

Nye, P. H. and M. H. Bertheux. 1957. The distribution of phosphorus in forest and savanna soils of the Gold Coast and its agricultural significance. *J. Agr. Sci.* **49**:141–149.

Nye, P. H. and D. J. Greenland. 1960. The soil under shifting cultivation. *Commonwealth Bur. Soils Tech. Commun.* 51. 156 pp.

Nye, P. H. and W. N. M. Foster. 1961. The relative uptake of phosphorus by crops and natural fallow from different parts of their root zone. *J. Agr. Sci.* **56**:299–306.

Nye, P. H. and D. Stephens. 1962. Soil fertility. Pp. 127–143. In J. B. Wills (ed.), *Agriculture and Land Use in Ghana*. Oxford University Press, London.

Nye, P. H. and D. J. Greenland. 1964. Changes in the soil after clearing a tropical forest. *Plant and Soil* **21**:101–112.

Odum, H. T. 1970. Summary, an emergent view of the ecological system at El Verde. Pp. 1191–1281. In H. T. Odum (ed.), *A Tropical Rain Forest*. U.S. Atomic Energy Commission, Washington.

Ofori, C. S. 1973. Decline in fertility status of a tropical forest Ochrosol under continuing cropping. *Exptal. Agr.* **9**:15–22.

Ovington, J. D. and J. S. Olson. 1970. Biomass and chemical content of El Verde lower montane rain forest plants. Pp. H53–H75. In H. T. Odum (ed.), *A Tropical Rain Forest*. U.S. Atomic Energy Commission, Washington.

Pendleton, R. L. 1954. The place of tropical soils in feeding the world. *Ceiba* **4**:201–222.

Pereira, H. C., E. M. Chenery, and W. R. Mills. 1954. The transient effects of grasses on the structure of tropical soils. *Emp. J. Exptal. Agr.* **22**:148–160.

Phillips, J. 1974. Effects of fire in forest and savanna ecosystems of sub-Saharan Africa. Pp. 435–482. In T. J. Kowlowski (ed.), *Fire and Ecosystems*. Academic Press, New York.

Popenoe, H. L. 1957. The influence of the shifting cultivation cycle on soil properties in Central America. *Proc. 9th Pacific Sci. Congr. (Bangkok)* **7**:72–77.

Popenoe, H. L. 1960. Effects of shifting cultivation on natural soil constituents in Central America. Ph.D. Thesis, University of Florida, Gainesville. 156 pp.

Ramsay, J. M. and R. Rose-Innes. 1963. Some observations on the effects of fire on the Guinea Savanna vegetation of northern Ghana over a period of 11 years. *Afr. Soils* **8**:41–85.

Reed, W. E. 1951. Reconnaissance soil survey of Liberia. *U.S. Dept. Agr. Agr. Inf. Bull.* 66. Washington.

Richardson, H. L. 1968. The use of fertilizers. Pp. 137–154. In R. P. Moss (ed.), *The Soil Resources of Tropical Africa*. Cambridge University Press, London.

Russell, E. J. 1950. *Soil Conditions and Plant Growth*. p. 635. Longmans, London.

Ruthenberg, H. 1971. *Farming Systems in the Tropics*. Oxford University Press, London.

Sanchez, P. A. 1973. Soil management under shifting cultivation. Pp. 46–67. In P. A. Sanchez (ed.), "A Review of Soils Research in Tropical Latin America." *North Carolina Agr. Exp. Sta. Tech. Bull.* **219**.

Sanchez, P. A. and M. A. Nureña, 1972. Upland rice improvement under shifting cultivation systems in the Amazon Basin of Peru. *North Carolina Agr. Exp. Sta. Tech. Bull.* **210**.

Sanchez, P. A., C. E. Seubert, E. J. Tyler, et al. 1974. Investigaciones en manejo de suelos tropicales en Yurimaguas, Selva Baja del Peru. Pp. II-B-1–II-B-36. In *Reunion Internacional sobre Sistemas de Produccion para el Trópico Húmedo*. Zona Andina. Lima, Peru.

Seubert, C. E. 1975. Effect of land clearing methods on crop performance and changes in soil properties in an Ultisol of the Amazon Jungle of Peru. M.S. Thesis, North Carolina State University, Raleigh. 152 pp.

Siband, P. 1972. Etude de l'evolution des sols sous culture traditionelle en Haute-Casamance: principaux resultats. *Agron. Tropicale (France)* **27**:574–591.

Singh, K. 1961. Value of bush, grass or legume fallow in Ghana. *J. Sci. Food Agr.* **12**:160–168.

Snedaker, S. C. 1970. Ecological studies on tropical moist forest succession in eastern lowland Guatemala. Ph.D. Thesis, University of Florida, Gainesville.

Snedaker, S. C. and J. F. Gamble. 1969. Compositional analysis of selected second growth species in lowland Guatemala and Panama. *Bio. Sci.* **19**:536–538.

Sollins, P. and G. Drewry. 1970. Electrical conductivity and flow rates of water through the forest canopy. Pp. H135-H154. In H. T. Odum (ed.), *A Tropical Rain Forest.* U.S. Atomic Energy Commission, Washington.

Stephens, D. 1960a. Fertilizer trials on peasant farms in Ghana. *Emp. J. Exptal. Agr.* **28**:1–15.

Stephens, D. 1960b. Some fertilizer trials with phosphorus, nitrogen and sulfur in Ghana. *Emp. J. Exptal. Agr.* **28**:154–164.

Stephens, D. 1960c. Three rotation experiments with grass fallows and fertilizers. *Emp. J. Exptal. Agr.* **28**:165–178.

Stephens, D. 1967a. The effects of different nitrogen treatments and of potash, lime and trace elements on cotton on Buganda clay loam soil. *East Afr. Agr. For. J.* **23**:320–325.

Stephens, D. 1967b. Effects of grass fallow treatments in restoring fertility of Buganda clay loam in South Uganda. *J. Agr. Sci.* **68**:391–403.

Stephens, D. 1969. The effects of fertilizers, manure and trace elements in continuous cropping rotations in Southern and Western Uganda. *East Afr. Agr. For. J.* **34**:401–417.

Suarez de Castro, F. 1957. Las quemas como práctica agricola y sus efectos. *Fed. Nac. Cafetaleros de Colombia Bol. Téc.* 2.

Suarez de Castro, F. and A. Rodriguez. 1955. Equilibrio de materia orgánica en plantaciones de café. *Fed. Nac. Cafeteros de Colombia Bol. Tec.* 15.

Tergas, L. E. and H. L. Popenoe. 1971. Young secondary vegetation and soil interactions in Izabal, Guatemala. *Plant and Soil* **34**:675–690.

Touré, E. H. O. 1964. Maintenance of the fertility of savanna zone soils. Effects of annual or biennial mineral fertilizer applications under conditions of continuous cropping in the Sudan zone of West Africa. *Afr. Soils* **9**:221–246.

Uribe, A., F. Suarez de Castro, and A. Rodriguez. 1967. Efectos de las quemas sobre la productividad de los suelos. *Cenicafé* **18**:116–135.

Urrutia, V. M. 1967. Corn production and soil fertility changes under shifting cultivation in Uaxatún, Guatemala. M.S. Thesis, University of Florida, Gainesville. 101 pp.

Van Beukering, J. A. 1947. Het Ladangvraagstuk, een bedrijsten sociaal economish probleem. *Landbouw* **19**:241–285.

Van der Weert, R. 1974. The influence of mechanical forest clearing on soil conditions and resulting effects on root growth. *Trop. Agr. (Trinidad)* **51**:325–331.

Van der Weert, R. and K. J. Lenselink. 1972. The influence of mechanical clearing of forest on some physical and chemical soil properties. *Surinaamse Landbouw* **20**:2–14.

Van Parijs, A. 1959. Mantien de la productivite des sols sons cultures continues en Ituri (Congo Belgue). Pp. 857–863. *III Conf. Interafricains des Sols.*

Vine, H. 1953. Experiments on the maintenance of soil fertility at Ibadan, Nigeria, 1922–1951. *Emp. J. Exptal. Agr.* **21**:65–85.

Vine, H. 1954. Is the lack of fertility of tropical African soils exaggerated? *Proc. 2nd Int. Afr. Soils Conf.* (*Leopoldville*) **2**:389–412.

Vine, H. 1968. Developments in the study of soils and shifting agriculture in tropical Africa. Pp. 89–119. In R. P. Moss (ed.), *The Soil Resources of Tropical Africa.* Cambridge University Press, London.

Watters, R. F. 1966. The shifting cultivation problem in the American Tropics. Pp. 1–16. In *Reunion Internacional sobre Problemas de la Agricultura en los Trópicos Humedos de America Latina.* Lima, Belem do Pará.

Watters, R. F. 1968. *La Agricultura Migratoria en Venezuela.* Instituto Forestal Latinoamericano de Investigación y Capacitación, Mérida. 134 pp.

Watters, R. F. 1971. Shifting cultivation in Latin America. *FAO For. Dev. Paper* 17. 305 pp.

Watters, R. F. and L. Bascones. 1971. The influence of shifting cultivation on soil properties at Altamira-Calderas, Venezuelan Andes. *FAO For. Dev. Paper* 17, pp. 291–299.

Webster, C. C. 1954. The ley and soil fertility in Britain and Kenya. *East Afr. Agr. For. J.* **20**:71–74.

Weigert, R. G. and P. Murphy. 1970. Effect of season, species and location on the disappearance rate of leaf litter in a Puerto Rican rain forest. Pp. H101–H104. In H. T. Odum (ed.), *A Tropical Rain Forest.* U.S. Atomic Energy Commission, Washington.

Woods, F. W. and C. M. Gallegos. 1970. Litter accumulation in selected forests of the Republic of Panamá. *Biotropica* **2**:46–50.

Zinke, P. J., S. Sabhasri, and P. Kundstadler. 1970. Soil fertility aspects of Lua forest fallow system of shifting cultivation. Unpublished paper, University of California School of Forestry and Conservation, Berkeley.

11

SOIL MANAGEMENT IN RICE CULTIVATION SYSTEMS

Rice is the food crop produced in largest quantities in the tropics and the one occupying the largest area planted as well. According to FAO statistics, over 170 million tons of rice was produced in 94 million ha in the tropics during 1970. More than 90 percent of this production comes from tropical Asia. Nevertheless, rice is an extremely important food crop in the African and American tropics also.

Rice is the only major food crop capable of growing in flooded soils, because of its ability to oxidize its rhizosphere. Flooding brings about a series of physical, chemical, and biological changes that provide a completely different set of soil–plant relationships from those observed in other crops. Substantial advances have been made in the last two decades on the management of flooded soils in relation to rice production. This chapter describes the physical and chemical changes brought about by flooding and puddling, the unique physiological properties of the rice plant, and the management of water and nutrients in the various rice cropping systems found in the tropics.

The vast majority of tropical rice is produced on relatively fertile Entisols, Inceptisols, Vertisols, Mollisols, and Alfisols with predominantly layer silicate ion exchange systems. Consequently this chapter will focus on such soils. A major exception is upland or dry land rice, which is also grown on Oxisols and Ultisols.

EFFECTS OF FLOODING ON SOIL PHYSICAL PROPERTIES

When a dry soil is suddenly and thoroughly flooded, its aggregates become saturated with water. The air in the soil pores is compressed by the advancing water until small air explosions occur, causing the breakdown of larger aggregates or clods into smaller ones (Yoder, 1936).

Swelling

Montmorillonite minerals then begin to swell; in order for aggregates to swell freely, their cementing agents undergo certain transformations. Kita and Kawaguchi (1960) found that hydration of reducible iron and manganese oxides, silicates, and organic matter takes place in paddy soils when flooded and that this facilitates swelling. The result is reduced cohesion within soil aggregates and increased cohesion between aggregates. In soils containing fixed ammonium between layer silicate lattices, this process also results in a quick release of NH_4^+ ions into the soil solution, independently of microbial nitrogen transformations (Bhattacharyya, 1971).

Permeability

Just after flooding, percolation losses increase rapidly, soon decrease, and gradually become negligible, as illustrated in Fig. 11.1. The increase in permeability during the first 10 to 20 days after flooding is attributed to the release of entrapped air and the rapid production of CO_2. The gradual decrease that follows is due to a slow disintegration of aggregates and to

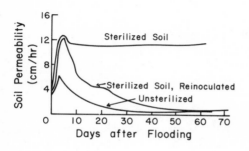

Fig. 11.1 Dynamics of soil permeability under flooded (unpuddled) conditions. *Source:* Allison (1947).

clogging of soil pores by microbial wastes. Allison (1947) and other workers have found through sterilization techniques that this process is primarily microbiological, as shown in Fig. 11.1. After drying and rewetting, permeability is usually restored to the original aerobic value.

With continuously flooded rice cultivation, soil permeability decreases further. Mikklesen and Patrick (1968) reported that in certain permeable California soils percolation rates decreased to one-third or one-fifth of the original values after several years of continuous rice cultivation. They attributed this effect to decreases in aggregate stability. Most soils used for rice have very low permeabilities to start with, because of high clay or silt contents, high sodium saturation, high water tables, or the presence of impermeable layers in the subsoil. Consequently the changes in permeability upon flooding may be less marked than in the example from California.

Aggregate Stability

The stability of soil aggregates generally decreases with flooding because of swelling, hydration, and increased solubility of some cementing agents. The magnitude of this phenomenon varies greatly with soil properties and water quality, ranging from no to almost complete aggregate breakdown (in the absence of puddling). In general, pure layer silicate systems with high pH or sodium contents show marked aggregate breakdown upon flooding. At the other extreme, soils high in organic matter, iron, and aluminum oxides often retain their structure while flooded, although the aggregates may be quite soft and can easily be destroyed if force is applied in the flooded state. A series of studies conducted by Kawaguchi et al. (1956) and Kawaguchi and Kita (1957) in Japan and by Ahmad (1963) in Trinidad confirmed that flooding gradually decreases aggregate stability because of organic matter decomposition and the reduction of iron and manganese oxide coats to soluble forms. Upon drying and reoxidation, aggregate stability increases because of the precipitation of ferric and manganic compounds as oxide coats around clay particles. This process is particularly important in oxide and oxide-coated layer silicate systems.

PUDDLING

Puddling can be defined as the process of breaking down soil aggregates into a uniform mud, accomplished by applying mechanical force to the soil at high moisture contents. In most cropping systems puddling is an unintentional effect of tillage at the wrong moisture content and usually results in

severe yield decreases or delays in planting. In lowland rice systems puddling is an important soil management practice, conducted with great care
for the purpose of destroying the topsoil structure. Puddling is almost
synonymous to lowland rice culture in Asia.

The process of puddling is accomplished by a series of tillage operations,
beginning at soil moisture contents above saturation and ending at moisture
contents closer to field capacity. This process is best understood by
considering the changes in soil strength within an aggregate and between
aggregates, as illustrated in Fig. 11.2. The cohesion *within* soil aggregates
decreases with increasing soil moisture contents. The individual aggregates
become soft and may or may not disintegrate, depending on their stabilities.
The cohesion *between* aggregates is very low at low moisture contents,

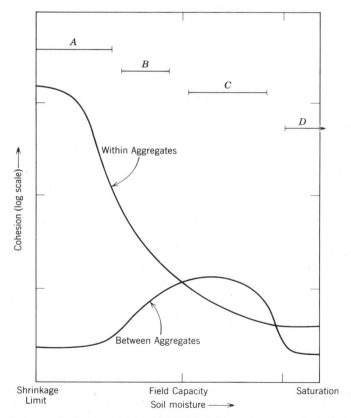

Fig. 11.2 Cohesion of soil aggregates as affected by soil moisture contents. *Source:* Adapted
from Koenigs (1961) and Koenigs (personal communication).

Plate 14 The final puddling operation involving land leveling: Ilocos Sur, Philippines.

increases rapidly with increasing moisture, peaks at about field capacity, and decreases sharply as moisture contents approach saturation (Koenigs, 1961). The cohesion between aggregates depends primarily on the number of contact points between them. The number of contact points is minimal in a dry soil and approaches a maximum at about field capacity because of the increased thickness of water films and the swelling of the aggregates themselves. At higher moisture contents the thick moisture films act as lubricants and decrease the number of contact points between aggregates. In Fig. 11.2 range *A* is too dry for effective tillage, *B* is the optimum for conventional tillage, and *C* indicates where maximum soil puddling is accomplished. At this range the cohesion within aggregates is very low, and the cohesion between aggregates is maximum. When force is applied by a plow or a man's foot, the aggregates are easily destroyed because of the multiplying effect of the high friction and the low internal strength of the aggregates.

In practice, farmers start their puddling operations at moisture contents above saturation. In this range, D in Fig. 11.2, both the cohesion within and the cohesion between aggregates are minimal. Applying force at this range does not cause maximum puddling, but it incorporates weeds and starts the aggregate destruction process. Farmers then continue to puddle the soil at progressively lower moisture contents until they reach maximum puddling at tensions close to field capacity.

The degree of puddling attained varies with soil type and management. High clay contents facilitate puddling and produce more aggregate destruction. Nevertheless, sandy soils with low clay contents can be puddled. Montomorillonitic families are puddled more easily and thoroughly than kaolinitic or oxidic families. Sodium-saturated soils are the easiest to puddle. In general, the higher the soil organic matter content or the iron and aluminum oxide contents, the more difficult it is to puddle the soil. Andepts and Oxisols are extremely difficult to puddle, and the degree of aggregate breakdown is probably less than in other soils.

The consequences of puddling are aggregate destruction, elimination of noncapillary pore space, increased capillary porosity, increased soil moisture retention, decreased evaporation and percolation losses, and soil reduction without flooding.

Aggregate Destruction

The primary consequence of puddling is the destruction of soil aggregates. A puddled soil consists essentially of a two-phase solid–liquid system. Individual clay particles or clusters of them are oriented in parallel and are surrounded by capillary pores saturated with water. Sand and silt particles and some remaining aggregates are also part of the matrix. The degree of aggregate destruction is difficult to quantify because drying is necessary to measure aggregation. Kawaguchi et al. (1956), Chaudhary and Ghildyal (1969), and others provide evidence of aggregate destruction after puddling and subsequent drying.

Changes in Porosity

Noncapillary pores are essentially eliminated in the process of puddling. Bodman and Rubin (1948) found that 91 to 100 percent of the volume occupied by such pores was destroyed by puddling a silt loam. Capillary porosity increases drastically upon puddling—by 223 percent, in an observation by Jamison (1953). Since most of these pores are smaller than 0.2 μ in effective radii, water may move through pores as a liquid but is lost only as vapor.

Changes in Bulk Density

After puddling, apparent specific gravity decreases because of the larger total pore volume occupied by water. With time, however, puddled soils increase their bulk densities while still flooded, probably because of a slow settling of the clays. When dried, however, puddled soils shrink dramatically and increase in bulk density by large, measurable amounts (Kawaguchi et al., 1956; Sanchez, 1968).

Increased Soil Moisture Retention

As a consequence of the destruction of noncapillary pores, the increase in water-saturated capillary pores, and the decrease in initial bulk density, puddled soils hold more water at a given soil moisture tension than in the unpuddled state. This effect, illustrated in Fig. 11.3, is measurable within a range of 0 to 10 bars of soil moisture tension.

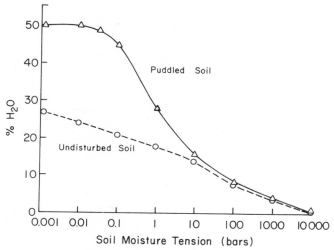

Fig. 11.3 Soil moisture retention curves for a puddled and undisturbed clay soil. *Source:* Croney and Coleman (1954).

Decreased Moisture Losses

The changes in porosity and water retention result in sharply reduced soil moisture loss patterns in puddled soils. Table 11.1 shows the effects of puddling on drainage rates of six Philippine rice soils ranging widely in texture and mineralogy. Puddling decreased percolation losses by a factor of 1000 times, regardless of soil properties. Among the puddled soils in this table, the sandy Entisol and the clayey Andept had higher percolation losses because of the difficulty in puddling such soils. Other clayey soils percolated at the rate of 0.5 mm/day, a finding that agrees with field observations.

The benefits of puddling in slowing water movement have been previously attributed to the formation of relatively impermeable "plow pans" just below the puddled layer. These plow pans are found in loamy soils grown to rice for many years, in well-drained Oxisols, and as concretionary materials in Andepts. Plow pans are absent in sandy and clayey soils, Vertisols, and young alluvial and calcareous soils (Sanchez, 1973). Puddling, however, is almost universally practiced in lowland rice farms. With the dikes surrounding the paddy fields and the puddled soil at the bottom, the rainfed fields hold the maximum amount of water possible for the rice crop growing during the rainy season.

Another consequence of puddling is the slow rate of drying. The drying process may take several months in puddled clayey soils, as opposed to well-structured ones (Sanchez, 1973).

Table 11.1 *Effects of Puddling on the Drainage Rates of Six Flooded Philippine Soils in Pot Conditions*

			Drainage Rate (cm/day)	
Soil	Mineralogy	Clay (%)	Granulated Condition	Puddled Condition
Psamment	Siliceous	9	267	0.45
Fluvent	Mixed	24	215	0.17
Aquept	Montmorillonitic	30	183	0.05
Aqualf	Montmorillonitic	40	268	0.05
Ustox	Kaolinitic	64	155	0.05
Andept	Allophanic	46	214	0.31
Mean			217	0.18

Source: Sanchez (1973).

Soil Reduction

Because of the absence of air, the soil reduction process takes place as soon as the soil is puddled (Breazeale and McGeorge, 1937). Puddled soils remain reduced, regardless of whether or not they are flooded, until cracks begin to form. The chemical consequences of soil reduction discussed in the next section can occur as a result of puddling.

Regeneration of the Structure

Puddling is not an irreversible process. The original structure can be regenerated through alternate wetting and drying. The puddled soil must be dried first; then aggregates are reformed by alternate wetting and drying. Tillage at the appropriate moisture content facilitates this process. Soils high in organic matter or iron and aluminum oxides are easier to regenerate than others (Koenigs, 1961, 1963; Sanchez, 1968).

CHEMICAL CONSEQUENCES OF FLOODING

When a soil is flooded, its oxygen supply decreases to zero in less than a day. The rate of atmospheric oxygen diffusion is 10,000 times slower through water layers or water-filled pores than through air or air-filled pores. Aerobic microorganisms quickly consume the remaining oxygen and become dormant or die. Anaerobes or facultative anaerobes multiply rapidly and take over the organic matter decomposition process, using, instead of oxygen, oxidized soil components as electron acceptors. These products are reduced in the following thermodynamic sequence: nitrates, manganic compounds, ferric compounds, intermediate products of organic matter decomposition, sulfates, and sulfites. The main reactions are shown in Table 11.2. For greater detail the reader is referred to comprehensive reviews by Ponnamperuma (1965, 1972).

The result is to change the soil from the oxidized to the reduced state. The oxidation–reduction potential (*Eh* values corrected to pH 7) is a useful parameter for measuring the intensity of soil reduction and for identifying the predominant reactions taking place. The change in *Eh* as a function of time after flooding is illustrated in Fig. 11.4. Nitrates become unstable at *Eh* values between $+400$ and $+300$ mV and are denitrified. After all the nitrates are consumed, anaerobic microorganisms reduce Mn^{4+} to Mn^{2+} compounds. This occurs at *Eh* values of about $+200$ mV. Manganic com-

Table 11.2   Principal Reduction Reactions Occurring in Flooded Soils in a Roughly Thermodynamic Sequence

Stage	Eh_7 (mv)	Reaction
0	800	$O_2 + 4H^+ + 4e^- \rightleftarrows 2H_2O$
1	430	$2NO_3^- + 12H^+ + 10e^- \rightleftarrows N_2 + 6H_2O$
2	410	$MnO_2 + 4H^+ + 2e^- \rightleftarrows Mn^{2+} + 2H_2O$
3	130	$Fe(OH)_3 + e^- \rightleftarrows Fe(OH)_2 + OH^-$
4	−180	Organic acids (lactic, pyruvic) $+ 2H^+ + 2e^- \rightleftarrows$ alcohols
5	−200	$SO_4^{2-} + H_2O + 2e^- \rightleftarrows SO_3^{2-} + 2OH^-$
6	−490	$SO_3^{2-} + 3H_2O + 6e^- \rightleftarrows S^{2-} + 6OH^-$

Source: Simplified from Ponnamperuma (1965, 1972).

pounds, however, are usually not abundant in soils; since only specific bacteria can reduce them, their role is relatively minor in most soils.

After the manganic ions are reduced, Fe^{3+} ions are reduced to Fe^{2+} ions, with a corresponding decrease in *Eh* to about +120 mV. This is probably the most important reduction reaction that takes place in flooded soils because iron compounds are usually more abundant in soils than nitrates, manganic hydroxides, or sulfates.

Several organic acids, such as lactic and pyruvic acid, are reduced to alcohols at *Eh* values of approximately −180 mV. Sulfate ions are reduced to SO_3^{2-} and S^{2-} ions at redox potentials of about −150 mV. Other reduction reactions take place in more intensively reduced soils but at *Eh* values not normally found in flooded rice soils.

The intensity of the reduction process depends on the amount of easily decomposable organic matter (the substrate of microorganisms) and the soil temperature. Figure 11.5 illustrates the decrease in *Eh* with time of submergence, followed by an equilibrium level. The higher the soil organic matter content, the greater is the intensity of reduction. The actual amounts of reduced substances can be measured as oxidizable matter. In most soils the concentration of reduced substances reaches a peak 2 to 4 weeks after flooding and decreases gradually to an equilibrium level (Ponnamperuma, 1965).

Soil reduction per se is not detrimental to the rice plant, except possibly at potentials greater than −300 mV, where sulfides may be produced at toxic levels (Patrick and Mahapatra, 1968). Ferrous iron is the most ₐabundant inorganic reduced product; consequently, *Eh* values are equilibrated by the Fe^{3+}–Fe^{2+} system at about +100 mV. *Eh* values low enough

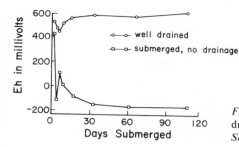

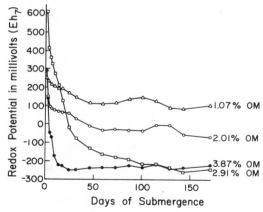

Fig. 11.4 Changes in redox potential of a well-drained and a submerged Inceptisol with time. *Source:* Ponnamperuma (1965).

to reduce sulfate are seldom attained, except in soils high in organic matter and low in iron.

The indirect consequences of soil reduction, however, are of fundamental importance to rice culture.

Oxidized and Reduced Zones in Flooded Soils

The profile of a flooded soil is not completely reduced. Several oxidized zones exist in which the processes described above do not take place. This is illustrated in Fig. 11.6. A superficial layer, ranging from 1 mm to 1 cm in depth, remains oxidized because it is in equilibrium with the oxygen dissolved in the water layer. This layer is easily identified in flooded rice

Fig. 11.5 Changes in the redox potentials of four Philippine soils with different organic matter contents when submerged. *Source:* Ponnamperuma (1965).

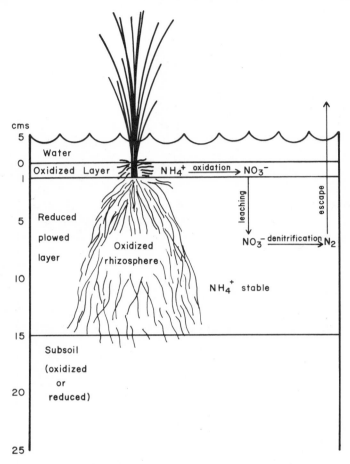

Fig. 11.6   Redox profile and nitrogen dynamics in a constantly flooded soil. *Source:* Sanchez (1972a).

fields because it maintains the aerobic soil color, whereas the reduced layer below changes to the grayish–bluish or black color typical of reduced iron compounds. The rest of the plowed layer is reduced except for the rhizosphere of active rice roots, which is oxidized because of the exudation of oxidized compounds by the roots. This can also be visually recognized by the presence of yellowish red root coatings, caused by the precipitation of ferric compounds on part of the root surfaces.

The subsoil may be reduced or oxidized, depending on its organic matter content and the presence or absence of a perched water table. Most subsoils of flooded soils may be slightly reduced, but in some cases the organic matter content is too low for the microorganisms to cause soil reduction. In cases of perched water tables, the subsoil may remain oxidized if flooding does not proceed below the plowed layer.

Changes in pH

Regardless of their original pH values, most soils reach pHs of 6.5 to 7.2 within 1 month after flooding and remain at that level until dried. Figure 11.7 shows this dramatic consequence of flooding in soils ranging from pH 3.4 to 8.1 in their aerobic stages. The overall effect of submergence is to increase the pH in acid soils and decrease it in alkaline soils. These changes are a function of the Fe^{2+} ion concentration and the partial pressure of CO_2. The pHs of acid soils increase because of the release of OH^- ions when $Fe(OH)_3$ and similar compounds are reduced to $Fe(OH)_2$ or $Fe_3(OH)_8$. Other reduction reactions described in Table 11.2 also show a decrease in H^+ ions or an increase in OH^- ions with reduction. The pH values of alkaline soils decrease to about 7 because of the increase in partial pressure of CO_2, which results in a net release of H^+ ions. In alkaline soils the effect of iron reduction is less important because they contain little iron, whereas in acid soils the increase in the partial pressure of CO_2 is less important than the reduction of iron compounds. In neutral soils the pH changes little because these two factors tend to balance each other.

Exceptions to this generalization are rare. Figure 11.7 shows one of the few recorded, an acid sulfate soil with an aerobic pH of 3.4, which increased only to 5. This soil happened to be extremely low in iron (0.08 percent reducible Fe) and had no reducible manganese. Consequently, not much reduction was taking place.

This remarkable consequence of flooding amounts to a self-liming operation and results in an optimum pH range for the availability of most nutrients. Aluminum toxicity is quickly eliminated in acid soils when they

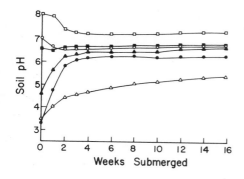

Fig. 11.7 Kinetics of the pH values of some submerged soils. *Source:* Ponnamperuma (1972).

are flooded, because exchangeable aluminum is precipitated at a pH of 5.5. Consequently, liming is of little value in flooded rice production. It should be noted, however, that this pH increase requires about 2 weeks of flooding. Rice may die from aluminum toxicity if transplanted at the start of flooding in some acid sulfate soils (Nhung and Ponnamperuma, 1966). Delaying transplanting for 2 or 3 weeks after flooding may eliminate this danger without lime applications.

Nitrogen Transformations

After oxygen is consumed, virtually all the nitrates present in the soil are denitrified and lost to the atmosphere. Losses on the order of 20 to 300 kg N/ha have been recorded within a month after flooding. High values occur when upward NO_3^- movement during the dry season results in nitrate accumulation in the topsoil, and when nitrate fertilizer sources are used. This reaction occurs very rapidly and completely. No nitrite accumulates in flooded soils (Ponnamperuma, 1955). The "flushes" of nitrogen mineralization associated with the first rains at the beginning of the rainy season may be completely lost when the soil is subsequently flooded.

Unlike nitrates, the ammonium ion is in the reduced state and therefore is stable in anaerobic conditions. The mineralization of soil organic nitrogen stops at the ammonification state. Consequently NH_4^+ ions accumulate in flooded soils either as exchangeable NH_4^+ or in the soil solution. Figure 11.8 shows the accumulation of exchangeable and soluble NH_4^+ in several rice soils. The magnitude of these increases depends primarily on the soil organic matter content, except in Andepts, which have lower mineralization rates; the soil with 6.2 percent organic matter in Fig. 11.8 is an Andept. When rice is grown, the accumulation of NH_4^+ ions can be depleted by plant uptake.

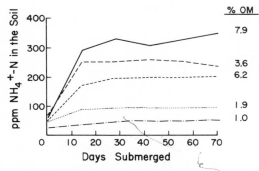

Fig. 11.8   Kinetics of NH_4 release in six submerged soils. *Source:* Ponnamperuma (1965).

A high concentration of exchangeable Fe^{2+} may displace considerable quantities of NH_4^+ ions from exchange sites into the soil solution. Leaching losses of NH_4^+ ions are an important mechanism in flooded soil with substantial water movement (Patrick and Mahapatra, 1968).

The presence of a thin oxidized layer over the reduced topsoil has profound implications in nitrogen movement. Nitrification takes place in the oxidized layer, either from organic matter mineralization or the addition of ammonium fertilizer sources. Nitrate ions may move down into the reduced layer either by diffusion or by downward water movement. In the reduced layer, denitrification quickly occurs and the N_2 gases produced escape to the atmosphere. This mechanism is shown in Fig. 11.6. The practical implication is that ammonium sources should be mixed into the reduced layer in order to avoid this problem, and not be broadcast on the soil surface.

Organic Matter Decomposition

Organic matter decomposition proceeds at a slower rate in flooded than in aerobic soils. The anaerobic bacteria involved are less efficient than the more diversified aerobic microflora. Anaerobic decomposition proceeds at lower energy levels and therefore, requires less nitrogen. Consequently, soil nitrogen mineralization can proceed at higher C:N ratios in flooded than in anaerobic soils (DeDatta and Magnaye, 1969). Organic materials not suitable for mineralization under aerobic conditions can be mineralized under flooded conditions. This is particularly true of rice straw, which has a high C:N ratio. In spite of the slower rate of organic carbon mineralization, the rate of organic nitrogen mineralization is often higher in flooded than in aerobic soils because of the utilization of higher C:N ratio materials (Patrick and Mahapatra, 1968).

The end products of organic matter decomposition are also different in flooded soils. Ponnamperuma (1972) points out that, whereas in a normal, well-drained soil the main end products are CO_2, NO_3^-, SO_4^{2-}, and resistant humified materials, in flooded soils they are CO_2, NH_4^+, methane, amines, mercaptans, H_2S, and partially humified residues. The pathways are similar in both conditions until the formation of pyruvic acid. In flooded soils these and other intermediate products are further reduced to several alcohols and organic acids, which eventually are reduced to CH_4 or CO_2 by strict anaerobes.

During the first few weeks of flooding, large quantities of CO_2, ranging from 1 to 3 tons/ha, are produced. These peaks are followed by a decline in CO_2 concentration, caused by escape, leaching, and precipitation as insoluble carbonates. The partial pressure of CO_2 can reach toxic levels, particularly in soils high in organic matter and low in iron.

Manganese and Iron Transformations

The solubility of manganese increases sharply upon flooding because of the reduction of Mn^{4+} compounds to the more soluble Mn^{2+}. Figure 11.9 shows that a peak concentration of Mn^{2+} in the soil occurs during the first month, followed by a gradual decrease. Ponnamperuma (1972) attributes this decrease to precipitation of Mn^{2+} as $MnCO_3$. Soils high in reducible manganese undergo the most pronounced changes in spite of original pH or organic matter levels. Acid soils high in manganese and organic matter develop peak concentrations of 90 ppm Mn^{2+} in the soil solution, followed by a decline and stabilization at about 10 ppm. Alkaline soils or soils low in manganese seldom contain more than 10 ppm Mn^{2+} in the soil solution.

Most flooded soils contain sufficient manganese for rice growth. No evidence of manganese toxicity has been reported in flooded conditions. In fact, flooding eliminates manganese toxicity in acid aerobic soils because the solubility of Mn^{4+} compounds is vastly decreased when the pH increases to about 7. The release of Mn^{2+} with flooding apparently is not sufficient to reach toxic levels for rice.

After nitrates and manganese compounds are reduced, the solubility of iron increases because of the reduction of Fe^{3+} compounds to more soluble Fe^{2+} compounds. A peak in the concentration of exchangeable Fe^{2+} or soil solution Fe^{2+} occurs normally within the first month after flooding and is followed by a gradual decline. The magnitude and intensity of these peaks vary substantially with soil properties. Figure 11.10 shows that the most pronounced peaks occur in acid soils high in organic matter. These are usually Oxisols or Ultisols with large quantities of reducible iron. High-pH soils show small increases in Fe^{2+} because of their lower contents of reducible iron. The percentage of total free iron reduced within a few weeks of

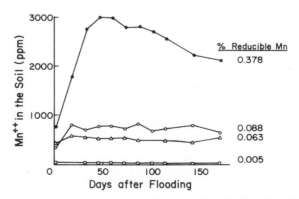

Fig. 11.9 Changes in concentration of Mn^{2+} in four soils with flooding. *Source:* Ponnamperuma (1965).

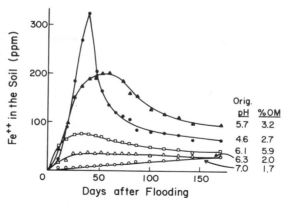

Fig. 11.10 Changes in concentration of Fe²⁺ in the soil solutions of five flooded soils. *Source:* Ponnamperuma (1965).

flooding ranges from 5 to 50 percent. More iron is reduced when the oxidized forms are less crystalline (Ponnamperuma, 1972). The grayish color of reduced soils is attributed to FeS, although hydrated magnetite [Fe₃(OH)₈], ferrous silicates, and vivianite may also be present.

Iron reduction is considered the most important reaction occurring in flooded soils because it raises the pH, increases the availability of phosphorus, and displaces cations from exchange sites. The increase in Fe²⁺ concentration is usually beneficial to rice in alkaline soils, reaching 20 ppm in the soil solution, a level that is usually sufficient to eliminate iron deficiency. On the other hand, increases in Fe²⁺ in acid Oxisols and Ultisols may reach levels of about 350 ppm Fe²⁺, which can cause iron toxicity to rice. Although iron toxicity is common in these soils, it can be avoided by drainage and sometimes by delaying transplanting until after the peaks of reduction have occurred.

Neither iron nor manganese can be considered a micronutrient in flooded rice culture. The rice plant assimilates large quantities of both elements, averaging 12 kg Fe or Mn/ha, a slightly lower amount than the uptake of phosphorus (Tanaka et al., 1964). Also, Fe²⁺ and Mn²⁺ become the dominant exchangeable cations in flooded soils and displace Ca²⁺, Mg²⁺, K⁺, and Na⁺ into the soil solution by mass action. The Fe²⁺ and Mn²⁺ ions in the soil solution may move toward oxidized parts of the soil profile, forming coprecipitates of Fe³⁺ and Mn⁴⁺ hydrous oxides in the form of stains, nodules, or concretions.

Phosphorus Availability

The concentration of phosphorus in the soil solution increases upon flooding. Such increases are due to (1) the reduction of ferric phosphates into

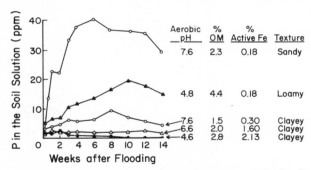

Fig. 11.11 Kinetics of water-soluble phosphorus in some submerged soils. *Source:* Ponnamperuma (1972).

more soluble ferrous phosphates; (2) the availability of reductant-soluble phosphorus compounds, caused by dissolving previously oxidized layers surrounding the phosphate particles; (3) the hydrolysis of some iron- and aluminum-bonded phosphates in acid soils, which results in a release of some fixed phosphorus at higher soil pH; (4) the increased mineralization of organic phosphorus in acid soils, caused by increasing the pH between 6 and 7; (5) the increased solubility of apatite in calcareous soils when the pH decreases to 6 or 7; and (6) the greater diffusion of $H_2PO_4^-$ ions in a larger volume of soil solution (Ponnamperuma, 1972; Turner and Gilliam, 1976).

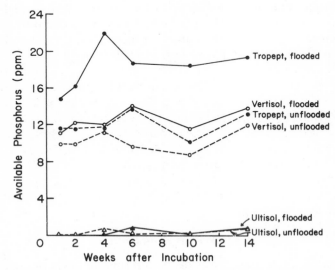

Fig. 11.12 Changes in the availability of Bray 2-extractable phosphorus in submerged and nonsubmerged soils. *Source:* Tusneem (1967).

Figure 11.11 shows that in some soils there are no clear peaks of soil solution phosphorus concentration, while others show a measurable peak and a subsequent decrease. These decreases have been attributed to refixation of soil solution phosphorus as calcium-bonded phosphates (Patrick and Mahapatra, 1968; Ponnamperuma, 1972).

The relative importance of these various mechanisms is not clear. What is clear is that in most soils there is a marked increase in soil solution phosphorus beyond the 0.1 to 0.2 ppm level, which is considered adequate for rice growth (Hossner et al., 1973).

The magnitude of this phenomenon varies with soil properties. One would expect it to be greater in soils high in ferric phosphates or reductant-soluble phosphates. Ponnamperuma's data in Fig. 11.11, however, indicate that the largest increases take place in sandy soils low in iron, while clayey soils high in iron do not produce large increases. The degree of crystallinity of iron compounds is probably the reason for this difference, assuming that more crystalline and therefore less soluble forms of iron are present in red clayey soils.

These changes can be detected by soil tests when provisions are made for analyzing the soil in the reduced state. Figure 11.12 shows the effect of flooding on available phosphorus extracted by the Bray 2 method shortly after sampling. The changes were more marked in the acid Inceptisol than in the Vertisol or the Ultisol used by Tusneem (1967).

Flooding also causes a marked change in the distribution of the different phosphorus fractions. The reductant-soluble and Ca-P fractions decrease, whereas the Fe-P fraction increases. Patrick and Mahapatra (1968) have shown this process in an acid soil in Fig. 11.13. Similar results have been observed for several Philippine soils by Tusneem (1967) and Singlachar and Samaniego (1973) and for a Mollisol of northern India by Gupta et al. (1972). The increase in iron-bonded phosphorus can be explained as precipitation of phosphorus by Fe^{3+} upon drying. Under subsequent flooding this newly formed $Fe^{3+}P$ will be reduced to $Fe^{2+}P$. In calcareous soils the long-term effect of flooding will be the transformation of Ca-P into Fe-P. This may result in lower phosphorus availability, not to flooded rice, but to subsequent aerobic crops growing in rotation with rice, because of the low solubility of Fe-P at the high pH levels of aerobic calcareous soils.

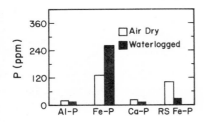

Fig. 11.13 Transformation of inorganic phosphate in an Alfisol as a result of waterlogging. *Source:* Patrick and Mahapatra (1968).

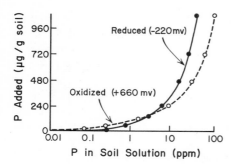

Fig. 11.14   Effect of flooding on phosphorus fixation after 24 hours equilibration in Crowley silt loam (Aqualf) from Louisiana. *Source:* Adapted from Patrick and Khalid (1974).

Flooding also alters the process of phosphorus fixation. A recent report on alluvial and rice soils of Louisiana indicates that under reduced conditions less phosphorus was sorbed at concentrations in the soil solution below 1 ppm P, and more phosphorus at high concentrations. The results are shown in Fig. 11.14. Patrick and Khalid (1974) concluded that the differences in phosphorus fixation and release pattern under aerobic and anaerobic conditions can be attributed to the reduction of ferric oxyhydroxide to gel-like ferrous compounds with greater surface area. All the soils in question have low phosphorus fixation capacities. In high-fixing Ultisols and Oxisols the magnitude of this phenomenon is probably different. Although no direct evidence is available, the lack of marked increase in water-soluble or extractable phosphorus upon flooding such soils observed by Ponnamperuma and others implies no major decrease in phosphorus fixation.

Sulfur Transformations

At very intense levels of soil reduction, sulfate ions are reduced to SO_3^{2-} and S^{2-} by bacteria of the genus *Desulfovibrio*. The relationships between these oxidized and reduced forms of sulfur are shown in Fig. 11.15. The

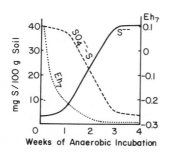

Fig. 11.15   Relationship between *Eh* and sulphate reduction. *Source:* Ponnamperuma (1965).

magnitude of sulfate reduction depends on soil properties. Acid soils first show an increase in soil solution SO_4^{2-} because of the release of sorbed SO_4^{2-} as pH increases. This is followed by a slow decrease, depending on the intensity of soil reduction. The availabilities of sulfur as SO_4^{2-} and as SO_3^{2-} are the same. The availability of sulfur decreases with the formation of S^{2-}, however, because it is mostly precipitated as FeS. In soils very low in iron, H_2S is formed, resulting in direct toxicity to the rice plants.

Nonreducible Cations

Because K^+, Ca^{2+}, Mg^{2+}, and Na^+ are already in the reduced state, they are not directly affected by soil reduction. The sheer increase in the volume of water in flooded soils, however, may accelerate the dissolution of solid compounds of these elements. The large quantities of NH_4^+, Fe^{2+}, and Mn^{2+} ions released upon flooding may displace substantial quantities of Ca^{2+}, Mg^{2+}, and K^+ from the exchangeable sites into the soil solution. An example is shown in Table 11.3, where the increase of soil solution K^+ is related to increases in Fe^{2+} and Mn^{2+}. Consequently, these ions have become more susceptible to leaching. On the other hand, flood waters usually carry considerable quantities of these exchangeable bases, which normally prevent potassium deficiencies in most lowland rice soils.

The forms of boron, copper, molybdenum, and zinc present in flooded soils do not undergo oxidation–reduction reactions. They are indirectly affected by pH changes, iron reduction, and the production of organic complexing agents. Prolonged submergence reduces the availability of zinc, partly because of pH increases. The exact mechanism, however, is not well understood (IRRI, 1970, 1971, 1972). A negative phosphorus–zinc interaction similar to that observed in aerobic soils is known to occur in rice soils

Table 11.3   Effects of Flooding on the Potassium Concentrations in the Soil Solutions of Several Philippine Soils (ppm)

Soil	Soil Solution $Fe^{2+} + Mn^{2+}$	Exchangeable K	Soil Solution K^+	
			Before Flooding	Peak at Flooding
Clay	342	165	2.3	7.9
Loamy loam	230	140	7.6	12.5
Loamy sand	72	100	3.2	5.2
Clay	39	60	1.6	1.9

Source: Adapted from Ponnamperuma (1965).

(Patrick and Mikklesen, 1971). The net result of flooding is probably an increase in the availability of boron, copper, and molybdenum (Patrick and Mikklesen, 1971; Ponnamperuma, 1972).

The concentration of monomeric silica [$Si(OH)_4$] in the soil solution increases slightly with flooding and decreases afterwards. The increase may be due to the release of occluded silicon by reduction of iron oxides and the effect of CO_2 on aluminum silicates. Although silica is for all practical purposes an element essential to the rice plant and responses to silica fertilization are known, these changes with flooding are not of sufficient magnitude to be of agronomic relevance.

Reversibility

All the above reactions are reversed when a flooded soil is dried and reoxidized. The pH level returns to its original value. The speed of reoxidization depends on the rate of water losses from the soils. In clayey puddled soils the process may last for several months, whereas in well-aggregated soils reoxidization begins within a few days of drainage.

Intermittent Flooding

The reactions just described apply to constantly flooded systems. In practice, however, over 80 percent of the tropical rice area is seldom constantly flooded, as will be discussed in the following section. Rainfed "lowland" fields may be flooded throughout crop growth or be alternately flooded or dried, depending on rainfall distribution. Many "upland" rice fields are temporarily flooded or waterlogged during periods of heavy rainfall. Deep-water rice starts as an aerobic system and becomes anaerobic at later stages of rice growth. Even in constantly flooded systems the land preparation process normally includes alternate flooding and drying at the beginning. The concepts developed in the preceding section, therefore, require modification when they are applied to a particular water regime.

The most important modifications are the nitrogen changes under alternate wetting and drying, not only because of their importance to fertilization practices but also because nitrification and denitrification thresholds occur at relatively high redox potentials. Under constant flooding nitrates are quickly lost and ammonium ions accumulate. Under intermittent flooding the following cycle develops. Right after flooding, nitrates quickly disappear and NH_4^+ contents increase. When the soil dries, a portion of the NH_4^+ ions is nitrified into NO_3^-. In the next flooding these

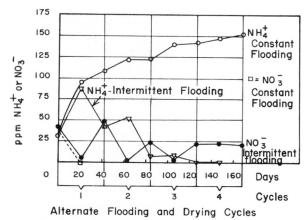

Fig. 11.16 Effects of constant and intermittent flooding on NH_4^+ and NO_3^- production under laboratory conditions. *Source:* Adapted from data of Patrick and Wyatt (1964).

NO_3^- ions are lost by denitrification or leaching. Alternate flooding and drying cycles reflect this pattern, which results in tremendous nitrogen losses (Fig. 11.16). Patrick and Wyatt (1964) showed that losses are greatest during the first cycle and decrease progressively afterwards.

The influence of intermittent flooding on phosphorus behavior and other nutrient transformations during crop growth has not been studied in detail. Patrick and Mikklesen (1971) have suggested, however, that phosphorus fixation is more intensive and less reversible under intermittent flooding than under either continuous flooding or continuous aerobic conditions. If drying is of sufficient length to cause reoxidation, aluminum or manganese toxicity may develop in soils that have low aerobic pH.

RICE CROPPING SYSTEMS

Tropical rice is grown in a variety of cropping systems. They can be grouped into five principal ones, described in Tables 11.4 and 11.5. The terms "upland" and "lowland," abundantly used in rice literature, refer to particular cropping systems and not to topography or elevation. For example, lowland rice is grown at 2000 m of elevation in the Philippine rice terraces, and upland rice at sea level throughout the Pacific coast of Central America. The term "paddy rice" is essentially synonymous to "lowland rice." The following is a description of the five principal rice cropping systems.

Table 11.4   Extent and Distribution of Rice Cultivation in the Tropics

Extent and Area Distribution	Tropical Asia	Tropical America	Tropical Africa	Total
Production (million tons)	153	11	5	169
Grain yield (tons/ha)	1.81	1.84	1.43	1.80
Area planted (million hectares)	84	6	3	94
Areal distribution (gross estimates) (%)				
Rainfed lowland (paddy)	50	0	20	46
Irrigated lowland (paddy)	20	4	18	19
Upland (dry land, secano)	20	75	72	25
Deep-water (floating)	10	0	—	9
Direct-seeded, irrigated	—	21	—	1

Source: FAO (1970), CIAT (1972), and unpublished sources.

Irrigated Lowland or Paddy System

Lowland or paddy rice accounts for about 65 percent of the total rice area in the tropics. The irrigated version, considered as the typical stereotype of rice production, is found in only 19 percent of the total tropical rice area. In ustic monsoon climates of Southeast Asia, the process begins after the advent of the first heavy rains marking the end of the dry season. Small paddy fields, usually less than 1 ha in size and often much smaller, are surrounded with dikes to trap as much rainfall as possible. After the soil has been saturated with water for several days, these fields are plowed by animal-drawn implements or small tractors to incorporate stubble and weeds. On a small, carefully tended area, the seedbeds are prepared and guarded by a family member. Two types of seedbeds are common: the conventional one, in which soaked, pregerminated seeds are sown directly on the soil at a rate of 1 to 2 tons seed/ha, and the "dapog" system, in which seeds are prevented from soil contact by a layer of banana leaves, plastic, or other barrier, and are sown at a density five times higher. In both cases water level is carefully controlled. Conventional seedbeds are pulled for transplanting at 15 to 30 days after seeding, while "dapog" seedbeds have to be transplanted between 8 and 12 days after seeding because by that time nutrient reserves in the seeds begin to be exhausted.

While the seedlings are growing in the seedbeds, the main fields are puddled by several harrowing operations conducted at progressively lower soil moisture contents until the topsoil is converted into a uniform mud. Basal fertilizer applications are broadcast and mixed into the puddled soil during

Table 11.5 Major Components of the Principal Rice Cropping Systems

Cropping System	Water Supply	Land Preparation	Crop Establishment	Field Moisture	Typical Location
Rainfed lowland	Rainfall	Puddling	Transplanting	Intermittent flooding	Philippines
Irrigated lowland	Irrigation	Puddling	Transplanting	Constant flooding	Philippines
Upland	Rainfall	Dry, no dikes	Direct seeding	Mostly unflooded	Brazil
Deep-water	Rainfall and floods	Dry, no dikes	Direct seeding	Flooded 5–400 cm deep	Thailand
Direct-seeded, irrigated	Irrigation	Dry, dikes	Direct seeding	Constant flooding	Nicaragua

the last puddling operation. The fields are water-leveled in the process, and extra mud is used to repair the dikes. Rice seedlings are then transplanted in "hills" or groups of three to six seedlings at spacings ranging from 20 × 20 to 50 × 50 cm. After a few days without flooding, the water level is raised to 5 or 10 cm above the soil surface and kept there until 2 or 3 weeks before harvest. One or two manual or chemical weed control operations are conducted during the first month after transplanting. A second nitrogen application may be top-dressed on the flooded soil at the panicle initiation stage. Rice is harvested between 100 and 150 days after seeding, usually by hand. The management of the residues is highly variable. A second or even a third crop may be planted during the year if irrigation water is available during the dry season. About half of the irrigated lowland area in Southeast Asia is double-cropped. These areas account for 25 percent of the rice production of tropical Asia (Barker, 1972). Most of the new short-statured rice varieties are grown according to this system. Also, the bulk of rice research has been conducted in lowland irrigated paddies.

Rainfed Lowland System

The rainfed lowland system is similar to the one just described, except for the lack of water control. It is the principal rice-growing system in the tropics and accounts for 46 percent of the total area. Rainfall dependence means that the timing of actual operations may be subject to tremendous variation, as rainfall variability during the rainy season is usually large. In a study of rainfed farms in Central Luzon, Philippines, Johnson (IRRI, 1966) found that most farmers do not start land preparation until 2 months after the rainy season begins, when rainfall is more reliable. After transplanting, the crop may suffer from water stress at several stages of growth, depending on the rainfall pattern. Only one crop can be grown per year in rainfed ustic regimes. A similar situation is also found in certain irrigated areas of Latin America, where the irrigation systems are very deficient and water stress often occurs. The soil moisture regime, unlike that for paddy rice, can best be described as intermittent flooding. Relatively little research has been conducted until recently on rainfed or intermittently flooded systems.

Upland Rice

"Upland rice" (equivalent terms: "dry land rice," "arroz secano," "arroz de sequeiro," "riz pluvial") refers to a system in which rice is grown like any other crop, without wetland preparation, transplanting, or dikes around the field. Approximately 25 percent of the tropical world's rice area is in upland rice. This system is characterized by conventional land preparation, direct seeding in a dry soil, and complete dependence on rainfall for moisture. Upland rice is grown in areas and seasons that average at least 150 mm of monthly rainfall. Most of the soils used are clayey; some of them are poorly drained, but many, as in central Brazil, are well drained. Upland rice is the predominant form of rice culture in tropical America and Africa. It accounts for 72 percent of the total rice area in tropical Africa, and 75 percent in tropical America.

Upland rice is cultivated under a wide range of management intensities. It is grown in intercropped shifting cultivation systems in the Amazon Basin, much of Africa, and the hill areas of Southeast Asia. It is also grown in small, settled subsistence farms throughout the tropics and in large, almost completely mechanized farms in the rolling country of central Brazil.

Almost universally, upland rice yields are generally lower than lowland rice yields. Water stress and the absence of substantial research on this system are probably the main reasons why upland rice yields are lower.

Deep-Water Rice

In parts of the floodplains of the Mekong, Chao Phya, Irrawaddy, and Ganges rivers in Asia, a unique system of rice cultivation has developed. Rice is broadcast on dry tilled land at the start of the rainy season. After the advent of the heavy rains, these fields are flooded to depths ranging from 50 cm to 4 m. Varieties adapted to these conditions keep elongating as the water rises. Often they have to be harvested with canoes. Yields are substantially lower than with lowland rice. Significant research advances have been made, primarily in Thailand, which may increase the yield potential of this system. Deep-water rice accounts for 9 percent of the total tropical rice area.

Direct-Seeded Irrigated System

This is the most mechanized rice cropping system, developed primarily in the United States and Australia and practiced in a few of the best-developed rice areas of the tropics, primarily in Latin America. Fields are large, usually well leveled and surrounded by dikes. Dry land preparation is mechanized. Preferminated seeds are broadcast on standing water from airplanes or drilled into the dry soil with large tractors. Water control is precise; a shallow flood is maintained from 20 days after seeding to 20 days before harvest. Nitrogen plus the required herbicides and insecticides are applied by airplane. Rice is harvested with combines. Yields are high, and local research is usually sufficient. This system is found in only 1 percent of the tropical rice area. Examples are the Piura Valley in northern Peru and parts of Nicaragua.

WATER MANAGEMENT

Adaptation of the Rice Plant to Flooded Conditions

Rice is one of the few crops capable of growing under flooded conditions because of its ability to oxidize its own rhizosphere. Oxygen is diffused from the leaves via the tillers and stems to the roots through lacuna or channels in the cortex tissues. It is not known how oxygen reaches the meristematic tissues devoid of lacuna. Nevertheless, this mechanism is sufficient not only to meet the oxygen requirements of respiring root cells but also to secrete oxygen or oxidized components into the rhizosphere (Alberda, 1953; Ponnamperuma, 1965; Luxmoore and Stolzy, 1972). Many other marsh plants also have this mechanism, while others can respire anaerobically.

Several other well-known crops have the capacity to oxidize their rhizosphere when flooded, but the difference is one of degree. For example, Yoshida (1967) measured the root oxidation capacities of common crop species in terms of the amount of naphthylamine oxidized per gram of root during 2 days of flooding. The results in Table 11.6 show that the root

Table 11.6   Relative Root Oxidation
Powers of Several Crops under Flooded
Conditions

Crop	Naphthylamine Oxidized in 48 Hours (mg/g dry root)
Rice	15–30
Soybean	7.1
Wheat	4.9
Sorghum	4.0
Oats	2.9
Corn	1.4

Source: Yoshida (1967).

oxidizing power of rice is an order of magnitude higher than that of other common crops. This table also explains why sorghum tolerates temporary flooding better than corn.

The oxidizing power of rice roots depends on several factors. Yoshida observed substantial differences among varieties. He also found that the oxidizing power is greater at earlier growth stages, and that it increases with solar radiation and decreases when the plant is deficient in nitrogen, phosphorus, or potassium but not in calcium, magnesium, and silica (IRRI, 1966).

As a result of the exudation of oxidized compounds, rice roots become coated with yellowish red precipitates of unknown composition. It is presumed that they are formed by ferric and manganic oxides and hydroxides. They are not formed around the active root tips, which remain white. The role of these root coats in ion and water uptake is not known.

Water Requirements of Rice Cultivation

The water requirements of rice plants per se are no different from those of other important crops. The transpiration ratio of rice (400 g H_2O/g dry matter) is similar to that of other crops (Sanchez, 1968). However, the total water consumption of lowland rice systems ranges from 6 to 16 mm/day with an average of 9 mm/day from transplanting to harvest (IRRI, 1963). This represents approximately 1000 mm for a 4 month rice crop. These figures are substantially higher than those for most other crops, which average about 5 mm/day or 600 mm in 4 months, according to a review by Kelley (1954). An additional 25 to 30 percent is also used in the process of wet land preparation, further decreasing the water use efficiency of lowland rice.

Water is consumed in lowland rice systems by evaporation, transpiration, percolation, and seepage. Average values for constantly flooded rice fields in Southeast Asia appear in Table 11.7, based on a survey by Kung et al. (1965). Most of the percolation losses are lateral because puddling decreases downward percolation (IRRI, 1965).

Water Stress Susceptibility

The reason why rice requires more water than other crops in spite of similar transpiration ratios is related to its susceptibility to water stress. Jana and DeDatta (1971) observed that rice suffers from moisture stress at a soil moisture tension as low as 0.3 bar, and that the so-called available moisture

Table 11.7   Average Water Consumption Figures for Paddy Soils of Southeast Asia

Use	Consumption per Crop (mm H_2O)
Land preparation	240
Evaporation	180–380
Transpiration	200–550
Percolation and seepage	200–700
Total	800–1200

Source: Kung et al. (1965).

range in rice may be between flooding and field capacity, rather than between 0.3 and 15 bars. Lal and Moomaw (1972) also showed this relationship with both an improved short-statured variety (IR8) and a traditional upland variety of Nigeria. Their results appear in Fig. 11.17. Furthermore, wilting symptoms are sometimes observed under flooded conditions on hot, sunny days, indicating that transpiration exceeds water uptake.

The reasons why rice is so susceptible to water stress are not well understood. Rice has a rather superficial root system, usually no deeper than 20 cm. The large amounts of air-filled voids in the roots probably retard water uptake. These changes probably reflect the result of crop evolution under flooded conditions, where there was no need for deep rooting or an efficient water uptake mechanism.

The most critical stages of water stress susceptibility are from panicle initiation to flowering. These are also the stages at which the water demand is the greatest. Table 11.8 shows the results of an experiment conducted during the dry season in the Philippines where water stress was applied at several growth stages. These and similar data indicate that severe water stress at any growth period can substantially decrease yields.

Varietal Differences in Water Stress Tolerance

Differences in the degree of water stress susceptibility occur among rice varieties. Varieties developed for upland systems are generally regarded as more tolerant of water stress than those developed for lowland conditions. This concept has been challenged, however, by new results from the Philippines, Senegal, and Peru. Chang et al. (1972) compared 25 "upland" and "lowland" varieties and found that the African and Asian upland varieties have low tillering capacity, poor early growth vigor, and long droopy leaves, which often roll when water stress begins. Moderately good drought tolerance is associated with varieties having deep, thick roots and low top:root ratios when water stress occurs. They have inherently low yielding capacity because of the limited number of panicles per unit area.

Several lowland varieties, however, have similar degrees of drought tolerance and much higher yield potentials because of improved plant type and higher tillering capacity (Nicou et al., 1970; Chang et al., 1972). Several IRRI varieties or selections have these properties: IR4, IR5, IR20, and IR442. In the Amazon Jungle of Peru, IR4-2, a lowland variety, consistently outyielded Carolino, the local upland variety, during an 18 month test period (Kawano et al., 1972; Sanchez, 1972b). Yields of both varieties were correlated with rainfall during crop growth, but, regardless of

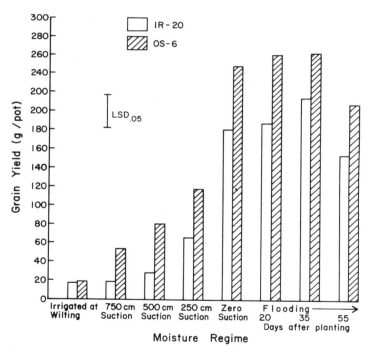

Fig. 11.17 Effect of soil moisture regime on grain yield of IR20 and OS-6 rice varieties. *Source:* Lal and Moomaw, (1972).

Table 11.8 *Grain Yields and Water Consumption by IR8 Rice under Various Water Management Practices in Los Baños, Philippines[a]*

Transplanting	Maximum Tillering	Panicle Initiation	Flowering	Maturity	Grain Yield (tons/ha)	Water Use (mm)
F ———————————————————→					7.16	1147
S →F ——————————→					5.84	1435
S ————→F ————————→					4.68	1438
S ———————————→F ——→					3.75	1121
Irrigated at visual water stress ——————————→					1.84	432
F →S ——————→F ——→					6.31	1178
F ————→S ————————→					5.87	730
F ———————————→S ——→					6.09	904

Source: DeDatta et al. (1973a).

[a] F = flooded to 2.5 cm; S = water stress (0.5 bar).

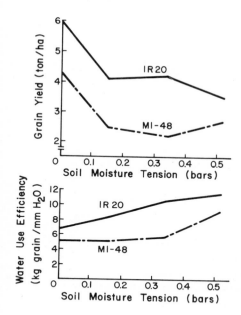

Fig. 11.18 Grain yield and water use efficiency as affected by soil moisture tension at time of irrigation of an "upland" (MI-48) and "lowland" (IR20) variety on a clayey Mollisol in Los Baños, Philippines. *Source:* DeDatta et al. (1973b).

rainfall regime, the short-statured IR4-2 consistently doubled the yield of the traditional upland variety. Figure 3.8 showed this relationship. In the Philippines, DeDatta et al. (1973b) has also shown that IR20 outyields the traditional upland variety MI-48 at several soil moisture tension levels (Fig. 11.18). Consequently, there is some evidence that the new plant type developed for lowland conditions may be applicable to upland conditions when combined with tolerance to water stress. This concept is poorly quantified, however, and its meaning unclear. DeDatta and Beachell (1972) have suggested that, rather than tolerance to prolonged stress, the ability to recover from relatively short stress periods may be more important. Other factors such as weed control and tolerance to disease and insect attacks are also important in selecting varieties for upland conditions (DeDatta et al., 1973).

Is Flooding Necessary for High Yields?

Rice generally yields better under flooded than unflooded conditions. Ample empirical evidence supports this statement, although very high rice yields are sometimes obtained without flooding. The flooded soil is usually a better medium for rice growth because (1) water stress is eliminated, (2) weed control is easier, and (3) the availability of certain nutrients, particularly phosphorus, increases as the pH approaches neutrality. Many publications emphasize the importance of one of these factors over the other two with evidence to support their claims. The abundant and conflicting literature on this subject has been reviewed periodically by Ponnamperuma (1955), Nojima (1963), and Sanchez (1968).

In this writer's opinion all three factors contribute to higher yields, and the relative importance of each one is site-specific. Since rice is quite susceptible to water stress, the physical elimination of this limitation is probably the most widespread factor. The highest yields occasionally obtained with upland rice (on the order of 6 to 7 tons/ha) have been grown in areas with high water tables and sufficient rainfall to prevent much water stress (DeDatta and Beachell, 1972).

Flooding certainly facilitates weed control because it eliminates aerobic weeds and increases the effectiveness of several herbicides in controlling aquatic weeds. However, there are other methods equally as effective in controlling weeds, although seldom as cheap. Saturation of the soil surface with water should be sufficient to eliminate water stress and produce the soil reduction. Under controlled conditions similar rice yields are obtained in a saturated and in a flooded soil. In actual field conditions, however, saturation is impractical because it is difficult to maintain and because it drastically increases weed growth.

The chemical benefits of flooding are of great practical importance to rice in certain soils, are of no consequence in others, and are extremely detrimental in a third group. In certain acid soils, particularly Oxisols, Ultisols, and Inceptisols, flooding eliminates aluminum and manganese toxicity and in some cases increases phosphorus availability (IRRI, 1970). Although liming and phosphorus applications may produce a similar effect, the automatic changes in the soil upon reduction do this at no cost. Flooding saline soils also decreases electrical conductivity and promotes the leaching of salts. Although rice is relatively susceptible to salinity, the dilution effect of flooding often permits rice to grow in soils where other crops will not, and at the same time helps to reclaim the soil through faster leaching (Pearson and Ayres, 1960). Flooding calcareous soils increases iron availability, which often results in marked yield increases (Ponnamperuma, 1965). In many paddy soils of alluvial origin, mostly Entisols, Inceptisols, Alfisols, Vertisols, and Mollisols with aerobic pHs between 6 and 7, flooding has no effect on nutrient availability. When water stress and weeds are eliminated, there is no difference in yields due to flooding in such soils.

Flooding causes detrimental effects in certain soils where reduction products may accumulate in toxic amounts. In certain Oxisols and oxidic families, ferrous iron toxicity may kill the plants or drastically decrease yields. In some sandy soils high in organic matter and low in iron, the concentration of organic acids or hydrogen sulfide may also reach toxic levels. The only practical solution in most instances is to drain the soil during short periods.

In summary, it is safe to assert that flooding usually contributes to high rice yields because of the elimination of water stress and the decrease in weed infestation in most soils, and improved soil chemical conditions in certain cases.

Is Puddling Necessary for High Yields?

The primary contribution of puddling is to decrease water losses, not to increase nutrient availability per se (Sanchez, 1973). However, puddling may indirectly increase nutrient availability by decreasing leaching losses of ions such as NH_4^+. Figure 11.19 shows the differences in water consumption between puddled and unpuddled Maahas clay and the increased yields due to the decrease in percolation. Consequently, puddling is generally beneficial to rice in soils subject to leaching losses in irrigated lowland systems. This generalization is of particular relevance for sandy soils and well-aggregated Oxisols and Andepts, where puddling is often essential for

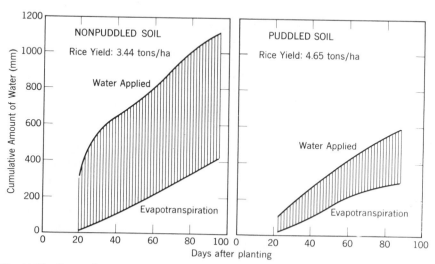

Fig. 11.19 Comparison of cumulative water applied, evapotranspiration, and percolation loss in a puddled and a nonpuddled clayey Inceptisol continually flooded at 5 cm at Los Baños, Philippines. *Source:* DeDatta and Kerim (1974).

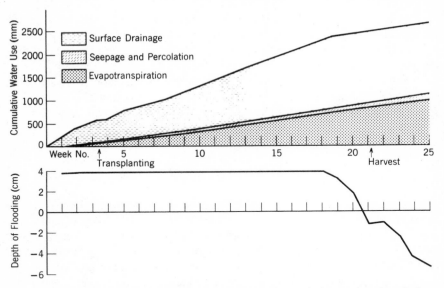

Fig. 11.20 Cumulative evapotranspiration, seepage, and percolation, and surface drainage (top) and computed flooding depth (bottom), in a constantly flooded paddy field of Central Luzon, Philippines. *Source:* Wickham (1971).

flooding such soils. At the other extreme, puddling is probably of little importance in soils having an impermeable layer in the subsoil or a constantly high water table, or in sodic soils that are already dispersed. In most rice soils puddling helps in decreasing percolation losses.

Puddling is a double-edged sword, however, in rainfed lowland systems. In most cases puddling atenuates the increases in soil moisture tension during temporary droughts and increases yields. However, when intense droughts take place early after transplanting, the puddled soil may shrink, crack, and impede root development to such a degree that the plants do not recover afterwards. DeDatta and Kerim (1974) reported rainfed rice yields of 2.8 tons/ha in unpuddled rainfed conditions and 4.0 tons/ha in puddled rainfed conditions. This difference is associated with less water stress in the puddled soil during a 20 day period at about the panicle initiation stage. In the same soil Sanchez (1973) observed that, when water stress occurred shortly after transplanting, severe root impedance took place and the yields reached only 1.8 tons/ha. When a similar water stress occurred at about the panicle initiation stage, rainfed puddled yields increased to about 4.5 tons/ha. The root system was already well developed, and stress at this later stage did not cause irreversible damage.

Another potentially detrimental effect of puddling is the time required for the soil to dry and be prepared for aerobic crops grown in rotation with rice. This time interval is extremely long in clayey montmorillonitic families but considerably shorter in clayey kaolinitic, allophanic, or oxidic families. In continuous lowland rice cropping, this effect is irrelevant.

Actual Water Management Practices

As previously mentioned, constant flooding is limited to about 20 percent of the tropical rice acreage. An example of an actual paddy field in Central Luzon, Philippines, is illustrated in Fig. 11.20, which shows that flooding was kept at a constant level before transplanting up to about 3 weeks before harvest. The actual water consumed by evaporation, seepage, and percolation was approximately 1000 mm. Figure 11.21 shows a typical rainfed lowland paddy characterized by alternate periods of flooding and unsaturation. The total water consumption in this case was about 2700 mm. These figures are typical of the higher water consumption under intermittent as opposed to constant flooding.

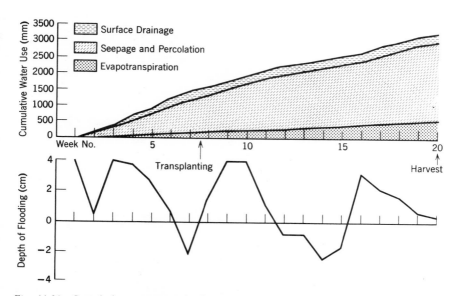

Fig. 11.21 Cumulative evapotranspiration, seepage, and percolation, and surface drainage (top) and computed flooding depth (bottom), in an intermittently flooded paddy field of Central Luzon, Philippines. *Source:* Wickham (1971).

NITROGEN MANAGEMENT

Rice responds almost universally to nitrogen applications, except on recently cleared land or in circumstances where other factors severely limit growth. In the principal tropical rice-producing countries—India, Indonesia, Thailand, the Philippines, and Brazil—the optimum responses are obtained at rates of 30 to 50 kg N/ha, with yields in the order of 2 to 3 tons/ha with tall-statured varieties. At higher rates the tall traditional *indica* varieties tend to lodge and decrease yields. The introduction of a new rice plant type by the International Rice Research Institute in 1966 and its rapid spread throughout the tropics have completely changed nitrogen management practices.

Rice responses to nitrogen depend primarily on nonsoil factors. The principal ones are plant type, solar radiation, water management, growth duration, and, lastly, soil properties.

Plant Type

The IR8 variety is the prototype of the new tropical rice plant. It is characterized by short stature, high tillering capacity, erect stems and

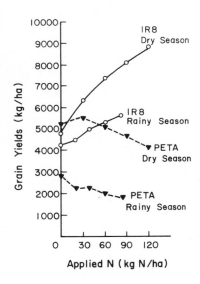

Fig. 11.22 Responses of two plant types (3 year averages) during seasons of different solar radiation in Los Baños, Philippines. Transplanting system under constant flooding. *Source:* DeDatta (1970).

Plate 15 Minabir-2, a tall-statured plant type (left), in contrast to IR8, a short-statured plant type (right): Lambayeque, Peru.

leaves, high grain:straw ratio, and resistance to lodging. The traditional plant types have vigorous growth, tall stature, low tillering capacity, weak stems and leaves, and low grain:straw ratios. These traditional varieties respond to nitrogen by increasing their height, which causes lodging and subsequent yield losses at high rates (DeDatta et al., 1966). Figure 11.22 illustrated the differences in responses between the short-statured plant type (IR8) and the traditional Peta variety. The negative responses of traditional varieties at high nitrogen rates are largely the result of lodging. Similar response curves have been obtained in other, widely different environments (Sanchez, 1972a).

The nitrogen responses of tall-statured and short-statured varieties are extremely different. The tall varieties increase their height dramatically with increasing nitrogen supply, whereas the short-statured varieties show only minor height increases (DeDatta et al., 1968; Sanchez et al., 1973a). Neither plant type shows significant changes in its growth duration with varying nitrogen rates, and nitrogen uptake at harvest is essentially identical. Among the three yield components, the number of panicles per unit area is more closely related to yield increases in short-statured varieties like IR8 than the number of filled grains per panicle or the individual grain weight. In tall-statured varieties, on the other hand, yield responses are related to both panicles per unit area and number of filled grains per panicle (Sanchez, 1972a). These results suggest that the new varieties respond

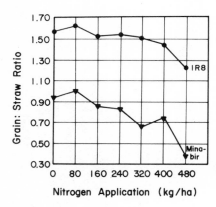

Fig. 11.23 Grain:straw ratio as a function of nitrogen rate in two plant types at Lambayeque, Peru. IR8 is short-statured; Minabir, tall-statured. *Source:* Sanchez et al. (1973a).

mainly by increasing the tiller number, while traditional varieties also respond in terms of panicle size. Tillering in rice is intimately associated with the nitrogen status of the plant (Tanaka et al, 1964).

The overall differences between the two plant types are summarized in Fig. 11.23, which shows the proportion of grain to straw produced as a function of nitrogen applied. The new varieties have the ability to convert more nitrogen and photosynthates into grain, using a smaller straw base to produce it than do the tall traditional varieties.

Solar Radiation

When rice is grown in constantly flooded soils, the differences in its nitrogen responses between rainy and dry seasons are associated mainly with differences in solar radiation, because the fluctuation of other climatic and agronomic factors is small (DeDatta, 1970). The greater solar radiation during the dry season provides more photosynthetic energy and allows larger nitrogen responses and yields in both plant types, the tall- and the short-statured, than during the rainy season. Nitrogen responses are greater in short-statured plant types subject to higher solar radiation. Positive responses occur in tall varieties at higher solar radiation, whereas only negative responses are observed at low solar radiation. This is shown in Fig. 11.24. The rainy season with its high degree of cloudiness provides less solar radiation and consequently lower yields and nitrogen responses. During the reproductive period, yields and solar radiation are highly correlated (DeDatta and Zarate, 1970). Nitrogen responses in the Peruvian Jungle are similar to those observed during the rainy season in the Philippines, because

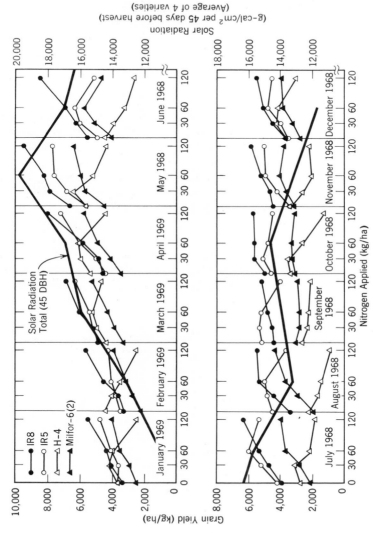

Fig. 11.24 Nitrogen responses of four indica rice varieties by month of harvest, plotted with the solar radiation total for 45 days before harvest (DBH), Los Baños, Philippines. Transplanted, constant flooding. *Source:* DeDatta and Zarate (1970).

of similar levels of solar radiation. The high responses obtained on the Peruvian coast result partly from solar radiation levels higher than those during the dry season in the Philippines, allowing high yields and optimum responses at 160 kg N/ha for the traditional varieties and about 300 kg N/ha for the short-statured types (Sanchez, 1972a). The intermittent flooding typical of this area explains in part the high rates of nitrogen used.

Water Management

Because of the large nitrogen losses caused by alternate flooding and drying, nitrogen responses are different under intermittent than under constant flooding. Figure 11.25 illustrates that the effect of progressively worse water management is strongly dependent on variety. Short-statured varieties like C4-63 show lower yields without added nitrogen as water management becomes worse, because of the depletion of available soil nitrogen. The responses to applied nitrogen are often lower with intermittent flooding, but in some cases yield levels are similar to those obtained with constant flooding. The situation with the tall-statured variety (Peta) is the opposite. Its yields increase at any nitrogen level with progressively poor water management because, when less nitrogen is available, less lodging occurs. This relationship probably reflects a natural adaptation of the tall varieties to poor water management. In spite of these differential responses, higher yields are

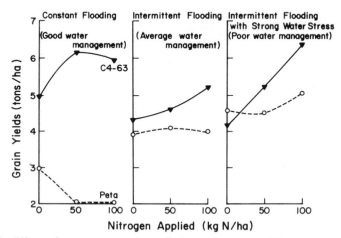

Fig. 11.25   Effects of water management and plant type on nitrogen response in Los Baños, Philippines. C4-63 is short; Peta, tall. *Source:* Adapted from DeDatta et al. (1973b).

usually attained with short-statured varieties at any water management level (Krupp et al., 1972; Sanchez, 1972b; DeDatta et al., 1973b).

Temperature and Growth Duration

The average growth duration of rice crops in the tropics ranges from 120 to 140 days from seeding to maturity for varieties relatively insensitive to photoperiodism. There are several large rice-producing areas in the tropics where low temperatures may affect rice growth, such as Bangladesh, Assam, Nepal, and parts of Madagascar and Peru. One result of low temperatures is an increase in growth duration. In such environments an interaction between growth duration, plant type, and nitrogen response exists. On the coast of Peru maximum yields and nitrogen responses are associated with a growth duration of around 180 days (Sanchez, 1972a). Earlier-maturing varieties fail to reach high yield levels because they do not accumulate sufficient dry matter. Later-maturing varieties also have lower yields because of lower grain:straw ratios and low-temperature-induced sterility.

Sources of Nitrogen for Constantly Flooded Systems

The nitrogen dynamics in flooded soils illustrated in Fig. 11.6 suggests that ammoniacal sources are superior to nitric ones. A review of the literature on this subject by DeDatta and Magnaye (1969), as well as a series of field experiments with ^{15}N conducted by the International Atomic Energy Agency in 15 countries (IAEA, 1970), confirms the general absence of differences between ammonium sulfate and urea in constantly flooded soils. For soils deficient in sulfur, ammonium sulfate is sometimes superior to urea, but the reverse is true in extremely acid soils with lower iron contents, where H_2S toxicity may occur. Urea hydrolysis into ammonium carbonate requires the same time in flooded soils as in well-aerated ones (Delaune and Patrick, 1970). Before hydrolysis, urea cannot be retained by clay particles; therefore it can move as fast as nitrates. This greater mobility and possible volatilization losses when urea is applied to the soil surface are the explanations most frequently advanced in cases where urea is inferior to ammonium sulfate (DeDatta and Magnaye, 1969).

The inefficiency of sodium nitrate is also obvious because of the denitrification process. Nitrate utilization increases when it is applied to the surface at the time when the rice plant has developed a superficial mat of roots capable of absorbing nitrates before they are leached to the reduced zone.

Even in these cases, however, nitrate is less efficient than the ammoniacal sources. Nitrification inhibitors used as additives have failed at the field level (DeDatta and Magnaye, 1969; IAEA, 1970). Anhydrous ammonia is an excellent nitrogen source for flooded rice, but mechanization difficulties and possible volatilization losses at the time of incorporation have prevented large-scale use. Organic manures of plant or animal origin have been employed for centuries in Asia. Although organic matter generally decomposes more slowly under flooded conditions, the responsible microorganisms function at higher C:N ratios than under aerobic conditions (DeDatta and Magnaye, 1969). With the nitrogen levels now recommended, the actual potential of organic fertilizers is limited. They may be used as possible supplements to inorganic sources.

Sources of Nitrogen for Intermittently Flooded Systems

Comparisons between nitrogen sources are very limited for intermittent flooding. Studies in Peru indicate that, when nitrogen is applied in split applications, there are no differences between urea and ammonium sulfate, in spite of the fact that the soil had an aerobic pH of 8.2. Sodium nitrate applications at panicle initiation proved extremely inefficient, because rice under alternate flooding and drying conditions does not develop a significant quantity of superficial roots able to absorb NO_3^- before it can be leached down to the reduced layer (Sanchez, 1972a). Several slow-release sources of nitrogen such as sulfur-coated urea have been studied in various countries. Experiments with sulfur-coated urea supplied by the Tennessee Valley Authority indicate that this source behaves similarly to conventional urea under constantly flooded conditions and low percolation rates (Englestad et al., 1972). However, under intermittent flooding, sulfur-coated urea incorporated before transplanting is superior to conventional sources applied in the same manner and, in some instances, to split applications of regular urea (Sanchez et al., 1973b). The potential of slow-release fertilizers for upland systems is being evaluated in several countries.

Placement of Nitrogen Fertilizers

Nitrogen is normally applied in either of two ways: incorporated in the soil before seeding or transplanting, or broadcast at different stages of growth. The need to incorporate ammoniacal sources into the reduced layer in systems with constant flooding is well known (Mikklesen and Finfrock, 1957; DeDatta, et al., 1968). Incorporations to 5 cm depth are sufficient for constant flooding conditions. For intermittent flooding a deeper application may be beneficial in order to escape reoxidation. The ^{15}N studies conducted by the International Atomic Energy Agency (IAEA, 1970) have provided additional information. The failure to observe the benefits of incorporation at 5 cm in soils with aerobic pHs of 4.7 or 8.1 has been attributed to inhibition of nitrification of ammoniacal sources applied to the superficial layer, where the pH does not change. At these pH extremes nitrification is minimal.

Under alternate redox conditions in Peru, incorporation at transplanting was inferior to that occurring with broadcast applications at advanced stages of growth. Sanchez (1972a) has attributed this difference to great nitrogen losses caused by frequent and pronounced flooding and drying cycles during the initial periods of growth.

Broadcast applications at the tillering or panicle initiation stages are more efficient in the presence of a thin layer of standing water. It is not recommended, therefore, that the soil be drained after nitrogen applications (AICRIP, 1969; DeDatta, 1970).

Timing of Nitrogen Applications under Constant Flooding

Because of the rapid changes that nitrogen undergoes in rice soils during short periods, the timing of nitrogen applications is an extremely critical management factor. Nitrogen uptake proceeds throughout the growth cycle of the rice plant, but its nitrogen content during two physiological stages is critical: at the beginning of tillering and at the panicle initiation stage (Matsushima, 1965). An adequate supply of available nitrogen during the beginning of tillering results in more tillers, which are closely correlated with yield in short-statured plant types. However, excessive supplies of available nitrogen after the maximum tillering stage and before panicle initiation may result in a large proportion of unproductive tillers and premature lodging of tall varieties. The nitrogen available between panicle initiation and flowering is closely correlated with the number of fertile grains per panicle. Excessive nitrogen after flowering may extend growth duration and increase susceptibility to certain diseases. The purpose of timing nitrogen applications is to synchronize the plant's requirements with the availability of this element in the soil throughout the growing season. As expected, there is great variability of experimental results in different localities, as well as in the same localities during different years.

Under constant flooding, a basal application entirely incorporated before seeding or transplanting is normally sufficient for soils with low percolation rates and for varieties resistant to lodging. In flooded soils with high perco-

lation rates, however, splitting the nitrogen application in two is more efficient, provided that the second half is applied at the panicle initiation stage
(Evatt, 1965; DeDatta, 1970). For varieties susceptible to lodging, applications at panicle initiation are advisable since they tend to reduce initial
excessive growth (DeDatta, 1970) and may prevent lodging altogether (Sims
et al., 1967).

Timing of Nitrogen Applications under Intermittent Flooding

When experiments are conducted on farmers' fields with intermittent water
management, the optimum rates as well as the optimum timing of applications are found to be completely different from experiment station results
(IRRI, 1969, 1971, 1972; Sanchez and Calderon, 1971; DeDatta et al.
1969). Using higher rates and splitting applications into two parts were
necessary to obtain high yields with tall and short plant types at several
solar radiation levels. In cases of alternate flooding and drying in Peru,
more than 90 percent of the nitrogen incorporated at transplanting may be
lost; whereas when nitrogen was applied at the panicle initiation stage,
yields increased (Fig. 11.26) and the efficiency of nitrogen utilization doubled. In upland rice, splitting nitrogen applications into two parts is defini-
tively superior to a single application, especially during the reproductive

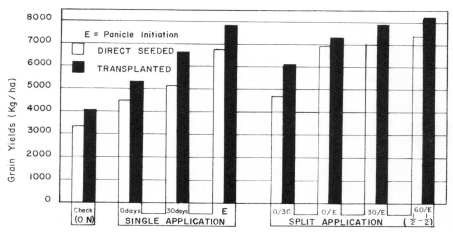

Fig. 11.26   Effects of timing of an application of 180 kg N/ha on IR8 rice under two planting
systems and intermittent flooding in Lambayeque, Peru. E = Panicle initiation. *Source:*
Sanchez and Calderon (1971).

phase (Cordero and Romero, 1972; Kass et al., 1974). In general, few additional benefits are obtained by more than two applications.

Recovery of Applied Nitrogen

The recovery of applied nitrogen is somewhat lower in flooded rice than in other crops. In the United States rice areas, Westfall (1969) estimated that fertilizer recovery ranges from 33 to 53 percent at rates of 40 to 120 kg N/ha. Racho and DeDatta (1968) reported a maximum of 33 percent recovery for applications of 30 kg N/ha in the rainy season and 57 percent with rates of 90 kg N/ha in the dry season in the Philippines. The different loss mechanisms have not been evaluated in detail, although it is assumed that denitrification and leaching are the most important processes involved.

Under alternate oxidation–reduction conditions, nitrogen losses increase. Nitrogen fertilizer recovery at harvest time fluctuated between 20 and 30 percent with conventional management practices in Peru; but efficiency can be increased substantially by the selection of sources, placement, and timing practices most adequate for local situations (Sanchez and Calderon, 1971; Sanchez et al., 1973b).

PHOSPHORUS MANAGEMENT

Responses

Rice responds to phosphorus applications less frequently and less intensely than other cereal crops. Several reasons account for this situation. The increased availability of phosphorus in the soil solution upon flooding is often of such magnitude that no phosphorus applications are necessary for lowland rice, whereas aerobic crops growing on the same soil require added phosphorus for high yields. Very often, in soils testing less than 1 ppm of extractable phosphorus, rice does not respond to phosphorus applications. In trials on such soils, rice may show a dramatic visual phosphorus response at early growth stages, but the unfertilized plots gradually catch up, and at harvest no yield differences are observed (DeDatta et al., 1966; Davide, 1965).

The increased availability of phosphorus under flooding is soil-dependent. Figure 11.11 shows that there is little change in some soils, particularly in highly weathered ones. Consequently, flooding does not eliminate phosphorus deficiency in such soils, examples of which are Oxisols, Ultisols, Andepts, and Sulfaquepts. Phosphorus responses are common in these soils, even while flooded. Phosphorus availability is further affected by the type of water management used (Sanchez and Briones, 1973).

Varietal Differences

Under upland conditions, rice usually requires lower rates of phosphorus to obtain maximum yields than other crops such as corn. As discussed in Chapter 8, the phosphorus requirement of rice may be lower than that of other cereals. Recent work conducted in the Philippines shows that varietal differences exist in relation to tolerance of low available phosphorus. Ponnamperuma and Castro (1972) and Koyama and Chammek (1971) have proved that certain varieties perform better under low available phosphorus than others. Commonly used varieties such as IR20 and IR22 gave 50 percent higher yields than IR8 in low-phosphorus conditions. Salinas and Sanchez (1976) report that such varietal differences are related to faster phosphorus adsorption and translocation rates.

Sources, Placement, and Timing

Ordinary or triple superphosphate is an excellent phosphorus source for rice. Ammoniated superphosphates are also good. In soils with low aerobic pH values, rock phosphates of high citrate solubility have proved to be as efficient as superphosphates. Several rock phosphate sources of different citrate solubility were evaluated in acid sulfate soils in Thailand and in Oxisols of Colombia in cooperation with a Tennessee Valley Authority program (Englestad et al., 1972, 1974). Figure 8.7 showed some of the results. The economic implications are obvious in areas like Thailand, where locally available rock phosphates cost about $20 (United States currency) a ton, while triple superphosphate costs $105 a ton at the time the studies were conducted. Rock phosphates with high citrate solubilities produce an excellent residual effect.

The most practical method of phosphate application to rice soils is to broadcast and incorporate it in the puddled layer before transplanting. In upland rice areas, phosphorus is applied in bands close to the planting furrow (Novais and Defelipo, 1971; Freitas et al., 1973).

The Problem of Soil Testing in Rice

Soil tests for phosphorus are, in general, failures when applied to lowland rice production. Poor correlations are obtained in the field and even in pot tests (Chang, 1965). There are several reasons for this poor performance. None of the common soil extractants can detect reductant-soluble phosphorus, which may become available upon flooding. Many of the common extractants (Bray, Truog, dilute double acid, etc.) do not detect much Fe–P. The Olsen extractant is usually better but seldom satisfactory. It would seem logical to preflood soils before extracting in order to account for the increased availability generally observed upon flooding. Unfortunately, attempts in this direction have failed. Research on this subject has generally been haphazard, and systematic efforts should produce better results.

MANAGEMENT OF OTHER NUTRIENTS

The extremely wide range of soils and climatic conditions in which rice is grown in the tropics has produced a number of locally important nutritional disorders, ranging from straight toxicities or deficiencies to complex nutrient interactions with plant diseases. Tanaka and Yoshida (1970) published a survey of these disorders in Asia, which has significantly increased our awareness of the importance of some of these problems. Their results are shown in Fig. 11.27. They also developed tentative critical levels of these nutrients in rice tissue for each disturbance (Table 11.9) and provided excellent photographs of the various symptoms. In Latin America and Africa, more evidence is also accumulating on several nutritional disturbances.

Zinc Deficiency

Zinc deficiency is probably the most widespread micronutrient disorder in tropical rice. It occurs in parts of India, Pakistan, the Philippines, and Colombia under lowland conditions (Tanaka and Yoshida, 1970; Yoshida and Forno, 1971; CIAT, 1971; IRRI, 1971, 1972). It also occurs throughout the Cerrado of Brazil under upland conditions (DeSouza and Hiroce, 1970). In lowland rice areas, zinc deficiency is associated with calcareous soils and is accentuated by prolonged flooding. Zinc deficiency symptoms are more pronounced at early growth stages, and sometimes the

Table 11.9   Deficiency and Toxicity Critical Contents of Various Elements in the Rice Plant

Element	Deficiency (D) or Toxicity (T)	Critical Level	Plant Part Analyzed	Growth Stage[a]
N	D	2.5%	Leaf blade	Til
P	D	0.1%	Leaf blade	Til
	T	1.0%	Straw	Mat
K	D	1.0%	Straw	Mat
	D	1.0%	Leaf blade	Til
Ca	D	0.15%	Straw	Mat
Mg	D	0.10%	Straw	Mat
S	D	0.10%	Straw	Mat
Si	D	5.0%	Straw	Mat
Fe	D	70 ppm	Leaf blade	Til
	T	300 ppm	Leaf blade	Til
Zn	D	10 ppm	Shoot	Til
	T	1500 ppm	Straw	Mat
Mn	D	20 ppm	Shoot	Til
	T	2500 ppm	Shoot	Til
B	D	3.4 ppm	Straw	Mat
	T	100 ppm	Straw	Mat
Cu	D	6 ppm	Straw	Mat
	T	30 ppm	Straw	Mat
Al	T	300 ppm	Shoot	Til

Source: Tanaka and Yoshida (1970).
[a] Mat = maturity; Til = tillering.

plant completely recovers at later growth stages. Tanaka and Yoshida attribute this effect to large bicarbonate concentrations during the peaks of soil reduction, which result in an immobilization of zinc in the roots. Not all calcareous soils, however, are zinc deficient. Conventional soil tests are well correlated with plant zinc content and responses (IRRI, 1972). The critical level by the Lindsay method is 1.5 ppm Zn in the soil; it is associated with levels of 14 ppm Zn in plant tissues. It is conceivable that the increased availability of phosphorus with flooding can decrease the availability of zinc, according to preliminary studies by Giordano and Mortvedt (1972).

In acid Oxisols of Brazil, zinc deficiencies are widespread in upland rice, in spite of the fact that the solubility of zinc is highest at such low pHs. Apparently these soils are so low in zinc that deficiencies occur in spite of the high solubility.

Zinc deficiency can be corrected by applications of 5 to 15 kg Zn/ha as zinc sulfate or oxide incorporated into the soil before seeding or transplanting (Yoshida et al., 1970; Giordano and Mordvedt, 1973). Such rates increased yields from 0.5 to 7.5 tons/ha in a calcareous soil of Colombia (CIAT, 1971). Other alternatives include dipping the seedlings in 1 percent zinc oxide suspension before transplanting, and mixing zinc oxide with presoaked rice seeds before direct seeding (Yoshida and Forno, 1971; CIAT, 1972).

There are important varietal differences in tolerating zinc deficiency. Local Colombian varieties are very tolerant, whereas most IRRI varieties are not (CIAT, 1971). This is a result of selection by breeding without the breeders knowing that they were selecting for zinc tolerance.

Iron Deficiency

Iron deficiency is widespread in calcareous and alkali soils under both upland and lowland conditions and is also found in some flooded soils with organic matter contents too low to produce significant soil reduction. Iron

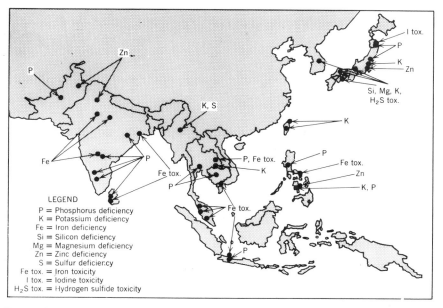

Fig. 11.27 Location and classification of nutritional disorders of rice in Asia. *Source:* Tanaka and Yoshida (1970).

Soil	Soil Condition	Disorder	Local Name
Very low pH	Acid sulfate soil	Iron toxicity	"Bronzing"
High in active iron	Low in organic matter	Phosphorus deficiency	
	High in organic matter	Phosphorus deficiency combined with iron toxicity	"Akagare Type III"
	High in iodine	Iodine toxicity combined with phosphorus deficiency	
	High in manganese	Manganese toxicity	
Low pH — Low in active iron and exchangeable cations	Low in potassium	Iron toxicity interacted with potassium deficiency	"Bronzing" "Akagare Type I"
	Low in bases and silica, with sulfate application	Imbalance of nutrients associated with hydrogen sulfide toxicity	"Akiochi"

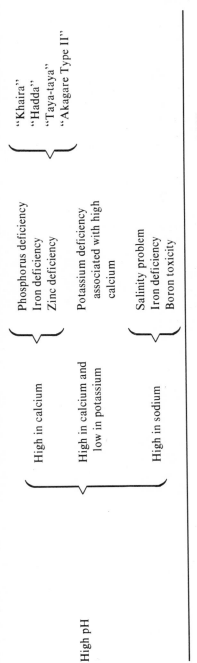

Fig. 11.27 Classification of Nutritional Disorders in Asia. *Source:* Tanaka and Yoshida (1970).

deficiencies are widespread in central and northern India, the Philippines, and parts of Japan and the United States (Tanaka and Yoshida, 1970; Okajima et al., 1970; Ponnamperuma and Castro, 1972). In flooded soils iron deficiency is associated with aerobic pHs higher than 7 or insufficient soil reduction. When rice is grown on calcareous soils, iron deficiency may occur in upland conditions but may be absent in flooded areas because of the generally increased availability of iron under flooding. Iron deficiency has been recorded in acid soils of pH around 5.0 both in the Philippines (DeDatta et al., 1975) and in Brazil (Costa and deSouza, 1972) in upland conditions during periods of water stress. Although the solubility of iron is high, apparently the crystallinity of the forms or sheer water stress may induce iron deficiency. Not all calcareous soils produce iron deficiency in rice. No iron deficiency symptoms and prominent iron coatings have been observed by this writer in calcareous soils of pH 8.3 throughout the coast of Peru.

Iron chlorosis symptoms may appear at early growth stages and disappear afterwards when the increase in Fe^{2+} has occurred because of flooding. Iron deficiency can be corrected by applying ferric sulfate, ferrous sulfate, and iron oxides to the standing water or incorporating them into the soil (Patrick and Mikklesen 1971). Sulfuric acid applied to decrease the pH in certain calcareous and sodic soils also eliminates iron deficiency. Varietal differences in tolerating iron deficiencies have been identified (Ponnamperuma and Castro, 1972; IRRI, 1971, 1972).

Iron Toxicity

Iron toxicity occurs when certain highly weathered Oxisols, Ultisols, or Sulfaquepts are flooded. The increased concentration of ferrous ions in the soil solution can reach toxic levels of 300 ppm Fe or more in the rice plants. Iron toxicity has recently been identified as the cause of "bronzing" disease in Sri Lanka and other parts of Southeast Asia, and as the primary cause of "anaranjamiento" disorder of Colombia (Tanaka and Yoshida 1970; CIAT 1971, 1972). Potassium deficiency and iron toxicity interact in "Akagare Type I" disease in Japan (Tanaka and Tadano, 1972).

The only practical solution for iron toxicity involves draining the soil to reoxidize the ferrous iron. In soils where there is a marked ferrous iron peak followed by a decrease, iron toxicity may be avoided by preflooding the soil for a month or so before sowing or transplanting. In iron-toxic Sulfaquepts, however, reoxidation is harmful because it results in aluminum toxicity. The best alternative for such soils is a combination of liming and manganese dioxide applications, prolonged submergence, and leaching to prevent the buildup of ferrous iron and to reduce aluminum toxicity (Nhung and Ponnamperuma, 1966; IRRI, 1972). In an Oxisol from Colombia, liming to pH 6.5 and avoiding organic matter additions depressed iron toxicity. The most practical solution for this soil, however, is to grow upland rice.

Varietal differences relative to iron toxicity also exist. New improved varieties like IR8 are very susceptible, whereas tall-statured local varieties and some short-statured ones are less sensitive to high Fe^{2+} levels (IRRI, 1972; CIAT, 1972).

Potassium Deficiency

Potassium responses in lowland rice are rare. Most lowland rice soils are usually high enough in exchangeable potassium. They also receive new supplies of potassium and other bases with the flood water every year. Potassium responses occur in some calcareous soils because of Ca:K imbalances. A common rice disease, *Helminthosporium* leaf spot, is accentuated by potassium deficiency and attenuated by potassium fertilization. When deficiencies occur, they can be corrected by application rates of about 40 kg K/ha (Tanaka and Yoshida, 1970). Some upland rice soils are low in potassium and respond to this element like any other crop (Kemmler, 1971; Kass et al., 1973). Soil tests are good tools for estimating potassium deficiency in rice soils.

Other Disturbances

In sandy soils high in sulfates and low in iron, H_2S can accumulate at toxic levels and kill the plants. The "akiochi" symptoms observed in Japan are a consequence of H_2S toxicity. Its occurrence in the tropics has not been documented (Tanaka and Yoshida, 1970). Japanese farmers frequently add iron-rich soils from the uplands to affected fields. This drastic practice causes precipitation of H_2S as FeS.

Aluminum and manganese toxicities are common in acid upland rice fields; liming and the use of tolerant varieties are the remedies. Manganese toxicity seldom occurs in flooded fields because rice is very tolerant of this element (up to 2500 ppm Mn in tissue). Manganese deficiencies have been occasionally reported in Southeast Asia.

Carbon dioxide and organic acids sometime accumulate in toxic levels in some flooded soils, usually in spots. The solution is drainage.

Silicon deficiency has been reported in Japan and Korea but not in the tropics (Okuda and Takahashi, 1965). Rice has a high requirement for silica

(Tanaka and Park, 1966). In upland rice areas of Brazil, some farmers apply burned rice hulls to Oxisols, a practice that may add silica to soils low in this element.

Salinity is a major problem in irrigated desert areas, where the rice yield potential is otherwise very high. Flooded rice may suffer from salinity stress at electrical conductivities of 4 mmho/cm or more. The fact that temporary drying increases salinity damage to rice poses a major problem in intermittently flooded areas. Drainage is the best practice. Varietal differences to salt tolerance exist; IR8 is considered relatively tolerant (Tanaka and Yoshida, 1970).

SUMMARY AND CONCLUSIONS

1. Rice, the number 1 crop produced in the tropics, is usually grown under a unique set of soil conditions caused by the physical and chemical changes resulting from flooding. Flooding per se causes entrapped air to explode in soil pores, decreasing the permeability and aggregate stability of soils. The magnitude of these changes varies with soil properties.

2. Puddling, the intentional process of breaking down soil aggregates into a uniform mud, is a common practice in lowland rice culture. Puddling causes almost complete aggregate breakdown, the destruction of macropores, increases in water-filled micropores, increased moisture retention at a soil moisture tension range from 0 to 10 bars, and decreased evaporation and percolation losses, and induces chemical reduction in the absence of flooding. The decrease in percolation losses is considered the main reason why farmers puddle. Puddling is not an irreversible process; the original structure can be restored by alternate wetting and drying and tillage at the right moisture content. The magnitude of these changes, as well as the ease of regenerating, depends on soil properties. Soils with layer silicate mineralogy require less effort to puddle, attain a greater degree of aggregate destruction, and require long periods of time for regeneration. Soils with oxidic or oxide-coated mineralogy (Oxisols, Ultisols, Andepts) are most difficult to puddle thoroughly, but structural regeneration is fairly simple.

3. The paramount chemical consequence of flooding is the microbiological transformation of an oxidized soil into a reduced one. Not all the profile of a flooded soil is reduced; a thin oxidized layer close to the surface and the rhizosphere of rapidly growing rice roots remain oxidized. The reduction processes start with nitrates, followed by manganic and ferric oxides and hydroxides, intermediate products of organic matter decomposition, and in

certain cases sulfates. The intensity of reduction is usually a function of soil organic matter content and can be measured as Eh (redox potential). The main consequences of soil reduction are change of the soil pH to between 6 and 7 regardless of aerobic pH, denitrification of nitrates and accumulation of ammonium, and increases in the solubility of iron, manganese, and in certain cases phosphorus. The magnitude of these changes also varies with soil properties, ranging from highly beneficial to neutral and then to highly toxic to the rice plant. Upon drying, these processes are also reversible.

4. Rice in the tropics is grown in five major cropping systems: the rainfed "lowland" or paddy system comprises 46 percent of the acreage; irrigated "lowland," 19 percent; "upland" or dry land, 25 percent; deep-water, 9 percent; and direct-seeded irrigated, 1 percent. The physical and chemical consequences of flooding vary according to the type of water management used. The chemistry of intermittent flooding is substantially different from that of continuous flooding.

5. Rice is capable of growing under flooded conditions because of its ability to oxidize its own rhizosphere. The water requirements of the rice plant per se are no different from those of other crops of comparable growth duration. The susceptibility to water stress requires insurance against this phenomenon through flooding. Wet land preparation also consumes additional water. The water consumption of an average flooded, puddled rice field is on the order of twice the amount of a crop of corn. The available moisture range for rice in clayey soils is on the order of 0 to 0.3 bar rather than the conventional 0.3 to 15 bars. Significant varietal differences in water stress susceptibility occur but are not related to plant type. Rice generally yields more under flooded conditions because of water stress elimination, easier weed control, and, in certain soils, the chemical benefits of soil reduction.

6. Rice responds to nitrogen applications almost universally except in recently cleared land or in situations where other factors severely limit growth. The magnitude of the responses, however, is minimal and often negative when tall-statured varieties are used because of their susceptibility to lodging. With the advent and widespread adaptation of short-statured, lodging-resistant varieties the responses are sharper and often highly profitable. In addition to soil properties other factors such as plant type, solar radiation, temperature, and water management affect the shape of the nitrogen response curve. The most effective sources of nitrogen are the ammoniacal ones, especially ammonium sulfate and urea. Under conditions of intermittent flooding, slow-release sources such as sulfur-coated urea have proved more efficient, but they are not advantageous for constantly flooded soils. The best method of application of conventional ammoniacal

fertilizers is the incorporation of a basal dose into the soil before seeding or transplanting, followed by a top-dressed application at the panicle initiation stage. Delaying the timing of the first application to the tillering stage appears to be a more efficient practice in intermittently flooded and upland systems. The efficiency of nitrogen utilization is lower in rice than in other crops and much lower under intermittent than under constant flooding.

7. Rice responds to phosphorus applications with less frequency than other crops grown in the same soil because of the increased availability of this element upon flooding. Responses are common under upland conditions. Varietal differences in low-phosphorus tolerance exist. Superphosphates and ammoniated phosphates are excellent sources for rice. In very acid soils rock phosphates behave very well. Soil tests for lowland rice have not been adequately developed.

8. Other nutritional problems are found throughout the tropical rice-growing region. Zinc deficiency is perhaps the most widespread one. Iron deficiency, iron toxicity, potassium deficiency, and disturbances caused by excessive concentrations of reduced products are also encountered.

REFERENCES

Ahmad, N. 1963. The effect of evolution of gases and reducing conditions in a submerged soil and its subsequent physical status. *Trop. Agr. (Trinidad)* **40**:205–209.

AICRIP. 1969. All India Coordinated Rice Improvement Project. *Progress Report Kharif,* Vol. 2. Indian Council of Agricultural Research, New Delhi.

Alberda, T. 1953. Growth and root development of lowland rice and its relation to oxygen supply. *Plant and Soil* **5**:1–28.

Allison, L. E. 1947. Effects of microorganisms on the permeability of soils under prolonged submergence. *Soil Sci.* **63**:439–450.

Barker, R. 1972. The economic consequences of the green revolution in Asia. Pp. 115–126. In *Rice, Science and Man.* International Rice Research Institute, Los Baños, Philippines.

Bhattacharyya, A. K. 1971. Mechanism of the formation of exchangeable ammonium nitrogen immediately after waterlogging rice soils. *J. Indian Soc. Soil Sci.* **19**:209–213.

Bodman, G. B. and J. Rubin. 1948. Soil puddling. *Soil Sci. Soc. Amer. Proc.* **13**:27–36.

Breazeale, J. F. and W. T. McGeorge. 1937. Studies on soil structure: some nitrogen transformations in puddled soils. *Arizona Agr. Exp. Sta. Tech. Bull.* 69.

Broadbent, F. E. and D. S. Mikklesen. 1968. Influence of placement on the uptake of tagged nitrogen by rice. *Agron. J.* **60**:674–677.

Broeshart, H. and H. Brunner. 1964. The efficiency of phosphate and nitrogen fertilization in rice cultivation. *Trans. 8th Int. Contg. Soil Sci. (Adelaide)* **4**:209–218.

Chang, S. C. 1965. Phosphorus and potassium tests for soils. Pp. 373–381. In International Rice Research Institute, *Symposium on the Mineral Nutrition of the Rice Plant.* Johns Hopkins Press, Baltimore.

Chang, T. T., G. C. Loresto, and O. Tagumpay. 1972. Agronomic and growth characteristics, of upland and lowland rice varieties. Pp. 645–661. In *Rice Breeding*. International Rice Research Institute, Los Baños, Philippines.

Cho, D. Y. and F. N. Ponnamperuma. 1971. Influence of soil temperature on the chemical kinetics of flooded soils and the growth of rice. *Soil Sci.* 112:184–194.

Chandrasekaran, S. and T. Yoshida. 1973. Effect of organic acid transformations in submerged soils on the growth of the rice plant. *Soil Sci. Plant Nutr.* 19:39–45.

Chaudhary, T. N. and P. B. Ghildyal. 1969. Aggregate stability of puddled soil during rice growth. *J. Indian Soc. Soil Sci.* 17:261–265.

CIAT. 1971, 1972. Annual Reports. Centro Internacional de Agricultura Tropical, Cali, Colombia.

CIAT. 1972. Políticas arroceras en America Latina. Información básica y antecedentes. Centro Internacional de Agricultura Tropical, Cali, Colombia.

Cordero, A. and A. Romero. 1972. *Estudios de Fertilización Nitrogenada del Arroz en el Pacífico Húmedo de Costa Rica*. Departamento de Agronomía, Ministerio de Agricultura y Ganadería, San José, Costa Rica. 14 pp.

Costa, A. S. and D. M. deSouza. 1972. *Clorose do Arroz Devida a Deficiência de Ferro em Solo Latosol Roxo de Campinas*. Instituto Agronômico de Campinas, São Paulo, Brazil.

Croney, D. and J. D. Coleman. 1954. Soil structure in relation to soil suction (pF). *J. Soil Sci.* 5:75–84.

Das Gupta, D. K. 1971. Effects of levels and time of nitrogen application and interaction between phosphorus and nitrogen on grain yield of rice varieties under tidal mangrove swamp cultivation in Sierra Leone. *Afr. Soils* 16:59–67.

Davide, J. G. 1965. The time and methods of phosphate fertilizer applications. Pp. 255–268. In International Rice Research Institute, *Symposium on the Mineral Nutrition of the Rice Plant*. Johns Hopkins Press, Baltimore.

DeDatta, S. K. 1970. Fertilizer and soil amendments for tropical rice. Pp. 106–145. In *Rice Production Manual*, 2nd ed. University of the Philippines at Los Baños.

DeDatta, S. K., J. C. Moomaw, V. V. Racho, and G. V. Simsiman. 1966. Phosphorus supplying capacity of lowland rice soils. *Soil Sci. Soc. Amer. Proc.* 30:613–617.

DeDatta, S. K., C. P. Magnaye, and J. C. Moomaw. 1968a. Efficiency of fertilizer nitrogen (N^{15} labelled) for flooded rice. *Trans. 9th Int. Congr. Soil Sci. (Adelaide)* 4:67–76.

DeDatta, S. K., A. C. Tauro, and S. N. Balaoing. 1968b. Effects of plant type and nitrogen level on the growth characteristics and grain yield of indica rice in the tropics. *Agron. J.* 60:643–647.

DeDatta, S. K. and C. P. Magnaye. 1969. A survey of forms and sources of fertilizer nitrogen for flooded rice. *Soil and Fertilizers* 32:103–109.

DeDatta, S. K., C. P. Magnaye, and J. T. Magbanua. 1969. Response of rice varieties to time of nitrogen application in the tropics. *Trop. Agr. Res. Ser.* 3:73–88. Ministry of Agriculture and Forestry, Tokyo, Japan.

DeDatta, S. K. and P. M. Zarate. 1970. Environmental conditions affecting growth characteristics, nitrogen response and yield of tropical rice. *Biometeorology* 4:71–89.

DeDatta, S. K. and H. M. Beachell. 1972. Varietal response to some factors affecting production of upland rice. Pp. 685–700. In *Rice Breeding*. International Rice Research Institute, Los Baños, Philippines.

DeDatta, S. K., H. K. Krupp, E. I. Alvarez, and S. C. Modgal. 1973a. Water management practices in flooded tropical rice. Pp. 1–18. International Rice Research Institute, In *Water Management in Philippine Irrigation Systems: Research and Operations.* Los Baños, Philippines.

DeDatta, S. K., W. P. Abilay, and G. N. Nalwar. 1973b. Water stress effects in tropical rice. Pp. 19–36. International Rice Research Institute, In *Water Management in Philippine Irrigation Systems: Research and Operations.* Los Baños, Philippines.

DeDatta, S. K., E. L. Aragon, and J. A. Malabuyoc. 1973c. Varietal differences in and cultural practices for upland rice. Unpublished paper. International Rice Research Institute, Los Baños, Philippines. 63 pp.

DeDatta, S. K., and M. S. A. A. Kerim. 1974. Water and nitrogen economy on rainfed rice on puddled and non-puddled soils. *Soil Sci. Soc. Amer. Proc.* **38**:515–518.

DeDatta, S. K., F. G. Faye, and R. N. Mallick. 1975. Soil–water relations in upland rice. Pp. 168–185. In E. Bornemisza and A. Alvarado (eds.), *Soil Management in Tropical America.* North Carolina State University, Raleigh.

Delaune, R. D. and W. H. Patrick, Jr. 1970. Urea conversion to ammonia in water-logged soils. *Soil Sci. Soc. Amer. Proc.* **34**:603–607.

DeSouza, D. M. and R. Hiroce. 1970. Diagnóstico e tratamiento de deficiência de zinco em solos con pH abaixo de 7. *Bragantia* **29**:91–103.

Englestad, O. P., J. T. Getsinger, and P. J. Stangel. 1972. *Tailoring Fertilizers for Rice.* Tennessee Valley Authority, Muscle Shoals, Ala. 56 pp.

Englestad, O. P., A. Jugsujinda, and S. K. DeDatta. 1974. Response by flooded rice to phosphate rocks varying in citrate solubility. *Soil Sci. Soc. Amer. Proc.* **38**:524–529.

Evatt, N. S. 1965. The timing of nitrogenous fertilizer applications on rice. Pp. 243–254. In International Rice Research Institute, *Symposium on the Mineral Nutrition of the Rice Plant.* Johns Hopkins Press, Baltimore.

FAO. 1970. *Production Yearbook.* Food and Agricultural Organization of the United Nations, Rome.

Feng, M. P. and J. L. Saldaña. 1973. The response of two paddy rice varieties to potash in the Dominican Republic. *Potash Rev.* **16**:63. 5 pp.

Freitas, J. A. C. de, J. M. Braga, S. S. Brandão, and F. B. Gomes. 1973. Adubação mineral (NPK) de arroz em solos da região do Maranhão. *Experientiae* **15**:291–313.

Ghildyal, B. P. 1971. Soil and water management for increased water and fertilizer use efficiency for rice production. *Proc. Int. Symp. Soil Fert. Eval.* (*New Delhi*) **1**:499–509.

Giordano, P. M. and J. J. Mordvedt. 1972. Rice response to zinc in flooded and non-flooded soil. *Agron. J.* **64**:521–523.

Giordano, P. M. and J. J. Mordvedt. 1973. Zinc sources and methods of application for rice. *Agron. J.* **65**:51–53.

Gupta, R. K., T. A. Singh, and B. P. Ghildyal. 1972. Fate of native and applied phosphorus under saturated and non-saturated soil moisture conditions. *Agrochimia* **16**:548–555.

Haddad, G. and L. Seguy. 1972. Le riz pluvial dans le Senégal Meridional. *Agron. Tropicale* (*France*) **21**:419–461.

Hossner, L. R., J. A. Freeouf, and B. L. Folsom. 1973. Solution phosphorus concentration and growth of rice in flooded soils. *Soil Sci. Soc. Amer. Proc.* **37**:405–408.

IAEA. 1970. Rice fertilization. *Tech. Repts. Ser.* 108. International Atomic Energy Agency, Vienna.

IRRI. 1963–1973. Annual Reports. International Rice Research Institute, Los Baños, Philippines.

Ishizuka, Y. 1965. Nutrient uptake at different stages of growth. Pp. 192–218. In International Rice Research Institute, *The Mineral Nutrition of the Rice Plant*. Johns Hopkins Press, Baltimore.

Ishizuka, Y. 1971. Physiology of the rice plant. *Adv. Agron.* **23**:241–315.

Jaggi, I. K. and M. B. Russell. 1973. Effect of moisture regimes and green manuring on ferrous iron concentration in soil and growth and yield of paddy. *J. Indian Soc. Soil Sci.* **21**:71–76.

Jamison, V. C. 1953. Changes in air–water relationships due to structural improvement of soils. *Soil Sci.* **76**:143–151.

Jana, R. K. and S. K. DeDatta. 1971. Effects of solar energy and soil moisture tension on the nitrogen response of upland rice. *Proc. Int. Symp. Soil Fert. Eval. (New Delhi)* **1**:487–497.

Jana, R. K. and B. P. Ghildyal. 1972. Effect of varying soil water regimes during the different growth phases on the yield of winter rice. *Il Riso* **21**:93–95.

Kampen, J. and G. Levine. 1970. Water losses and water balance studies in Philippine lowland rice irrigations. *Philipp. Agr.* **54**:283–301.

Kar, S. and S. B. Varade. 1972. Influence of mechanical impedance on rice seedling root growth. *Agron. J.* **64**:80–82.

Kass, D. L., J. Furlan, J. B. Pace, W. S. Couto, and E. de S. Cruz. 1973. *Adubação de Arroz de Sequeiro em Latosol Amaerlo na Zona Bragantina*. Instituto de Pesquisas Agropecuarias do Norte, Belém, Pará, Brazil. 25 pp.

Kass, D. L., N. T. DaPonte, J. Furlan, et al. 1974. *Some Agronomic and Economic Aspects of Rice Culture in the Brazilian Amazon*. Instituto de Pesquisas Agropecuarias do Norte, Belém, Pará, Brazil. 24 pp.

Kawaguchi, K., K. Kita, and K. Kyuma. 1956. A soil core sampler for paddy soils and some physical properties of the soil under waterlogged conditions. *Soil Plant Food* **2**:92–95.

Kawaguchi, K. and K. Kita. 1957. Mechanical and chemical constituents of water-stable aggregates of paddy soils in relationship to aggregate size. *Soil Plant Food* **3**:22–28.

Kawaguchi, K. and T. Kawachi. 1969. Cation exchange reactions in submerged soil. *J. Sci. Soil Manure (Japan)*. **40**:89–95.

Kawano, K., P. A. Sanchez, M. A. Nureña, and J. R. Velez. 1972. Upland rice in the Peruvian Jungle. Pp. 637–643. In *Rice Breeding*. International Rice Research Institute, Los Baños, Philippines.

Kelley, O. J. 1954. Requirement and availability of soil water. *Adv. Agron.* **6**:67–94.

Kemmler, G. 1971. Response of high yielding paddy varieties to potassium, experimental results from various rice growing countries. *Proc. Int. Symp. Soil Fert. Eval. (New Delhi)* **1**:391–406.

Kita, K. and K. Kawaguchi. 1960. The effects of both the reduction of the soil under waterlogged conditions and the dehydration of the reduced soil upon soil structure. *J. Sci. Soil Manure (Japan)* **31**:355–379, 495–498.

Koenigs, F. F. R. 1961. *The Mechanical Stability of Clay Soils as Influenced by Moisture Conditions and Some Other Factors*. Centre for Agricultural Publications and Documentation No. 67.7, Wageningen, Netherlands. 171 pp.

Koenigs, F. F. R. 1963. The puddling of clay soils. *Netherl. J. Agr. Sci.* **11**:145–156.

Koyama, T. and C. Chammek. 1971. Soil–plant nutrition studies on tropical rice. I. Studies on varietal differences in absorbing phosphorus from soils low in available phosphorus. *Soil Sci. Plant Nutr.* **17**:115–126.

Krupp, H. K., W. P. Abilay, and E. I. Alvarez. 1972. Some water stress effects on rice. Pp. 663–675. In *Rice Breeding,* International Rice Research Institute, Los Baños, Philippines.

Kumar, V., K. T. Mahajan, S. B. Varade, et al. 1971. Growth responses of rice to submergence, soil aeration and soil strength. *Indian J. Agr. Sci.* **41**:527–534.

Kung, P., C. Atthayodhin, and S. Kruthabandhi. 1965. Determining water requirements of rice by field measurement in Thailand. *Int. Rice Comm. Newsletter* **19**(4):5–18.

Lal, R., and J. C. Moomaw. 1972. Effect of different soil moisture regime on the growth and development of rice. International Institute for Tropical Agriculture, Ibadan, Nigeria.

Luxmoore, R. J. and L. H. Stolzy. 1972. Oxygen diffusion in the soil–plant system. VI. A synopsis with commentary. *Agron. J.* **64**:725–729.

MacRae, I. C., R. R. Ancajas, and S. Salandanan. 1968. The fate of nitrogen in some tropical soils following submergence. *Soil Sci.* **105**:327–334.

Mahapatra, I. C. and W. H. Patrick, Jr. 1971. Evaluation of phosphate fertility in waterlogged soils. *Proc. Int. Symp. Soil Fert. Eval. (New Delhi)* **1**:53–61.

Mandal, L. N. and S. K. Khan. 1972. Release of phosphorus from insoluble phosphatic materials in acidic lowland rice soils. *J. Indian Soc. Soil Sci.* **20**:19–25.

Matsushima, S. 1965. Nutrient requirements at different stages of growth. Pp. 219–242. In International Rice Research Institute, *The Mineral Nutrition of the Rice Plant.* Johns Hopkins Press, Baltimore.

Mikklesen, D. S. and D. C. Finfrock. 1957. Availability of ammoniacal nitrogen to lowland rice as influenced by fertilizer placement. *Agron. J.* **49**:296–300.

Mikklesen, D. S. and W. H. Patrick, Jr. 1968. Fertilizer use on rice. Pp. 403–432. In *Changing Patterns in Fertilizer Use.* Soil Science Society of America, Madison, Wisc.

Naphade, J. D. and B. P. Ghildyal. 1971. Influence of puddling and water regimes on soil characteristics, ion uptake and rice growth. *Proc. Int. Symp. Soil Fert. Eval. (New Delhi)* **1**:511–517.

Nhung, M. T. and F. N. Ponnamperuma. 1966. Effect of calcium carbonate, manganese dioxide and ferric hydroxide on chemical and electrochemical changes and the growth of rice on a flooded acid sulphate soil. *Soil Sci.* **102**:29–41.

Nicou, R., L. Seguy, and G. Haddad. 1970. Comparison de l'enracienment de cuatre varietes du riz pluvial en presence on absence de travail du sol. *Agron. Tropicale (France)* **8**:639–659.

Nojima, K. 1963. Irrigation and drainage. Pp. 399–423. In K. Matsubayashi et al. (eds.), *Theory and Practice of Growing Rice.* Japan Ministry of Agriculture and Forestry, Tokyo.

Novais, R. F. and B. V. Defelipo. 1971. Níveis ótimos de NPK na adubação de arroz de sequeiro em um solo de cerrado de Patos de Minas. *Experientiae* **11**:281–296.

Okajima, H., N. D. Mannikar, and M. J. Rao. 1970. Iron chlorosis of rice seedlings in calcareous soils under upland conditions. *Soil Sci. Plant Nutr.* **16**:128–132.

Okuda, A. and E. Takahashi. 1965. The role of silicon. Pp. 123–146. In International Rice Research Institute, *The Mineral Nutrition of the Rice Plant.* Johns Hopkins Press, Baltimore.

Pande, H. K. and N. K. Adak. 1971. Leaching loss of nitrogen in submerged rice cultivation. *Exptal. Agr.* **7**:329–336.

Pande, H. K., and B. N. Mittra. 1971. Effects of depth of submergence, fertilization and cultivation on water requirement and yield of rice. *Exptal. Agr.* **7**:241–248.

Patnaik, S. and F. E. Broadbent. 1967. Utilization of tracer N by rice in relation to the time of application. *Agron. J.* **59**:287–288.

Patrick, W. H., Jr., and R. Wyatt. 1964. Soil nitrogen loss as a result of alternate submergence and drying. *Soil Sci. Soc. Amer. Proc.* **28**:647–653.

Patrick, W. H., Jr., and I. C. Mahapatra. 1968. Transformation and availability to rice of nitrogen and phosphorus in waterlogged soils. *Adv. Agron.* **20**:323–359.

Patrick, W. H., Jr., and D. S. Mikklesen. 1971. Plant nutrient behavior in flooded soil. Pp. 187–215. In R. A. Olsen (ed.), *Fertilizer Technology and Use,* 2nd ed. Soil Science Society of America, Madison, Wisc.

Patrick, W. H., Jr., and R. D. Delaune. 1972. Characterization of the oxidized and reduced zones in flooded soil. *Soil Sci. Soc. Amer. Proc.* **36**:573–576.

Patrick, W. H., Jr., and M. E. Tusneem. 1972. Nitrogen loss from flooded soil. *Ecology* **53**:735–737.

Patrick, W. H., Jr., and R. A. Khalid. 1974. Phosphate release and sorption by soils and sediments: effect of aerobic and anaerobic conditions. *Science* **186**:53–55.

Pearson, G. A. and A. D. Ayres. 1960. Rice as a crop for salt-affected soils in process of reclamation. *U.S. Dept. Agr. Res. Service, Production Res. Rept.* 43.

Ponnamperuma, F. N. 1955. The chemistry of submerged soils in relation to the growth and yield of rice. Ph.D. Thesis, Cornell University, Ithaca, N.Y. 414 pp.

Ponnamperuma, F. N. 1965. Dynamic aspects of flooded soils and the nutrition of the rice plant. Pp. 295–328. In International Rice Research Institute, *The Mineral Nutrition of the Rice Plant.* Johns Hopkins Press, Baltimore.

Ponnamperuma, F. N. 1972. The chemistry of submerged soils. *Adv. Agron.* **24**:29–96.

Ponnamperuma, F. N. and R. U. Castro. 1972. Varietal differences in resistance to adverse soil conditions. Pp. 677–684. In *Rice Breeding.* International Rice Research Institute, Los Baños, Philippines.

Racho, V. V. and S. K. DeDatta. 1968. Nitrogen economy of cropped and uncropped flooded soils under field conditions. *Soil Sci.* **105**:417–427.

Robinson, D. O. and J. P. Page. 1951. Soil aggregate stability. *Soil Sci. Soc. Amer. Proc.* **15**:25–29.

Salinas, J. G. and P. A. Sanchez. 1976. Soil–plant relationships affecting varietal and species differences in tolerance to low available soil phosphorus. *Ciência e Cultura (Brazil)* **28**:156–168.

Sanchez, P. A. 1968. Rice performance under puddled and granulated soil cropping systems in Southeast Asia. Ph.D. Thesis, Cornell University, Ithaca, N.Y. 381 pp.

Sanchez, P. A. 1972a. Nitrogen fertilization and management of tropical rice. *North Carolina Agr. Exp. Sta. Tech. Bull.* 213.

Sanchez, P. A. 1972b. Tecnicas agronómicas para optimizar el potencial productivo de las nuevas variedades de arroz en America Latina. Pp. 27–43. In *Políticas Arroceras en America Latina.* Centro Internacional de Agricultura Tropical, Cali, Colombia.

Sanchez, P. A. 1973. Puddling tropical rice soils. I. Growth and nutritional aspects. II. Effects of water losses. *Soil Sci.* **115**:149–158, 303–308.

Sanchez, P. A. and M. V. Calderon. 1971. Timing of nitrogen applications for rice grown under intermittent flooding in the Coast of Peru. *Proc. Int. Symp. Soil Fert. Eval.* (*New Delhi*) 1:595–602.

Sanchez, P. A. and M. A. Nureña. 1972. Upland rice improvement under shifting cultivation systems in the Amazon Basin of Peru. *North Carolina Agr. Exp. Sta. Tech. Bull.* 210.

Sanchez, P. A. and A. M. Briones. 1973. Phosphorus availability of some Philippine rice soils as affected by soil and water management practices. *Agron. J.* **65**:266–278.

Sanchez, P. A., G. E. Ramirez, and M. B. de Calderon. 1973a. Rice responses to nitrogen under high solar radiation and intermittent flooding in Peru. *Agron. J.* **65**:523–529.

Sanchez, P. A., A. Gavidia, G. E. Ramirez, R. Vergara, and F. Minguillo. 1973b. Performance of sulfur-coated urea under intermittently flooded rice culture in Peru. *Soil Sci. Soc. Amer. Proc.* **37**:789–792.

Sanchez, P. A., G. E. Ramirez, and C. Perez. 1975. Influence of solar radiation on the varietal response of rice to nitrogen in the coast of Peru. Pp. 246–257. In E. Bornemisza and A. Alvarado (eds.), *Soil Management in Tropical America.* North Carolina State University, Raleigh.

Savant, N. K. and M. M. Kibe. 1971. Influence of continuous submergence on pH, exchange acidity and pH-dependent acidity in rice soils. *Plant and Soil* **32**:205–208.

Savant, N. K. and M. M. Kibe. 1972. Influence of added calcium hydroxide in submerged and subsequently acid lateritic rice soils. *Soil Sci. Soc. Amer. Proc.* **36**:529–531.

Sims, S. L., V. L. Hall, and T. H. Johnston. 1967. Timing of nitrogen fertilization of rice. I. Effects of application near midseason and varietal performance. *Agron. J.* **59**:63–66.

Singlachar, M. A. and R. Samaniego. 1973. Effect of flooding and cropping on the changes in the inorganic phosphate fractions in rice soils. *Plant and Soil* **39**:351–360.

Sreenivasan, P. S. and J. R. Banerjee. 1973. The influence of rainfall on the yield of rainfed rice at Karjat. *Agr. Meteorol.* **11**:285–292.

Stout, B. A. 1966. Equipment for rice production. *FAO Agri. Dev. Paper* 84. 169 pp.

Tanaka, A. 1969. Physiological basis for fertilizer response of rice varieties. *Japan Min. Agr. For. Trop. Agr. Res. Ser.* 3:37–43.

Tanaka, A., S. A. Navasero, C. V. García, F. T. Parao, and E. Ramirez. 1964. Growth habit of the rice plant in the tropics and its effect on nitrogen response. *IRRI Tech. Bull.* 3.

Tanaka, A., C. V. García, and N. T. Diem. 1965. Studies on the relationship between tillering and nitrogen uptake by the rice plant. *Soil Sci. Plant Nutr.* **11**:9–13.

Tanaka, A., K. Kawano, and J. Yamaguchi. 1966. Photosynthesis, respiration, and plant type of the tropical rice plant. *IRRI Tech. Bull.* 7.

Tanaka, A. and S. A. Navasero. 1966. Interaction between iron and manganese in the rice plant. *Soil Sci. Plant Nutr.* **12**:197–201.

Tanaka, A. and Y. D. Park. 1966. Significance of the adsorption and distribution of silica in the growth of the rice plant. *Soil Sci. Plant Nutr.* **12**:191–195.

Tanaka, A. and S. Yoshida. 1970. Nutritional disorders of the rice plant in Asia. *IRRI Tech. Bull.* 10.

Tanaka, A. and T. Tadano. 1972. Potassium in relation to iron toxicity of the rice plant. *Potash Rev. Sub.* 9 (Suite 20), pp. 1–12.

Taylor, H. M. 1972. Effect of drying on water retention of a puddled soil. *Soil Sci. Soc. Amer. Proc.* **36**:972–973.

Turner, F. T. and J. W. Gilliam, 1976. Diffusion as a factor affecting the availability of phosphorus in flooded soils. *Plant and Soil* (in press).

Tusneem, M. E. 1967. Availability and transformation of phosphorus fertilizers in submerged and non-submerged soils. M.S. Thesis, University of the Philippines, Los Baños. 151 pp.

Tusneem, M. E. and W. H. Patrick, Jr. 1971. Nitrogen transformation in water-logged soils. *Louisiana Agr. Exp. Sta. Bull.* 657.

Westfall, D. G. 1969. The efficiency of applied N in rice-producing soils. *Rice J.* 72(7):66–67.

Wickham, T. H. 1971. Water management in the humid tropics: A farm-level analysis. Ph.D. Thesis, Cornell University, Ithaca, N.Y. 269 pp.

Williams, B. C. and W. H. Patrick, Jr. 1973. Dissolution of complex ferric phosphates under controlled *Eh* and pH conditions. *Soil Sci. Soc. Amer. Proc.* 37:33–36.

Yoder, R. E. 1936. A direct method of aggregate analysis of soils and a study of the physical nature of soil erosion losses. *J. Amer. Soc. Agr.* 20:337–351.

Yoshida, S. 1967. Some problems of iron nutrition in higher plants. Unpublished seminar presented at the Soils Department, College of Agriculture, University of the Philippines, March 15, 1967. 7 pp.

Yoshida, S., G. W. McLean, M. Shafi, and K. E. Mueller. 1970. Effects of different methods of zinc application on growth and yields of rice in a calcareous soil, West Pakistan. *Soil Sci. Plant Nutr.* 16:147–149.

Yoshida, S. and D. A. Forno. 1971. Zinc deficiency of the rice plant on calcareous and neutral soils in the Philippines. *Soil Sci. Plant Nutr.* 17:83–87.

Yoshida, T. and R. R. Ancajas. 1973. Nitrogen fixing activity in upland and flooded rice fields. *Soil Sci. Soc. Amer. Proc.* 37:42–46.

12

SOIL MANAGEMENT IN MULTIPLE CROPPING SYSTEMS

Multiple cropping, the growing of two or more crops on the same land during a year, is a widespread form of agriculture in the tropics. For centuries farmers have taken advantage of year-round adequate temperature and solar radiation, as well as water availability, in udic soil moisture regimes. Multiple cropping is also practiced in ustic and aridic areas during the rainy season or throughout the year with irrigation.

Growing a mixture of crops in a seemingly random arrangement has been scorned by agricultural scientists as primitive and disorderly (Jolly, 1958; Grimes, 1963). Some of these systems are so diametrically opposed to the evenly spaced rows of single-crop stands that agronomists unfamiliar with them seek immediately to replace mixtures with single stands. In an early review of cropping systems in the tropics, Wood (1934) recognized that crop mixtures rather than single stands are predominant in small farming systems. Multiple cropping is also nature's way, since it is rare to find a pure stand of one species in native vegetation (Trouse, 1975). Wood reported that the crop mixtures used by these small farmers are more productive because of a better utilization of space and time. However, difficulties in mechanizing operations is an important limiting factor. Wood theorized that crop mixtures may have been the original form of agriculture; but as harvesting time became critical, innovations such as the scythe and eventually mechanical harvesters favored the growing of single-crop stands. The establishment of a definite succession of single-crop stands led eventually to the development of crop rotations, a basic principle of crop production in the temperate region.

In spite of some early studies emphasizing the importance of multiple cropping (Aiyer, 1949; Anderson, 1950; Iso, 1954; Abeyratne, 1956), agricultural scientists essentially ignored multiple cropping until the work of Bradfield (1964, 1969, 1970, 1972) in the Philippines attracted international attention. Since then, a veritable research explosion has taken place throughout the tropics and even in the United States. Taking advantage of the shorter growth duration and photoperiod insensitivity of the new "green revolution" varieties of wheat and rice, areas in India that grew only one crop a year are now growing two or three (Kanwar and Krishnamoorthy, 1971; IARI, 1972). The more complex intercropped systems are being studied and improved at many research institutions.

Soil management is different for multiple cropping than for single cropping. This chapter examines what is known about managing the soil in the various forms of multiple cropping systems and its relationship to small farming operations. The various forms of multiple cropping are described in detail in order to clarify their differences and the management implications thereof.

MULTIPLE CROPPING AND SMALL FARMING SYSTEMS

Harwood and Price (1976) showed that the extent and importance of multiple cropping increase as farm size decreases. Also, the smaller the farm, the more complex are the crop combinations. An examination of the concept of "small farming systems" is, therefore, in order.

Small Farming Systems

A small farm is best defined as a farm operation based primarily on family manual and animal labor, where a considerable proportion of the farm output is consumed by the family, but a significant proportion is sold or bartered at nearby markets. This definition has three key components. First, the bulk of the labor is not mechanized, although some mechanization can take place, such as contract tractor plowing or the use of hand rototillers, which partially replaces animal power. Second, a *significant* amount of what is grown is traded for other goods in the market. This writer has yet to see a truly subsistence farm in the tropics where the family consumes only what it grows and does not trade with others. On the other hand, farms where only a minor proportion of the output is consumed by the family do not fit this definition regardless of size. The third component is the presence of nearby markets to which the products can be delivered by foot or animal power. This excludes the large cattle ranches in relatively isolated areas like the Llanos Orientales of Colombia, which otherwise would have fit the definition. Thus small farming systems are defined not by a specific size of farm but by the nature of farm operations.

With the exception of the Asian rice-growing areas, the contribution of small farming systems to national food production has been considered marginal. The more visible commercial farms have been traditionally regarded as the major food producers. Many commercial farms produce export crops or nonfood crops such as sugarcane, cacao, and cotton, which are primary sources of foreign exchange or of raw materials. Studies conducted in Colombia and Central America, however, show that about 70 percent of the food consumed in those countries is produced on small farms (CATIE, 1974; Pinchinat et al., 1976).

Small farms are by far the most numerous type in the tropics. In tropical Asia 75 percent of all farms are smaller than 2 ha (Harwood and Price, 1976), and in Central America 69 percent are smaller than 5 ha, according to calculations from data presented by CATIE (1974). The average farm size for 20 tropical African countries reporting such data in the 1973 FAO *Production Yearbook* was 5.4 ha.

The number of multiple cropping combinations in actual use in the tropics must be on the order of several thousand. Many of them have been described in reviews by Blencowe and Blencowe (1970) for Malaysia; IARI (1972) for India; ASPAC (1974) for Taiwan; IRRI (1975) and Harwood and Price (1976) for tropical Asia; Pinchinat et al. (1976) for tropical America; Okigbo and Greenland (1976) for tropical Africa, and Dalrymple (1971), Hart (1974), and Andrews and Kassam (1976) on a worldwide basis. Annual crops are combined with other annuals and with perennials. A "dominant" or principal food crop is generally well defined for broad ecological regions. It may be rice, cassava, or cacao for the udic lowlands, corn for the ustic lowlands and at moderate elevations, wheat in the cooler highlands, and sorghum or millet in the aridic areas. Exceedingly complex and sophisticated decisions appropriate for linear programming analysis are being made daily by small farmers (Harwood and Price, 1976).

Classification of Multiple Cropping Systems

The essence of multiple cropping is the intensification of crop production into a third and a fourth dimension (the first dimension: area of production, the second: yield per unit area). Multiple cropping introduces time as a third and space as a fourth dimension when two crops share the same given area of land at the same time. Multiple cropping systems can be classified according to the degree of intensification in time and space. Table 12.1 shows a proposed classification system. A major distinction is made between intercropping, where two or more crops are grown simultaneously, and sequential cropping, where two or more crops are grown one after

Table 12.1   Definitions of the Principal Multiple Cropping Patterns

Multiple Cropping: The intensification of cropping in time and space dimensions. Growing two or more crops on the same field in a year.

Intercropping: Growing two or more crops *simultaneously* on the same field per year. Crop intensification is in both time and space dimensions. There is intercrop competition during all or part of crop growth. Farmers manage more than one crop at a time in the same field.

   Mixed intercropping: Growing two or more crops simultaneously with no distinct row arrangement.

   Row intercropping: Growing two or more crops simultaneously with one or more crops planted in rows.

   Strip intercropping: Growing two or more crops simultaneously in different strips wide enough to permit independent cultivation but narrow enough for the crops to interact agronomically.

   Relay intercropping: Growing two or more crops simultaneously during part of each one's life cycle. A second crop is planted after the first crop has reached its reproductive stage of growth but before it is ready for harvest.

Sequential Cropping: Growing two or more crops in *sequence* on the same field per year. The succeeding crop is planted after the preceding one has been harvested. Crop intensification is only in the time dimension. There is no intercrop competition. Farmers manage only one crop at a time in the same field.

   Double cropping: Growing two crops a year in sequence.

   Triple cropping: Growing three crops a year in sequence.

   Quadruple cropping: Growing four crops a year in sequence.

   Ratoon cropping: Cultivating crop regrowth after harvest, although not necessarily for grain.

Source: Andrews and Kassam (1976).

Table 12.2   Related Terminology Used in Multiple Cropping Systems

---

Single Stands: The growing of one crop variety alone in pure stands at normal density. Synonymous with "solid planting," "sole cropping." Opposite of "multiple cropping."

Monoculture: The repetitive growing of the same crop on the same land.

Rotation: The repetitive growing of two or more sole crops or multiple cropping combinations in the same land.

Cropping Pattern: The yearly sequence and spatial arrangement of crops, or of crops and fallow, on a given area.

Cropping System: The cropping patterns used on a farm and their interactions with farm resources, other farm enterprises, and available technology that determine their makeup.

Mixed Farming: Cropping systems that involve the riasing of crops and animals.

Cropping Index: The number of crops grown per annum on a given area of land × 100.

Relative Yield Total (RYT):   The sum of the intercropped yields divided by yields of sole crops. The same concept as land equivalent ratios. "Yield" can be measured as dry matter production, grain yield, nutrient uptake, energy, or protein production, as well as by the market value of the crops.

Land Equivalent Ratio (LER):   The ratio of the area needed under sole cropping to the one under intercropping to give equal amounts of yield at the same management level. The LER is the sum of the fractions of the yields of the intercrops relative to their sole-crop yields. It is equivalent to RYT, expressed in commercial yields.

Income Equivalent Ratio (IER):   The ratio of the area needed under sole cropping to produce the same gross income as is obtained from 1 ha of intercropping at the same management level. The IER is the conversion of the LER into economic terms.

Source: Adapted from Andrews and Kassam (1976).

another during a year. These differences are fundamental enough to require separate discussion. Table 12.2 defines some additional terms used in relation to multiple cropping systems.

INTERCROPPING SYSTEMS

Intercropping is the simultaneous growing of two or more crops in the same field at the same time. Unlike sequential cropping, where intensification is in the time dimension, intercropping involves intensification in both space and time. Farmers practice intercropping when they can obtain higher yields by growing a mixture than from dividing an equal area into separate single-crop stands.

Basic Concepts of Intercrop Competition

An examination of some concepts of how plants react in mixtures is an appropriate first step in understanding intercropping. Botanists define "plant interference" as the response of an individual plant or species to its environment as modified by the presence of another individual plant or another species (Hall, 1974a,b; Trenbath, 1974). Such interference can be noncompetitive, competitive, or complementary. Noncompetitive interference occurs when different plants share a growth factor (light, water, nutrients) that is present in sufficient amounts so that it is not limiting. Plant yields are not affected by this type of interference. Competitive interference, or simply competition, occurs when one or more growth factors are limiting. In such cases the plant or species better equipped to utilize a growth factor increases its yield at the expense of the other plant or species, which suffers a yield decrease. Complementary interference, or simply complementarity, occurs when one plant helps another, as in the case of legumes supplying nitrogen to grasses via symbiotic fixation. Interference occurs among plants of the same species in single stands and among plants of the same and different species in intercropped systems.

An understanding of plant interference has developed primarily through the study of grass–legume pastures and varietal mixtures of crop species. Reviews by Hall (1974a,b), Hart (1974), and Trenbath (1974) provide insights relevant to intercropping. Trenbath compiled the results of 572 comparisons of varietal or species mixtures with single-crop stands that did not include legume–grass mixtures. His results, shown in Fig. 12.1, indicate that two-thirds of the cases had relative yield totals (RYT) of dry matter production close to 1.0, which suggest no advantage of intercropping. Figure 12.1 also shows that 20 percent of the mixtures has RYT values ranging from 1.1 to 1.7, which indicate advantages in intercropping, and that 14 percent had RYT values lower than 0.9, indicating clear disadvantages. This is an almost normal frequency distribution, with the vast majority of the cases showing little or no advantage to intercropping.

Farmers have obviously selected combinations at the higher end of this frequency distribution. A review by Andrews and Kassam (1976) shows that the RYT values obtained in farmer fields with two-crop mixtures measured as economic yield (not total dry matter) ranged from 1.2 to 1.6. In experi-

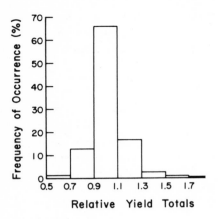

Fig. 12.1 Distribution of the relative yield totals of mixtures, based on dry matter production in 572 published experiments, excluding legume–grass mixtures. *Source:* Adapted from Trenbath (1974).

ment stations RYT values as high as 2.4 have been obtained. Some examples of successful intercropping combinations are summarized in Table 12.3, where RYTs are expressed in terms of crop yields (LER) and gross income (IER). These examples show that a hectare of intercropping produced from 24 to 82 percent more total yield than two half-hectares of single crops, and similar increases in gross income. When three or four crops are intercropped, RYT values increase even more (Soria et al., 1975).

The yield advantages of successful intercropping systems are probably related to minimizing interspecific competition for light, water, and nutrients. Some mixtures of tall permanent crops with shade-loving shorter crops, such as rubber over coffee, probably approach this ideal because they do not compete for light and apparently not significantly for water and nutrients. Hart (1974) cites several examples where RYTs exceed 2.0.

The most common situation, however, involves some degree of competition between species, which results in a decrease in yields per hectare of both crops relative to single cropping, although the RYTs are above 1.0. Competition is minimized by growing compatible species and manipulating time of planting, population, spacing, and other variables in the different forms of intercropping, which are discussed in the following section.

Mixed Intercropping

Mixed intercropping encompasses a wide array of apparently random arrangements of several crops in a field. The term "apparently" deserves emphasis because such arrangements actually reflect farm experience and, in many cases, sophisticated plant-spacing concepts. Mixed intercropping is

Table 12.3 Examples of the Intercropping Effect Recorded at Several Locations

Type	Crops A–B	Location	Yields (tons/ha) Single Stands Crop A	Single Stands Crop B	Intercropped Crop A	Intercropped Crop B	Land Equiv. Ratio (LER)	Income Equiv. Ratio (IER)
Relay	Corn–beans	Mexico	2.05	0.84	1.53	0.48	1.33	1.34
Mixed	Corn–peanuts	Tanzania	1.73	0.83	1.68	0.42	1.48	—
Row	Corn–mung beans	Philippines	2.43	1.17	1.85	0.90	1.53	1.45
Row	Corn–soybeans	North Carolina, U.S.A.	6.77	2.69	5.27	1.69	1.40	1.34
Row	Sugarcane–cowpeas	Brazil	1.41	1.36	1.41	1.35	2.00	—
Row	Corn–cotton	Kenya	3.05	1.51	2.65	0.56	1.24	1.06
Row	Corn–pigeon peas	Trinidad	3.13	1.87	2.61	1.85	1.82	—
Mixed	Cacao–rubber	Costa Rica	—	1.28	0.93	0.69	—	1.35

Source: Lepiz (1971), Evans (1960), Cordero and McCollum (unpublished), IRRI (1972), Krutman (1968), Dalal (1974), and Hunter and Camacho (1961).

485

Plate 16   Mixed intercropping of yams, cassava, corn, and some vegetables near Ibadan, Nigeria.

common when cereals, grain legumes, and root crops are grown together and no tillage is practiced. For example, farmers in the Amazon region plant simultaneously upland rice, cassava, and plantains. Rice, which is harvested first, is planted approximately at 50 cm intervals, which is the appropriate spacing for the tall, lodging-susceptible varieties used. Earlier-maturing vegetables may be planted closer and harvested even before the rice. Cassava is planted at 1 to 2 m spacings, which permit the development of a satisfactory crop canopy after the rice harvest. Plantains are planted at 3 to 5 m intervals, resulting in an adequate plantain canopy after the cassava harvest. Shade-tolerant vegetables or grain legumes may be planted after the rice or cassava harvests to utilize the available space under the larger crops.

The differences in plant size and growth duration probably decrease the competition for solar radiation. The rice canopy develops with little interference from the slow-growing cassava and plantains. After the rice is harvested, cassava becomes dominant; after its harvest the plantain canopy takes over. In essence this system mimics natural successions of forest regrowth. Competition for water and nutrients is also minimized by planting the most demanding crop (rice) first. When the availability of some nutrients is low, farmers increase the spacing between plants in order to reduce competition. Hart (1974), working with a bean–corn-cassava mixed cropping system in Costa Rica, obtained RYT values of 1.37 for yields and 1.30 for net income, as compared with growing the three crops in single stands.

In parts of West Africa, mixed intercropping is practiced in mounds or ridges of soil constructed with hoes. Several crops are planted on different parts of the mounds (Okigbo and Greenland, 1976). Mounding is thought to be beneficial because it increases the volume of soil available for root crops in Alfisols, where the subsoil is often gravelly or plinthitic. Cassava and yams are planted on top of the mound; corn and other crops that require good drainage are planted on the sides of the mounds; while rice, which thrives on poor drainage, occupies the areas between the mounds, where water tends to accumulate. Although few quantitative data exist on such systems, these arrangements follow the concepts of minimizing competition and altering the root environment in ways most favorable for the different crops.

Row Intercropping

This form occurs when one or more crops are planted at about the same time in rows that are close to each other. Row intercropping is common in tilled areas and is perhaps the central concept of intercropping. Competition between species for light, water, and nutrients is on the row basis except when two crops are planted in the same row. Corn and upland rice are commonly grown in this fashion in the Asian and Central American lowlands. Corn and wheat are grown in terraced rows in the highlands of Guatemala; sorghum and peanuts, in the ustic and aridic areas of Africa.

Row intercropping is most advantageous when a tall-statured crop is grown with a short-statured one, and when the crops have different growth durations. Competition for light is minimized when the crops have different canopy arrangements, particularly when the tall crops has a more erect leaf habit and the shorter crop more horizontal leaf angles (Trenbath, 1974). These mixtures intercept more solar radiation over time and thus have higher potentials for photosynthesis than single-crop stands (Alvim and Alvim, 1969; Bantilan et al., 1974). Table 12.4 shows the greater light interception of tall–short crop combinations when the short-statured crops are planted in solid stands and the tall crop (corn) is planted in rows of varying spacings. In addition to the advantage of intercepting more solar radiation for photosynthesis, less light is available for weed growth (Bantilan et al., 1974).

Plate 17 Row intercropping of corn, cotton, and castor beans near Campina Grande, Paraíba, Brazil.

When the crops have different growth durations, the advantages of row intercropping increase further. The stages of maximum demand for light, water, and nutrients occur at different times, even though both crops are planted at about the same time. A classic example is corn–mung bean intercropping in the Philippines (IRRI, 1973). Mung bean (*Phaseolus aureus*) reaches its flowering stage about 35 days after planting, before it is shaded by corn. It is harvested at 60 days, when corn demands are at a maximum. There is little shadowing at the seedling stage of either crop, which is an advantage because both are very susceptible to light stress. A similar situation takes place when a tall crop is havested first, as in sweet corn–upland rice intercropping, or when two tall crops of widely different growth durations are used. Examples of the latter are row intercropping of 128 day corn with 168 day pigeon peas in Trinidad (Dalal, 1974), or 75 day millet with 195 day sorghum in Nigeria (Andrews, 1974). In such cases the later-maturing crop must be tolerant of shading and have limited nutrient requirements at the peak periods of the first crop's growth.

Row intercropping of annual crops under perennials is very common. Tall-growing crops such as corn, cassava, and bananas are planted on young coffee or rubber plantations to provide shade and produce income while the permanent crops develop (Pushparajah and Wong, 1970; Pushparajah and Tang, 1970). The same situation occurs with sugarcane, where corn, beans, soybeans, peanuts, sweet potatoes, upland rice, or even tobacco is planted simultaneously and harvested before the sugarcane developes a full canopy (Chang, 1965; Bains et al., 1970; Streeter, 1974). At normal spacings and planting densities the yields of intercropped cowpeas and sugarcane were identical to monoculture yields (RYT of 2.0) in northeast Brazil (Krutman, 1968), and in Taiwan sugarcane yield reduction ranged from 0 to 10 percent when the sugarcane was intercropped with several species (Chang, 1965).

Fully grown permanent crops are commonly intercropped with shade-tolerant permanent crops. Cacao, which is shade-tolerant, is grown under mature rubber (Hunter and Camacho, 1961; Blencowe and Templeton, 1970). Under the more open-canopied permanent crops such as coconuts, many crops can be grown (Nelliat et al., 1974). Permanent grass–legume pastures under coconuts support a thriving cattle industry in parts of Asia without decreasing coconut production (Santhirasegaram, 1967; MacEvoy, 1974). Varieties of pasture species have been selected and bred for this purpose (DeGuzman, 1974; Javier, 1974). One of the most sophisticated row intercropping systems this writer has observed is a three-canopy combination of pineapples, papayas, and coconuts in volcanic soils near Lake Taal in the Philippines.

Table 12.4   Light Interception of Row-Intercropped Canopies in Comparison with Single Crop Stands at 44 days after Seeding in Los Baños, Philippines

| | | Light Interception (%) | |
| | | Intercropped with Corn Rows at | |
Crop	Single Stands	1 m	2 m
Corn		68[a]	43[a]
Sweet potatoes	55	83	89
Peanuts	79	89	92
Mung beans	78	94	90

Source: Adapted from Bantilan et al. (1974).

[a] Corn single stands at 1 and 2 m row spacings.

Relay Intercropping

In the two systems discussed above, crops are planted at roughly the same time but are harvested at different times because of differences in growth duration. When a second crop is planted after the first crop has entered the reproductive growth phase but before harvest, the system is called "relay intercropping." A very common example is the almost ubiquitous corn–beans (*Phaseolus vulgaris*) system in most of Central America and much of tropical South America as well. Corn is planted in rows, usually at the beginning of the rainy season, and even before in parts of Guatemala. The tall, photoperiod-sensitive corn varieties require 6 to 10 months to mature. When the ears are well formed but not mature (33 to 55 percent water in the kernels), farmers break the stalks just below the ear and plant climbing bean varieties. The corn matures normally and suffers no major physiological damage in terms of grain yield or seed germination if harvested as long as 10 weeks after breaking (Byrd, 1967). This is in effect a form of storage in which the corn ears are protected from rain, mold, and diseases by hanging upside down. The beans grow quickly and increase their leaf areas by climbing up the doubled corn stalk. Beans mature when the

Plate 18 Relay intercropping of corn and *Phaseolus vulgaris* beans in Turrialba, Costa Rica. Corn stalks are doubled over.

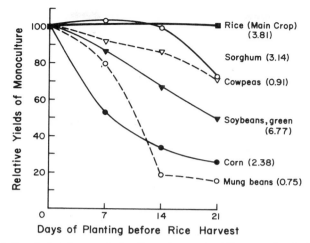

Fig. 12.2   Relative shade tolerance during the early growth of five crops in relation to the time of relay intercropping with rice in Los Baños, Philippines. Numbers in parenthesis indicate yields (tons/ha) when grown as sole crops. *Source:* Calculated from data by Herrera and Harwood (1973).

weather is drier and the attack of pathogens is less. Farmers then harvest both crops at the beginning of the dry season.

Relay intercropping is also very common in rice-based systems in Taiwan (Shen, 1964; ASPAC, 1974). Up to five crops per year may be harvested by two relays successions: rice–melons, followed by planting rice again and relaying with cabbages and corn. A very intensive relay system involving seven crops a year has been developed by Hildebrand (1976) and is practiced on small farms of El Salvador.

Pastures are often established by relay intercropping on standing cereal crops. This practice is common in newly cleared forested areas, where the higher nutrient availability after burning is used by a cash crop in the process of pasture establishment. Shelton and Humphreys (1975) successfully established *Stylosanthes guyanensis* in a growing crop of upland rice without decreasing rice yields. Relay intercropping of pastures with corn is commonly practiced in the Amazon Jungle and in East Africa (Poultney, 1963).

Competition is minimized in relay intercropping by reducing the time during which the two crops are grown together. For this system to be successful, the crop planted during the reproductive stage of another must be tolerant to the shading by the first crop. Figure 12.2 shows that some crops—for example, corn, mung beans, and soybeans—are very sensitive to

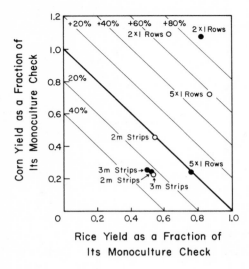

Fig. 12.3 Effect of row arrangement on productivity of corn–rice intercrop (number of rice rows × number of corn rows). Corn population: 60,000 plants/ha (white dots) and 15,000 plants/ha (black dots). *Sources:* Herrera and Harwood (1974) and IRRI (1973).

shading at their early growth stages, whereas others, such as sorghum and cowpeas, are less sensitive. Experiments in Costa Rica have shown that *Phaseolus vulgaris* beans actually benefit from shading at early growth stages (Sanabria, 1975).

Because of the lower period of competition, relay intercropping generally can provide higher relative yield totals than mixed or row intercropping. Herrera and Harwood (1975) obtained higher RYT values with relay intercropping of corn and sweet potatoes with an overlap period of 35 days than by planting both at the same time, which resulted in an overlap period of 95 days.

Strip Intercropping

This pattern occurs when individual crops are grown in the same field in strips wide enough to permit independent cultivation but close enough to produce some agronomic interaction. This practice is more common in the temperate region, where it is used primarily to provide barriers that reduce wind speed and improve water conservation (Radke and Hagstrom, 1976). There is little competition between crops except in the border rows, but the benefits accrued in wind protection and water conservation may give positive RYT values. Figure 12.3 shows examples of a negative effect of strip intercropping of corn and upland rice in contrast with a sharp positive effect from certain row intercropping combinations.

Row Arrangement and Planting Density

Many studies have been conducted on different planting patterns, spacing, and seeding density in an attempt to understand the superior performance of intercropping systems in farmers' fields and also to improve their productivity. Amazingly, only one comparison between mixed and row intercropping has been found in the available literature. Dalal (1974) reported that row intercropping of corn and pigeon peas produced a higher RYT (1.82), as compared with a mixed cropping RYT of 1.57. Most of the available information consists of comparing different patterns of row or relay intercropping.

Experimental work conducted under well-fertilized conditions suggests that the productivity of intercropping increases as the planting densities of the different crops increase. Figure 12.3 shows this effect with upland rice and corn. When both crops were grown in 2 or 3 m wide strips, the values were below 1.0, indicating a negative effect of strip intercropping. As the crops were planted closer together, RYT values increased, reaching very high levels of over 1.7 when two rows of rice were intercropped between one corn row spaced at 1.4 m. Studies in the Philippines and Nigeria have also shown that skipping rows of the tall crops or widening their spacing within certain limits does not decrease their yields (IRRI, 1972; Andrews, 1974). The increased yield per plant is due to border effects and compensates for the decrease in plant population. Changing corn row spacing from 1 to 2 m did not vary the LERs of corn–soybean row intercrops, shown in Figure 12.4.

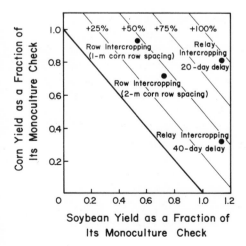

Fig. 12.4 Effect of intercropping corn and soybeans with different row spacings and delays in planting corn into established soybeans. *Source:* IRRI (1973).

This figure also emphasizes one limitation of LER or other relative yield totals: it confounds the proportions of each crop. At 1 m row spacings, corn yields were almost identical to those from single corn stands, while soybean yields decreased by half. At 2 m spacing corn yields were about 70 percent of single-stand values, and soybeans 80 percent. Both LER values were 1.5, but the proportion of the crops varied significantly. Similar results have been obtained with other intercrop combinations by Osiru and Willey (1972), Willey and Osiru (1972), Pinchinat and Oelsligle (1974), and Francis et al. (1976).

Figure 12.4 also shows the advantages of some relay intercropping combinations. In this example soybeans flowered about 30 days after planting. Consequently, in both the 20 and the 40 day relays, corn was planted after the soybeans entered their reproductive phase. Figure 12.4 shows that relaying corn at 20 days after planting the soybeans increased LER values up to 1.9, but relaying at 40 days was not as productive. Francis et al. (1976) in Colombia found that LER values of corn–beans increased from 1.3 to 1.7 as corn planting was delayed from 0 to 15 days after planting the beans. When the planting of beans was delayed from 0 to 15 days after planting the corn, however, the LER dropped from 1.3 to 1.1. These results suggest that the optimum timing combination should be determined locally.

Comparisons between alternate rows, a mixture of both crops in the same row, and double rows suggest that little advantage is obtained by more complicated arrangements than alternate rows (Evans, 1960; Krutman, 1968; Pinchinat and Oelsligle, 1974; Francis et al., 1976). When one crop is used to support another, as with corn and climbing beans, however, special row arrangements designed to provide maximum support, such as double rows, are beneficial (Hildebrand, 1976).

In comparing intercropped with single-stand combinations, the choice of plant density will affect the outcome. When the same planting density is used, the results may be biased in favor of the single stands if the normal single-stand density is used, or in favor of the intercropped treatments if the single stands are planted at densities above or below the optimum. An example of the latter case appears in the work of Soria et al. (1975), where very high RYT values were obtained when the single stands were planted at uncommonly low or high densities. The most practical comparisons are those in which both single stands and intercropped combinations are planted at their optimum densities and spacings. If these parameters are not known, research should be conducted to establish them.

All the above information has been collected under controlled conditions where the competition for water and nutrients is thought to be eliminated by sufficient fertilization and/or irrigation. Few studies are available in which

Table 12.5   Influence of Bean Growth Habit on the Performance of Corn–Bean Intercropping Systems in Chapingo, Mexico

Plant Population (1000 plants/ha)		Bean Type	Grain Yield (tons/ha)		Land Equiv. Ratio	Gross Income (U.S. $/ha)
Corn	Beans		Corn	Beans		
40	0	—	2.45	—		74
0	110	Climbing	—	1.46		90
20	20[a]	Climbing	1.67	0.95	1.33	124
20	60	Climbing	1.47	1.20	1.42	149
20	90	Climbing	1.40	1.30	1.46	158
30	90	Climbing	1.86	1.27	1.62	189
0	110	Bush	—	1.17	1.00	53
20	20	Bush	1.69	0.35	0.98	55
20	60	Bush	1.86	0.55	1.23	91
20	90	Bush	1.85	0.71	1.35	112
30	90	Bush	2.41	0.71	1.58	139
$LSD_{0.05}$			0.41	0.16		33

Source: Adapted from Lepiz (1971).
[a] Farmer's practice.

various increases in plant density are compared under less than optimal conditions. One such work from Mexico, shown in Table 12.5, shows that tripling the density of climbing beans and increasing that of corn by 50 percent more than the farmer's practice increased LER values and almost tripled gross income.

The Effect of Varieties

Using varieties of widely different plant types can affect the productivity of intercropped systems. Climbing-type beans were more productive than bush-type beans at low soil fertility levels in the example shown in Table 12.5. At a high fertility level in Palmira, Colombia, on the other hand, bush beans were superior to climbing beans, perhaps because the former compete less with corn (Francis et al., 1976). The effects of changing the plant type of beans and/or corn are shown in Table 12.6. The best combination was

Table 12.6   Effects of Shift in Plant Type of Both Corn and Bean Varieties on the Productivity of Row Intercropping in Palmira, Colombia

Corn Plant Type	Bean Plant Type	Corn Grain Yield (tons/ha)	Beans Grain Yield (tons/ha)	LER
Normal	Bush	7.6	0.84	1.65
Dwarf	Bush	8.8	0.67	1.44
Normal	Climbing	7.3	0.43	1.30
Dwarf	Climbing	8.1	0.22	1.08

Source: Adapted from Francis et al. (1976).

normal, tall-statured corn with bush beans. Other combinations with dwarf (brachytic) corn or with climbing beans produced lower bean yields, presumably because of less distance between the two crop canopies and consequently more competition.

Crop varieties have traditionally been bred and selected for single stands. Breeding programs aimed specifically at developing improved varieties for intercropped systems have been started in the Philippines (IRRI, 1972), Tanzania (Finlay, 1974), Colombia (Francis et al., 1976), and Nigeria (Andrews, 1974; Okigbo and Greenland, 1976). The specific objectives of such programs as outlined in Francis' paper are as follows: more erect leaf habits of tall crops to reduce competition, earlier maturity, photoperiod insensitivity, and flexibility in responding to different plant populations. The existence of these breeding programs is very encouraging and may have a major impact in improving the productivity of intercropped systems.

SOIL-PLANT RELATIONS IN INTERCROPPING SYSTEMS

The preceding section showed how competition for light among different species can be minimized by sound choices of crop combinations, time of planting, planting density, and row arrangement. Much less is known about relationships below the ground, although a consensus exists that competition for water and nutrients is more frequent and severe than competition for light (Hall, 1974; Trenbath, 1974). Agronomic experience also indicates that water and nutrients are generally more limiting than solar radiation in the tropics. Much of the experimental work described above was conducted under conditions where water and nutrients were not limiting—thus in the apparent absence of competition below ground.

Above-ground and below-ground competition affect each other. A heavily shaded plant will develop a weaker root system that will absorb less water and nutrients than a light-saturated plant. Conversely, a root system that is poorly developed because of the presence of aluminum toxicity or other limiting factors will produce a weak crop canopy.

Very little is known about root development when two or more crops are grown together . Nelliat et al. (1974) observed that roots of coconuts, cacao, cinnamon, and pineapple in mixed intercropping systems of southern India occupy different soil volumes. These crops were spaced widely enough so that little overlap of root systems probably occurred. In row intercropping systems at conventional spacings, the situation is probably different. Studies with crops and weeds (Pavlychenko, 1937) show that parts of the root systems of different species tend to mutually avoid each other, particularly at lower depths, but that they do intermingle close to the surface. Studies with single stands of row crops show a similar degree of mutual avoidance between rows but not within a row (Kurtz et al., 1952; Raper and Barber, 1970; Baldwin and Tinker, 1972). This separation is often related to soil compaction in the furrows, according to Trouse (1975). Intercropping with species of widely different root system configurations may result in stratification of the two root systems at different depths. Harwood and his coworkers at IRRI (1972) observed this effect in corn–peanut row intercroppings. Peanut roots were concentrated in superficial soil layers, whereas corn roots developed at greater depths. Trenbath (1974) mentions that such stratification of root systems may be an expression of root avoidance, which forces the root system that develops later to use greater soil depths.

The possibility of horizontal or vertical root system stratification suggests that this may be another fundamental advantage of intercropping. With roots at different soil layers, water and nutrient uptake may increase in relation to single-crop stands. Earlier work in Illinois (Kurtz et al., 1947, 1952; Bray, 1954) suggested that, when water and mobile nutrients such as nitrate or sulfate are present in sufficient amounts, competition between root systems will be minimal; but if these elements are limiting, competition for them will be strong. Trouse (1975) reported that corn and soybean roots developed freely near each other in well-aerated soils sufficiently provided with water and nutrients; but when the soil was compacted, the root systems became restricted and competition was intense.

Relay intercropping is also advantageous in separating the root systems in terms of time. Most root systems stop developing after flowering, and the uptake of nutrients and water decreases from that point to harvest (Trouse, 1975). Having a second crop developing its root system while the first one is essentially phasing out its activities is a logical way to diminish competition.

Nutrient Uptake

Several studies have compared intercropped combinations with single-crop stands in terms of yields, dry matter, and nutrient uptake: Ibrahim and Kabesh (1971) in Egypt; Kassam and Stockinger (1973) in Nigeria; IRRI (1973, 1974), Palada and Harwood (1974), and Liboon and Harwood (1975) in the Philippines; Dalal (1974) in Trinidad; and North Carolina State University (1974) and Oelsligle et al. (1976) in Costa Rica. The unanimous finding was that the intercropped mixtures extracted more nutrients from the soil than did single stands per unit area of land. The most comprehensive data were obtained by Dalal (1974), who compared corn and pigeon peas (*Cajanus cajan*) in single stands, mixed intercropping, and row intercropping. The corn and pigeon peas had different growth durations, the corn being harvested 112 days after planting and the pigeon peas 168 days. The dry matter and nutrient accumulation patterns shown in Fig. 12.5 illustrate that the difference in growth duration tended to minimize competition and produce smooth sigmoid growth curves for the intercropped mixture. Yield and uptake data at harvest (Table 12.7) indicate that a

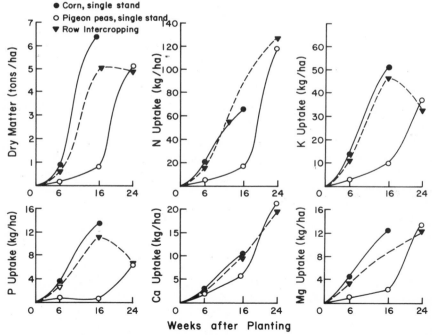

Fig. 12.5 Growth and nutrient uptake patterns of corn and pigeon peas in monoculture and row intercropping in Trinidad. *Source:* Adapted from Dalal (1974).

Table 12.7 Effects of Mixed and Row Intercropping on Yields and Nutrient Uptake of Corn (C) and Pigeon Peas (PP) on River Estate Loam (pH 5.2) in St. Augustine, Trinidad

Parameter	Single Stands		Mixed Intercropping			Row Intercropping		
	C	PP	C	PP	RYT	C	PP	RYT
Grain yields (tons/ha)	3.1	1.9	2.0	1.7	1.54	2.6	1.8	1.78
Total dry matter (tons/ha)	6.4	5.1	4.2	3.8	1.40	5.0	4.9	1.74
N uptake (kg/ha)	66	119	48	100	1.56	54	127	1.88
P uptake (kg/ha)	13	6	9	5	1.52	11	7	2.01
K uptake (kg/ha)	51	37	37	32	1.59	46	33	1.79
Ca uptake (kg/ha)	10	22	10	15	1.68	9	19	1.76
Mg uptake (kg/ha)	12	14	9	8	1.32	9	12	1.61

Source: Adapted from Dalal (1974).

hectare of the mixed intercropping system accumulated from 30 to 63 percent more nutrients than two half-hectares of single stands. Row intercropping was more efficient than mixed intercropping, with RYTs for yield, dry matter, nitrogen, phosphorus, and potassium uptake on the order of 1.8, as compared with RYT values on the order of 1.5 for these parameters in mixed intercropping. Similar results for nitrogen uptake were obtained by Kassam and Stockinger (1973) with a row-intercropped mixture of early-maturing millet and late-maturing sorghum in northern Nigeria.

Higher nitrogen uptake with intercropping has also been recorded for row-intercropped mixtures of crops with similar growth durations, such as corn–upland rice, corn–soybeans, corn–beans, and wheat–horsebeans in the Philippines, Costa Rica, and Egypt. Palada and Harwood (1974; IRRI, 1973) observed that they could not correlate the increased nitrogen uptake of a corn–upland rice mixture with greater light interception, because the light interception was similar to that of the rice single stand. This suggests that increased nutrient uptake of mixtures may be due to favorable interaction of the root systems. Mixed crops may draw nutrients from larger soil volumes than single-crop stands because of some degree of mutual avoidance of root systems or perhaps more root–soil contact in the same volume of soil. If this is the case, the greater efficiency of highly mobile nutrients such as nitrates could be accounted for by root development at greater depths, which could rescue nutrients that would otherwise be leached.

Since many subsoils in the tropics present limitations in water-holding capacity, high levels of exchangeable aluminum, and low levels of available phosphorus, mixtures of crops with different degrees of tolerance to these factors may increase the efficiency of nutrient uptake further. Research on this subject is needed.

The increased efficiency of nutrient uptake is about the same for "mobile" and "less mobile" nutrients. Dalal's results in Table 12.7, as well as those of Ibrahim and Kabesh (1971), indicate simiar RYT values for the more mobile nutrients like nitrogen and potassium and the less mobile elements like phosphorus, calcium, and magnesium. Although there was competition in all cases, as indicated by the decrease in yields of both species in mixtures, this lack of distinction gives further support to the theory that different root systems exploit different soil volumes, as stated in the mobility-competition concepts of Bray (1954) and Kurtz et al. (1947, 1952).

Intercropping with Legumes

The value of mixed intercropping of grass and legume pasture species is well known (see Chapter 13) and leads to the assumption that a similar beneficial effect could be obtained by intercropping cereal or grass crops with grain or forage legumes. Studies on this topic indicate, however, that intercropping with legumes can be beneficial or detrimental, depending on the nitrogen-fixing capacity of the legume, the degree of compatibility or competition between the species, the manner of planting—whether at the same time or relay intercropped, and the fertility level of the soil.

These effects can be better understood by evaluating mixed pasture data in terms of the RYTs of the different components. Hall (1974a) analyzed the results of Vallis et al. (1967) on mixtures of *Chloris gayana* and *Stylosanthes humilis* grown in different plant proportions (Table 12.8). When the grass:legume proportion was 85:15, RYT values were very low denoting little if any beneficial effect of mixed intercropping. When species were grown in equal proportions, RYT values for dry matter and nitrogen uptake increased to about 1.2. When 85 percent of the plants were legumes, maximum RYT values were observed for these two parameters. At this proportion the legume was supplying ample amounts of fixed nitrogen, which produced the maximum intercropping effect in terms of dry matter and protein content. The RYT values from nitrogen uptake derived from the soil (using ^{15}N) suggest strong competition for this factor, unlike total nitrogen, for which there was little competition since the legume was using atmospheric nitrogen. Hall's "relative crowding coefficient products" give a measure of the degree of overall competition. The larger this parameter, the less competition there is for the different factors. Consequently there was strong competition for soil nitrogen ($NH_4^+ + NO_3^-$) but less competition for total nitrogen uptake due to symbiotic fixation, and consequently less competition in terms of dry matter production.

A second study by Hall (1974b) shows the interaction between two other pasture species, in relation to the supply of another nutrient. The grass

*Table 12.8   Effects of Different Proportions of a Grass (*Chloris gayana*)— Legume (*Stylosanthes humilis*) Mixture on Dry Matter, Total Nitrogen Uptake, and the Uptake of Soil Nitrogen Measured with* ^{15}N

Parameter	Proportion of Grass Legume Plants (RYT)			Relative Crowding Coefficient Product (K)
	85:15	50:50	15:85	
Dry matter	1.05	1.20	1.32	3.5
Total N uptake	1.05	1.24	1.45	7.8
N uptake from soil	1.01	1.05	1.12	1.5

Source: Adapted from Hall (1974a).

Setaria anceps and the legume *Desmodium intortum* were grown, with and without potassium fertilization, on a soil strongly deficient in potassium. The results shown in Table 12.9 are given at the optimum proportions of the two species. When potassium was limiting, RYT values for dry matter and nutrient uptake were low, except for nitrogen due to legume symbiosis. In the presence of sufficient quantities of potassium, all RYT values increased to about 1.4. The relative crowding coefficients show that competition for all factors decreased when the potassium supply became adequate. These studies indicate that in order for the legume to be beneficial it must be grown in adequate proportions and in the presence of sufficient available nutrients for both crops.

With these concepts in mind, the apparent contradictions in the literature on legume intercropping can be clarified. The possibility of intercropping clovers, alfalfa, and other legumes with corn was studied in the United States Midwest and discarded. Triplett (1962) found little benefit from intercropping legumes on continuous corn or corn–soybean rotations in Ohio. At low nitrogen levels, intercropping corn with alfalfa increased yields as opposed to single stands of corn. These increases, however, were not enough to reach the desired yield level of 100 bu/acre (6 tons/ha) of corn, which was achieved with heavy nitrogen fertilization. Intercropping corn and alfalfa produced about half the corn yields obtained with the classic corn–wheat–alfalfa rotation used in the area. Kurtz et al. (1947), working in Illinois, observed similar effects with corn–clover intercropping. These were cases where the nitrogen demands for high corn yields could not be met by growing a legume during the short growing season of the northern temperate region.

In sharp contrast, a four-season experiment conducted on udic Alfisols in Ife, Nigeria, by Agboola and Fayemi (1971, 1972) showed sustained corn

*Table 12.9   Effects of Potassium Deficiency or Sufficiency on the Growth and Nutrient Uptake of a Grass (*Setaria anceps*)–Legume (*Desmodium intortum*) Mixture at 15:85 Proportion*

Parameter	Relative Yield Total (RYT)		Coefficient Product	
	K Deficiency	K Sufficiency	K Deficiency	K Sufficiency
Dry matter	1.04	1.40	1.1	5.3
N uptake	1.26	1.44	2.7	6.8
P uptake	1.06	1.30	1.3	4.0
K uptake	0.98	1.38	0.9	4.7

Source: Adapted from Hall (1974b).

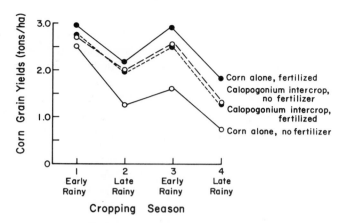

Fig. 12.6 Effects of intercropped legumes on corn yields in four consecutive seasons on an Alfisol from Ife, Nigeria (fertilization: 55 kg h/ha, 10 kg P/ha, 55 kg K/ha). *Source:* Adapted from: Agboola and Fayemi (1972).

yield increases when shade-tolerant legumes like *Calopogonium muconoides,* cowpeas, and mung beans were row-intercropped with corn and incorporated into the soil after the corn and grain legume harvest. The results with *Calopogonoium* are shown in Fig. 12.6, where a fertilizer application of 55-10-55 kg NPK/ha per crop was compared with legume intercropping. The intercropped treatment produced yield levels that were not significantly different from those obtained with the fertilized single stands of corn. Thus the farmers were able to essentially save the investment in fertilizer, at least for the first four crops of corn. The RYT values of the grain yields during the fourth planting—3.11 without fertilization and 1.62 with fertilization—suggest little competition between corn and *Calopogonium* and the other shade-tolerant legumes. Agboola and Fayemi also compared other legumes (*Mucuna utilis* and *Phaseolus lunatus*), but these competed aggressively with corn.

In this region of Africa, researchers have recommended the use of *Mucuna* as a green manure crop between corn plantings as a means for stabilizing yields (Vine, 1953). Farmers have not adopted the practice, however, because they do not appreciate planting a crop with no direct cash value. Intercropping is obviously a better solution because no time is wasted. Agboola and Fayemi (1971) compared corn–green manure sequences with intercropping and found yield benefits from both, but only two corn crops could be grown with sequential cropping, as compared to four with intercropping, during the 2 year study.

In ustic soil moisture regimes where two cereal crops cannot be grown in sequence, intercropping corn with green manures has proved to be even more beneficial. Relay intercropping is the predominant form practiced in these regions in order to decrease competition early and to allow the legume crop to grow into the dry season, utilizing residual moisture. Viegas et al. (1960) conducted four long-term experiments (ranging from 6 to 10 years) in São Paulo, Brazil, with corn–velvet bean (*Stizolobium* spp.) relay intercropping in the presence and the absence of phosphorus, potassium, and lime, but without fertilizer nitrogen applications. They observed positive advantages of velvet bean, particularly with phosphorus and potassium fertilization, on overall average corn yields in three out of the four cases. The negative result was obtained in an area recently cleared of grass–legume pastures, which was probably adequate in available nitrogen.

The contribution of intercropped legumes in terms of nitrogen was estimated in another relay-intercropped experiment conducted in the highlands of Mexico by Peregrina (1965). Three legumes of the genera *Melilotus* and *Vicia* were planted in the corn when it was 2 months old, and were allowed to grow on residual moisture after the corn grain and stover were harvested during the dry season. Right before the rains, the legumes were incorporated into the soil and the following corn crop was planted. Peregrina compared these treatments with annual nitrogen fertilization. His results appear in Fig. 12.7, where the yield of corn intercropped with legumes is plotted along the nitrogen response curves. Intercropping corn with *Melilotus alba* or *Vicia villosa* provided yields similar to those obtained with nitrogen rates of 170 and 94 kg N/ha to the following crop of corn. Although it cannot be ascertained whether this was a direct effect of the nitrogen contribution of legumes, the fact is that excellent results were

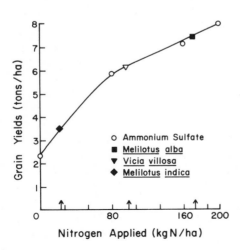

Fig. 12.7 Effects of relay intercropping three annual legumes and their incorporation on yields of the subsequent corn crop, as compared to annual nitrogen applications in La Piedad, Michoacán, Mexico. Arrows indicate the contribution of the legumes, expressed in equivalent rates of ammonium sulfate. *Source:* Adapted from Peregrina (1965).

obtained. The importance of species selection is underscored by the relatively poor behavior of *Melilotus indica,* which produced corn yield increases equivalent to those obtained with only 35 kg N/ha as ammonium sulfate. Similar effects were observed by Pathak et al. (1968) in India, where corn–mung bean intercropping produced higher corn yields than a rate of 90 kg N/ha.

Intercropping and incorporating *Crotolaria juncea* as green manure in sugarcane rows was found to provide as high sugar yields and juice quality as a rate of 168 kg N/ha of ammonium sulfate in Madras, India (Gowda and Mariakulandai, 1972).

In summary, intercropping legumes with cereal crops can be beneficial in terms of growth and nitrogen uptake if the legume species does not seriously compete with the cereal, like the shade-tolerant ones in row intercropping or others with relay intercropping. It is also important that the growth period of the legume be sufficiently long to accumulate considerable amounts of nitrogen fixed from the atmosphere. In some cases the apparent nitrogen contribution is sufficient for high crop yields; in other examples intercropping was successful only at relatively low yield levels. The answer is site-specific and involves selection of crops, timing, and yield expectations.

Response to Fertilizer Applications

Does the more efficient nitrogen uptake of the intercropped system result in a more efficient use of fertilizer? The few studies designed to answer this question suggest that fertilizer nitrogen utilization is indeed more efficient in row-intercropped systems than in comparable single stands. The relative advantage of intercropping varies, however, with different rates of nutrient application. Figure 12.8 shows an example of corn–upland rice row intercropping in the Philippines where the benefits of intercropping increased with increasing rates of nitrogen applications. At the rate of 180 kg N/ha the RYT values reached 1.5, and the return per dollar invested in nitrogen also attained a much higher level than in single-stand plantings.

Other cases show decreasing RYT values with increasing nitrogen rate. Figure 12.8 illustrates this situation with beans intercropped in 2 m corn rows, but not in 1 m rows, in Costa Rica. When corn row spacing was shortened to 1 m, RYT values increased with increasing nitrogen rate. The most efficient return on fertilization, however, was obtained with the 2 m intercrop. This example shows that the shape of the RYT–nitrogen curve depends on spacing configurations and probably other factors, such as variety and plant density. The profitability of the systems will depend on the

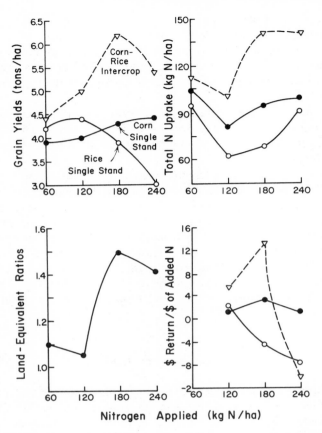

Fig. 12.8 Comparison between an upland rice–corn intercrop and single-crop stands on yields, nitrogen uptake, and nitrogen response in Los Baños, Philippines. *Source:* Adapted from Palada and Harwood (1974).

price ratios of both crops, as well as their relative yields. Consequently LER values or other physical RYTs may give misleading information. Economic interpretation based on actual crop and fertilizer practices is probably the best guideline for decision. A higher RYT value at a certain nitrogen rate may not be as profitable as applying the same amount of nitrogen at different rates to single-stand crops, as was the case with corn in Fig. 12.9.

The nitrogen response of intercropped mixtures of high-responsive and low-responsive crops is likely to give decreasing RYT values with increasing fertilizer rates. Liboon and Harwood (1975; IRRI, 1974) found that the RYT value of a corn–soybean row intercrop was very high without nitrogen

(1.47) but decreased to about 1.10 when nitrogen was applied. Their nitrogen uptake data indicate that soybeans fixed about 125 kg N/ha when no nitrogen was added. A nitrogen application of 60 kg ha stopped nitrogen fixation, resulting in lower LER values. Similar legume growth depression was observed by Pathak et al. (1968) in India and by Bantilan et al. (1974) in the Philippines. This situation did not occur in the corn–beans example of Fig. 12.9, however, since *Phaseolus vulgaris* is a very poor symbiotic nitrogen fixer in Latin America.

Very little work has been conducted with nutrients other than nitrogen. An experiment in liming corn–cowpea intercrops on acid Ultisols of the Amazon Jungle of Peru showed RYT values increasing with liming rates up to the apparently optimum level of 3 tons/ha and decreasing afterwards (Table 12.10). In this case there was no benefit of intercropping without liming, but a substantial benefit at the optimum rate of lime.

The available information is far too limited to determine the reasons why fertilization enhances or diminishes the effect of intercropping. Research comparing the fertilizer response of single stands versus intercrops needs to be increased. The limited data, however, indicate that the benefits of inter-cropping are not restricted to low fertility levels.

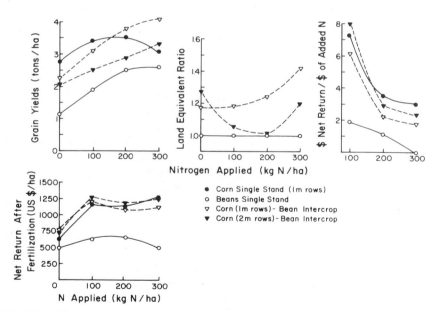

Fig. 12.9 Nitrogen response of corn–beans (*Phaseolus vulgaris*) row intercrops on a Dystrandept in Turrialba, Costa Rica. *Source:* Adapted from Oelsligle and Pinchinat (1973), and North Carolina State University (1974).

Table 12.10   Effects of Liming on the Productivity of Corn–Cowpeas Row Intercrops in Acid Ultisols of Yurimaguas, Peru

| Lime Rate (tons/ha) | Grain Yield (tons/ha) | | | | |
| | Single Stands | | Intercropped | | |
	Corn	Cowpeas	Corn	Cowpeas	RYT
0	0.02	0.36	0.01	0.17	0.93
1	1.26	1.07	0.47	0.63	0.96
2	2.24	1.20	0.89	0.69	0.92
3	1.99	1.16	1.47	0.73	1.37
4	2.41	1.40	1.28	0.72	1.11

Source: North Carolina State University (1974).

Determining Rates of Fertilizer Application

Oelsligle et al. (1976) reviewed the literature on this subject and found no adequate methods in practice for determining fertilizer rates for intercropped systems. The problem is a simple lack of information. They found that many researchers recommend applying the sum of the optimum fertilizer requirements of the individual crops, based on data regarding single-crop responses. This practice is likely to promote good growth, but it is also likely to result in inefficient fertilization. Empirical determinations of fertilizer recommendations have been developed in Mexico and Guatemala for corn–bean intercrops. The recommended amounts range from 50 to 120 kg N/ha and 0 to 80 kg P_2O_5/ha, depending on the soil. The trials in Guatemala were conducted as row intercropping (DelValle 1975). Unfortunately, farmers in this area use relay intercropping practices, which are likely to give different results. Observations in Costa Rica indicate that the fertilizer response of beans relay-intercropped with corn is almost identical to that of single stands of beans (Oelsligle et al., 1976). The lack of much information on this subject is one of the most important gaps in understanding tropical soil management.

When short-term crops are intercropped with a plantation crop like sugarcane or rubber to produce some food for the laborers, the plantation economy dictates that little if any yield reduction of the plantation crop occur. In effect an RYT of about 2 is needed. This can be achieved with crops that differ greatly in growth habits and duration. The approach used is to determine how much additional fertilizer or nutrient is needed to completely eliminate competition. Bhoj and Kapoor (1970) reported that inter-cropping corn with sugarcane was not recommended in parts of India be-cause it decreased the yields of sugarcane. They found that adding an addi-tional 112 kg N/ha and an extra irrigation to the sugarcane–corn intercrop produced the same sugarcane yields and juice quality and increased net pro-fits by 41 percent because of the value of the corn crop. In immature rubber plantations in Malaysia, fertilizer recommendations for corn and peanuts grown between young rubber trees are the same as for single stands of these crops (Pushparajah and Wong, 1970). There is little competition between the slow-growing trees and the food crops.

Sources, Timing, and Placement

There is very little information on how to manage fertilizers in intercrop-ping systems. To determine the appropriate fertilizer sources, timing, and placement, knowledge about how the root systems interact is needed. This writer has been able to find only one study in which this effect of fertilizer placement was studied. Chang et al. (1969) placed ^{32}P- and ^{86}Rb-tagged fertilizers in bands under different crops grown in double-row intercropping in a sandy acid soil of Chianam, Taiwan. In the sugarcane–sweet potato system, when the fertilizers were placed under one crop, the other crop benefited very little from them (Table 12.11). In the sugarcane–peanut system, however, a completely different situation ensued. Regardless of whether the tagged fertilizers were placed under the sugarcane or the peanut rows, the sugarcane absorbed several times more ^{32}P and ^{86}Rb than the peanuts. These results suggest limited competition for band-placed phos-phorus and potassium fertilizers in the sugarcane–sweet potato combina-tion, and an overpowering effect of sugarcane over peanuts. The authors observed a clear case of root stratification in the sugarcane–sweet potato case, with the sweet potato roots remaining superficial, but provided no such observations for the sugarcane–peanut comparison. The latter experi-ment was conducted during a warmer season than the former. Cold weather may have limted root development and, therefore, competition in the sugarcane–sweet potato experiment. Localized placement may be of value in fertilizing one crop in preference to another when the root systems are not competitive, but this practice is not likely to work when one crop is able to effectively utilize fertilizers placed directly under the companion crop.

Common sense dictates that fertilizer applications be timed and placed in such a way that an adequate nutrient supply is available at periods of expected high demand. Unfortunately the review by Oelsligle et al. (1976) mentions no investigations on this subject. Researchers have been making

*Table 12.11   Effects of the Placement of ^{32}P and ^{86}Rb Fertilizers on the
Uptake of Sugarcane, Sweet Potatoes, and Sugarcane–Peanuts at 60 Days
after Seeding on a Sandy Soil (pH 5.3) of Taiwan*

Double rows of sugarcane alternate with two rows of sweet potatoes and
four rows of peanuts.

		^{32}P or ^{86}Rb in Dry Matter (%)	
Element	Placement under rows of:	Sugar cane (2 rows)	Sweet potatoes (2 rows) or Peanuts (4 rows)
^{32}P	Sugarcane	0.035	0.008
	Sweet potatoes	0.002	0.027
^{86}Rb	Sugarcane	0.168	0.026
	Sweet potatoes	0.009	0.171
^{32}P	Sugarcane	0.069	0.012
	Peanuts	0.042	0.015
^{86}Rb	Sugarcane	0.032	0.005
	Peanuts	0.025	0.016

Source: Adapted from Chang et al. (1969).

obvious guesses as to adequate timing, such as applying nitrogen after harvesting the early-maturing crop in order to stimulate the growth of the
later-maturing crop, which then has no further competition. Also, when
nitrogen-fixing legumes such as soybeans are intercropped with cereals, the
sensible practice is to apply most of the nitrogen along the rows of the
cereal crops in order not to inhibit symbiotic nitrogen fixation by the
legumes.

Common sense also indicates that applications of phosphorus or lime
with a considerable residual effect could be incorporated before planting.
For nutrients that are easily leached, such as nitrogen and potassium,
however, top-dressed applications should be used.

More careful reflection on the basic concepts of fertilizing single-crop
stands should make the reader quite uncomfortable with foregoing
generalizations. What happens when one crop is sensitive to aluminum
toxicity, whereas the companion crop is tolerant? How should nitrogen be
applied to a combination of cereal and root crops, such as corn and sweet
potatoes, in which the adequate rate for corn will decrease the tuber:shoot
ratio in sweet potato? Could this problem be avoided by relay intercropping? Banding phosphorus fertilizers in soils very deficient in phosphorus
promotes root proliferation only around the band. Would this increase or

decrease competition? Could yields be further increased by mixing species with different tolerances to a limiting factor? What is the role of slow-release fertilizers?

These questions cannot be answered by extrapolating our knowledge of the fertilizer responses of single crops. The interactions when two or more crops are grown simultaneously are fundamental enough so that answers can be obtained only by doing research on the system in question. The analogy to the mixed pasture system is obvious. If we had pooled our knowledge of how to fertilize grass pastures and legume pastures separately (not only with nitrogen, but with the other nutrients as well) and then guessed at what the mixed system would do, the results would have been pretty poor when compared to our knowledge of mixed grass–legume systems.

Advantages and Limitations of Intercropping

This discussion has shown that certain intercropping systems are more productive than single-crop stands covering the same area because they are capable of more efficient utilization of solar radiation and available nutrients from the soil. In addition, intercropping poses fewer problems of weed control, pests, and diseases. Work in the Philippines shows that certain intercrop combinations repel insects that attack one of the crops (IRRI, 1973, 1974). Intercropping is also believed to decrease the peaks of labor demand, to increase farm income, and to improve the nutritional diet of the farm family (Andrews and Kassam, 1976; Harwood and Price, 1976; Okigbo and Greenland, 1976).

One of the main reasons for intercropping is to reduce the risk of crop failure. If one crop fails, the farmer will have two or three more to harvest. Studies in the Philippines confirm that this is the case when crop failures occur at early stages. The remaining crop then will act as a single stand. If, however, one crop fails at later growth stages because of attack by a disease after flowering, the yield of the companion crop is not able to increase because competition has already reduced its potential (IRRI, 1974).

The primary disadvantage of intercropping is the difficulty in mechanization. This may restrict its widespread use in large farming systems. Relay intercropping, however, presents fewer problems than row intercropping. Some degree of mechanization, primarily with small multiple-use hand tractors, appears feasible as a supplement but not as a substitute for manual labor. Agricultural engineers, however, have not taken a serious look at what can be done. In view of the development of machines for harvesting tobacco, tomatoes, and grapes, it appears unduly pessimistic to assume that intercropped systems too cannot be fully mechanized.

Intensifying crop production has its limits. The more complex five- or six-crop a year systems require so much labor that small farmers may use them only on part of their land (Hildebrand, 1976). Moreover, the most intensive systems may not be the most profitable (IRRI, 1974).

SEQUENTIAL CROPPING SYSTEMS

Double, Triple, or Quadruple Cropping

Sequential cropping is the simplest form of multiple cropping: one crop is planted after the preceding one is harvested. The number of crops per year defines the various forms, such as double cropping, triple cropping, and quadruple cropping. The definition of a calendar year, however, it is not always adequate. In a common sequential system in some irrigated paddy rice areas of tropical Asia, five crops are grown in 2 years.

Sequential cropping is practiced on the more commercialized farms, particularly those with irrigation, although most still fit the small farm definition. Sequential cropping has recently been introduced on large farms of the southern United States, where two crops a year are now harvested on lands that previously produced one crop (Lewis and Phillips, 1976). In tropical regions where wheat grows well during the cooler season and there is enough moisture from rainfall or irrigation to grow other crops potatoes are double- or triple-cropped with corn and grain legumes (Dalrymple, 1971; IARI, 1972). Paddy rice is commonly double cropped with wheat in the northern fringes of ropical Asia on loamy or sandy soils, where a transformation from the puddled to the well-structured aerobic state is feasible in short periods of time (Harwood and Price, 1976). Many kinds of sequential double or triple cropping are also practiced in tropical America in udic soil moisture regimes or in ustic or aridic regimes with irrigation. Bernal (1972) and Pinchinat et al. (1976) list several double-crop combinations, such as corn–corn and corn–soybeans, practiced in Colombia. Triple cropping of upland rice–corn–soybeans appears to be the most promising sequential multiple cropping system for shifting cultivation areas in the Amazon Jungle of Peru (North Carolina State University, 1974). In irrigated lowland rice farms throughout the tropics, where neither water nor temperature is limiting, double or triple cropping of transplanted rice is practiced. Such systems permit utilization of the higher solar radiation during the dry season.

The basic precept of sequential cropping is that the farmer manages only one crop at a time. From the soil management point of few, experience from single-crop stands is not entirely applicable to sequential cropping systems because of the influence of a previous crop on soil physical properties, water, and nutrient availability to the succeeding crop. Soil management practices should be geared to the crop sequence or rotation rather than to individual crops.

Soil Physical Properties

The greater the number of months the soil is protected by a crop canopy, the lower will be the need for tillage operations. Tillage is practiced to improve soil structure and to control weeds. Studies near New Delhi, India, showed that keeping a continuous crop canopy during the year with triple cropping improved soil structure. Table 12.12 indicates the effects of increasing the number of crops per year on the infiltration rates of slowly permeable alluvial soils. Studies at IARI (1972) also showed that the number of necessary tillage operations decreased as the number of crops grown during the year increased. The effects of increasing crop intensity on soil structure will depend on soil properties and management. In strongly aggregated soils such as Oxisols, such effects may be of less importance than in poorly aggregated soils such as the alluvial soils low in organic matter in Table 12.12. The quality of management will also exert a marked influence.

A major advantage of increasing the number of crops per year is the decrease in weed infestation. Weeds have less time to grow and compete when the time between harvest and subsequent planting is minimized. Minimum or no tillage had been used advantageously in double cropping

Table 12.12 Effects of Increasing the Number of Crops per Year on the Infiltration Rates of an Alluvial Soil near New Delhi, India: Tillage after Land Preparation is Indicated

Crops per Year	Infiltration Rate (cm/hour)
2: Corn-wheat (cultivated)	0.6
3: Mung–corn–wheat (cultivated)	1.2
3: Mung–corn–wheat (no additional cultivation)	1.7
4: Mung–corn–toria[a]–wheat (relay intercropped, no additional cultivation)	2.7

Source: IARI (1972).

[a] *Brassica campestris.*

systems. Sanford et al. (1973), working on a Mollisol from Mississippi, found that planting sorghum or soybeans after wheat with a device that disturbed only a 5 cm wide and 10 cm deep strip of soil per row produced yields of all crops similar to or better than those obtained with conventional cultivation, provided that weed control was adequate. Contact herbicides with short activity periods and no residual effects are successfully used in no-till systems (Lewis and Philipps, 1976). No-till systems are likely to fail even in the absence of weeds, however, if the machine harvesting of the first crop compacts the soil severely (Trouse, 1975). In such cases additional tillage is required.

Sequential cropping in areas where a second crop depends almost entirely on residual soil moisture also benefits from decreasing the number of tillage operations. In calcareous sandy loams of the Punjab, India, growing a legume before planting cotton was beneficial, but plowing the legume residues as green manures was detrimental because a significant quantity of moisture was lost in the process. Gautam et al. (1964) recommended cutting the crop residue and using it in other fields where moisture was not limiting. Singh (1967) found it more practical to apply extra nitrogen to compensate for the effect of incorporating green manures than to plow the soil again.

When sequential systems include a paddy rice crop, the management of physical properties consists of destroying aggregation during the puddling process and restoring the structure for growing other crops. The choice of crop species to follow paddy rice in double cropping systems depends on their relative tolerance to the poor aeration caused by puddling, and on the texture and mineralogy of the soil. Harwood (1975) developed a soil classification system that defines the multiple cropping potentials of puddled rice soils. Table 12.13 shows that the increase in bulk density with puddling is a useful parameter to determine the double cropping potential of soils. Peanuts and corn can be double-cropped after puddled rice only in coarse-textured soils, whereas soybeans, mung beans, and cowpeas can tolerate a wider range of poor soil physical conditions. Vine crops such as watermelon and squashes are the most tolerant when they are transplanted on the puddled soil with mounds made around each plant to improve soil aeration.

Yields and Soil Chemical Properties

An example of intensive sequential cropping is given in Table 12.14, based on the work of Nair and Singh (1971) and Nair et al. (1973 a, b) in northern India. Three crops were grown per year with little time between crops. The fact that very little difference was observed between upland and puddled rice indicates an efficient change in soil structure after rice growth. Estimates of photosynthetic efficiency (i.e., the actual amount of solar radiation

Table 12.13   Soil Classification Categories that Indicate Multiple Cropping Potential in Puddled Rice Soils Having Limited Water

	Soil Textural Class			
	1	2	3	4
(2:1 Clay type)	Sandy loam	Silt loam	Clay loam	Clay
(1:1 Clay type)	Silt loam	Clay loam	Clay	
	Increase in Bulk Density When Puddled (%)			
	< 4	4.1–8	8.1–12	> 12
	Crop Potential[a] After Paddy Rice (Subject to Water Availability)			

Peanut
Corn
Sorghum
Soybeans
Rice (transplanted)
Mung
Cowpea
Intensive
  vegetables
Vine crops

Source: Harwood (1975).

[a] —— Grows well; – – – grows with difficulty.

used in photosynthesis) were on the order of 2 percent, which is similar to the level attained in the best single-crop systems in the temperate region. Nair and his coworkers attained this level with three crops a year instead of one. In spite of the high amounts of nutrient removal there were no appreciable changes in soil organic carbon, total nitrogen, and available phosphorus and potassium. Nair et al. (1973a) attribute the stable levels of organic matter to annual additions of 4 to 6 tons/ha of roots and stubble. Annual fertilization rates were quite high, as shown in Table 12.14, but were closely balanced by crop removal of nitrogen and phosphorus. This was possible because of the probably low phosphorus fixation capacity of this soil. In the case of potassium, crop removals exceeded annual additions, but the rate of release of nonexchangeable potassium to available forms was believed to be high in this Mollisol.

The preceding example illustrates almost ideal conditions in terms of excellent soils, good management as judged by photosynthetic efficiency, and intensive use of fertilizers. A much longer study on a light-textured alluvial soil from Taiwan shows similar results at a lower level of management. Table 12.15 summarizes the data of Lin et al. (1973) on the double cropping

Table 12.14 Annual Production and Use of Solar Radiation in Heavily Fertilized Triple Cropping Systems in a Mollisol of Pantnagar, U.P., India: Average Figures of 2 years
Nutrient data are the average of puddled and upland rice systems.

Cropping Sequence	Grain or Tuber Yield (tons/ha)			Total Cropping Duration (days)	N (kg/ha)		P (kg/ha)		K (kg/ha)	
	Crop 1	Crop 2	Crop 3		Fertilizer Added	Crop Removal	Fertilizer Added	Crop Removal	Fertilizer Added	Crop Removal
Upland rice–wheat–millet	5.1	6.4	1.3	339	310	323	74	84	98	296
Puddled rice–wheat–millet	4.8	6.2	1.5	343						
Upland rice–wheat–mung bean	5.2	5.7	1.3	342	337	401	94	89	83	290
Puddled rice–wheat–mung bean	5.2	5.6	1.4	345						
Upland rice–potato–wheat	5.4	11.7	4.5	311	360	383	96	98	150	342
Puddled rice–potato–wheat	4.9	18.5	4.4	315						

Source: Adapted from Nair et al. (1973a, b).

Table 12.15 *Effects of Continuous Double Cropping of Lowland Rice for 48 Years in an Alluvial Soil of Taiwan on Yields and Soil Chemical Properties*

Fertility Treatment[a]	Average Rice Yield per Crop (1924–1972) (tons/ha)	pH	Organic C (%)	Total N (%)	N Mineralized in 10 weeks (ppm)	Bray 1 P (ppm)	Exch. K (ppm)
					Topsoil Properties in 1972		
Check	1.63	5.2	2.1	0.15	143	34	36
NPK	2.55	5.3	2.2	0.17	156	102	48
NPK + lime	2.70	5.8	2.3	0.15	157	81	42
Animal manure	2.75	5.3	2.4	0.20	181	55	39
Animal manure + P	2.85	5.4	2.6	0.21	182	97	43
Green manure	2.56	5.2	2.3	0.11	190	40	40
Green manure + P	2.59	5.4	2.8	0.19	157	81	42
LSD$_{0.05}$	0.15						

Source: Adapted from Lin et al. (1973).
[a] 95 kg N/ha, 41.5 kg P/ha, 79 kg K/ha per crop. Nitrogen contents in animal and green manures were also 95 kg N/ha per crop.

of paddy rice in Taiwan for 48 years with 88 crops (no cropping during World War II). Average rice yields were similar between fertility treatments, with the same amount of nitrogen added. The effects on chemical soil properties were also similar and suggest an equilibrium level without major differences among treatments. There is little question that adequately managed sequential cropping systems can maintain production almost indefinitely.

On the other hand, double or triple cropping sequences involving sweet potatoes, taro (*Colocasia esculenta*), peanuts, sorghum, and cowpeas, conducted in New Britain on volcanic alluvial soils, showed progressive yield decreases with time. Bourke (1974) reports that these decreases were related to decreases in soil fertility parameters. No fertilizers were used; but when cropping was alternated with 1.5 years of legume green manure, the fertility decline was delayed. Intensive cropping of this nature requires fertilization to sustain long-term sequential cropping.

Residual Effects of Fertilization

In sequential cropping systems the nature one crop and the fertilizers applied to it will probably affect the performance of the crop that follows.

Raheja et al. (1971) compiled the long-term residual effects of uniform corn–wheat double cropping trials conducted throughout India. Approximately one-third of the 66 kg N/ha applied to corn was utilized by the succeeding wheat crop. In a triple-cropped wheat–mung bean–corn sequence near New Delhi, Oza and Subbiah (1973) found with the use of ^{15}N that wheat recovered approximately 40 percent of the added nitrogen but that the succeeding mung and corn crops recovered only about 1 percent of the nitrogen added to the wheat crops. Their study suggests that most of the remaining nitrogen remained in the soil as immobilized organic nitrogen. A third example from India on the residual effects of nitrogen applications to rice on a succeeding soybean crop was reported by Reddi et al. (1973) for a sandy soil from Tirupati. They showed that soybean yields increased from 1.3 to 1.9 tons/ha when the nitrogen applications to the preceding rice crop increased from 0 to 180 kg N/ha. The residual fertilizer nitrogen, however, decreased nodulation in the soybeans, although the overall effect was positive; a 40 percent yield increase.

These three examples suggest that the residual effect of nitrogen fertilization is affected by many variables and is, therefore, site-specific. The nitrogen rate applied, the recovery of added fertilizer by the first crop, leaching, immobilization, denitrification, and rainfall pattern are likely to affect the magnitude of the residual effect. It appears, however, that some residual effects can be expected and should be considered in fertilizing the succeeding crop.

The residual effects of other mobile nutrients are likely to be as varied as those of nitrogen, but the residual effects of liming and phosphorus can be predicted with reasonable accuracy on the basis of the discussions in Chapters 7 and 8. For example, sequential double and triple cropping experiments in an acid Ultisol in Yurimaguas, Peru, showed that initial applications of 60 kg K_2O/ha lasted for about 1 year, after which extreme potassium deficiency reduced rice, corn, and soybean yields drastically. In contrast initial applications of 4 tons/ha of lime or 172 kg P/ha had a completely residual effect for the first six or seven consecutive harvests (North Carolina State University, 1974).

Effects of Preceding Crops

The preceding crop species per se can have a beneficial or detrimental effect on the performance of the succeeding crop. The well-known beneficial effects of a preceding legume crop on a cereal are also found in these forms of multiple cropping, but the magnitude of the effects varies with the way legumes are harvested and with legume species.

In a review of legume research in Colombia, Bernal (1972) reported the results of two sequential cropping experiments conducted on an alluvial soil near Medellin and on a Mollisol in the Cauca Valley. The average corn yields of the Medellin experiment are shown in Fig. 12.10, where double-cropped corn is compared with corn–soybeans (harvested for grain) and corn–*Dolichos lablab,* a legume incorporated as green manure before flowering. Corn in monoculture responded positively to nitrogen applications. Corn preceded by soybeans yielded about 0.5 to 1.0 ton/ha more than corn preceded by corn and also responded positively to nitrogen applications. Corn preceded by the incorporated green manure crop produced the highest average yields without any fertilizer nitrogen applications. This is a good example of the effect of green manuring, which in this case appeared to be equivalent to 100 kg N/ha applied to corn. The savings in nitrogen, however, must be weighed against the profit from growing a second crop of corn or a crop of soybeans.

The temporary effect of green manure incorporations, such as the *Dolichos lablab* example, should be emphasized. In a thorough review of the effect of green manures in South Africa, Haylett (1961) concluded that the beneficial effects last only for one or two succeeding crops.

In the Cauca Valley experiments, the effects of soybean or alfalfa crops were evaluated for the first and second successive corn crops fertilized with

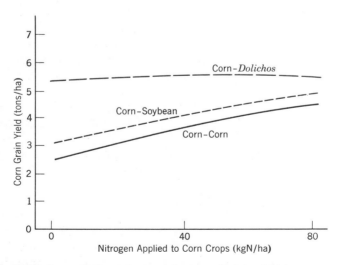

Fig. 12.10 Effect of preceding crop (corn, soybeans, or *Dolichos lablab* green manure) on the nitrogen response of corn in an alluvial soil of Medellin, Colombia (means of nine years). *Source:* Bernal (1972).

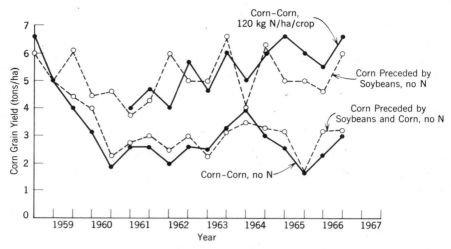

Fig. 12.11 Long-term effects of preceding crops and nitrogen applications on double-cropped crop yields in a high-base-status soil of the Cauca Valley of Colombia. *Source:* Bernal (1972).

0 or 120 kg N/ha in a soil where only nitrogen was deficient. The results shown in Fig. 12.11 indicate a sharp yield decline in unfertilized double-cropped corn monoculture within the first 2 years and a leveling off at yields of 2.8 tons/ha afterwards. When a corn–soybean or a corn–alfalfa (harvested for hay) sequence was used, the corn yields without nitrogen averaged about 5.4 tons/ha—almost identical to those for corn monoculture fertilized at 120 kg N/ha per crop (Table 12.16). When two corn crops were grown successively, the effect of the soybean crop disappeared, but that of alfalfa was still evident, although at a lower yield level. Bernal's report shows that soybean residue decomposition could supply the equivalent of 120 kg N/ha of nitrogen fertilizer to one, but not two, subsequent corn crops, while the effect of alfalfa could last longer but lacked the advantage of a cash crop like soybeans.

The effects of preceding crops are not limited to sequential multiple cropping systems. Jones (1974) observed striking differences in Samaru, Nigeria, where only one crop a year is possible because of the strong ustic soil moisture regime. Table 12.17 shows the corn yields at three nitrogen rates, as affected by whether the previous rainy season crop was peanuts, cowpeas, or sorghum. Corn yields were higher when preceded by peanuts than by cowpeas, and also higher when preceded by cowpeas than by sorghum. The differences were larger without nitrogen applications and decreased at the apparently optimum application level of 84 kg N/ha. Table 12.18 shows that the preceding peanut crop produced a higher nitrogen

Table 12.16   Effects of Preceding Crop and Nitrogen Fertilization on Corn Yields in Double Cropping Experiments Conducted for 9 years on Mollisols of the Cauca Valley of Colombia

Preceding Crop(s)	Average Corn Yield (tons/ha)	
	0 kg N/ha Added to Each Crop	120 kg N/ha Added to Each Crop
Corn	2.8	5.7
Soybeans	5.3	6.5
Soybeans and corn	3.0	5.2
Alfalfa	5.5	6.1
Alfalfa and corn	4.5	6.4

Source: Bernal (1972) and Gomez (1968).

uptake by corn, which was related to higher soil nitrogen mineralization rates, presumably because of residue decomposition. The difference between peanuts and cowpeas is probably related to the greater amount of root residues in the former crop. Similar differences have been observed in experiments in progress at IITA, where the beneficial effect of the grain–legume preceding crop is a function of the amounts of residues left after the grain harvest.

The effect of the preceding crop can be negative when it is a fast-growing graminea which depletes the soil of inorganic nitrogen or exhausts soil moisture reserves. Detrimental effects have also been observed with grain legumes. Harwood and his coworkers in the Philippines observed that

Table 12.17   Effects of Previous Crops on Yields and Nitrogen Responses of Corn in Ustalfs from Samaru, Nigeria

Crop Grown during Previous Rainy Season	Corn Yield (tons/ha)		
	Nitrogen Added to Corn (kg/ha)		
	0	84	168
Peanuts	3.4	5.4	5.2
Cowpeas	2.7	5.2	4.1
Sorghum	2.4	4.9	4.0

Source: Adapted from Jones (1974).

Table 12.18   Nitrogen Uptake by Corn at 12 Weeks of Growth and Soil Nitrogen Mineralized after 5 Weeks' Incubation in the Zero Nitrogen Treatments of Table 12.17

Crop Grown during Previous Rainy Season	Plant N Uptake (kg/ha)	Soil N Mineralized (kg/ha)
Peanuts	78	62
Cowpeas	50	50
Sorghum	48	40

Source: Adapted from Jones (1974).

farmers seldom planted corn or cowpeas after mung beans, or even a second mung bean crop. Experiments proved that mung beans have a depressing effect on yields, particularly at low levels of nitrogen (IRRI, 1973; Herrera and Harwood, 1975). The results are shown in Fig. 12.12. Apparently mung beans secrete certain toxins yet to be identified which depress growth. Many of these relationships are known by farmers, but the reasons should be quantified by researchers.

Ratoon Cropping

Ratoon cropping is another form of sequential cropping. It consists of cultivating the regrowth of a crop previously harvested; thus there is no need for additional land preparation, and the regrowth takes partial or total advantage of an existing root system. This system has been thoroughly reviewed by Plucknett et al. (1970). Regrowth originates from basal buds in the stem or crown and is encouraged by certain cutting heights, irrigation, and nitrogen fertilization. Although the cultivation of sugarcane, pineapples, pasture species, and even bananas is based on this principle, ratoon cropping fits the multiple cropping concept in the case of short-term crops such as sorghum, rice, millet, and cotton, in which more than one crop can be grown in a year.

In rice, a ratoon crop growing from a healthy first crop normally produces two-thirds of the first-crop yields in about two-thirds of the time. Cutting the straw of the original crop close to the ground and either removing it or arranging it along the rows is essential for successful ratoon growth. Nitrogen fertilization before or shortly after cutting is also essential to produce good ratoon rice yields, since the time lag between tillering and panicle initiation is very short.

The success of ratoon cropping depends on an adequate temperature and moisture regime following the original crop harvest, the way the crop is harvested, and, in most cases, the degree of insect, disease, and weed con-

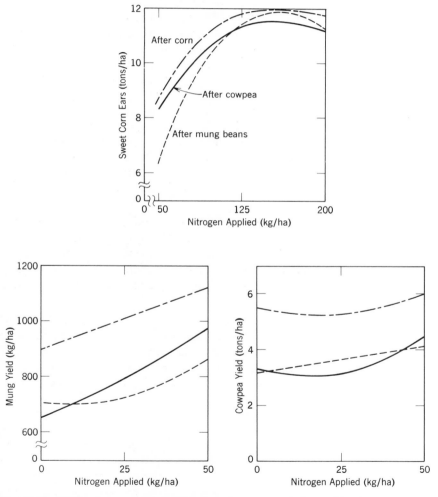

Fig. 12.12 Yields of sweet corn, mung beans, and cowpeas as influenced by nitrogen level and preceding crop. *Source:* IRRI (1973).

trol of the first crop. Insect, disease, and weed problems are compounded in the ratoon crops.

The potential of ratoon cropping in the tropics can be visualized by the work of Bradfield (1970) in a trial conducted on a paddy field on a fertile montmorillonitic soil in Los Baños, Philippines. Bradfield harvested four crops a year by planting rice in unpuddled condition at the start of the rainy season, followed by planting a grain–sorghum hybrid, out of which he obtained one main crop and two ratoon crops with irrigation during the dry season. The four crops yielded a total of 22 tons/ha a year of grain. The sorghum ratoon crops matured 10 to 15 days earlier than the main crop. To this writer's knowledge this annual grain production figure has been exceeded only by a 28 ton/ha a year yield produced with quadruple cropping of irrigated, puddled rice, also at the same location (IRRI, 1969).

SUMMARY AND CONCLUSIONS

1. The bulk of food consumed in most tropical countries is produced in small farming systems characterized by landholdings of a few hectares, limited mechanization, and a wide variety of multiple cropping systems whereby several crops, often grown simultaneously in the same field, are harvested in a year.

2. Multiple cropping systems are divided into simultaneous and sequential cropping. Intercropping is defined as simultaneous cropping where crop intensification is in time and space. Four main forms of intercropping are mixed, row, relay, and strip intercropping. Sequential cropping systems involve intensification in time, such as double, triple, quadruple, or ratoon cropping.

3. Many intercropping systems are more productive than growing the same crops in single stands because higher yields per hectare are obtained than with two half-hectares of single stands. Relative yield totals define the relative advantage or disadvantage of intercropped systems. Successful intercropping systems are more efficient because of (1) better utilization of available solar radiation, (2) higher efficiency in utilizing soil or fertilizer-applied nutrients because of unknown reasons involving root competition versus complementarity, (3) fewer problems with weeds, pests, and disease control, and (4) better use of available manual labor and other low-energy technology. These successful systems have been selected by farmers from a wide variety of combinations, most of which are not more productive than single-crop stands. The main limitation of intercropping appears to lie in the difficulty of mechanization.

4. When more than one crop is grown at the same time, soil–plant relationships are not well understood. The limited evidence available suggests a higher uptake of nutrients when two separate root systems tap different soil areas or do not compete aggressively with each other for nutrients. Competition for easily mobile nutrients like nitrate is believed to be more critical than for nutrients like phosphorus. Intercropping green manures with cereal crops appears to be an effective way for growing green manures at the same time that a cash crop is grown. The interactions between crops of different nutritional demands are not well understood. Little is known and much needs to be learned about how to fertilize effectively intercropped systems.

5. Sequential cropping systems also make better use of available solar energy, high temperatures, and available moisture in tropical areas by growing two or more crops a year, one after another. Most of the basic knowledge obtained from single-crop stands is applicable for the individual crops, but the effects of the preceding crop and the short time interval between harvesting and the next planting affect soil management practices.

6. The number of tillage operations needed to promote good soil structure and control weeds decreases as the number of days that the soil is protected by a crop canopy increases. In the case of puddled rice, the kinds of crops that can be grown in succession depend on the degree of soil structure regeneration and the crop tolerance to poor soil aeration.

7. The effects of intensive cropping on soil chemical properties are usually beneficial when the crops are properly fertilized. The residual effects of nitrogen fertilization of preceding crops vary according to species, soil properties, and management. The residual effects of phosphorus and liming are along conventional knowledge.

8. Alternating a cereal and a grain legume provides positive results and considerable savings in nitrogen fertilizers in double-cropped systems. The beneficial effect of the legume residues after the grain is harvested is usually limited to one succeeding crop. There are certain incompatible successions associated with toxins secreted by some grain legumes, such as mung beans.

REFERENCES

Abeyratne, E. L. F. 1956. Dryland farming in Ceylon. *Trop. Agriculturalist* **112**:191–229.

Agboola, A. A. and A. A. Fayemi. 1971a. Preliminary trials on the intercropping of maize with different tropical legumes in Western Nigeria. *J. Agr. Sci.* **77**:219–225.

Agboola, A. A. and A. A. Fayemi. 1971b. Effects of interplanted legumes and fertilizer treatments on the major soil nutrients. *Int. Symp. Soil Fert. Eval. Proc.* (*New Delhi*) **1**:529–540.

Agboola, A. A. and A. A. Fayemi. 1972. Effects of soil management on corn yields and soil nutrients in the rainforest zone of Western Nigeria. *Agron. J.* **64**:641–644.

Aiyer, A. K. Y. N. 1949. Mixed cropping in India. *Indian J. Agr. Sci.* **19**:439–543.

Alvim, R. and P. T. Alvim. 1969. Efeito da densidade de plantio no aproveitamento de energía luminosa pelo milho (*Zea mays*) e pelo feijão (*Phaseolus vulgaris*) em culturas exclusivas e consorciadas. *Turrialba* **19**:383–393.

Anderson, E. 1950. An indian garden in Santa Lucía, Guatemala. *Ceiba* **1**:97–103.

Anderson, E. and L. O. Williams. 1954. Maize and sorghum as a mixed crop in Honduras. *Ann. Missouri Bot. Garden* **41**:213–215.

Andrews, D. J. 1972. Intercropping with sorghum in Nigeria. *Exptal. Agr.* **8**:139–150.

Andrews, D. J. 1974. Responses of sorghum varieties to intercropping. *Exptal. Agr.* **10**:57–63.

Andrews, D. J. and A. H. Kassam. 1976. The importance of multiple cropping in increasing world food supplies. In R. I. Pappendick et al. (eds.), *Multiple Cropping Symposium*. American Society of Agronomy (in press).

ASPAC. 1974. *Multiple Cropping Systems in Taiwan*. ASPAC and Food Fertilizer Technology Center, Taipei, Taiwan. 77 pp.

Bains, S., S. Dayanand, and K. N. Singh. 1970. A note on the relative performance of different intercrops in sugarcane. *Indian J. Agron.* **15**:86.

Baldwin, J. P. and P. B. Tinker. 1972. A method for estimating the lengths and spacial patterns of two interpenetrating root systems. *Plant and Soil* **37**:209–213.

Bantilan, R. T., M. C. Palada, and R. R. Harwood. 1974. Integrated weed management. 1. Key factor affecting crop–weed balance. *Weed Sci. Bull. Philip.* (in press).

Bernal, J. 1972. Las leguminosas como fuentes de nitrógeno en pastos y rotaciones. *Suelos Ecuatoriales* **4**:175–194.

Bhoj, R. L. and P. C. Kapoor. 1970. Intercropping of maize in spring planted sugarcane gives high profits with adequate nitrogen use. *Indian J. Agron* **15**:242–246.

Blencowe, E. K. and J. W. Blencowe (eds.). 1970. *Crop Diversification in Malaysia*. Incorporated Society of Planters, Kuala Lumpur. 300 pp.

Blencowe, J. W. and J. K. Templeton. 1970. Establishing cocoa under rubber. pp. 286–295. In E. K. Blencowe and J. W. Blencowe (eds.), *Crop Diversification in Malaysia*. Incorporated Society of Planters, Kuala Lumpur

Bourke, R. M. 1974. A long term rotation trial in New Britain, Papua, New Guinea. Unpublished paper, Lowlands Agricultural Experiment Station, Keravat, East New Britain, Papua, New Guinea. 18 pp.

Bradfield, R. 1964. Some unconventional views of rice culture in Southeast Asia. Unpublished seminar paper, International Rice Research Institute, Los Baños, Philippines. 25 pp.

Bradfield, R. 1969. Training agronomists for increasing food production in the humid tropics. *ASA Spec. Publ.* **15**:45–64. American Society of Agronomy, Madison, Wisc.

Bradfield, R. 1970. Increasing food production in the tropics by multiple cropping. Pp. 229–242. In D. G. Aldrich (ed.), *Research for the World Food Crisis*. American Association for the Advancement of Science, Washington.

Bradfield, R. 1972. Maximizing food production through multiple cropping systems centered on rice. Pp. 143–163. In *Rice, Science and Man*. International Rice Research Institute, Los Baños, Philippines.

Bray, R. H. 1954. A nutrient mobility concept of soil-plant relationships. *Soil Sci.* **78**:9–22.

Byrd, H. W. 1967. Effects of breaking over corn plants in Brazil on dry matter accumulation, germination and vigor of kernels. *Fitotecnia Latinoamer.* **4**:109–123.

CATIE. 1974. Conferencia sobre sistemas de producción agrícola para el trópico. Informe final. Centro Agronómico Tropical de Investigación y Enseñanza, Turrialba, Costa Rica.

Chang, H. 1965. Rotations and intercropping systems of sugarcane in Taiwan. *Taiwan Sugar* **12**:1–6.

Chang, H. 1971. Relationships between net radiation, soil temperature and sugarcane growth in an intercropped field. *Taiwan Agr. Quart.* **7**:123–138.

Chang, H., C. H. Chang, and F. W. Ho. 1969. Competition between sugarcane and intercrops for fertilizer tagged with P^{32} and Rb^{86}. *J. Agr. Assoc. China* **63**:43–49.

Chundawat, G. S. 1972. Note on the effect of phosphate fertilization and legume, non-legume component on nitrogen reserve of soil. *Indian J. Agr. Res.* **6**:167–168.

Cordero, A. and R. E. McCollum. Unpublished data. Soil Science Department, North Carolina State University, Raleigh.

Dalal, R. C. 1974. Effects of intercropping maize with pigeon peas on grain yields and nutrient uptake. *Exptal. Agr.* **10**:219–224.

Dalrymple, D. G. 1971. Survey of multiple cropping in less developed nations. *U.S. Dept. Agr. Foreign Economic Devel. Service FEDR* 12. Washington.

DeGuzman, M. R. 1974. Pasture and fodder production under coconuts. *ASPAC Ext. Bull.* 45. Taipei, Taiwan. 29 pp.

DelValle, R. 1975. Efecto de fertilización con NPK en el sistema maíz-frijol asociado bajo condiciones del Valle de Monjas. Tesis Ing. Agr., Universidad de San Carlos, Guatemala. 41 pp.

Enyi, B. A. C. 1973. Effects of intercropping maize or sorghum with cowpeas, pigeon peas and beans. *Exptal. Agr.* **9**:83–90.

Evans, A. C. 1960. Studies of intercropping. I. Maize or sorghum with groundnuts. *East Afr. Agr. For. J.* **26**:1–10.

FAO. 1973. *Production Yearbook.* P. 8. Food and Agricultural Organization of the United Nations, Rome.

Finlay, R. C. 1974. Intercropping research and the small farmer in Tanzania. In *Field Staff Symposium.* International Research and Development Centre, Ottawa.

Francis, C. A., C. A. Flor, and S. R. Temple. 1976. Adapting varieties for intercropping systems in the tropics. In R. I. Pappendick et al. (eds.), *Multiple Cropping Symposium.* American Society of Agronomy (in press).

Gautam, O. P., V. H. Shah, and K. P. M. Nair. 1964. Agronomic investigations with hybrid maize. II. Studies of intercropping, row spacing and method of phosphorus application with hybrid maize. *Indian J. Agron.* **9**:247–254.

Gehrke, M. R. 1962. Distribution of absorbing roots of coffee and rubber in mixed plantings in two ecological zones of Costa Rica. M.S. Thesis, Instituto Interamericano de Ciencias Agrícolas, Turrialba, Costa Rica. 105 pp.

Gomez, J. 1968. Rotación y rendimiento de maíz. *Agr. Tropical (Colombia)* **24**:204–220.

Gowda, B. K. L. and A. Mariakulandai. 1972. Intercropped green manures vs. green leaf manuring on sugarcane yields. *Madras Agr. J.* **59**:312–317.

Grimes, R. C. 1963. Intercropping and alternate row cropping of cotton and maize. *East Afr. Agr. For. J.* **28**:161–163.

Hall, R. L. 1974a. Analysis of the nature of interference between plants and species. I. Concepts and extension of the DeWit analysis to examine effects. *Aust. J. Agr. Res.* **25**:739–747.

Hall, R. L. 1974b. Analysis of the nature of interference between plants and species. II. Nutrient relations in a Nandi *Setaria* and Greenleaf *Desmodium* association with particular reference to potassium. *Aust. J. Agr. Res.* **25**:749–756.

Hart, R. D. 1974. The design and evaluation of a bean, corn and manioc polyculture cropping system for the humid tropics. Ph.D. Thesis, University of Florida, Gainesville. 158 pp.

Hart, R. D. 1975. A bean, corn and manioc polyculture cropping system. I. The effect of interspecific competition on crop yield. *Turrialba* **25**:294–301.

Harwood, R. R. 1975. Farmer-oriented research aimed at crop intensification. Pp. 12–32. In *Proceedings of the Cropping Systems Workshop*. International Rice Research Institute, Los Baños, Philippines.

Harwood, R. R. and E. C. Price. 1976. Multiple cropping in tropical Asia. In R. I. Pappendick et al. (eds.), *Multiple Cropping Symposium*. American Society of Agronomy.

Haylett, D. G. 1961. Green manuring and soil fertility. *South Afr. J. Agr. Sci.* **4**:363–378.

Herrera, W. A. T. and R. R. Harwood. 1973. *Crop Interrelationships in Intensive Cropping Systems*. Saturday Seminar, International Rice Research Institute, Los Baños, Philippines. 26 pp.

Herrera, W. A. T. and R. R. Harwood. 1974. Effect of plant density and row arrangement on productivity of corn-rice intercrop. Unpublished paper, International Rice Research Institute, Los Baños, Philippines. 7 pp.

Herrera, W. A. T. and R. R. Harwood. 1975. *Agronomic Results in the Intensive Upland Rice Area in Batangas*. Saturday Seminar, International Rice Research Institute, Los Baños, Philippines. 40 pp.

Hildebrand, P. E. 1976. Multiple cropping makes "dollars and sense" agronomy. In R. I. Pappendick *et al* (eds.), *Multiple Cropping Symposium*. American Society of Agronomy (in press).

Hildebrand, P. E. and E. C. French. 1974. *Un Sistema Salvadoreño de Multicultivos: su Potencial y sus Problemas*. Departamento de Economía Agrícola, CENTA, Santa Tecla, El Salvador. 23 pp.

Hunter, J. R. and C. Camacho. 1961. Some observations on permanent mixed cropping in the humid tropics. *Turrialba* **11**:26–33.

IARI. 1972. Recent research on multiple cropping. *Indian Agr. Res. Inst. Res. Bull.* 8 (New Series).

Ibrahim, N. E. and M. O. Kabesh. 1971. Effect of associated growth on the yield and nutrition of legume and grass plants. I. Wheat and horsebeans mixed for grain production. *United Arab Republic J. Soil Sci.* **11**:271–283.

IICA. 1974. Bibliografía sobre sistemas de agricultura tropical. IICA-CIDIA Documentación e Información Agricola No. 27, Instituto Interamericano de Ciencias Agrícolas, Turrialba, Costa Rica.

IRRI. 1969. Agronomy section. P. 109. Annual Report. International Rice Research Institute, Los Baños, Philippines.

IRRI. 1972. Multiple cropping. Pp. 21–35. Annual Report. International Rice Research Institute, Los Baños, Philippines.

IRRI. 1973. Multiple cropping. Pp. 14–34. Annual Report. International Rice Research Institute, Los Baños, Philippines.

IRRI. 1974. Multiple cropping. Annual Report. International Rice Research Institute, Los Baños, Philippines (in press).

IRRI. 1975. *Proceedings of the Cropping Systems Workshop.* International Rice Research Institute, Los Baños, Philippines. 396 pp.

Iso, E. 1954. *Rice and Crops in Its Rotation in Subtropical Zones.* Japan FAO Association, Tokyo. 611 pp.

Javier, E. Q. 1974. Improved varieties of pastures under coconuts. *ASPAC Ext. Bull.* 37. Tapei, Taiwan.

Jolly, A. J. 1958. Mixed farming in the tropics. *Turrialba* 8:52–54.

Jones, M. J. 1974. Effects of previous crop on yield and nitrogen response of maize at Samaru, Nigeria. *Exptal. Agr.* 10:273–279.

Kamath, M. B. and B. V. Subbiah. 1971. Phosphorus uptake pattern by crops from different depths. *Int. Symp. Soil Fert. Eval. Proc. (New Delhi)* 1:281–291.

Kanwar, J. S. and C. Krishnamoorthy. 1971. Arid soils-multiple cropping and soil fertility problems in India. *FAO World Soil Resources Rept.* 41, pp. 60–67.

Kashirad, A. and H. Marschner. 1974. Iron nutrition of sunflower and corn plants in mono and mixed culture. *Plant and Soil* 41:91–101.

Kassam, A. H. and K. Stockinger. 1973. Growth and nitrogen uptake of sorghum and millet in mixed cropping. *Samaru Agr. Newsletter* 15:28–33.

Koli, S. E. 1970. Agronomy on cereal crops. *Afr. Soils* 15:157–164.

Koregave, B. A. 1964. Effect of mixed cropping on the growth and yield of suram (Elephant yam, *Amorophallus campanulatus* B.). *Indian J. Agron.* 9:255–260.

Krutman, S. 1968. Cultura consorciada cana x feijoeiro; primeros resultados. *Pesq. Agropec. Bras.* 3:127–134.

Kurtz, T., M. D. Appleman, and R. H. Bray. 1947. Preliminary trials with intercropping corn and clover. *Soil Sci. Soc. Amer. Proc.* 11:349–355.

Kurtz, T., S. W. Melsted, and R. H. Bray. 1952. The importance of water and nitrogen in reducing competition between intercrops and corn. *Agron. J.* 44:13–17.

Lepiz, I. R. 1971. Asociación de cultivos maíz-frijol. *Agr. Técnica (Mexico)* 3:98–101.

Lewis, W. and J. A. Phillips. 1976. Double cropping in southern United States. In R. I. Pappendick et al., *Multiple Cropping Symposium.* American Society of Agronomy. (in press).

Liboon, S. P. and R. R. Harwood. 1975. Nitrogen response in corn–soybean intercropping. Unpublished paper, International Rice Research Institute, Los Baños, Philippines. 12 pp.

Lin, C. F., T. S. Lee Wang, A. H. Chang, and C. Y. Cheng. 1973. Effects of some long-term fertilizer treatments on the chemical properties of soil and yield of rice. *J. Taiwan Agr. Res.* 22:241–262.

MacEvoy, M. G. 1974. Establishment and management of pastures in coconut plantation. *ASPAC Ext. Bull.* 38. Tapei, Taiwan. 17 pp.

Mahapatra, I. C. and N. Sadanandan. 1973. Effects of multiple cropping on some of the physical and chemical properties of upland alluvial rice soils. *Int. Rice Comm. Newsletter* 22:26–34.

Nair, P. K. R. and A. Singh. 1971. Production potential, economic feasibilities and input requirements of five high intensity crop rotations. *Indian J. Agr. Sci.* 41:805–815.

Nair, P. K. R., A. Singh, and S. C. Modgal. 1973a. Maintenance of soil fertility under intensive cropping in northern India. *Indian J. Agr. Sci.* **43**:250–255.

Nair, P. K. R., A. Singh, and S. C. Mogdal. 1973b. Harvest of solar energy through intensive multiple cropping. *Indian J. Agr. Sci.* **43**:983–988.

Nelliat, E. V., K. V. Bavappa, and P. K. R. Nair. 1974. Multi-storeyed cropping—a new dimension in multiple cropping for coconut plantations. *World Crops* November–December 1974:260–265.

North Carolina State University. 1973. *Agronomic–Economic Research on Tropical Soils.* Annual Report. Soil Science Department, North Carolina State University. Raleigh. 190 pp.

North Carolina State University. 1974. *Agronomic-Economic Research on Tropical Soils.* Annual Report. Soil Science Department, North Carolina State University. Raleigh. 240 pp.

Oelsligle, D. D. and A. M. Pinchinat. 1975. Effect of varying nitrogen levels on grain yields, energy and protein production and economic returns of corn and beans when grown alone and in different combinations. Unpublished paper, Soil Science Department, North Carolina State University, Raleigh. 16 pp.

Oelsligle, D. D., R. E. McCollum, and B. T. Kang 1976. Soil fertility management in multiple cropping systems in the tropics. In R. I. Pappendick et al. (eds.), *Multiple Cropping Symposium* American Society of Agronomy (in press).

Okigbo, B. D. and D. J. Greenland. 1976. Intercropping systems in tropical Africa. In R. I. Pappendick et al. (eds.), *Multiple Cropping Symposium.* American Society of Agronomy (in press).

Osiru, D. S. O. and R. W. Willey. 1972. Studies of mixtures of dwarf sorghum and beans with particular reference to plant population. *J. Agr. Sci.* **71**:531–540.

Oza, A. M. and B. V. Subbiah. 1973. Residual fertilizer nitrogen in the soil under multiple cropping conditions using nitrogen-15. *Indian Soc. Nucl. Tech. Agr. Biol. Newsletter* **2**:55–56.

Palada, M. C. and R. R. Harwood. 1974. The relative return of corn–rice intercropping and monoculture on nitrogen application. Unpublished paper, International Rice Research Institute, Los Baños, Philippines. 7 pp.

Pan, Y. C., T. T. Sun, and K. C. Lin. 1963. The nitrogen requirement for interplanting tobacco with sugarcane. *Soils Fert. Taiwan* **1962**:74.

Pathak, R. D., T. K. Ghosh, and V. C. Srivatsava. 1968. Response of interseeding legumes in maize to different dates and nitrogen levels. *Ranchi Univ. J. Agr. Res.* **3**:4–6.

Pavlychenko, T. K. 1937. Qualitative study of the entire root system of weed and plants under field conditions. *Ecology* **18**:62–79.

Peregrina, R. P. 1965. La magnitud de la aportación de nitrógeno por diferentes leguminosas en siembras asociadas con maiz. *II. Congreso Sociedad Mexicana de Ciencia de Suelo* **1**:135–141.

Pinchinat, A. M. and D. D. Oelsligle. 1974. Combining corn and soybeans for increased production in the tropics. *Agron. Abs.* **1974**:46.

Pinchinat, A. M., J. Soria, and R. Bazan. 1976. Multiple cropping in tropical America. In R. I. Pappendick et al. (eds.), *Multiple Cropping Symposium.* American Society of Agronomy (in press).

Plucknett, D. L., J. P. Evenson, and W. G. Sanford. 1970. Ratoon cropping. *Adv. Agron.* **22**:205–326.

Poultney, R. G. 1963. A comparison of direct seeding and undersowing on the establishment of grass and the effect on the cover crop. *East Afr. Agr. For. J.* **29**:26–30.

Pushparajah, E. and S. Y. Tang. 1970. Tapioca as an intercrop in rubber. Pp. 128–138. In E. K. and J. W. Blencowe (eds.), *Crop Diversification in Malaysia.* Incorporated Society of Planters, Kuala Lumpur.

Pushparajah, E. and P. W. Wong. 1970. Cultivation of groundnuts and maize as intercrops in rubber. Pp. 53–65. In E. K. and J. M. Blencowe (eds.), *Crop Diversification in Malaysia.* Incorporated Society of Planters, Kuala Lumpur.

Radke, J. K. and R. T. Hagstrom. 1976. Intercropping for wind protection. In R. I. Pappendick et al. (eds.), *Multiple Cropping Symposium.* American Society of Agronony (in press).

Raheja, S. K., R. Prasad, and H. C. Jain. 1971. Long-term fertilizer studies in crop rotations. *Int. Symp. Soil Fert. Eval. Proc. (New Delhi)* **1**:881–903.

Raper, C. D. and S. A. Barber. 1970. Rooting systems of soybeans. I. Differences in root morphology among varieties. *Agron. J.* **62**:581–584.

Reddi, G. H. S., Y. Y. Rao, and Y. P. Rao. 1973. Residual effect of N, P and K applied to IR8 rice on succeeding soybean crop. *Indian J. Agr. Res.* **7**:177–187.

Sadanandan, N. and I. C. Mahapatra. 1970. The influence of multiple cropping on the bulk density of upland alluvial rice soils. *Agr. Res. J. Kerala* **8**:98–100.

Sadanandan, N. and I. C. Mahapatra. 1972. Study on the soil available phosphorus as affected by multiple cropping. *J. Indian Soc. Soil Sci.* **20**:371–374.

Sadanandan, N. and I. C. Mahapatra. 1973. A study of the nitrogen status of the soil as affected by multiple cropping. *J. Soc. Soil Sci.* **21**:173–175.

Sanabria, E. 1975. Producción de biomasa, nutrición mineral y absorción de agua en la asociación frijol-maiz cultivada en solución nutritiva. M.S. Thesis, Instituto Interamericano de Ciencias Agrícolas, CATIE, Turrialba, Costa Rica. 74 pp.

Sanford, J. O., D. L. Myhre, and N. C. Merwine. 1973. Double cropping systems involving no tillage and conventional tillage. *Agron. J.* **65**:978–982.

Santhirasegaram, K. 1967. Intercropping of coconuts with special reference to food production. *Ceylon Coconut Planters Rev.* **5**:12–24.

Sarma, V. and R. V. Patil. 1971. Residual effect of sorghum and maize fertilization on succeeding crop of groundnut. *J. Indian Soc. Soil Sci.* **19**:313–316.

Sharma, B. M. and D. L. Deb. 1974. Copper status of soils of the Union Territory of Delhi with special reference to cropping sequence. *J. Indian Soc. Soil Sci.* **22**:145–150.

Shelton, H. M. and L. R. Humphreys. 1975. Undersowing rice (*Oryza sativa*) with *Stylosanthes guyanensis. Exptal. Agr.* **11**:89–112.

Shen, T. H. 1964. *Agricultural Development in Taiwan since World War II.* Pp. 155–164. Comstock, New York.

Singh, S. 1967. Cotton yield as influenced by the preceding legumes raised with and without phosphorus in combination with nitrogen application to cotton. *Indian J. Agr. Sci.* **37**:57–68.

Singh, S. D., D. K. Misra, D. L. Vyas, and H. S. Davlay. 1971. Fodder production in association with different legumes under different levels of nitrogen. *Indian J. Agr. Sci.* **41**:172–176.

Soria, J., R. Bazán, A. M. Pinchinat, G. Páez, N. Mateo, R. Moreno, J. Fargas, and W. Forsythe. 1975. Investigación sobre sistemas de producción agrícola para el pequeño agricultor del trópico. *Turrialba* **25**:283–293.

Streeter, C. P. 1974. *Reaching the Developing World's Small Farmers.* Pp. 63–68. Rockefeller Foundation, New York.

Trenbath, B. R. 1974. Biomass productivity of mixtures. *Adv. Agron.* **24**:177–210.

Triplett, G. B., Jr. 1962. Intercrops in corn and soybean cropping systems. *Agron. J.* **54**:106–109.

Trouse, A. C. 1975. Below-ground reactions in multiple cropping systems. Unpublished paper. USDA-ARS, Auburn, Alabama.

Vallis, I., K. P. Haydock, P. J. Ross, and E. F. Henzell. 1967. Isotopic studies on the uptake of nitrogen by pasture plants. 3. The uptake of small additions of N^{15} labeled fertilizers by Rhodes grass and Townsville Lucerne. *Aust. J. Agr. Res.* **18**:865–877.

Van Parijs, A. 1957. Rotations des plantes vivrieres dan la region de Nioka (Haut-Ituri). *Bull. Agr. Congo Belgue* **48**:1515–1544.

Viegas, G. P., E. S. Freire, and C. G. Fraga. 1960. Adubação do milho XIV. Ensaios com múcuna intercalada e adubos minerais. *Bragantia* **19**:909–941.

Vine, H. 1953. Experiments on the maintenance of soil fertility at Ibadan, Nigeria 1922–1951. *Emp. J. Exptal. Agr.* **21**:65–85.

Willey, R. W. and D. S. O. Osiru. 1972. Studies of mixtures of maize and beans (*Phaseolus vulgaris*) with particular reference to plant population. *J. Agr. Sci.* **79**:519–529.

Wood, R. C. 1934. Rotations in the tropics. *Trop. Agr.* (*Trinidad*) **11**:44–46.

13

SOIL MANAGEMENT FOR TROPICAL PASTURE PRODUCTION

Meat and milk production from forage-consuming animals is a very important component of tropical agriculture. Approximately half of the world's permanent pastures and half of the cattle population are located in the tropics (Table 13.1). However, by standards that have become accepted in temperate climates, tropical livestock productivity is low. Only one-third of the world's meat and one sixth of its milk products are produced in this region (Jones, 1972).

The low productivity of forage-consuming animals in the tropics has been attributed to several factors such as heat stress and animal diseases. Although animal production involves many complex aspects, most animal scientists agree that feed supply is a greater limiting factor than environmental or health aspects (McDowell, 1972). Many of the factors limiting pasture production are related to soil properties.

Unlike crops, pastures are an intermediate step toward the desired output: meat, milk, fiber, or work. The results of soil management practices for pasture production should be measured in terms of animal production. Soil research with animal measurements, however, is extremely expensive

Table 13.1 Importance of Pastures and Forage-Consuming Animals in the Tropics (millions)

Region	Hectares in Permanent Pastures	Cattle	Sheep	Goats	Buffaloes	Forage-Consuming Animal Units[a]	Animals/ha of Pasture
Tropical America	297	179	62	33	—	214	0.9
Tropical Africa	642	130	73	94	—	170	0.3
Tropical Asia	33	230	47	91	76	321	9.7
Australia and Oceania[b]	455	29	137	—	—	57	0.1
Total	1427	568	319	218	76	762	0.5
Percent of world	47	49	31	55	61	57	—

Source: Based on FAO 1973 figures.

[a] Expressed in cattle equivalent-animal units as calculated by Russel et al. (1974): Cattle = 1; Buffaloes, horses, and camels = 0.8; sheep and goats = 0.2.

[b] Includes temperate Australia.

534

because of the large areas of land required, as well as the large number of animals needed to account for animal variability. The usual procedure is to screen a wide range of treatments in small-plot clipping experiments, and then to select a few of the more promising possibilities for larger-plot trials with grazing animals.

From the soil management point of view, three principal systems of pasture production can be identified: extensive grazing on native savannas, extensive grazing on improved grass–legume mixtures, and intensive pasture or forage production based on fertilized grass species. Although this is somewhat of an oversimplification, each of these categories represents distinct soil management systems. Animals are also fed via nomadic grazing in aridic areas, but in this case the range, rather than the soil, is the management objective. On small farms primarily in Asia, animal production is heavily dependent on crop residues, but in this case soils are managed for crop production purposes.

The range in productivity between these three soil management systems represents different orders of magnitude. Beef cattle production on native savannas results in annual liveweight gains on the order of 20 to 50 kg/ha. On improved grass–legume mixtures with minimum fertilizer inputs, annual liveweight gains are on the order of 100 to 300 kg/ha. Intensively fertilized grass pastures may produce from 500 to over 1000 kg/ha of liveweight gains per year.

In most of the tropics animals do not compete with human beings for grain consumption. The argument about the luxury of meat consumption via supplemental grain feeding is not very relevant in the tropics. Actually, beef is most often produced on soils that have serious limitations for crop production because of low native fertility, steep slopes, or poor road connections. Milk production based on supplemental grain feeding is more likely to be economical than is grain finishing of beef animals. It also allows more intensive use of labor and capital, if dairy animals of high genetic merit are available and other management practices are sound. However, even with dairy animals one seldom encouters much use of cereal grains as feed in tropical settings. By-products such as molasses, milling residues, and oil meals are more prevalent in dairy supplements.

An exception to these generalizations is encountered in certain intermountain valleys of Central and South America, where prime valley bottom land is grazed by cattle while surrounding steep hillsides are cultivated to intercropped systems. This is caused by a land tenure pattern whereby wealthy ranchers bought the best lands from small farmers, who then retrenched up the slopes.

For simplicity, data on beef production by steers are used in this chapter. Animal scientists can adjust for other species, milk production, or different age of animals. The term "grazed pastures" denotes direct harvesting by the animals, whereas "forages" is used when the pasture is cut by hand and fed to animals away from the field where it is produced.

NATURAL GRASSLANDS

The vast majority of permanent pasture land in the tropics consists of natural grasslands, mainly in the lowland savanna areas but also in the highlands. In tropical South America, Australia, and Africa natural grasslands cover more than 90 percent of the permanent pasture areas (Crowder, 1974; Jones, 1972; McIlroy, 1972). Animal production is low; one animal unit requires on the order of 5 to 25 ha of grassland to feed itself. The dominant natural species are adapted grasses of the genera *Themeda, Trachypogon, Heteropogon, Paspalum, Axonopus, Hyparrhenia,* and, in some highland areas, *Pennisetum.* With the exception of tropical America and certain East African highlands, few native legume species are found in the natural grasslands.

Limiting Factors

The two factors principally responsible for the low animal productivity in natural grasslands are inadequate soil moisture and soil fertility. Most of these areas have ustic soil moisture regimes characterized by a strong dry season. In parts of Africa and Australia tropical grasslands may approach the aridic soil moisture regime.

The interplay between these two factors has been accurately described by Crowder (1974) and may be paraphrased as follows: "Pasture is generally abundant during the rainy season, when new shoots or seedlings develop and grow rapidly. Such leafy material is usually consumed in large amounts and is more digestible than mature plants. The crude protein of some immature indigenous species may reach 8 to 10 percent on a dry matter basis; hence cattle usually gain weight during the rainy season. Bunch-type grasses dominate, and stem elongation and flowering begin before the onset of the dry season. As physiological maturity approaches, the leaf:stem ratio decreases and consequently the nutritive value declines. Plants become progressively lower in protein, minerals, and soluble carbohydrates and higher in fiber and lignin. These changes diminish the palatability, intake, and digestibility of the plants, resulting in decreased energy and protein consumption. In the mature tropical grasses crude protein may fall to a critical level of 3 to 5 percent on a dry matter basis. A deficiency of certain minerals in the forage, particularly phosphorus and calcium, may further accentuate the problem. During the dry season, the digestibility of the grasses decreases further because soluble minerals, energy, and protein are lost through plant respiration, leaching, and microbial fermentation. With intake lowered and the digestibility of the grasses impaired, particularly during drought, cattle go hungry for long periods. The nutritional level drops below that needed for maintenance; cattle lose weight, conception is delayed, animal fertility decreases, and maturity is prolonged so that slaughter often occurs when animals are 6 to 7 years old. Under nutritional stress, breaks in breeding cycles often occur that lower yearly calving percentages unduly; thus it becomes difficult to maintain herd replacements and to provide cattle for slaughter."

The result is a zigzag pattern of liveweight increases and decreases. Figure 13.1 illustrates this pattern, as measured in the Northern Territory of Australia. Cattle gain weight rapidly during the rainy season, but lose a significant proportion of these gains during the dry season. Under these conditions, 4 to 5 years is needed to attain a market-size weight of 400 to 450 kg. According to McDowell (1972), as long as 10 years is required to reach this size in the drier parts of Africa. Without these limiting factors cattle should be ready for slaughter in 2 years.

Fire as a Management Tool

Periodic burning is an almost universal management practice for cattle raising in native savannas, and scattered fires are a common feature of the landscape during the dry season. Although such fires may be caused by lightning or by discarding lit cigarettes along the road, ranchers inten-

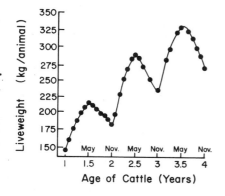

Fig. 13.1 Seasonal fluctuations in cattle weights as a function of wet and dry seasons in native pastures of Katherine, N. T., Australia. *Source:* Adapted from Norman (1966).

Plate 19   Annual burning in savannas of the Llanos Orientales of Colombia.

tionally burn their native savannas. This practice is easily condemned by people unfamiliar with the system, citing ecological damage as the main reason. An examination of the available data, however, indicates that periodic burning is essential for maintaining the stand and productivity of many natural grasslands and that the frequency and timing of burning operations are important management parameters.

The primary reason for burning is to destroy the dry, unpalatable grass stands and to promote the regrowth of younger, more palatable grass shoots. Burning also helps to destroy insects and pests that attack both pastures and animals, and provides moderate amounts of phosphorus and bases to the soil. Most of the carbon, nitrogen, and sulfur content of the vegetation is lost to the atmosphere. Burning is a rapid process: a curtain of fire travels quickly along the grass. It is likely, therefore, that its effects on soil properties are less marked than those of forest burning in shifting cultivation areas. Native grasses of the genera *Andropogon, Paspalum, Hyparrhenia, Trachypogon, Themeda,* and others are well adapted to this process. Some introduced grass species such as guinea grass (*Panicum maximum*) tolerate burning, whereas others, for example, molasses grass (*Melinis minutiflora*), do not.

A long-term experiment conducted by Van Rensburg (1952) on "red loam" soils of the Tanzanian highlands showed that, when native grasses were protected from burning, they developed very poor grazing properties. The pasture had poor growth vigor, and large quantities of unpalatable dead material smothered new growth. When pastures were burned periodically, however, they exhibited vigorous regrowth during the rainy season. In Ngong, Kenya, Edwards (1942) observed that the native dominant grass species, *Themeda trianda,* disappeared from an area that was protected against burning for 8 years, after which a violent fire took place. An undesirable species, *Digitaria abyssinica,* took its place. Edwards noted that annual burning maintained stable *Themeda* stands in good condition for grazing.

The timing of burning during the year can cause shifts in botanical composition. Ramsay and Rose-Innes (1963) compared burning early and late in the dry season in northern Ghana savannas for 11 years. The found that, when pastures were burned early in the dry season while some leaves were still green, regrowth of the palatable *Andropogon gayanus* grass was promoted. Burning late in the dry season, on the other hand, resulted in very hot fires and caused the unpalatable *Loudetia acuminata* grass to dominate the savanna. Herdsmen in these areas burn early in the dry season. Similar results were obtained by Van Rensburg (1952) in Tanzania. In more aridic areas, however, burning late in the dry season is favored. Norman (1963) found late burning superior in Katherine, Australia, probably because the dry matter accumulation may have been insufficient to cause much damage. He also observed that frequent burning early in the dry season followed by grazing is a good way to eliminate the native pasture before planting improved species.

The optimum frequency of burning appears to be related to moisture supply during the rainy season. In areas of more than 1200 mm rainy season rainfall, grass growth is abundant, and thus there is a considerable amount of dried material to be eliminated during the dry season. Annual burning is indicated for areas with this rainfall regime, such as the Llanos of Colombia, the Cerrado of Brazil, and the "Guinea" or derived savanna of West Africa. In areas of lower rainy season rainfall, such as East Africa and northern Australia, burning once every 2 years was found to be best (Edwards, 1942; Van Rensburg, 1952; Norman, 1963). Apparently there is less dead material to be eliminated.

Attempts to improve the productivity of native tropical grasslands can be grouped into three major categories: improved burning and grazing practices, fertilization, and oversowing legumes with fertilization. These three approaches are examined in the following sections.

Improved Burning and Grazing Management

A series of experiments conducted by CIAT (1972, 1973) on an Oxisol from Carimagua, Colombia, illustrates several attempts to improve the productivity of a native savanna dominated by the grasses *Trachypogon vestitus* and *Paspalum pectenatum*. By simply increasing the stocking rate from 0.18 to 0.31 animal/ha, the annual liveweight gains were increased from 5 to 12 kg/ha. When rotational grazing was compared with the traditional continuous grazing, liveweight gains decreased sharply (CIAT, 1972). In the following year, the traditional burning at the beginning of the dry season was compared with sequential burning of different paddocks every 2 months. Annual liveweight gains increased from 18 to 24 kg/ha with sequential burning at a stocking rate of 0.2 animal/ha. When the stocking rate was increased to 0.35 animal/ha, the traditional annual burn system produced 33 kg/ha of liveweight gains, while sequential burning produced 39. These figures are extremely low and reflect the low beef productivity of native Oxisol savannas, including the rapid weight losses during the dry season. Although all yields are low, it is clear that production can be increased significantly within this range.

Fertilizing the Native Grasslands

The second alternative, direct fertilization of the native grasslands, has produced dubious results. Research on this subject shows that annual dry matter production increases from two to four times with annual applications of 60 to 150 kg/ha of nitrogen, phosphorus, or potassium (Norman, 1962; Lotero et al., 1965; Walker, 1969; Keya, 1973; Olsen, 1974; Olsen and Santos, 1975). In the only experiment of this group that measured annual liveweight gains (Walker's in Uganda), the results were a function of dry matter increases. Walker, however, concluded that fertilization was too expensive to be economical. It appears that the native grasses are well adapted to the limited soil fertility, and that they are inefficient in recovering fertilizer nitrogen and phosphorus and in transferring these elements to the grazing animal. For example, Norman (1962) observed an apparent recovery of added nitrogen of less than 10 percent in trials with native *Themeda* spp. pastures in northern Australia, in comparison with improved grass species that recover from 40 to 70 percent of the nitrogen applied.

Oversowing Legumes

Most native grasslands in the tropics are low in available nitrogen, and few contain significant quantities of legume species. The most successful way to improve the productivity of native pastures has been to plant legumes in them, without eliminating the native grasses.

In a considerable proportion of the grasslands of northern Australia, productivity has been increased by oversowing Townsville lucerne (*Stylosanthes humilis*), an annual legume that reseeds itself and is tolerant of drought, acidity, and low available phosphorus levels. The soils of the areas (Ultisols, Alfisols, and Inceptisols) are particularly deficient in nitrogen, phosphorus, sulfur, and molybdenum but are not high in exchangeable aluminum. The native spear grass (*Heteropogon contortus*) savanna is burned during the dry season. Stylo seeds are mixed with single superphosphate enriched with molybdenum, are either drilled or broadcast and then are incorporated into the soil with a disk harrow. Seeding is done when heavy rains are expected. After the rains start, the pasture is heavily grazed to depress the growth of the native grass and allow more light to reach the slow-growing legume. At the seedling stage, stylo is unpalatable and cattle allow it to grow. Adequate grass–legume pastures are established within a year.

The Australian work indicates that fertilization is indispensable for the successful establishment and persistence of *Stylosanthes humilis*. The molybdenum-enriched simple superphosphate is applied at rates on the order of 220 kg/ha, which provides 22 kg P, 22 kg S, and 1 kg Mo/ha. Triple superphosphate is not used because it does not provide the sulfur needed in these soils. A maintenance application of one-half to one-fourth of the establishment rate is added annually at the start of the rainy season.

The results of this practice are shown in Table 13.2. Although planting stylo alone doubled stocking rates and liveweight gains per hectare, the use of superphosphate increased production even further, particularly during an extremely dry year (1964–1965) and during an extremely wet year (1968–1969). Graham and Meyer (1972) also found that a larger number of steers attained the desired market weight in 2.5 years when fed with the fertilized grass–legume pasture. An extra year was needed for animals grazing other treatments to attain market weight. The value of steers marketed was also higher with the fertilized legume–grass pasture. These practices have proved to be very profitable.

Research conducted by Keya and his coworkers on Alfisols of Kitale, Kenya, showed that legumes can be successfully established in native *Hyparrhenia* grasslands by broadcasting without machinery incorporation. Their results, shown in Table 13.3, indicate that a well-adapted legume, *Desmodium uncinatum,* did not increase forage dry matter or protein production without superphosphate applications. When 50 kg P/ha as superphosphate was added, the proportion of legumes in the sward increased from 12 to 54 percent, total dry matter doubled, and protein

Table 13.2 *Effects of Introducing a* Stylosanthes humilis *in a Native* Heteropogon contortus *Savanna on Beef Production during 4 Years in Marlborough, Queensland, Australia*

Treatment	Stocking Rate (animals/ha)	Animal Liveweight Gain (kg/ha)					Market Value of Individual Steers at 2.5 Years (A $)
		1964–65	65–66	67–68	68–69	Mean	
Native savanna	0.05	26	38	37	34	34	81
Stylo, unfertilized	0.10	40	78	76	63	64	93
Stylo + superphosphate[a]	0.10	50	127	90	68	84	120

Source: Graham and Mayer (1972).

[a] 220 kg/ha of molybdized simple superphosphate initially, plus 110 kg/ha annual maintenance.

Table 13.3   Effects of Broadcast Legumes and Phosphorus Fertilization on the Botanical Composition, Dry Matter, and Protein Production of Hyparrhenia *Grasslands in Kitale, Kenya, 16 months after seeding*
    Soil is a red loam (Alfisol) with pH 5.7.

Legume Broadcast	Super-phosphate Applied (kg/ha)	Total Dry Matter (tons/ha)	Proportion of Legumes in Dry Matter (%)	Total Protein Production (kg/ha)
None	0	3.6	—	266
	500	4.4	—	325
Desmodium uncinatum	0	4.1	12	270
	500	8.2	54	840

Source: Keya et al. (1971a).

production tripled. Another study (Keya et al., 1972) showed no differences between methods of introducing *Desmodium uncinatum* into the native grassland. Prior burning, grazing, cultivation, raking, or rotovating did not affect stand or yields. In the most recent study by this group Keya and Kalangi (1973) found that the superphosphate rate could be reduced to 25 kg P/ha without yield decreases. They also found that broadcasting improved grass species such as *Chloris gayana* was not successful, whereas most legumes were easily established.

IMPROVED PASTURES BASED ON GRASS–LEGUME MIXTURES

The improvement of native grasslands is limited by the low productivity of native grass species. To increase beef production and stocking rates to one or more animals per hectare, it is necessary in most cases to replace the native grasses with improved species. The establishment of pastures in areas other than native grasslands, such as forests or cropland, also requires the planting of improved pasture species. Considerable research has been conducted in the tropics on species selection and on the establishment and maintenance of improved pastures.

Species Selection and Adaptation to Soil Conditions

Australian scientists have worked on the premise that, despite climate limitations, plant species of low productivity, and the inherent low soil fertility of most natural tropical grasslands, the climate should permit higher pasture production. Pasture improvement has been based on a change of species, together with the use of appropriate fertilizers to raise soil fertility (Davies and Shaw, 1964; Shaw and Norman, 1970). Because of the high cost of fertilizers relative to the price of beef in most of tropical Australia, a mixture of adapted grass–legume species is used, the legume being the main source of nitrogen to the system. Applications of other nutrient elements are kept at a minimum by using species and varieties well adapted to specific soil moisture regimes and certain soil fertility limitations such as high exchangeable aluminum, low exchangeable calcium, and low available phosphorus. Selection of the appropriate grass–legume mixture is, therefore, site-specific.

The characteristics of the most widely used improved grass and legume pasture species have been described in several review articles by Williams (1967), Davies and Hutton (1970), Hutton (1970), Williams and Andrews (1970), Jones (1972), McIlroy (1972), and Crowder (1974). Although the botanical and agronomic properties of these species are beyond the scope of this book, their adaptation to soil conditions is an extremely important component of soil management. Much has been learned through the work of Andrew and his colleagues in Australia and Spain in Colombia about how species differ in their tolerances to adverse soil conditions.

Tropical grasses. A tentative list summarizing the adaptation of the principal grass species to certain soil conditions appears in Table 13.4. The many blank spaces reflect the limited knowledge about certain species. In general, these grasses can be grouped according to their adaptability to ustic lowlands, udic lowlands, the cooler highlands, and aquic soil moisture regimes.

Among the species well adapted to ustic soil moisture regimes, *Hyparrhenia rufa, Melinis minutiflora, Paspalum plicatulum,* and *Brachiaria decumbens* are considered tolerant of acid soil conditions (Spain, 1975; Spain et al., 1975) and of low available soil phosphorus (Andrew and Robins, 1971a). The term "tolerant" indicates that the species may respond to lower rates of lime or phosphorus than the "susceptible" grasses, which usually reqire lower levels of exchangeable aluminum and higher levels of available soil phosphorus to produce maximum yields. Other grasses adapted to the ustic or aridic soil regimes do not tolerate high levels of exchangeable aluminum (e.g., *Sorghum almum*) or low available phosphorus (e.g., *Paspalum dilatatum*). These tolerances or susceptibilities can be related to the soil properties of the regions where the grasses originally evolved.

Several improved species are well adapted to udic soil moisture regimes. Some of them, like *Panicum maximum* and *Digitaria decumbens,* are also

Table 13.4   List of Commonly Used Improved Tropical Grass Species and Their General Adaptabilities to Soil Conditions

		Preferred Soil Moisture Regime	Preferred Temperature[a]	Tolerance[b,c] to		
Scientific Name	Common Names			Soil Acidity	Low P Availability	Fire
Hyparrhenia rufa	Jaragua, Puntero	Ustic	Hot	T	T	T
Melinis minutiflora	Molasses, Gordura	Ustic	Hot	T	T	S
Paspalum dilatatum	Dallis grass	Ustic	Hot	T	S	T
Paspalum plicatulum	Brown seed paspalum	Ustic	Hot	T	T	T
Setaria sphacelata	Setaria, bristle grass, Nandi[d]	Ustic	Hot, cool	?	?	?
Brachiaria decumbens	Brachiaria, signal grass	Ustic, udic	Hot	T	T	?
Digitaria decumbens	Pangola	Udic, ustic	Hot	M	T	?
Panicum maximum	Guinea, Castilla, Colonião[d]	Udic, ustic	Hot	T	M	T
Cynodon plectostachyum	Star grass	Udic	Hot, cool	?	?	?
Pennisetum purpureum	Elephant, Napier, Merker[d]	Udic	Hot, cool	T	?	?
Paspalum notatum	Bahia grass	?	Cool	?	?	?
Cynodon dactylon	Bermuda, Dhub	Udic	Cool	?	?	?
Chloris gayana	Rhodes grass	Ustic, aridic	Cool	S	S	?
Cenchrus ciliaris	Buffel grass	Ustic, aridic	Cool	S	S	?
Pennisetum clandestinum	Kikuyu	Ustic, udic	Cool	?	S	?
Sorghum almum	Perennial sorghum	Aridic, ustic	Hot	S	M	?
Brachiaria mutica	Para grass	Aquic	Hot	T	S	?

[a] "Hot" refers to isohyperthermic lowlands; "cool", to highlands or the fringes of the tropics.
[b] Adapted from Andrew and Robins (1971), Spain (1975), Spain et al. (1975), and personal observations.
[c] T = tolerant, S = susceptible, M = moderately tolerant.
[d] Commonly used names of varieties.

tolerant of soil acidity and low available phosphorus. Others such as *Pennisetum purpureum,* although tolerant of aluminum, require higher quantities of other nutrients, particularly nitrogen and potassium, to approach maximum yields. The use of these grasses is generally restricted to soils of somewhat higher fertility than those suitable for the aluminum-tolerant species adapted to ustic soil moisture regimes.

Table 13.4 also lists several species that favor either the tropical highlands or the cooler areas on the fringes of the tropics. Many of them are less tolerant of low available phosphorus than those that are better adapted to the ustic or udic lowlands. Finally, certain species such as *Brachiaria mutica* prefer the aquic soil moisture regime and are thus restricted to poorly drained soils.

Tropical legumes.   A list of the most widely used tropical legumes is shown in Table 13.5. Legume species can also be grouped according to their adaptation to ustic lowlands, udic lowlands, cooler highlands, and aquic soil moisture regimes. Several species seem equally well adapted to both udic and ustic soil moisture regimes. The nutritional requirements of the tropical legumes are much better understood than those of tropical grasses.

Plate 20   A pure stand of *Stylosanthes guyanensis* grown for seed production in Oxisols of Matão, São Paulo, Brazil.

Several legume species listed in Table 13.5 are extremely well adapted to acid, infertile soil conditions. Most of them enter into symbiosis with the slow-growing cowpea-type *Rhizobium*, which is found in most acid tropical soils, and therefore require no innoculation. According to Norris (1967), these slow-growing *Rhizobium* strains release alkaline substances, whereas the fast-growing strains typical of temperate legumes release acid substances. Legume species possessing the alkali-releasing mechanism include both *Stylosanthes* and *Desmodium* spp., *Pueraria phaseoloides, Centrosema pubescens, Calopogonium muconoides, Macroptilium lathyroides, Phaseolus atropurpureus,* and *Lotononis bainesii.* All are tolerant of high levels of aluminum in the soil solution, according to Andrew et al. (1973), Andrew and Vanden Berg (1973), and Spain (1975). Some species, however, are susceptible to high levels of exchangeable manganese, thus limiting their suitability to acid soils low in this element. According to Andrew and Hegarty (1969), *Desmodium uncinatum* and *Phaseolus atropurpureus* are sensitive to high manganese levels while still tolerant of aluminum. All species of this aluminum-tolerant group have considerably lower requirements for calcium and magnesium than those susceptible to aluminum. Small quantities of lime are used to provide calcium and magnesium rather than to neutralize the exchangeable aluminum.

Among the legumes tolerant of soil acidity factors, some are also tolerant of low levels of available soil phosphorus. Table 13.5 lists the species with a critical plant phosphorus content of 0.18% P or less as "tolerant" and those with more than 0.20 % P as "susceptible," as suggested by the work of Andrew and Robins (1971). The following species combine tolerance to high exchangeable aluminum and manganese with tolerance to low calcium, magnesium, and phosphorus levels: *Stylosanthes guyanensis, Stylosanthes humilis, Pueraria phaseoloides, Centrosema pubescens,* and *Calopogonium muconoides.* A similar study, conducted by Andrew and Robins (1969) with potassium, established critical levels for this element. *Stylosanthes humilis, Centrosema pubescens,* and *Desmodium uncinatum* have critical levels of less than 0.8 percent K in plant tissue, while those grouped as "susceptible" have critical levels of 1.0 percent K or more. Varietal differences in tolerance to low available phosphorus and potassium have been observed in *Stylosanthes guyanensis* by Jones (1974) and Brolmann and Sonoda (1975).

In contrast to the foregoing group, Table 13.5 lists several legume species that are susceptible to most of the factors associated with soil acidity: *Leucaena leucocephala, Glycine wightii, Trifolium* spp., and *Medicago sativa.* These species developed on and are better adapted to high-base-status soils. They thrive in such soils in the tropics, particularly the clovers and alfalfa in calcareous soils with cool temperatures. Although *Centrosema pubescens*

Table 13.5 List of Commonly Used Improved Tropical Legume Species, Their General Adaptabilities to Soil Conditions, and the types of Rhizobium Used

Scientific Name	Common Names	Preferred Moisture or Temperature[a] Regime	Tolerance[b,c] to High Al	High Mn	Low Ca + Mg	Low P	Low K	Rhizobium Type[a]
Stylosanthes guyanensis	Stylo	Ustic-udic	T	T	T	T	?	Cowpea[e]
Stylosanthes humilis	Townsville lucerne, Townsville Stylo	Ustic-aridic	T	T	T	T	T	Cowpea
Pueraria phaseoloides	Kudzu, Puero	Udic	T	T	T	T	?	Cowpea
Centrosema pubescens	Centro	Udic-ustic (aquic)	T	T	T	T	T	Specific
Desmodium intortum	Greenleaf desmodium, pega-pega	Udić-ustic	T	M	T	S	M	*Desmodium*
Desmodium uncinatum	Silverleaf	Udic (cooler)	T	S	T	S	T	*Desmodium*
Calopogonium muconoides	Calopo	Udic-ustic	T	T	T	T	?	Cowpea
Macroptilium lathyroides	Murray lathyroides	Ustic	T	M	T	S	T	Cowpea
Lotononis bainesii	Miles lotononis	Ustic	T	S	T	T	S	Specific
Phaseolus atropurpureus	Siratro	Ustic (cooler)	T	S	?	T	T	Cowpea
Leucaena leucocephala	Ipil-ipil, Koa-haole, Leucaena	Ustic	S	S	?	?	?	Specific
Glycine wightii	Perennial soybean	Ustic	S	S	?	S	M	Cowpea
Trifolium repens[f]	White clover	Highlands	S	S	?	S	S	Clover
Medicago sativa[f]	Alfalfa, lucerne	Highlands	S	S	?	S	S	Alfalfa

[a] Lowland (isohyperthermic) unless otherwise specified.

[b] Based on Andrew and Robins (1969, 1971), Andrew and Hegarty (1969), Andrew et al. (1973), Andrew and Vanden berg (1973), and Spain et al. (1974).

[c] T = tolerant, S = susceptible, M = moderately tolerant.

[a] From Hutton (1970).

[e] Except Oxley fine stem and IR11022 varieties, which require specific strains.

[f] Also adapted to temperate climates.

is well adapted to udic and ustic regimes, it tolerates wet soil conditions and is often grown together with *Brachiaria mutica*.

Grass–legume mixtures. As previously stated, most improved pasture systems rely on a grass–legume mixture rather than single stands of grass or legume species. The role of the legumes in such mixtures is to contribute nitrogen to the grass and to improve the overall nutritional content of the pasture, particularly protein, phosphorus, and calcium. The grasses are expected to provide the bulk of the energy to cattle because of their larger dry matter production.

The compatibility of grass and legume species is related to their growth habits and similar adaptation to a specific climatic, soil moisture, and soil fertility regime. Tall grass species that grow in bunches, such as *Hyparrhenia rufa, Panicum maximum,* and *Pennisetum purpureum,* are compatible with low-lying or crawling legumes such as *Stylosanthes, Centrosema pubescens,* and *Pueraria phaseoloides.* Grass species that grow and cover the ground with a thick mat (e.g., *Brachiaria decumbens*) tend to outcompete the legumes unless growth is controlled by management. Some legume species such as *Stylosanthes guyanensis* have the advantage of being very unpalatable to cattle at early growth stages. This favors their establishment and permits their consumption at later stages when they are palatable.

The proper grass–legume combination is, therefore, site-specific. Such combinations reflect adaptation to finer climatic or edaphic distinctions than those presented in Tables 13.4 and 13.5. Some examples are shown in Table 13.6. In acid, infertile Oxisols of the Llanos of Colombia, a mixture of *Melinis minutiflora* and *Stylosanthes guyanensis* is successful. Both species are tolerant of the various soil fertility limitations and are adapted to the ustic moisture regime. In the jungle of Peru, characterized by a udic environment and acid, infertile Ultisols, a mixture of *Panicum maximum*

Table 13.6 Some Examples of Successful Grass–Legume Mixtures Used in the Tropics

Location	Soil	Grass Species	Legume Species
Colombia	Ustox	*Melinis minutiflora*	*Stylosanthes guyanensis*
Peru	Udults	*Panicum maximum*	*Pueraria phaseoloides*
Australia	Ustalfs	*Setaria sphacelata*	*Desmodium intortum*
Australia	Usterts	*Sorghum almum*	*Medicago sativa*
Kenya	Udalfs	*Pennisetum clandestinum*	*Desmodium uncinatum*
Australia	Aquic soils	*Brachiaria mutica*	*Centrosema pubescens*

Sources: Spain (1975), Davies and Hutton (1970), McIlroy (1972), and Bryan (1970).

and *Pueraria phaseoloides* is successful because of the adaptability of both species to the high rainfall and generally low fertility, and their compatability. In ustic high-base-status soils of northern Australia, two species that require somewhat higher native fertility levels are being used: *Setaria sphacelata* and *Desmodium intortum*. In calcareous Vertisols of Queensland, two species that prefer high-base-status soils and cooler temperatures, *Sorghum almum* and *Medicago sativa,* are used. For the highlands of Kenya with high-base-status Alfisols, two species adapted to the better fertility and the cool climates, *Pennisetum clandestinum* and *Desmodium uncinatum*, are mixed. Finally, in wet soils of Queensland, Australia, two species tolerant of poor drainage, *Brachiaria mutica* and *Centrosema pubescens*, are mixed. These are some examples of ·viable associations. Many others are described in the literature and used by livestock growers.

Establishing Improved Pastures

Soil management practices can be divided into those related to the establishment of a sward and those used to maintain production after establishment.

Establishment methods depend on the initial situation: forest, cropland, native savanna, or a worn-down, unproductive pasture. Some of the current methods are examined in this section.

Forests.   The clearing of tropical forests for establishing pastures is taking place at a fast rate in the Amazon Jungle. Within the past decade more than a million hectares of pastures have been planted in the Amazon Jungle of Brazil and in Peru and Colombia (Serrão and Neto, 1975; Santhirasegaram, 1975). The vast majority of the area has been planted to two grass species: guinea grass (*Panicum maximum*) in the higher rainfall areas, and yaragua (*Hyparrhenia rufa*) in the areas with ustic soil moisture regimes or close to them. In Australia smaller areas of rainforests have been cleared and planted to grass–legume pastures for beef fattening. The procedure is to cut and burn the forest during the dry season. Seeds of *Panicum maximum* mixed with *Stylosanthes humilis, Centrosema pubescens,* or *Pueraria phaseoloides* are simply sown into the ashes (Bryan, 1970). The mixture is fertilized with molybdenized simple superphosphate at rates on the order of 37 kg P/ha, followed by maintenance applications. Better legume establishment is usually attained by drilling the seed with machinery equipped with "stump jumps," which can operate well around stumps and roots.

In most of the Amazon, grasses are planted right after burning, using either botanical seed or, more commonly, vegetative propagation. Unlike the Australian practice, legumes are not normally planted and fertilizer is not commonly used. After a few years the guinea or yaragua grass disappears, and coarse, unpalatable species such as *Paspalum conjugatum* and *Axonopus compressus* dominate, along with spontaneous growth of legumes such as centro, *Desmodium intortum,* and calopo. In udic areas, where land clearing involves burning and cattle management practices are good, guinea grass pastures persist for more than 20 years with a carrying capacity of 1 animal/ha (Sanchez, 1973; Serrão and Neto, 1975). The presence of spontaneous legumes such as centro provides much of the nitrogen, while good grazing management allows its recycling.

In ustic areas that have been mechanically cleared, yaragua grass pastures degenerate after 4 to 5 years. Santhirasegaram (1975) attributed this decline to soil fertility depletion and compaction problems in Ultisols of Pucallpa, Peru. The fields are eventually abandoned to the regrowth of the secondary forest. The planting of adapted legumes such as *Stylosanthes guyanensis,* centro, and kudzu in established yaragua pastures with annual applications of 10 to 50 kg P/ha as ordinary superphosphate, however, has reversed this trend in Pucallpa. Santhirasegaram has shown that such practices, plus mineral supplements to cattle, double the stocking rate, dou-

Plate 21 A 20 year old guinea grass pasture, never fertilized, in acid Ultisols of Yurimaguas, Peru. Native legumes are abundant between the guinea grass clumps. With adequate management this pasture had a carrying capacity of 1 animal/ha.

ble the calving rate, triple the annual liveweight gain, and reduce by half the age at which the animals attain slaughter weight. When the pastures reach equilibrium, they contain 25 to 33 percent of the dry matter as legumes. Table 13.7 summarizes some of the results obtained during the first year of legume introduction. The maintenance of a comparable level over time is another matter. Australian experience suggests annual applications of ordinary superphosphate and potassium in soils deficient in this element.

The boost in soil nutrient status caused by the ash after burning tropical forests has been used profitably. Australian ranchers plant the more nutritious molasses grass (*Melinis minutiflora*) immediately after burning, and graze it heavily for 1 year. When its productivity begins to decline, they burn it, along with remaining stumps and logs. Subsequently the permanent grass–legume mixture is planted, and superphosphate applied. Some farmers in the Amazon region of Peru usually plant a cash crop such as upland rice or corn, and after the last weeding interplant sprigs of guinea grass. When the crop is harvested, the grass is already established. A similar practice has been introduced into northeast Thailand by Shelton and Humphreys (1975), who established *Stylosanthes guyanensis* pastures by broadcasting the legume in upland rice fields. When 2 kg/ha of stylo seeds was broadcast 10 days after rice planting, 3 tons/ha of *Stylosanthes* dry matter was obtained after the rice harvest without appreciable reduction in rice yields.

A more expensive method was satisfactorily used in the jungles of Hawaii by Motooka et al. (1968). Large areas were sprayed with tree-killing her-

Table 13.7   Effects of Different Methods of Pasture Establishment in Jungle Areas on the Initial Annual Liveweight Gains of Nellore Heifers Grazing at the Stocking Rate of 2 animals/ha on an Ultisol from Pucallpa, Peru

Treatment	Annual Liveweight Gain (kg/ha)	Pregnant Heifers (%)
Yaragua alone	79	—
Yaragua + kudzu	120	25
Yaragua + kudzu + 100 kg/ha superphosphate	159	38
Yaragua + kudzu + mineral supplement	321	88
Yaragua + kudzu + 100 kg/ha superphosphate + mineral supplement	352	88

Source: Adapted from Santhirasegaram (1975).

bicides by airplane and subsequently burned. Then guinea grass and *Desmodium intortum* seeds were mixed with granulated triple superphosphate and broadcast by airplane.

Savannas. Establishing improved pastures in native savannas involves heavy burning during the latter part of the dry season, followed by disk harrowing to incorporate the ashes, the fertilizers, and the seed. Broadcasting seeds without harrowing produces unreliable results, particularly with legumes. If the native grass has not been eliminated by burning and harrowing, high stocking rates after planting are recommended to suppress regrowth. Using these techniques, Winks (1973) changed a sown pasture that consisted of 90 percent native grass to 90 percent improved legumes in 3 years.

The transformation of native savannas into improved pastures can result in dramatic increases in beef production. When Oxisols of Colombia were planted to unfertilized molasses grass, the stocking rate increased from 0.2 to 1 animal/ha and the annual liveweight gain rose from 8 to 64 kg/ha (CIAT, 1972; Spain, 1975). Molasses grass alone, regardless of phosphorus or potassium fertilization, however, did not prevent serious weight losses during the dry season (CIAT, 1973). The addition of phosphorus and the planting of *Stylosanthes guyanensis* are expected to offset this situation and improve the yield and nutritive value of the pasture during the dry season.

Establishment techniques involving the planting of a crop to offset clearing costs are traditionally practiced in savanna areas of East Africa, particularly those with higher annual rainfall. Poultney (1963) observed that, when molasses grass and other species were planted into corn after the last weeding, corn yields were not affected, and grass dry matter and protein production were the same as when the grass was planted without corn. The results are shown in Table 13.8 and reflect the best combinations found. Unsuccessful combinations are commonly reported; the best technique is site-specific.

Improverished pastures. The regeneration of run-down pastures involves similar practices. The sowing of legumes with superphosphate has been successful in Australia (Bryan, 1967) except when persistent dry weather is encountered (Grof et al., 1970). Moderate nitrogen applications are commonly needed to give the newly planted grass–legume mixture a good start. Many of these pastures are severely compacted by animal trampling. Pereira and Beckley (1953) report that contour cultivation was needed for the successful establishment of star grass (*Cynodon plectostachyum*) pasture in overgrazed areas of Kenya. This practice doubled water infiltration rates into the soil and permitted the adequate growth of the grass.

*Table 13.8   Example of a Successful Practice of Interplanting
Grass in a Corn Crop in High-Rainfall Savanna Areas of near
Kitale, Kenya: Mean of 3 Seasons (tons/ha)*

System	Corn Grain Yield	Molasses Grass	
		Dry Matter	Protein
Corn alone	6.0	—	—
Corn–molasses grass	5.9	23.8	2.16
Molasses grass alone	—	24.1	2.10

Source: Adapted from Poultney (1963).

The need for some degree of harrowing is illustrated by Table 13.9, which shows the typical beneficial effect of this practice. The depth of fertilizer incorporation with harrowing is shallow, usually less than 10 cm.

Legume Inoculation and Seed Pelleting

A common method of establishing legume species on acid soils in the temperate region is to inoculate the seed with the appropriate *Rhizobium* strain and form a "pellet" around the seed with the inoculum, adhesive agent, lime, and phosphorus. This practice is considered sound because it promotes nodulation, protects the seed and inoculum against soil acidity, provides nutrients close to the seed, and reduces fertilization costs. The value of pelleting for tropical areas, however, has been severely questioned by Norris

*Table 13.9   Effects of Land Preparation and Ordinary Superphosphate
Additions on the Establishment of* Stylosanthes guyanensis *after Close
Grazing of Established* Hyparrhenia rufa *in Ultisols of Pucallpa, Peru
(Yields in tons dry matter/ha 24 weeks after sowing the legume)*

Land Preparation	Superphosphate Added (kg/ha)	*Stylosanthes guyanensis*	*Hyperrhenia rufa*
None	0	0.64	0.96
	200	1.24	1.07
Light disk harrowing	0	2.04	1.19
	200	2.51	1.35

Source: Santhirasegaram (1975).

(1967); this, in turn, has provoked considerable controversy among tropical soil microbiologists (Graham and Hubbell, 1975a, b).

Most tropical legume species that are tolerant to soil acidity and low phosphorus availability enter into symbiosis with slow-growing cowpea-type *Rhizobium* strains, as shown in Table 13.5. Extensive work conducted in Australia on soils with pH values of 5.2 to 5.9 indicates no need to inoculate species such as stylo, kudzu, and calopo (Norris, 1967). In the more acid Oxisols of Colombia with a pH of 4.3 and 77 percent aluminum saturation, the cowpea-type *Rhizobium* population is very low and inoculation of *Stylosanthes guyanensis* has been effective (Morales et al., 1973). Graham and Hubbell (1975a) attributed this effect to the extreme acidity and high temperatures encountered in such soils. They also noted significant differences among *Stylosanthes guyanensis* varieties and among cowpea-type strains. No need for inoculating kudzu and calopo, unlike stylo, was observed, because the former species are apparently able to establish effective symbioses under such extreme conditions.

Inoculation is definitely needed, however, for the aluminum-tolerant species that require specific strains, such as *Centrosema pubescens, Desmodium intortum, Desmodium uncinatum,* and *Lotononis bainesii,* as well as two improved varieties of *Stylosanthes guyanensis*: Oxley fine-stem and IRI 1022. Although nodulation is observed without inoculation, it is often ineffective, according to Graham and Hubbell (1975a). Inoculation is almost universally needed for aluminum-sensitive legume species such as alfalfa, the clovers, *Glycine wightii,* and *Leucaena leucocephala.* The use of local *Rhizobium* strains often results in poor nodulation (Lopes et al., 1971).

Commercial inoculants are produced in few tropical countries, such as Australia and Brazil. Importation from other countries often renders the inoculum inactive because of long delays in shipping and customs clearance. Although the relatively expensive adherents such as gum arabic can be replaced with cheaper products available in most developing countries (Graham et al., 1974), there is little doubt that the lack of commercial inoculants is an important limiting factor, particularly in countries with large grazing areas of acid soils.

Lime pelleting appears attractive even for species that are quite tolerant to exchangeable aluminum because of their response to low levels of calcium and magnesium fertilization. Work by Norris (1971, 1972) has shown little beneficial effect of lime pelleting on nodulation, because the cowpea-type *Rhizobium* secretes alkaline substances around its immediate environment. In fact, negative responses to lime pelleting are commonly observed. In sharp contrast, lime pelleting is extremely beneficial for establishing aluminum-sensitive species, such as alfalfa, the clovers, and *Leucaena leucocephala* in acid soils (Norris, 1967, 1973). The *Rhizobium* strains associated with these species exude acid substances that further aggravate the situation. In alfalfa and the clovers lime pelleting also lengthens the viability of the inoculum in storage. No such effect has been found in the acid-tolerant legume species (Norris, 1971, 1972), except when the soil contains toxic levels of manganese. Döbereiner and Abramovitch (1965) observed better nodulation of *Centrosema pubescens* with lime pelleting in a manganese-toxic soil. Apparently lime precipitated the excess manganese around the seed, without altering the soil pH.

The practice of pelleting with superphosphate or rock phosphate has noteworthy advantages and is recommended by Norris (1967, 1971, 1972) and by Graham and Hubbell (1975a,b) for acid-tolerant tropical legumes. Many of these species have higher initial internal phosphorus requirements for establishment than for subsequent growth. Fox et al. (1974) gives an example, shown in Table 8.6, of this situation. The mixing with superphosphate must be done shortly before planting because of possible ill effects from salt injury or extreme acidity under storage. Simple superphosphate will also provide some calcium and sulfur, which help to stimulate establishment. The use of rock phosphate dust is perhaps a more promising alternative because it can be mixed and stored without danger. The addition of molybdenum oxide to the pellets has shown promise, but ammonium and sodium molybdate cause injury (CIAT, 1973).

The adaptability of some grass and legume species to low phosphorus availability may be related to the presence of vesicular-arbuscular mycorrhiza in their roots. Mosse et al. (1973) found that *Centrosema pubescens* and *Paspalum notatum* roots infected with mycorrhizae utilized phosphorus at extremely low concentrations in the soil solution which uninfected roots were not capable of utilizing. Mycorrhiza inoculations have increased the growth of siratro in Colombia (CIAT, 1973) and both the growth and the nodulation of stylo and centro in Australia (Crush, 1974). Mycorrhiza inoculation may prove to be an economically viable practice in the future.

Much needs to be done toward quantifying the relationships among *Rhizobium,* mycorrhizae, tropical pasture species, and soils conditions common to tropical grasslands. These considerations, together with the possibility of significant nitrogen fixation in certain tropical grasses, observed by Döbereiner and Day (1974) and discussed in Chapter 6, have made tropical soil microbiology a very exciting area of research.

Nitrogen-Supplying Power of Legumes in Mixed Pastures

The range in annual nitrogen fixation by pasture legumes in the tropics is thought to be similar to that in the temperate region. According to reviews by Henzell and Norris (1962) and Jones (1972), *Rhizobium* fixation is on the order of 100 to 300 kg N/ha a year. A more recent compilation by Whitney (1975), however, gives an annual range of 47 to 905 kg N/ha, the higher figure was obtained for pure stands of an improved variety of *Leucaena leucocephala* in Hawaii. This implies that the upper limit of nitrogen fixation in the tropics exceeds the value of 603 kg N/ha a year reported for the temperate region by Henzell and Norris (1962). The year-round high temperatures and rainfall in udic regimes should explain why greater nitrogen fixation could be obtained in the tropics.

It has been established that the amount of nitrogen fixed by legumes is highly correlated with the dry matter content of the legume tops. Figure 13.2 shows two examples compiled by Jones (1972). In the highly productive udic environment of the Hawaiian example total nitrogen fixed and dry matter yields were high. Approximately 4 tons/ha of legume dry matter was needed to fix 100 kg N/ha a year. In the ustic regime of Queensland both dry matter and nitrogen fixed were lower, but only 3 tons/ha of legume dry matter was required to fix 100 kg N/ha a year. These differences however, could be due to different methods for calculating nitrogen fixation from nitrogen uptake data. Jones (1972) found that differences among adapted species within a specific environment were closely related to dry matter production. This suggests that there is little difference in the capacity of legumes to fix nitrogen as long as they are adapted to the environment. Factors affecting dry matter production, such as moisture or nutrient stress,

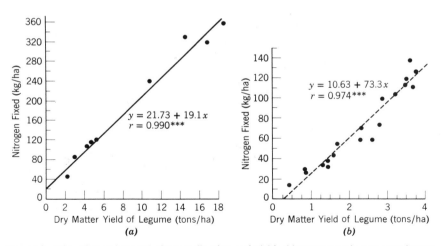

Fig. 13.2 The relation between nitrogen fixation and yield of legume tops in two experiments. *Source:* Compilation by Jones (1972). (*a*) Waikea, Hawaii (udic): 1 unit N/52 units dry matter. *Source:* Whitney et al. (1967). (*b*) Samford, Queensland (ustic): 1 unit N/27 units dry matter. *Source:* Jones et al. (1967).

solar radiation, diseases, compatibility with the grass species, and grazing management will, therefore, also affect nitrogen fixation.

The direct transfer of nitrogen fixed by the legumes to the associated grass species ranges from 0 to 53 percent, calculated by differences in nitrogen uptake of the grasses in the presence or absence of legumes (Jones et al., 1967; Vallis et al., 1967; Whitney et al., 1967; Whitney, 1970). The presumed mechanisms are legume leaf fall and root and nodule decomposition. The higher values reported, 53 percent in a *Digitaria decumbens–Desmodium intortum* mixture (Whitney, 1970) and 43 percent in a *Paspalum plicatulum–Phaseolus atropurpureus* mixture (Jones et al., 1967), were obtained with longer cutting intervals and greater height of cutting and after 3 years of establishment. These factors increase the top and root decomposition of the legume. Indirect transfer via the urine and feces of grazing animals is probably more important since about 80 percent of the nitrogen ingested by adult beef cattle is returned to the field as excrements (Vicente-Chandler et al., 1974).

Productivity of Improved Grass–Legume Mixtures

The productivity of properly selected and established grass–legume mixtures is well documented for the initial establishment period and the first 2 years. Beef production normally increases by a factor of two to four times (Jones, 1972). The question that immediately arises is whether these increases can be sustained with time. Few experiments provide reliable long-term data on this aspect. Figure 13.3 shows the changes in beef production for 7 years as a result of establishing a legume with and without fertilization in Queensland, Australia. In spite of the year-to-year fluctuation, caused primarily by rainfall variability, the introduction of *Stylosanthes humilis* without fertilization increased the average annual liveweight gain from 24 to 93 kg/ha and doubled the carrying capacity from 0.3 to 0.7 animal/ha. When the mixed pasture received modest annual applications of 10 kg P/ha as simple superphosphate and 40 kg K/ha plus molybdenum, the average annual liveweight gain reached 148 kg/ha, with a stocking rate of 0.95 animal/ha. It is interesting to observe that the fertilized treatments showed much greater yearly fluctuation in beef production, probably because of sharp fertilizer responses during years of adequate rainfall and minimal responses in years of poor or excessive rainfall. Applying the same fertilizer rate to pure grass stands produced less than half as much beef per hectare as application to the grass–legume mixture.

A direct relationship exists between the proportion of the legume in the mixture and beef production. This is illustrated in Fig. 13.4, which shows

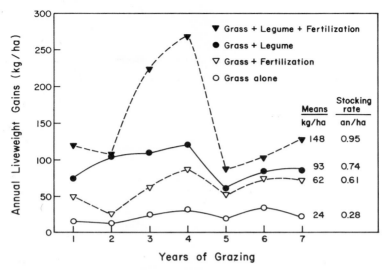

Fig. 13.3 Long-term effects of a legume species (*Stylosanthes humilis*) and annual fertilization (10 kg P/ha as single superphosphate, 36 kg K/ha, and 0.04 kg Mo/ha) on beef production and stocking rates on *Heteropogon contortus* grasslands on solodic soils (Alfisols?) in Rodd's Bay, Queensland, Australia. *Source:* Adapted from Shaw and Mannetje (1970).

progressive increase in liveweight gain as the percentage of legumes increased from 13 to 30. The 30 percent legume mixture approached the production levels of a high annual rate of nitrogen. Increasing the proportion of legumes much beyond 50 percent, however, is likely to decrease beef production, because cattle depends on grass consumption for the bulk of its energy requirements. However, well-managed, pure legume stands can also meet these energy requirements.

Rainfall distribution will affect the relative animal intake of grasses or legumes throughout the year. During the rainy season, when grass growth is more vigorous and the grass of good quality, cattle are likely to consume more grass than legumes. The reverse occurs during the dry season as the grasses become less palatable, and most legumes maintain in a higher protein content and general nutritional quality, at least during part of the dry season.

Nutritional Quality of Improved Species

The nutritional quality of pastures is a function of many factors, some plant- and others animal-related. The plant-related factors can be sum-

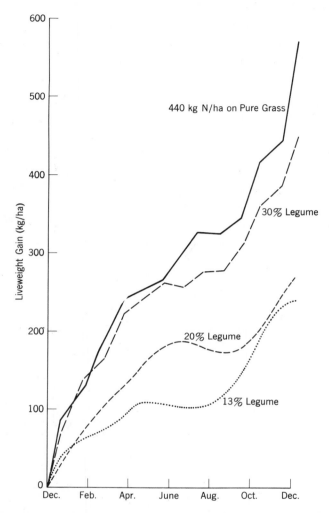

Fig. 13.4 Effect of legume content of tropical pastures on cumulative liveweight gains of beef cattle at Beerwah, Queensland, Australia: means of 2 to 7 years. *Source:* Bryan (1970).

marized under the headings of nutrient concentration (chemical composition), ability to be consumed by the animal (related to nondigestible fiber content, as well as to palatability), and degree of utilization of consumed material by the animal (digestibility of dry matter). The principal animal-related factor is the level of productivity of the animal in question. For example, a forage that is incapable of supporting high levels of milk production and therefore is of low nutritive value for a high-yielding dairy cow may be perfectly adequate for the maintenance of adult beef cattle.

The effect of stage of growth (harvest interval) on a few common measures of nutritive value are shown for guinea and pangola grasses in Table 13.10, and for kikuyu grass versus a kikuyu–red clover mixture in Table 13.11. These tables are a summarization of data compiled by McDowell et al. (1974). They demonstrate that variation due to physiological growth stage is usually much greater than variation among the commonly used species.

Digestible energy content at early vegetative stages ranges from 2.18 to 2.68 Mcal/kg of dry matter in the grasses and from 2.40 to 2.76 Mcal/kg in the legumes, according to other calculations from McDowell et al. This relatively narrow range is probably a result of selecting these species as palatable and nutritious. Digestible protein content at early vegetative stages varies from 7 to 12 percent among the grasses and from 13 to 22 percent in the legumes, according to the Latin American feed tables (McDowell et al., 1974).

The nutritional quality of tropical grasses varies more with management than among species. As plants grow older, the proportion of leaves decreases, the lignin content increases, and the protein, calcium, and phosphorus contents decrease. The effect of these factors on the energy and protein digestible by cattle is illustrated in Table 13.10 for guinea and pangola grasses. Digestible energy and digestible protein contents decrease with plant age or age after cutting. *In vitro* digestibility data show the same trend. The decline in phosphorus and calcium concentrations is present but less marked.

The decline in nutritional quality for tropical legumes, unlike that of the grasses, is small during similar time periods. Jones (1972) has assembled data which show that the digestibility and the protein and phosphorus contents of *Centrosema pubescens, Desmodium intortum, Glycine wightii,* and *Phaseolus atropurpureus* remain essentially constant during a 3 to 4 month period.

In mixed grass–legume pastures the situation is intermediate. Table 13.11 shows an example with kikuyu grass alone and kikuyu mixed with red clover. Although digestible energy and protein decreased with age, the decrease was less marked in the grass–legume mixture.

The use of species tolerant of low available phosphorus results in a lower concentration of this element in the pasture. When a combination of tolerant grass and legume species is grown, such as *Melinis minutiflora* and *Stylosanthes humilis,* both with a critical plant phosphorus content of about 0.18 percent, the total phosphorus available to cattle is lower than the amount obtained with a mixture of two species with high critical phos-

Table 13.10 Effect of the Stage of Growth of Guinea Grass (Panicum maximum) and Pangola (Digitaria decumbens) Grass on Their Nutritional Quality for Cattle Production

Days after Cutting	Guinea Grass					Pangola Grass				
	Digestible Energy (Mcal/kg)	Digestible Protein (%)	In vitro Digestibility (%)	P (%)	Ca (%)	Digestible Energy (Mcal/kg)	Digestible Protein (%)	In vitro Digestibility (%)	P (%)	Ca (%)
1–14	2.41	11.2	—	0.31	0.69	2.70	10.5	83	—	—
15–28	2.61	7.8	81	0.36	0.86	2.74	8.1	77	0.22	0.43
29–42	2.28	6.2	54	0.32	0.84	2.28	8.2	76	—	—
43–56	1.91	3.8	47	0.28	0.72	2.01	4.6	74	0.25	0.63
57–70	2.35	5.6	58	0.22	0.78	2.55	3.6	72	0.18	0.34
71–84	2.18	6.3	—	—	—	1.94	3.3	68	0.28	0.35

Source: Adapted from McDowell et al. (1974).

562

Table 13.11 Effect of a Legume (Trifolium pratense) *on Maintaining the Nutritional Quality of Kikuyu Grass* (Pennisetum clandestinum) *Pasture for Cattle*

Days after Cutting	Kikuyu Alone		Kikuyu + Red Clover	
	Digestible Energy (Mcal/kg)	Digestible Protein (%)	Digestible Energy (Mcal/kg)	Digestible Protein (%)
15–28	2.14	15.5	2.59	16.9
29–42	2.31	11.5	2.49	11.3
43–56	2.00	6.1	2.37	8.7

Source: Adapted from McDowell et al. (1974).

phorus contents, such as *Chloris gayana* and *Glycine wightii,* both with 0.23 percent P (as shown in Table 8.7). Direct phosphorus supplementation to cattle may be necessary in cases where phosphorus-tolerant mixtures are used (CIAT, 1973; Santhirasegaram, 1975). The same relationships may hold for calcium and magnesium, and have been documented for nitrogen (Minson, 1971).

Fertilization markedly affects the nutritive value of grasses. In studies in Puerto Rico by Vicente-Chandler et al. (1974) heavy applications of nitrogen, phosphorus, potassium, and lime increased considerably the digestible dry matter, protein, phosphorus, and calcium contents of grasses. Vicente-Chandler et al. (1974) also observed that the nutritive quality of grasses can be adversely affected by the available silicon content of the soil. High silica contents in grasses decrease their digestibility. A 60 day old pangola grass pasture grown on a young Inceptisol had 3.3 percent Si, whereas the same species, harvested at the same time, contained 0.7 percent on an Oxisol with lower available silicon content because of more advanced weathering.

Management of Grass-Legume Pastures

To maintain productivity, mixed tropical pastures require different grazing pressures at different times of the year. Appropriate weed and brush control and other practices similar to those developed in the temperate region are also needed. There is one aspect, tolerance to frequent defoliation, in which several tropical legumes differ significantly from alfalfa and the clovers. Several of the creeping species such as *Desmodium* spp., siratro, and perennial soybean are very sensitive to frequent defoliation and require more time to fully recover from grazing than does alfalfa (Jones, 1972). Other species such as *Stylosanthes humilis* withstand heavy grazing. The effects of these and other management practices can be appreciated from the data in Table 13.12. The total dry matter and the protein production decreased sharply when the pangola–*Desmodium* mixtures were cut frequently or close to the ground. Annual nitrogen fixation by the legume decreased from 313 to 85 kg N/ha when intensive and frequent cutting was practiced.

Jones (1972) observed that the effects of increased grazing pressure vary with the sensitivity of the legume to defoliation. When a setaria-siratro pasture that produced maximum liveweight gains at a stocking rate of 2.4 animals/ha was grazed with 3 animals/ha, the beef production decreased sharply and the legume component of the pasture fell from 25 to 12 percent. On the other hand, intensifying the grazing pressure from 1.2 to 2.5 animals/ha on a *Urochloa mozambicensis–Stylosanthes humilis* pasture actually doubled the liveweight gain and the legume composition of the pasture. Siratro is sensitive to frequent defoliation, but *Stylosanthes humilis* is not.

In all these cases the effect of grazing management on beef production can be related to the amount of legume dry matter present in the pasture, which is well correlated with nitrogen fixation. The optimum proportion of legume will depend on the productivity of the associated grass.

Fertilization of Grass–Legume Pastures

After the establishment phase, mixed pastures enter into a fairly efficient nutrient cycling through the grazing animal. More than 80 percent of the nitrogen, phosphorus, and potassium consumed by the animals is excreted in their urine and feces and is fairly well distributed if the animals are allowed to move freely around the pasture (Vicente-Chandler et al., 1974; Mott, 1974). Net nutrient losses take place when growing animals are removed from the pasture, or are due to urea volatilization from feces or to nutrient leaching through the soil. The fertilization strategy is geared toward replacing these losses in order to maintain a stable soil–plant–animal nutrient cycle.

Nitrogen. In most cases nitrogen fixation by legumes supplies enough nitrogen to offset annual losses, with appropriate grazing management. Inorganic nitrogen fertilization usually results in a sharp decrease in legume nitrogen fixation because the grass outgrows and shades the legume and reduces its presence drastically. These effects are shown in Table 13.12,

Table 13.12 Effects of Harvest Interval, Cutting Height, and Nitrogen Fertilization on the Productivity and Nitrogen Fixation of Pangola Grass–Desmodium intortum Mixtures in an Inceptisol from Hawaii

Species	N Fertilizer (kg/ha)	Cutting Interval (weeks)	Cutting Height (cm)	Dry Matter Production		Crude Protein Production		N Fixed by Legume (kg/ha)
				Total (tons/ha)	Legume (%)	Total (tons/ha)	Legume (%)	
Grass alone	0	10	5	5.4	—	0.4	—	—
	410	10	5	22.8	—	2.3	—	—
	410	5	5	14.3	—	1.8	—	—
Grass–legume	0	5	5	6.2	40	0.9	58	85
	0	5	13	7.2	50	1.2	66	125
	0	10	5	13.7	55	2.3	67	275
	0	10	13	14.7	58	2.4	69	313
Grass–legume	410	5	5	16.4	4	2.2	7	39
	410	5	13	14.6	6	2.0	9	34
	410	10	5	23.0	9	2.8	15	84
	410	10	13	22.0	12	2.7	18	98

Source: Whitney (1970).

where the legume component dropped from more than 40 to less than 12 percent when the pasture was fertilized with 410 kg N/ha a year. Small initial nitrogen applications, however, may have a priming effect in starting growth, but if continued the legume component may be reduced or even eliminated (Chaverra et al., 1967.; Caro and Vicente-Chandler, 1963).

Phosphorus. The maintenance phosphorus requirements will depend on the nutritional requirements of the grass–legume mixture, the phosphorus fixation capacity of the soil, and the amounts added during the process of establishment. In most of tropical Australia very small quantities of phosphorus are added. Initial applications on the order of 10 to 35 kg P/ha are followed by annual maintenance applications of 5 to 10 kg P/ha. These low rates are a result of the use of species tolerant to low levels of available phosphorus in the soil and the low phosphorus-fixing capacity of most of this region. Modest maintenance applications top-dressed to the pastures have been as effective as large initial applications in low-fixing soils of Australia (Jones, 1964; Fisher and Campbell, 1972; Bruce, 1972).

In soils with high phosphorus fixation capacities the initial application may need to be an order of magnitude higher. Experiments on high-phosphorus-fixing Ultisols and Oxisols in Hawaii reported lowest initial application rates on the order of 335 kg P/ha (Younge et al., 1964; Younge and Plucknett, 1965). High initial applications of 602 kg P/ha maintained maximum yields of a grass–legume mixture on an Oxisol of Hawaii for 11 years without annual maintenance applications. (The results are shown in Fig. 8.5.)

It seems doubtful that high initial phosphorus applications are economical for soils with high phosphorus fixation capacities, although they have been profitable when fertilizers were cheap. In view of the time lag in recuperating fertilizer investments through beef production, it seems more reasonable to use tolerant species plus modest initial and annual applications, even in Oxisols.

The cost of phosphorus fertilization can be reduced by applying cheaper sources such as rock phosphates, particularly in acid soils. This approach appears promising for Oxisols of Brazil with extremely high phosphorus fixation capacities (North Carolina State University, 1974). Some species differences in the relative response to superphosphate *versus* rock phosphate have been observed by Bryan and Andrew (1971). *Stylosanthes guyanensis* and *Lotononis bainesii* were able to extract more phosphorus from rock phosphate than could *Desmodium uncinatum* and *Microptilium lathyroides.* The reasons for this behavior are not known. In most cases, however, an initial application of a highly soluble source like superphosphate is sensible because of the low initial availability of rock phosphate and the relatively higher initial phosphorus requirements of some tropical pasture legumes. Such applications are likely to be more effective if incorporated to about 15 cm.

Liming. When both grass and legume species are acid-tolerant, they respond to small amounts of lime, usually on the order of 0.15 to 1.0 ton $CaCO_3$/ha, in soils that normally require 4 to 6 tons/ha to neutralize the exchangeable aluminum and raise the pH to 5.5 (Jones and Freitas, 1970; Spain, 1975). Figure 13.5 shows some examples from the Llanos Orientales of Colombia. The acid-tolerant grass and legume species that are normally grown in mixtures apparently responded only to calcium or magnesium fertilization because the 150 kg lime/ha rate was too small to affect soil properties. The residual effect of such small applications has not been

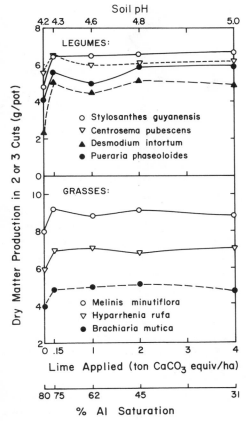

Fig. 13.5 Lime responses of several acid-tolerant grass and legume species grown on an Oxisol from the Llanos Orintales of Colombia. *Source:* Compiled from CIAT (1973).

determined, nor is it known whether the maintenance requirements are different from these establishment requirements. The maintenance applications of superphosphate mentioned in the preceding section are probably sufficient to provide the calcium needs of these legumes.

The lime needs of acid-susceptible legumes such as *Leucaena leucocephala* and *Glycine wightii* are totally different (Jones and Freitas, 1970; Hill, 1971; Lee and Wilson, 1972). Rates on the order of 10 tons lime/ha were needed to obtain maximum yields of *Leucaena leucocephala* in acid soils of New Guinea (Hill, 1970).

The vast majority of liming trials on tropical pastures have been conducted at rates that bypass the effect shown in Fig. 13.5 for the acid-tolerant legumes. A compilation of Colombian work by Lotero et al. (1971) shows that most experiments started with rates of 2 to 4 tons/ha, and consequently acid-tolerant species approached maximum yields at these levels. There is considerable need for field work on lime requirements for maintaining adapted grass–legume mixtures in acid soils. Depth of liming is seldom considered; it is common to incorporate lime to less soil depth in pastures than in field crops.

Potassium.   Although there is a major distinction among tropical legumes with respect to whether they require high or low levels of potassium for maximum growth (Andrew and Robins, 1969b), the knowledge of potassium requirements for the establishment and maintenance of adapted grass–legume pastures is very slight. The response to potassium in most of the Australian work has been sporadic or nonexistent (Shaw and Mannetje, 1970; Teitzel, 1969; Bryan and Evans, 1973), as well as in Brazil (Jones and Freitas, 1970). Even in Oxisols and Ultisols of Puerto Rico, which are known to be low in this element, potassium applications are not recommended for kudzu–molasses grass mixtures (Vicente-Chandler et al., 1964). In the absence of definite research information, reliance on soil test and plant analysis values may give an indication as to whether or not this element should be applied. In Florida, Brolmann and Sonoda (1975) found the critical level for stylo to be 0.8 percent K in tissues. Since about 80 percent of the potassium consumed by animals is returned to the soil via excreta in grazing systems, the need for maintenance applications is likely to be lower. One problem is the usual practice of adding generous quantities of potassium in experiments designed to evaluate the effects of other nutrients. Considering the extreme potassium deficiency of many Oxisols, Ultisols, and many sandy soils, research in this area is seriously needed.

Sulfur.   Responses to sulfur in mixed pastures in tropical Australia are well documented (Jones et al., 1971; Robinson and Jones, 1972). The reason why ordinary superphosphate is used instead of triple superphosphate in this region is its sulfur content. Sulfur responses are also common in Oxisols of Brazil. Jones and Quagliato (1973) observed that *Stylosanthes guyanensis* responded to sulfur immediately after establishment, whereas *Centrosema pubescens* and *Glycine wightii* did not respond until the third cut. Alfalfa showed the largest responses. The actual requirement seems to be on the order of 20 to 40 kg S/ha, according to McClung and Quinn (1959).

Micronutrients. The occurrence of micronutrient deficiencies depends on soil properties. The classic example is the widespread molybdenum deficiency observed in Australia, which led to the almost universal use of molybdenized simple superphosphate for both establishment and maintenance applications (Moore, 1970). The recommended rate is less than 1 kg/ha. Iron deficiencies have been recorded in kudzu growing on calcareous soils in Puerto Rico (Vicente-Chandler et al., 1964), while boron and molybdenum deficiencies have been observed in the same species in the Amazon Jungle of Peru (North Carolina State University, 1973). Additions of silica have suppressed manganese toxicities in *Sorghum sudanense* in Hawaii (Bowen, 1972). The occurrence of secondary and micronutrient problems is best identified by soil, plant, and animal analyses. Beeson et al. (1972) and Sutmoller et al. (1966) have made such studies in the Amazon Jungle of Peru and Brazil. Moody (1974), Teitzel and Bruce (1972, 1973), and many others in Australia have demonstrated the value of conducting soil fertility evaluation studies in the greenhouse before opening new areas for pasture production.

Effect of Legumes on Soil Properties

Legumes are considered to be "soil improvers." The available evidence indicates, however, that increases in total soil nitrogen are not appreciable until the legume is well established. Crack (1972) reported no changes in total soil nitrogen or in inorganic nitrogen during the first 2 years after oversowing *Stylosanthes humilis* on native savannas of Australia. After the third year net increases in topsoil organic nitrogen were observed with generous rates of superphosphate applications. Nitrate nitrogen increased similarly. The topsoil nitrate levels during the dry season and during the "flushes" at the beginning of the rainy season increased severalfold in the presence of legumes. When the legume was fertilized, the soil nitrate levels increased even further. These results, shown in Table 13.13, suggest mineralization of the increased organic nitrogen.

The annual increases in soil organic nitrogen range from 44 to 90 kg/ha

Table 13.13   Effects of the Introduction of Stylosanthes humilis *on Native* Heteropogon contortus *Savannas and Ordinary Superphosphate Applications on Changes in the Top 7.5 cm of a Solodic Soil near Townsville, Australia (initial % N = 0.08)*

Management	Changes in Organic N (kg N/ha)		NO$_3$-N (ppm)	
	0–2 Years	0–4 Years	End of Dry Season	"Flush" in Early Rainy Season
Native grass	−18	+ 10	+ 2	+ 2
Native grass + legume	−19	+ 68	+ 4	+ 8
Native grass + legume + 18 kg P$_2$O$_5$/ha a year	+ 3	+179	+10	+14

Source: Adapted from Crack (1972).

in savanna areas of southern Queensland. Vallis (1972) reported that dry years resulted in no increases or in actual decreases in total soil nitrogen. In udic areas with relatively high-base-status soils of northern Queensland, Bruce (1965) compared the long-term changes in organic carbon and nitrogen in unfertilized guinea grass and in mixed guinea grass–centro pastures. His comparison of pastures of known ages sampled at the same time is shown in Fig. 13.6. The pure stands of guinea grass suffered a decrease in nitrogen and carbon but approached a new equilibrium at about 8 years of grazing. The mixed grass–legume pasture, on the other hand, maintained soil organic matter levels close to those before clearing. The difference between the two was calculated to represent an annual topsoil organic matter accretion of 103 kg N/ha, which is the contribution of the legume.

The establishment and maintenance of such equilibrium soil organic nitrogen and carbon levels is remarkable. In cropping systems soil organic matter depletion is much more pronounced than in these grazing systems. The return of a large proportion of the nutrients consumed by the animals to the pastures via excreta is a major reason for this equilibrium. Figure 13.6 may explain why unfertilized guinea grass pastures remain productive for many years in udic environments. The role of the legume in maintaining levels of soil organic carbon and nitrogen similar to those before the forests were cleared lends further support to its important contribution to the maintenance of desirable soil chemical properties.

When the introduction of legumes is accompanied by adequate annual fertilization, soil properties actually improve. In the example shown in Table 13.14, significant increases in organic carbon, total nitrogen, availa-

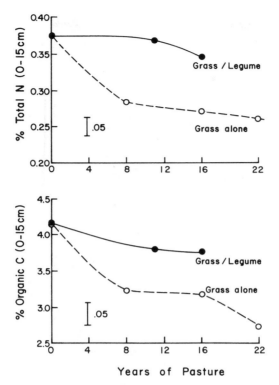

Fig. 13.6 Long-term effects of unfertilized guinea grass (*Panicum maximum*) pastures, with and without *Centrosema pubescens,* on the topsoil organic matter of an Alfisol of pH 5.7 after clearing a rainforest in South Johnstone, Australia. *Source:* Adapted from Bruce (1965).

ble phosphorus, exchangeable potassium, exchangeable calcium, and total cation exchange capacity were recorded after 8 years of grazing. The improved grass–legume pasture of this example was fertilized annually with 12 kg P/ha as single superphosphate and 10 kg K/ha as KCl.

INTENSIVE PASTURE AND FORAGE PRODUCTION BASED ON GRASS FERTILIZATION

The third major division of soil management systems for animal production involves the intensive use of fertilizers in the absence of legumes. The basis for this approach is that legumes seldom can provide sufficient nitrogen for grasses to achieve their maximum yield potential. Annual dry matter

Table 13.14 Changes in Topsoil Properties of an Ultisol after 8 Years of Chloris gayana, Desmodium intortum, Desmodium uncinatum, *and* Lotononis bainesii *Pastures in Southeast Queensland, Australia*

Treatment	pH	Organic C (%)	Total N (%)	Available P (ppm)	Exchangeable (meq/100 g)			CEC at pH 7
					Ca	Mg	K	
Virgin soil	5.4	0.56	0.035	3	0.75	0.44	0.38	3.4
Pasture + 125 kg/ha a year of ordinary superphosphate and KCl	5.4	1.99	0.053	33	1.80	0.44	0.92	5.4

Source: Adapted from Bryan and Evans (1973).

production on the order of 45 tons/ha has been achieved commercially with intensive fertilization of elephant or guinea grass in Puerto Rico (Vicente-Chandler et al., 1964, 1974). Such production supports a carrying capacity of up to 10 animals/ha with cut forage or 5 animals/ha under grazing, which results in over 1100 kg/ha of annual beef production or 7400 kg of milk per cow per year on steep Ultisols and Oxisols with the animals on an all-grass ration. Legume–grass mixtures do not produce sufficient nitrogen to supply such extremely high quantities of pasture.

Intensive grass fertilization is economically sound in areas where the ratio of fertilizer costs to animal product prices is low. This is the case for many dairying operations, as well as for beef production near urban areas with high market demands. Small irrigated areas are sometimes intensively fertilized to supply high-quality forage during the dry season or for fattening purposes.

Semi-intensive Pasture Fertilization

In areas where legumes have not been successfully introduced and price:cost ratios do not justify heavy investment in fertilizers, a semi-intensive system is practiced. An example is shown in Fig. 13.7, adapted from the work conducted by the IRI Research Institute in São Paulo, Brazil (Quinn et al., 1961, 1962; Mott et al., 1970; Mott, 1974). The improved Colonião variety of *Panicum maximum* received an annual application of

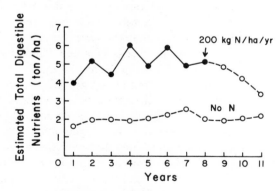

Fig. 13.7 Effect of annual nitrogen applications for 8 years in guinea grass pastures on the estimated total digestible nutrients consumed by grazing steers, and the residual effect for 3 years on a sandy Alfisol (pH 6) in Araçatuba, São Paulo, Brazil. Both treatments involved applications of 200 kg P_2O_5/ha and 60 kg S/ha for the first 2 years. *Source:* Adapted from Mott (1974).

200 kg N/ha for 8 consecutive years and 86 kg P/ha and 60 kg S/ha during the first 2 years. This more than doubled annual digestible energy yields and increased annual liveweight gains from 300 to 700 kg/ha. The economic return from nitrogen fertilization was $42 (United States currency) per year at 1963 prices. Although phosphorus and sulfur applications were terminated during the second year, the stable production figures suggest a strong residual effect of these elements. The annual nitrogen applications were terminated during the eighth year, after which production declined appreciably but still remained above the control levels. This suggests a strong residual effect of nitrogen, which Mott (1974) attributed to the efficient nitrogen recycling between the soil plants and animals.

Grazing versus Cut Forages

A fundamental division exists among intensive systems: those in which the animals harvest the pasture, and those in which man cuts the pasture and feeds it to animals elsewhere. In practice most intensively managed systems involve a combination of the two types, but for soil management purposes the differences are crucial.

In grazed pastures the animal recycles most of the nutrients back to the soil, so the fertilizer requirements are lower. Vicente-Chandler et al. (1974) calculated that intensively managed grazing animals return annually 176 kg N/ha, 20 kg P/ha, and 115 kg K/ha to the soil via excreta. The distribution of the feces is not uniform, and excessive growth in "dung spots" may require clipping. In heavily fertilized pastures this is not a problem, however, because the grasses are growing near their maximum potential.

Puerto Rican experience indicates that the fertilizer requirements of cut forages are twice those of grazed pastures. Vicente-Chandler et al. (1974) recommend annual applications of 370 kg N/ha, 53 kg P/ha, and 197 kg K/ha for cut forages but half that amount for grazed pastures in Oxisols and Ultisols of Puerto Rico. The difference between these two recommended rates corresponds very closely to the calculated amounts of nutrients recycled via excreta. Stephens (1967) compared cutting versus grazing elephant grass in Uganda and found higher nitrogen removals under cutting than under grazing. He also observed strong responses to potassium and magnesium applications in cut forages but none under grazing management in the same soil.

In addition, grazing animals cause changes in soil properties, primarily compaction by trampling when the soil is wet, and erosion along the trails they develop. Animals also select the most nutritious plant parts and reject older stems or even certain species.

In forage systems there is no nutrient cycling through the animals. Excreta are accumulated in barns and other confined areas and may or may not be returned to the same field, depending on management. There is little compaction or trampling. More dry matter with lower protein content is produced and consumed since the animal has little choice of plant parts or species. In terms of productivity, cut forages produce more dry matter and animal product per hectare per year than grazing systems of similar harvest intervals, but at a higher cost.

Excellent comparisons of the two systems, including the economic implications at the farm management scale, are given in the two review bulletins of the Puerto Rican work (Vicente-Chandler et al., 1964, 1974). A summary of these comparisons in terms of grass and animal production is shown in Table 13.15. The average values show that grazing produced only about 60 percent of the total digestible energy obtained by cutting. Grazed pastures, however, contained more than twice the crude protein percentage of the cut forages because of selective feeding by the animal and more frequent grazing.

Nitrogen Fertilization

Rates. Just as grass species differ in their tolerances to soil acidity and low available phosphorus, they also differ in their responses to nitrogen applications. Grass species can be grouped into high and low nitrogen response categories. The high-response species include elephant grass (*Pennisetum purpureum*), guinea grass (*Panicum maximum*), pangola grass (*Digitaria decumbens*), star grass (*Cynodon plectostachyum*), paragrass (*Brachiaria mutica*), Congo grass (*Brachiaria ruziziensis*), kikuyu (*Pennisetum clandestinum*), setaria (*Setaria sphacelata*), Rhodes grass (*Chloris gayana*), and Dallis grass (*Paspalum dilatatum*). These grasses require annual nitrogen applications of 400 to 900 kg/ha to reach maximum yields as cut forage, according to studies conducted in widely different soils throughout the tropics (Vicente-Chandler et al., 1964, 1974; Crowder et al., 1964; Herrera et al., 1967; Lotero et al., 1968; Whitney and Green, 1969; Evans, 1969; Olsen, 1975). In the typical response pattern shown in Fig. 13.8, elephant grass produced 44 tons/ha a year of dry matter with 880 kg N/ha, while some of the other species mentioned approached maximum annual dry matter yields of 30 tons/ha with 440 kg N/ha. The literature is full of experiments showing a linear response to nitrogen applications in these grasses when the highest rates used were less than 600 to 800 kg N/ha. This high nitrogen requirement supports the concept that maximum production of these grasses cannot be accomplished by mixing with legumes

Table 13.15 Comparison of the Annual Productivities of Intensively Managed Ultisols and Oxisols in the Udic Region of Puerto Rico under Forage and Grazing Management (C = cut forages, G = grazed pastures)

Forages received annual applications of 370 kg N/ha, 53 kg P/ha and 197 kg K/ha and were limed to pH 6; grazed pastures were fertilized at half of those rates.

Grass Species	Dry Matter Production (tons/ha)		Total Dry Matter Consumed by Cattle (tons/ha)		Crude Protein (%)		Carrying Capacity for 270 kg Steers (animals/ha)		Liveweight Gain (kg/ha)
	C	G	C[a]	G	C	G	C[1]	G	G
Elephant	37	13	14	8	7.2	16.0	10.0	5.4	1232
Guinea	30	13	12	8	7.7	17.4	8.3	5.4	1232
Pangola	28	12	11	7	7.4	16.0	7.6	5.0	1008
Molasses[b]	14	7	6	4	8.9	—	3.9	2.7	616

Source: Adapted from Vicente Chandler et al. (1974).
[a] One 270 kg steer requires 1425 kg total dry matter/year = 3600 kg/year of cut forage, assuming 20% waste in feeding and 50% digestibility of forage consumed.
[b] Not a recommended species.

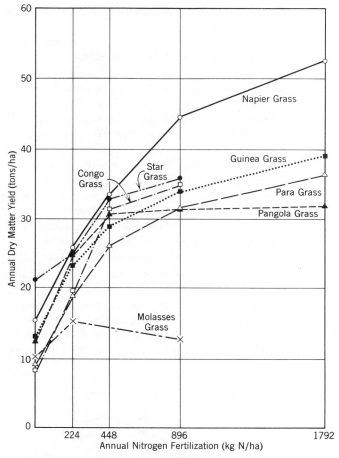

Fig. 13.8 Dry matter response to nitrogen fertilization by tropical grasses cut for forage in Ultisols of the udic mountain regions of Puerto Rico. *Source:* Vicente-Chandler et al. (1974).

because the maximum contribution of legumes in such mixtures is on the order of 100 to 300 kg N/ha a year (Henzell, 1968).

Grass species that respond only to moderate nitrogen applications include molasses grass (*Melinis minutiflora*), yaragua (*Hypharrenia rufa*), Bahia grass (*Paspalum notatum*), and native grassland species such as *Heteropogon contortum, Andropogon* spp., and *Paspalum plicatulum,* according to studies by Vicente-Chandler et al. (1964, 1974), Bastidas et al. (1967), Perez (1970), and Mannetje and Shaw (1972). The last authors demonstrated that these species are less efficient users of applied nitrogen:

20 kg dry matter/kg N at initial rates, in contrast to 40 to 50 kg dry matter/kg N at similar rates with the more nitrogen-responsive grasses. Figure 13.8 includes molasses grass as an example of the limited response species. The maximum dry matter yield of about 15 tons/ha a year was obtained with the optimum annual application of 200 kg N/ha. It is no coincidence that such species are used in grass–legume mixtures. They probably approach their maximum nitrogen response potentials with the rate that a legume can provide (100 to 300 kg N/ha a year) in mixtures. These grass species, therefore, are seldom used in intensively fertilized systems.

Annual additions of 400 to 800 kg N/ha usually require additional phosphorus and potassium plus liming to overcome the high residual acidity when ammoniacal sources of nitrogen are used. Economic considerations usually reduce the optimum rates of application to about 400 kg N/ha a year for cut pastures, as illustrated in Table 13.15. When these intensively fertilized pastures are grazed instead of cut for forage, nutrient recycling lowers the optimum nitrogen application rate to half, but the carrying capacity is correspondingly reduced (Vicente-Chandler et al., 1974). An example of the relationship among nitrogen rates, liveweight gains, and carrying capacities is illustrated in Fig. 13.9 from a trial conducted under almost ideal conditions with irrigated pangola grass on Mollisols in the Cauca Valley of Colombia. In this case the optimum level was 500 kg N/ha, which produced annual liveweight gains of 1000 kg/ha with a carrying capacity of 6.7 animals/ha. Optimum nitrogen rates will also depend on soil

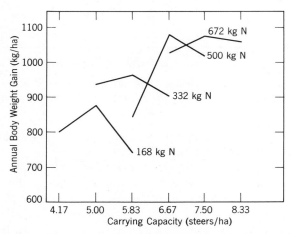

Fig. 13.9 Relationship between annual nitrogen applications, carrying capacities, and annual liveweight gains on irrigated pangola grass on high-base-status Mollisols of Palmira, Colombia. *Source:* CIAT (1973).

conditions, estimated leaching losses, grazing management, and many other factors. The invasion of weeds and less desirable grass species is less of a problem at high nitrogen rates than at lower rates because of competition (Gartner, 1967, 1969).

The response of grasses to nitrogen fertilization is not uniform throughout the year. The highest dry matter production is attained during periods of both high temperature and high rainfall. Most of the nitrogen-responsive grasses are quite susceptible to low temperatures (Vicente-Chandler et al., 1964, 1974; Wollner, 1968, Whitney and Green, 1969) and to water stress (Oakes, 1967; Guerrero et al., 1970). Mannetje and Shaw (1972) correlated nitrogen response and efficiency of fertilizer utilization directly with rainfall. Most of the intensively fertilized grasses are grown in udic environments or with supplemental irrigation during the dry season.

Since the desired dry matter and protein production can be reasonably well controlled with nitrogen applications, one way to regulate production at different times of the year is to increase or decrease nitrogen rates, in addition to the influence of temperature and rainfall.

Sources. A review of the Latin American literature on pasture fertilization provided no evidence of differences in efficiency among the commercial nitrogen sources (Sanchez, 1973). Urea is generally recommended because of its lower unit cost and moderate acidifying effect, except when conditions are favorable for volatilization losses. When sulfur deficiencies are encountered, ammonium sulfate is recommended unless the sulfur requirements are met by simple superphosphate. Although sulfur-coated urea has been found more effective for fertilizing grasses under high leaching conditions in the temperate region (Allen and Mays, 1974), little is known about its effect under tropical conditions. Initial results in Ultisols from the Amazon Jungle of Peru indicate promise for this slow-release source. Guinea grass forage attained the same dry matter production with an annual rate of 400 kg N/ha as urea split in six applications as with 100 kg N/ha a year as sulfur-coated urea in two applications (North Carolina State University, 1974).

Timing. Studies on the timing of application indicate that best results can be attained by dividing the annual rates into equal amounts, applied after each cut or grazing period; this normally ranges from four to eight times a year (Crowder et al., 1967; Herrera et al., 1967; Vicente-Chandler et al., 1964, 1974). Some benefits may be obtained by applying lower rates during periods of expected cool weather or water stress. A more important timing issue, however, is the frequency of cutting or grazing. Harvesting every 90 days produces more annual dry matter than harvesting every 40 days, but the protein content is much higher with 40 day than with 90 day intervals (Vicente-Chandler et al., 1974). When both energy demand and protein demand are considered, the optimum timing is probably between these two points. It is commonly recommended that nitrogen be applied after every cut or grazing.

Nitrogen is commonly applied by broadcasting. Lotero et al. (1968) in Colombia observed no benefits from banding or concentrating nitrogen applications around elephant grass bunches, as compared with broadcasting over the entire area. Foliar spray applications are ineffective (Uribe and Grisales, 1960; Crespo, 1972).

Fertilizer Recovery. Because of the importance of estimating protein content, nitrogen analysis is commonly conducted in pasture research that permits the calculation of the apparent recovery of added nitrogen by uptake differences. Tropical grasses are more efficient in recovering applied nitrogen than most other crops. At recommended nitrogen rates apparent recoveries range from 43 to 75 percent in Puerto Rico, Colombia, and Australia (Henzell, 1971; Sanchez, 1973). Recovery has been as high as 77 to 83 percent in Australia when calculated with the use of ^{15}N (Vallis et al., 1973). In grazed pastures it would be reasonable to assume a higher recovery through recycling. Henzell (1972), however, working on Ultisols in Queensland, was able to account, for only 40 percent of the total 374 kg N applied/ha a year, by measurements of the amount used by cattle, the amount accumulated in standing herbage, and the amount present in the top 75 cm of the soil. The recovery of nitrogen applied to cut pastures decreases with increasing rates of application (Henzell, 1971; Vicente-Chandler et al., 1974), but the magnitude of the decrease is much lower than with crops. Actual recovery values depend on soil properties and management. Higher fertilizer recoveries are obtained with species classified as responsive to high nitrogen applications.

Phosphorus Fertilization

The phosphorus requirements of intensively fertilized grasses, unlike those for nitrogen, depend more on soil properties than on the grass species present. Although some of the highly nitrogen-responsive species are considered relatively tolerant of low available soil phosphorus (Table 13.4), their high growth rate, caused by heavy nitrogen fertilization, increases their phosphorus requirements. According to Vicente-Chandler et al. (1974), these grasses remove about 53 to 72 kg P/ha annually (Table 13.16). The optimum levels in Puerto Rican Oxisols and Ultisols are on the order of 75

Table 13.16   Annual Nutrient Extraction by Intensively Fertilized Cut Forages in an Ultisol from Puerto Rico (same fertilization as in Table 13.15)

Grass Species	Dry Matter Production (tons/ha)	Nutrient (kg/ha)					
		N	P	K	Ca	Mg	S
Elephant	28	338	72	565	107	71	84
Guinea	25	322	50	406	167	110	50
Pangola	26	335	53	401	122	75	50

Source: Adapted from Vicente-Chandler et al. (1974).

kg P/ha a year except for pangola grass, which requires less phosphorus. Figure 13.10 shows a typical response. A review of the Colombian literature on this subject by Michelin et al. (1974) reports optimum rates of 43 kg P/ha a year for elephant and pangola grass in volcanic and alluvial soils of the highlands of that country. Because of the higher annual dry matter production and phosphorus removal in intensively fertilized grasses, the recom-

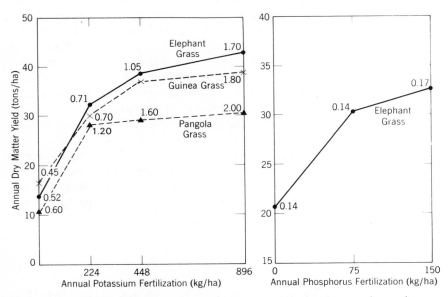

Fig. 13.10 Responses of tropical grasses to fertilization with phosphorus and potassium on a typical Ultisol in Puerto Rico. Numbers show percent potassium and phosphorus content of forage on a dry weight basis. *Source:* Vicente-Chandler et al. (1974).

mended rates of annual phosphorus application are considerably higher than those used in cut forages in the temperate region (Kamprath, 1973).

Studies on phosphorus sources indicate a higher recovery of superphosphate than of rock phosphates (Michelin et al., 1974). Considering the lower cost of rock phosphates, Werner et al. (1968) recommended a mixture of both sources for pastures in Brazil. Aspects related to phosphorus fixation and residual effects discussed in the section on legume-based pastures are applicable to intensive forage production. As previously mentioned, the phosphorus requirements can be decreased by half in grazing systems because of recycling.

Potassium Fertilization

Intensively managed grasses extract enormous amounts of potassium from the soil. Table 13.16 shows an annual range of 401 to 565 kg K/ha for three species. The fast growth due to heavy nitrogen fertilization requires correspondingly rates of potassium. Figure 13.10 shows that annual rates of 225 to 450 kg K/ha are sufficient to approach maximum yields. The optimum rate seems to be assocoated with a 1.0 percent K content in dry matter; higher rates can result in the classic luxury consumption of this element. Vicente-Chandler et al. (1974) observed little response to potassium during the first 2 years after establishment in about 50 soils of Puerto Rico. Beyond 2 years the average amount of potassium available per year was 60 kg/ha (as measured by nutrient uptake), with only small differences noted among soil types. Soil tests correlated well with potassium uptake. No differences between KCl and K_2SO_4 were observed in Puerto Rico. The slow-release sulfur-coated KCl has not been tested under these conditions.

Potassium should be applied along with nitrogen after every cut or grazing. Considering a 75 percent rate of recovery, Vicente-Chandler et al. (1974) recommended an annual application rate of 200 kg K/ha for Oxisols and Ultisols for cut forages. Similar levels have been recommended for cut pangola grass in Nigeria (Chheda et al., 1973). Under grazing the recommended amounts in Puerto Rico are decreased by half because of nutrient recycling.

Liming

The nitrogen-responsive tropical grasses are tolerant of relatively high levels of exchangeable aluminum. Liming is used primarily to counteract the residual acidity of high rates of nitrogen applications. In the process any calcium and magnesium deficiency encountered is corrected—an important benefit, considering the high annual uptake values of these two elements (107 to 167 kg Ca/ha and 71 to 110 kg Mg/ha in the examples of Table 13.16). When calcitic lime is used, magnesium additions must be made separately if the soil is deficient in this nutrient.

The residual acidity of heavy nitrogen applications in tropical forages is one of the most extreme recorded. Annual rates of 400 to 500 kg N/ha decreased soil pH from 5.4 to 4.2 in 5 years in Colombian Andepts (Villamízar and Lotero, 1967). Heavy nitrogen applications also decrease the pH and base saturation of the subsoil; an example based on the work of Abruña et al. (1958) was shown in Fig. 6.9. In Puerto Rico an annual application of 1 ton/ha of lime is recommended to counteract the acidity caused by the recommended rate of 1 ton/ha of 15-5-10 fertilizer containing ammonium sulfate at the rate of 370 kg N/ha. Liming can be as effective when broadcast on the soil surface as when incorporated into 15 cm, according to Vicente-Chandler et al. (1974). The residual effect of such lime applications was shown in Fig. 7.12. In grazing systems lime needs are reduced by half because nitrogen fertilizer applications are reduced by 50 percent.

Sulfur and Micronutrients

Heavily fertilized grasses extract large quantities of sulfur (50 to 84 kg S/ha in the example of Table 13.16). When ammonium sulfate or simple superphosphate is used, these requirements are usually satisfied. With other fertilizer sources, however, sulfur deficiencies are common, and annual gypsum applications are required for their correction (McClung and Quinn, 1959; Figure 8.10).

Micronutrient deficiencies are not usually encountered in intensively fertilized, well-managed grass pastures. One reason may be the amounts of impurities present in the heavy applications of fertilizers and lime. Adequate levels of zinc, iron, manganese, and copper have been reported by Vicente-Chandler et al. (1974) in intensively fertilized grasses. If micronutrient deficiencies arise, it is probably economically sound, considering the high level of investment, to supply the missing elements. Soil testing and plant analysis can best detect these problems.

Nutritional Quality of Intensively Fertilized Grasses

The nutritive values of tropical grasses are considered to be lower than those of temperate grasses. Minson and McLeod (1970) compiled data from a large number of *in vivo* dry matter digestibility studies and obtained mean values of 52 percent for the tropical grasses and 70 percent for the temperate grasses. High temperature and evaporation rates and harvesting at more mature stages are cited as the main reasons for these differences. A decreased voluntary intake by the animals may intensify the problem. Table 13.10 showed how increasing time after cutting severely affects nutritional quality. In unfertilized grasses the decline in digestibility is rather sharp after 12 weeks of growth. Annual nitrogen application rates of 200 kg/ha did not markedly affect dry matter digestibility in several grasses in Brazil (Gomide et al., 1969).

Intensive nitrogen fertilization results in improved nutritive value, mainly with respect to crude protein content. In fact, protein yields per unit of land from grasses heavily fertilized with nitrogen often surpass those obtained from grass–legume combinations. Annual protein production at recommended fertilizer rates in Puerto Rico ranged from 1.9 to 2.6 tons/ha, either as cut forage or pastures consumed by grazing. Vicente-Chandler et al. (1974) observed that a kudzu–elephant grass mixture receiving identical fertilization except no nitrogen produced only 1.1 tons/ha a year of protein, or about half the amount produced by intensively fertilized grasses (Fig. 13.11). The annual dry matter production was only 13 tons/ha with the grass–legume mixture, or less than half that obtained from grasses alone, as shown in Table 13.15.

Digestibility studies conducted in several countries indicate that nitrogen applications beyond those required to produce a 6 percent protein content do not increase or decrease digestibility or voluntary intake (Chicco et al., 1971; Ford and Williams, 1973; Minson, 1973). There is also little difference among the species adapted to heavy fertilization. The principal factor affecting digestibility is the interval between cuttings or grazings. Figure 13.12 shows that *in vitro* digestibility decreased at an approximate rate of 0.5 percent per day from 1 to 9 weeks of growth. At the recommended intervals of 6 to 8 weeks *in vitro* digestibility ranged from 62 to 75 percent, which is within the normal range of temperate-region grasses found by Minson and McLeod (1970).

The phosphorus and calcium contents of well-fertilized grasses range from 0.18 to 0.22 percent P and from 0.28 to 0.78 percent Ca in forages cut at 60 day intervals (Vicente-Chandler, 1975). These levels are sufficient for meeting cattle requirements at the high stocking rates used. The beef production figures of Table 13.15 were obtained by feeding grasses exclusively; no concentrates or mineral supplements were needed. Milk production with all-grass rations averaged over 10 liters per day with a butterfat content of 3.8 percent, for a total of 3000 liters per lactation, with an average calving interval of 13.5 months over 5 years (Vicente-Chandler, 1975).

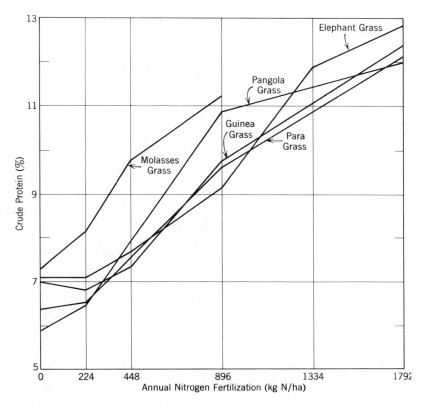

Fig. 13.11 Effect of intensive nitrogen fertilization on the protein contents of several tropical grasses cut every 60 days. *Source:* Vicente-Chandler et al. (1974).

Effects of Intensively Fertilized Grass Production on Soil Properties

The high stocking rates used in intensively fertilized grass pastures would be expected to cause severe soil compaction. Studies in Puerto Rico have shown that his effect depends on soil properties. Vicente-Chandler and Silva (1970) observed little compaction in a clayey Ultisol even between bunches of guinea grass at a stocking rate of 6 animals/ha for 4 years. No responses to tillage were observed, along with changes in porosity and bulk density. The Puerto Rican group has observed excellent permeability in the Oxisols and Ultisols, but severe soil compaction occurred in Vertisols under intensive management (Vicente-Chandler et al., 1974). Similar problems can be expected in coarser-textured Ultisols and other soils susceptible to compaction.

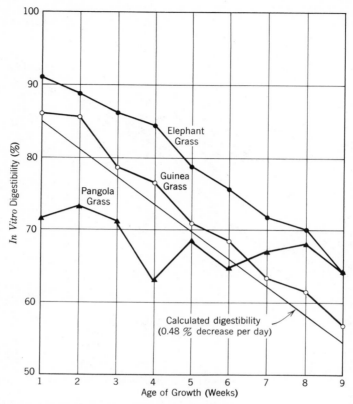

Fig. 13.12 Digestibility of intensively fertilized grasses in Puerto Rico as a function of cutting interval. *Source:* Vicente-Chandler et al. (1974).

Table 13.17 Recovery of Fertilizer Nitrogen by Guinea Grass and Effect of 5 Years of Heavy Nitrogen Fertilization on the Topsoil of an Ultisol from Puerto Rico

Annual N Applied (kg/ha)	Fertilizer Recovery (%)	Soil Organic Matter (%)	Soil Organic N (%)
0	—	3.0	0.17
800	46	3.4	0.21

Source: Vicente-Chandler et al. (1964)

586

Table 13.18   Effect of 4 Years of Heavy Fertilization (900-80-670 kg/ha NPK) and Liming (10 tons/ha) on the Profile Properties of an Ultisol from Puerto Rico (same as in Table 13.17)

Soil Depth (cm)	Unlimed			Limed (10 tons/ha a year)		
	pH	Exch. Ca + Mg (meq/100 g)	Al Saturation (%)	pH	Exch. Ca + Mg (meq/100 g)	Al Saturation (%)
0–8	3.9	1.9	77	5.4	11.2	4
8–15	4.2	1.6	78	4.6	5.0	31
15–30	4.4	2.7	65	4.6	4.2	38
30–45	4.6	3.0	62	4.8	5.0	39

Source: Adapted from Vicente-Chandler (1974).

Intensive fertilization has actually improved soil chemical properties. In the example shown in Table 13.17 soil organic matter and total nitrogen increased significantly after 5 years. This suggests that a considerable proportion of the applied nitrogen not recovered by the crop can be converted into soil nitrogen. Heavy fertilization and liming improved the base status of the top 45 cm of these Ultisols, as shown in Table 13.18.

Perhaps the most dramatic effect is the soil conservation aspect of these systems. Annual soil erosion losses in these steep Ultisols and Oxisols were reduced from 27 tons/ha with cropping to 1.5 tons/ha in heavily fertilized pastures (Vicente-Chandler et al., 1974). In less intensive pasture systems soil erosion would be a major hazard, but with excellent grass cover throughout the year erosion is practically eliminated.

APPROACHES FOR PRODUCING FEED DURING THE DRY SEASON

There is little doubt that one of the most seriously limiting factors for beef production in ustic areas is the scarcity of feed during the dry season. This problem could be solved by storing excess grass produced during the rainy season as hay or silage, by supplementary feeding, or by transferring the animals to areas that have no moisture limitations. Hay production from intensively fertilized grasses or from grass–legume pastures is not commonly practiced in the tropics because of problems of curing hay in the field during the rainy season, and the low nutritive quality of the grasses when they are harvested at mature stages of growth (McDowell, 1972; Blue and Tergas, 1969). Also, it is difficult for coarse-stemmed species to dehydrate quickly enough to prevent damage from mold and fungi. In certain situations mature grass can be used as "standing hay" to supply energy for part of the dry season. Norman and Stewart (1964) reported that cattle needed a supplement of 10.5 kg protein/day to prevent weight losses during the dry season.

Pure stands of legumes such as *Stylosanthes humilis* have been harvested as hay and stored satisfactorily during the dry season in northern Australia (Wright, 1971). This species does not lose its nutritive quality appreciably and can be harvested at the beginning of the dry season as soon as the seed matures. Ensilage of intensively fertilized corn or sorghum is another alternative. McDowell (1972) considers that silage making has limitations in the tropics because of the unfavorable ratio of soluble carbohydrates to protein found in tropical forages, especially at high levels of nitrogen fertilization, which cause an unfavorable kind of fermentation. However, storage of heavily fertilized forage sorghum grown during the rainy season in inexpensive trench silos has been adopted as a viable solution for small dairy farmers in northern El Salvador (North Carolina State University, 1973).

Attempts have been made to improve the nutritional quality of grasses during the dry season by late applications of nitrogen at the end of the rainy season. Application of 75 and 150 kg N/ha to *Hyparrhenia rufa* 1 month before the dry season maintained dry matter production and protein and phosphorus contents at adequate levels for 1 to 2 months into the dry season in Guanacaste, Costa Rica (Tergas and Blue, 1971; Tergas et al., 1971). Pasture regrowth in the following rainy season was stimulated by these applications, suggesting a strong residual effect. Similar positive results have been observed in Brazil (Werner, 1970), but the negative results obtained in Venezuela (Chacón et al., 1971) imply that the success of this practice may depend on local conditions. If heavily fertilized grasses are grown during the rainy season, a similar positive carryover effect may take place.

Another possibility from the soil management point of view is a combination of the extensive legume-based system in large areas and intensive fertilization in small areas. The intensive system could be established on soils that are more fertile or can hold moisture better during the dry season. Excess production during the rainy season could be used for hay if conditions permitted. In the Cerrado of Brazil, where economic considerations suggest the use of extensive legume-based systems, about 5 to 10 percent of the area could be irrigated throughout the year by relatively inexpensive gravity systems (North Carolina State University, 1974). Intensively fertilized areas could maintain a carrying capacity up to 5 or 10 times higher than the extensive areas, judging from the data reviewed in this chapter. By devoting these small irrigated areas to intensively fertilized grasses, annual weight losses could be eliminated without the use of expensive concentrates. Research based on these concepts and economic tradeoffs between the extensive and intensive systems could help to solve the dry season feeding problem.

SUMMARY AND CONCLUSIONS

1. Animal production is a major component of tropical agriculture. Low intake of energy and protein by grazing animals is considered the main limiting factor preventing high beef and milk production in the tropics. Pasture production is strongly affected by soil moisture and fertility limitations. Tropical pastures are produced in three distinct soil management systems: natural grasslands, improved grass–legume pastures, and intensively fertilized grasses used for grazing or cut forage.

2. Over 90 percent of the pasture area used for animal production in the tropics consists of natural grasslands. Cattle gain weight during the rainy season but suffer severe losses during the dry season. This zigzag pattern of weight gains delays slaughtering to an average age of 4 to 10 years, decreases animal fertility rates, and results in annual liveweight gains on the order of 20 to 50 kg/ha with stocking rates of only 0.05 to 0.3 animal/ha.

3. Periodic burning is essential for maintaining the productivity of natural grasslands, even at such low levels. Annual burning is required in areas of high rainy season rainfall to eliminate the mature, unpalatable grasses and allow fresh regrowth. In areas of lower rainy season rainfall, burning every 2 or three years is more effective. Fertilizing natural grasslands is seldom beneficial because native grasses do not respond efficiently to added nutrients. The best alternative known to present technology is to oversow adapted legume species with moderate fertilization.

4. Improved grass and legume species may be planted on natural grasslands, newly cleared forests, and/or cropland that it is desirable to transform into pastures. Several grass and legume species are well adapted to moisture stress, as well as to high exchangeable aluminum and manganese, and to low phosphorus, potassium, calcium, and magnesium levels. Many of the tropical legume species tolerant of these soil fertility limitations enter into symbiosis with the slow-growing, "cowpea-type" *Rhizobium* strains, which are also well adapted to these chemical limitations. Good nodulation is thus obtained. In contrast, other legume species and certain

specific strains of rhizobia are very sensitive to these limiting factors. Inoculation with certain acid-tolerant species may be needed, and pelleting with rock phosphate, but not lime, is often beneficial.

5. In grass–legume systems the grass is expected to supply most of the energy needs, while the legume supplies a major part of the protein and produces nitrogen, which when cycled through the animal fertilizers the grasses. The level of annual nitrogen fixation by tropical legumes is comparable to and in some cases exceeds the amounts fixed by temperate-region legumes. The annual nitrogen contribution of the legumes to the pasture is directly correlated with the dry matter content of legume tops, when adapted species are used and are well managed. Many tropical legumes are very sensitive to defoliation and require longer grazing intervals in order to fix high quantities of nitrogen. Nitrogen fertilization reduces nitrogen fixation by the legumes, as well as their relative proportion in the sward. The requirements for nutrients other than nitrogen in grass–legume pastures are moderate because of the efficient recycling by animals, except in soils with high phosphorus fixation capacities. Improved grass–legume pastures with modest fertilization produce about 100 to 300 kg/ha of annual liveweight gains at stocking rates of 0.5 to 1 animal/ha.

6. Intensively fertilized grass pastures are grown where cost:price ratios justify relatively large investments per unit of land. No legumes are used because grasses require rates of nitrogen on the order of 400 to 900 kg N/ha a year, which are beyond the potential contribution of the legumes. Only grasses that have a high capacity to respond to nitrogen and to utilize it efficiently are used. The annual fertilization requirements for cut forages are on the order of 400-50-200 kg NPK/ha a year plus 1 ton lime/ha to counteract the residual acidity of nitrogen fertilizers. Under grazing the above rates are reduced by half because of nutrient recycling through the animals. The protein production of intensively fertilized grasses and their digestibility are high as long as they are cut at appropriate intervals. Intensively managed grass pastures support annual liveweight gains of 500 to over 1000 kg/ha with stocking rates on the order of 5 to 10 animals/ha.

7. The basic problem of feed supply during the dry season in ustic areas can be solved by storing surplus rainy season production as hay or silage, although both systems involve management problems. An alternative in regions where a small proportion of the land area can be irrigated is to grow intensively fertilized grasses to supplement the grass–legume pastures grown on the bulk of the land area.

8. The choice among the three management systems or combinations thereof is simply a matter of which one has the comparative advantage, given the locally prevalent cost: price ratios.

REFERENCES

Abruña, F., R. W. Pearson, and C. Elkins. 1958. Qualitative evaluation of soil reaction and base status changes from field applications of residually acid forming nitrogen fertilizers. *Soil Sci. Soc. Amer. Proc.* **22**:539–542.

Allen, S. E. and D. A. Mays. 1974. Coated and other slow-release fertilizers for forages. Pp. 559–582. In D. A. Mays (ed.), *Forage Fertilization.* American Society of Agronomy, Madison, Wisc.

Andrew, C. S. and M. P. Hegarty. 1969. Comparative responses to manganese excess of eight tropical and four temperate pasture species. *Aust. J. Agr. Res.* **20**:687–696.

Andrew, C. S. and M. F. Robins. 1969. The effect of potassium on the growth and chemical composition of some tropical and temperate pasture legumes. *Aust. J. Agr. Res.* **20**:999–1021.

Andrew, C. S. and W. H. J. Pieters. 1970. Effect of potassium on the growth and chemical composition of some pasture legumes. 3. Deficiency symptoms of 10 tropical pasture legumes. *CSIRO Tech. Paper Div. Trop. Pastures* **5**:2–11.

Andrew, C. S. and M. F. Robins, 1971. Effects of phosphorus on the growth, chemical composition and critical phosphorous percentages of some tropical pasture grasses. *Aust. J. Agr. Res.* **22**:693–706.

Andrew, C. S., A. D. Johnson, and R. L. Sandland. 1973. Effect of aluminum on the growth and chemical composition of some tropical and temperate pasture legumes. *Aust. J. Agr. Res.* **24**:325–339.

Andrew, C. S. and P. J. Vanden Berg. 1973. The influence of aluminum on phosphate sorption by whole plants and exised roots of some pasture legumes. *Aust. J. Agr. Res.* **24**:341–351.

Bastidas, A., J. Bernal, J. Lotero, et al. 1967. Frecuencia de corte y aplicación de nitrogeno a cuatro gramíneas en climas calientes. *Agr. Tropical (Colombia)* **23**:747–756.

Beeson, K. C., A. Goytendía, and G. G. Gomez. 1972. Selection of soils for pastures using common chemical measures. *Agron. J.* **64**:58–59.

Bernal, J. 1972. Las leguminosas como fuentes de nitrógeno en pastos y rotaciones. *Suelos Ecuatoriales* **4**:175–194.

Blue, W. G., L. Andrade, E. Rey, M. T. Ramirez, et al. 1963. Investigations on the potential for pasture development in the Atlantic zone of Costa Rica. *Proc. Soil Crop Sci. Soc. Fla.* **23**:208–221.

Blue, W. G. and L. E. Tergas. 1969. Dry season deterioration of forage quality in the wet-dry tropics. *Proc. Soil Crop. Sci. Fla.* **29**:224–238.

Boughey, A. S., P. E. Munro, J. Mieklejohn, R. M. Strang, and M. J. Swift. Antibiotic reactions between African savanna species. *Nature* **203**:1302–1303.

Bowen, J. E. 1972. Manganese–silicon interaction and its effect on growth of sudan grass. *Plant and Soil* **37**:577–588.

Brolmann, J. B. and R. M. Sonoda. 1975. Differential response of three *Stylosanthes guyanensis* varieties to three levels of potassium. *Trop. Agr. (Trinidad)* **52**:139–142.

Bruce, R. C. 1965. Effect of *Centrosema pubescens* on soil fertility in the humid tropics. *Queensl. J. Agr. Anim. Sci.* **22**:221–226.

Bruce, R. C. 1972. The effect of topdressed superphosphate on the yield and botanical composition on a *Stylosanthes guyanensis* pasture. *Trop. Grassl.* **6**:135–140.

Bruce, R. C. and I. J. Bruce. 1972. Correlation of soil phosphorus analyses with response of tropical pastures to superphosphate on some North Queensland soils. *Aust. J. Exptal. Agr. Anim. Husb.* **12**:188–194.

Bryan, W. 1967. Botanical changes following application of fertilizer and seed on run-down paspalum, kikuyo and matgrass pastures on scrub soil at Mateny, southeast Queensland. *Trop. Grassl.* **1**:167–170.

Bryan, W. W. 1970. Tropical and subtropical forests and heaths. Pp. 101–111. In R. M. Moore (ed.), *Australian Grasslands.* Australian National University Press, Canberra.

Bryan, W. W. and C. S. Andrew. 1971. Value of Nauru rock phosphate as a source of phosphorus for some tropical pasture legumes. *Aust. J. Exptal. Agr. Anim. Husb.* **11**:532–535.

Bryan, W. W. and T. R. Evans. 1973. Effects of soils, fertilizers and stocking rates on pastures and beef production on the Wallum of southeast Queensland. 1. Botanical composition and chemical effects on plants and soils. *Aust. J. Exptal. Agr. Anim. Husb.* **13**:516–529.

Caro, R., J. Vicente-Chandler, and J. Figarella. 1960. The yield and composition of five grasses growing in the humid mountains of Puerto Rico as affected by nitrogen fertilization, season and harvest procedure. *J. Agr. Univ. Puerto Rico* **44**:107–120.

Caro, R. and J. Vicente-Chandler. 1961. Effect of fertilization on carrying capacity and beef produced by Napier grass. *Agron. J.* **53**:204–205.

Caro, R. and J. Vicente-Chandler. 1963. Effect of liming and fertilization on productivity and species balance of tropical kudzu–molasses grass pasture under grazing management. *J. Agr. Univ. Puerto Rico* **47**:236–241.

Caro, R., J. Vicente-Chandler and J. Figarella. 1965. Productivity of intensively managed pastures on five grasses on steep slopes in the humid mountains of Puerto Rico. *J. Agr. Univ. Puerto Rico* **49**:99–111.

Caro, R., F. Abruña and J. Vicente-Chandler. 1972a. Comparison of heavily fertilized pangola and stargrass pastures in terms of beef production and carrying capacity in the humid mountain region of Puerto Rico. *J. Agr. Univ. Puerto Rico* **56**:104–109.

Caro, R., F. Abruña and J. Figarella. 1972b. Effect of nitrogen rates harvesting interval and cutting heights on yield and composition of stargrass in Puerto Rico. *J. Agr. Univ. Puerto Rico* **56**:267–279.

Caro, R., J. Vicente-Chandler and F. Abruña. 1972c. Effect of four levels of fertilization on beef production and carrying capacity of pangola grass pastures in the humid mountain region of Puerto Rico. *J. Agr. Univ. Puerto Rico* **56**:219–222.

Caro, R. and J. Vicente-Chandler. 1972. Effect of heavy rates of fertilization on beef production and carrying capacity of Napier grass pastures over five consecutive years of grazing under humid tropical conditions. *J. Agr. Univ. Puerto Rico* **56**:223–227.

Caro, R., F. Abruña and J. Vicente-Chandler. 1973. Comparison of heavily fertilized pangola grass and stargrass pastures under humid tropical conditions. *Agron. J.* **65**:132–133.

Caro, R., and J. Vicente-Chandler. 1974. Milk production of young Holstein cows fed only grass from steep, intensively managed tropical grass pasture over three successive lactations. *J. Agr. Univ. Puerto Rico* **58**:18–26.

Cassady, J. T. 1973. The effect of rainfall, soil moisture and harvesting intensity on grass production in two rangeland sites in Kenya. *East Afr. Agr. For. J.* **39**:26–36.

Cassidy, G. J. 1971. Response of a mat-grass-paspalum sward to fertilizer applications. *Trop. Grassl.* **5**:11–22.

Chacón, E., S. Rodriguez y C. F. Chicco. 1971. Efecto de la fertilización tardía con nitrógeno sobre el valor nutritivo del pasto pangola. *Agron. Tropical (Venezuela)* **21**:503–509.

Chaverra, H., S. Echeverri y L. V. Crowder. 1967. Aplicación de nitrógeno a mezcla de gramíneas y leguminosas. *Agr. Tropical (Colombia)* **23**:226–232.

Chheda, H. R. and J. O. Akinola. 1971. Effects of cutting frequency and level of applied nitrogen on crude protein production and nitrogen recovery by three *Cynodon* strains. *West Afr. J. Biol. Appl. Chem.* **14**:31–38.

Chheda, H. R., M. Saleem and M. A. Mohamed. 1973. Effects of nitrogen and potassium fertilizers on *Cynodon* IB.8 and on soil potassium in southern Nigeria. *Exptal. Agr.* **7**:249–255.

Chesney, H. A. D. 1969. Fertilizer studies with pangola-grass on Tiwiwid fine sand, Guyana. *Agr. Res. Guyana* **3**:136–138.

Chesney, H. A. D. 1972. Yield response to pangola-grass grown on Tiwiwid fine sand to magnesium and fitted micronutrients. *Agron. J.* **64**:152–154.

Chicco, C. F., S. Rodriguez, and C. E. Fuenmayor. 1971. Efecto de la fertilización con nitrógeno sobre el rendimiento, consumo y digestibilidad del heno de pangola. *Agron. Tropical (Venezuela)* **21**:215–227.

CIAT. 1972, 1973. Annual Reports. Centro Internacional de Agricultura Tropical, Palmira, Colombia.

Coaldrake, J. E. 1970. The Brigalow. Pp. 123–140. In R. P. Moore (ed.), *Australian Grasslands.* Australian National University Press, Canberra.

Colman, R. L. 1972. Factors affecting the response to nitrogen of temperate and tropical grasses. *J. Aust. Inst. Agr. Sci.* **38**:225–226.

Cortes, H. 1966. Niveles y frecuencia de aplicación de nitrógeno al pasto pangola. *Acta Agron. (Colombia)* **16**:101–131.

Crack, B. J. 1972. Changes in soil nitrogen following different establishment procedures for Townsville Stylo on a solodic soil in northeastern Queensland. *Aust. J. Exptal. Agr. Anim. Husb.* **12**:274–280.

Crespo, G. 1972. Efectos de tres niveles de urea y dos sistemas de aplicación en el rendimiento y contenido de nitrógeno de pasto pangola. *Rev. Cubana Cienc. Agr.* **6**:235–244.

Crowder, L. V. 1974. Pasture and forage research in tropical America. *Cornell Int. Agr. Bull.* 28.

Crowder, L. V. and Riveros. 1962. Resumen de las investigaciones en pastos y forrajes. *Agr. Tropical (Colombia)* **15**:35–51.

Crowder, L. V., J. Lotero, A. Michelin, et al. 1963. Fertilización de gramíneas tropicales y subtropicales en Colombia. *Div. Inv. Agrop. Bol. Divulg.* 12.

Crowder, L. A., A. Michelin, and A. Bastidas. 1964. The response of pangola grass to rate and time of nitrogen application in Colombia. *Trop. Agr. (Trinidad)* **41**:21–29.

Crush, J. R. 1974. Plant growth responses to vesicular-arbuscular mycorrhiza. 7. Growth and nodulation of some herbage legumes. *New Phytol.* **73**:743–749.

Davies, J. G. and N. H. Shaw. 1964. General objectives and concepts. *Commonwealth Bur. Pastures Field Crops Bull.* **47**:1–49.

Davies, J. G. and E. M. Hutton. 1970. Tropical and subtropical pasture species. Pp. 273–302. In R. P. Moore (ed.), *Australian Grasslands.* Australian National University Press, Canberra.

Dávila, V. and S. Echeverri. 1967. Aplicación de nitrógeno y riego en pasto kikuyo. *Agr. Tropical (Colombia)* **23**:744–746.

Deinum, B. and J. G. P. Dirven. 1975. Climate, nitrogen and grass. 6. Comparison of yield and chemical composition of some temperate and tropical grass species grown at different temperatures. *Netherl. J. Agr. Sci.* **23**:69–82.

Dirven, J. G. P. 1970. Yield increase in tropical grasslands by fertilization. Pp. 403–409. *Proc. Ninth Congr. Int. Potash Inst.*

Döbereiner, J. and S. Abramovitch. 1965. Efeito da calagem e da temperatura do solo la fixação do nitrogênio de *Centrosema pubescens* em un solo com toxides de manganés. *Proc. Ninth Int. Grassl. Congr.* (*São* Paulo) **2**:1121–1124.

Döbereiner, J., J. M. Day, and P. J. Dart. 1973. Fixação de nitrogênio na rizosfera de *Paspalum notatum* e da cana-de-açúcar. *Pesq. Agropec. Bras.* **8**:153–157.

Döbereiner, J. and J. M. Day. 1974. *Associative Symbiosis in tropical Grasses: Characterization of Microorganisms and Nitrogen-Fixing Sites.* Instituto de Pesquisa Agropecuaria do Centro Sul, EMBRAPA, Rio de Janeiro.

Downes, R. W. 1967. Establishment of legume in pastures of savanna woodlands in north Queensland. *Queensl. J. Agr. Anim. Sci.* **24**:23–29.

Edwards, D. C. 1942. Grass burning. *Emp. J. Exptal. Agr.* **10**:219–231.

Escobar, L., A. Ramirez, and J. Lotero. 1967. Dósis y frecuencias de aplicaciones de nitrógeno en tres gramíneas tropicales. *Agr. Tropical* (*Colombia*) **23**:726–737.

Escobar, L., A. Ramirez, and J. Lotero. 1969. Dósis y frecuencias de aplicación de nitrógeno al pasto bermuda de costa. *Rev. Inst. Colomb. Agropec.* **4**:269–276.

Evans, T. R. 1969. Beef production from nitrogen fertilized pangola grass on the coastal lowlands of southern Queensland. *Aust. J. Exptal. Agr. Anim. Husb.* **9**:282–286.

FAO. 1973. *Production Yearbook.* Food and Agricultural Organization of the United Nations, Rome.

Falade, J. 1973. Effect of phosphorus on the growth and mineral composition of four tropical forage legumes. *J. Sci. Food Agr.* **24**:795–802.

Falvey, L. 1975. Effect of mowing frequency and height of Townsville Stylo pastures on cleared Blain soil. *Trop. Agr.* (*Trinidad*) **52**:143–148.

Fernandes, A. P. M., J. A. Gomide, and J. M. Braga. 1970. Efeito da adubação potássica sôbre a produção e valor nitritivo de algumas gramíneas forrageiras tropicais. *Experientiae* **10**:185–208.

Figarella, J., F. Abruña, and J. Vicente-Chandler. 1972. Effect of five nitrogen sources applied at four rates to pangola grass under humid tropical conditions. *J. Agr. Univ. Puerto Rico* **56**:410–416.

Fisher, M. J. 1970. The effects of superphosphate on the growth and development of Townsville Stylo in pure ungrazed swards at Katherine, N. T. *Aust. J. Exptal. Agr. Anim. Husb.* **10**:716–724.

Fisher, M. J. and N. A. Campbell. 1972. The initial and residual responses of phosphorus fertilizers of Townsville Stylo in pure ungrazed swards at Katherine, N. T. *Aust. J. Exptal. Agr. Anim. Husb.* **12**:488–494.

Ford, C. W. and W. T. Williams. 1973. *In vitro* digestibility and carbohydrate composition of *Digitaria decumbens* and *Setaria anceps* grown at different levels of nitrogenous fertilizers. *Aust. J. Agr. Res.* **24**:309–316.

Fox, R. L., S. M. Hassan, and R. C. Jones. 1971. Phosphate and sulfate sorption by Latosols. *Int. Symp. Soil Fert. Eval. Proc.* (*New Delhi*) **1**:857–864.

Fox, R. L., R. K. Hashimoto, J. R. Thompson, and R. S. de la Peña. 1974. Comparative external phosphorus requirements of plants growing in tropical soils. *Tenth Int. Congr. Soil Sci.* (*Moscow*) **4**:232–239.

França, G. E. de, and M. M. de Carvalho. 1970. Ensaio exploratório de fertilização de cinco leguminosas tropicais em un solo de Cerrado. *Pesq. Agropec. Bras.* **5**:147–153.

Freitas, L. M. M. de, A. C. McClung, and W. L. Lott. 1960. Field studies of fertility problems in two Brazilian campos cerrados, 1958–59. *IBEC Res. Inst. Bull.* 21.

Gartner, J. A. 1966. The effects of different rates of fertilizer nitrogen on the growth, nitrogen uptake and botanical composition of tropical grass swards. Pp. 223–227. *Proc. 10th Int. Grassl. Congr.*

Gartner, J. A. 1967. Fertilizer response of green panic on the Atherton Tableland, Queensland. *Queensl. J. Agr. Anim. Sci.* **24**:345–352.

Gartner, J. A. 1969. Effect of fertilizer nitrogen on a dense sward of kikuyu, paspalum and carpet grass. 1. Botanical composition, growth and nitrogen uptake. *Queensl. J. Agr. Anim. Sci.* **26**:21–33.

Gates, C. T. 1974. Nodule and plant development in *Stylosanthes humilis:* symbiotic response to phosphorus and sulfur. *Aust. J. Bot.* **22**:45–55.

Gates, C. T. and J. R. Wilson. 1974. The interaction of nitrogen and phosphorus on the growth, nutrient status and nodulation of *Stylosanthes humilis. Plant and Soil* **41**:325–333.

Gibson, T. A. and L. R. Humphreys. 1973. The influence of nitrogen nutrition of *Desmodium uncinatum* on seed production. *Aust. J. Agr. Res.* **24**:667–676.

Gómez, G., T. Aguilar, and K. C. Beeson. 1966. Interrelaciones suelo-planta-nutrición. 2. Calidad nutritiva de los forrajes del valle del Huallaga Central. *Anal. Cient. (Perú)* **4**:147–161.

Gomide, J. A., C. H. Noller, G. O. Mott, J. H. Conrad, and D. L. Hill. 1969. Effect of plant age and nitrogen fertilization on the chemical composition and *in vitro* cellulose digestibility of tropical grasses. *Agron. J.* **61**:116–123.

Graham, T. G. and B. G. Meyer. 1972. Effect of method of establishment of Townsville Stylo and the application of superphosphate on the growth of steers. *Queensl. J. Agr. Anim. Sci.* **29**:289–296.

Graham, P. H., V. M. Morales, and R. Cavallo. 1974. Materiales exipientes y pegantes de posible uso en nodulación de leguminosas en Colombia. *Turrialba* **24**:47–50.

Graham, P. H. and D. H. Hubbell. 1975a. Interacciones suelo–planta–*Rhizobium* en agricultura tropical. Pp. 211–235. In E. Bornemisza and A. Alvarado (eds.), *Manejo de Suelos en America Tropical.* North Carolina State University, Raleigh.

Graham, P. H. and D. H. Hubbell. 1975b. Legume-*Rhizobium* relationships in tropical agriculture. Pp. 9–21. In E. C. Doll and G. O. Mott (eds.), "Tropical Forages in Livestock Production Systems." *ASA Spec. Publ.* 24. American Society of Agronomy, Madison, Wisc.

Grof, B., J. Courtice, and D. G. Cameron. 1970. Effect of renovation and nitrogen fertilization on an old stand of buffelgrass in subcoastal Central Queensland. *Queensl. J. Agr. Anim. Sci.* **26**:359–364.

Grof, B. and W. A. T. Harding. 1970. Dry matter yields and animal production of guinea grass on the humid tropical coast of north Queensland. *Trop. Grassl.* **4**:85–95.

Guerrero, R., H. W. Fassbender, and J. Blydenstein. 1970. Fertilización del pasto Elefante en Turrialba, Costa Rica. 1 y 2. *Turrialba.* **20**:53–63.

Haggar, R. J. 1969. A guide to the management and use of stylo (*Stylosanthes gracilis*). *Samaru Agr. Newsletter (Nigeria)* **11**:63–66.

Harding, W. A. T. and D. G. Cameron. 1972. New pasture legumes for the wet tropics. *Queensl. Agr. J.* **98**:394–406.

Harty, R. L. 1967. Effect of superphosphate on the germination of Townsville Lucerne. *Queensl. J. Agr. Anim. Sci.* **24**:235–236.

Hendy, K. 1972. Response of a pangola grass pasture near Darwin to the wet season application of nitrogen. *Trop. Grassl.* **6**:25–32.

Henzell, E. F. 1968. Sources of nitrogen for Queensland pastures. *Trop. Grassl.* **2**:1–17.

Henzell, E. F. 1970. Use of nitrogenous fertilizers on subtropical pasture in Queensland. *J. Aust. Inst. Agr. Sci.* **36**:206–213.

Henzell, E. F. 1971. Recovery of nitrogen from four fertilizers applied to Rhodes grass in small plots. *Aust. J. Exptal. Agr. Anim. Husb.* **11**:420–430.

Henzell, E. F. 1972. Loss of nitrogen from a nitrogen fertilized pasture. *J. Aust. Inst. Agr. Sci.* **38**:309–310.

Henzell, E. F. and D. O. Norris. 1962. The use of nitrogen fertilizers on pastures in the subtropics and tropics. *Commonwealth Agr. Bur. Bull.* **46**:161–172.

Herrera, G., J. Lotero, and L. V. Crowder. 1967. Influencia del nitrógeno y frecuencia de aplicación en la producción de forraje y proteina del pasto pangola. *Agr. Tropical* (*Colombia*) **23**:297–312.

Herrera, G., A. Ramirez, and J. Lotero. 1968. Dósis de nitrógeno y frecuencia de aplicación en sorgo forrajero. *Agr. Tropical* (*Colombia*) **24**:675–680.

Hill, G. D. 1970. Studies on the growth of *Leucaena leucocephala*. 1. Effect of clean weeding and nitrogen fertilizer on early establishment. 2. Effect of lime at sowing production from a low calcium status soils of the Sogeri Plateau. *Papua New Guinea Agr. J.* **22**:29–30, 69–71.

Hutton, E. M. 1970. Tropical pastures. *Adv. Agron.* **22**:1–73.

Jaramillo, R., H. Chaverra, and C. P. Oñoro. 1968. Frecuencia de aplicación y dósis de nitrógeno para *Festuca alta* y *F. media*. *Rev. Inst. Colomb. Agropec.* **3**:179–193.

Jones, M. B. and L. M. M. de Freitas. 1970. Respostas de quatro leguminosas a fósforo, potássio e calcário num Latossolo Vermelho-Amarelo de Campo Cerrado. *Pesq. Agrop Bras.* **5**:91–99.

Jones, M. B., J. L. Quagliato, and L. M. M. de Freitas. 1970. Respostas de alfalfa e algumas leguminosas tropicais a aplicaçóes de nutrientes minerais, em tres solos de Campo Cerrado. *Pesq. Agrop. Bras.* **5**:209–214.

Jones, M. B. and J. L. Quagliato. 1973. Response of four tropical legumes and alfalfa to varying levels of sulphur. *Sulphur Inst. J.* **9**:6–9.

Jones, R. J. 1972. The place of legumes in tropical pastures. *ASPAC Tech. Bull.* 9. Taipei, Taiwan. 69 pp.

Jones, R. J., J. G. Davies, and R. B. Waite. 1967. The contribution of some tropical legumes to pasture yields of dry matter and nitrogen at Samford, southeastern Queensland. *Aust. J. Exptal. Agr. Anim. Husb.* **7**:57–65.

Jones, R. K. 1968. Initial and residual effects of superphosphate on a Townsville Lucerne pasture in northeastern Queensland. *Aust. J. Exptal. Agr. Anim. Husb.* **8**:521–527.

Jones, R. K. 1973. Deep sandy soils in Cape York Peninsula, North Queensland. 2. Plant nutrient status. *Aust. J. Exptal. Agr. Anim. Husb.* **13**:89–97.

Jones, R. K. 1974. Phosphorus responses of a wide range of accessions from the genus *Stylosanthes*. *Aust. J. Agr. Res.* **26**:847–862.

Jon s, R. K., P. J. Robinson, K. P. Haydock, et al. 1971. Sulphur–nitrogen relationships in the tropical legume *Stylosanthes humilis*. *Aust. J. Agr. Res.* **22**:885–894.

Jones, R. M. 1970. Sulfur deficiency of dryland lucerne in the eastern Darling Downs of Queensland. *Aust. J. Exptal. Agr. Anim. Husb.* **10**:749–754.

Kalma, J. D. 1971. The annual course of air temperature and near surface soil temperature in a tropical savannah environment. *Agr. Met.* **8**:292–303.

Kamprath, E. J. 1973. Phosphorus. Pp. 138–161. In P. A. Sanchez (ed.), "A Review of Soils Research in Tropical Latin America." *North Carolina Agr. Exp. Sta. Tech. Bull.* **219**.

Kerridge, P. C., C. S. Andrew, and G. G. Murtha. 1972. Plant nutrient status of soils of the Atherton Tableland, north Queensland. *Aust. J. Exptal. Agr. Anim. Husb.* **12**:618–627.

Keya, N. C. O. 1973. The effect of N P fertilizers on the productivity of *Hyparrhenia* grassland. *East Afr. Agr. For. J.* **39**:195–200.

Keya, N. C. O. 1975. Grass-legume pastures in western Kenya. 1. A comparison of the productivity of cut and grazed swards. 2. Legume performance at Kitale, Kisii and Kakamega. *East Afr. Agr. For. J.* **39**:240–258.

Keya, N. C. O., F. J. Olsen, and R. Holliday. 1971a. The role of superphosphate in the establishment of oversown tropical legumes in natural grasslands of western Kenya. *Trop. Grassl.* **5**:109–116.

Keya, N. C. O., F. J. Olsen, and R. Holliday. 1971b. Oversowing improved pasture legumes in natural grasslands of the medium altitudes of western Kenya. *East Afr. Agr. For. J.* **37**:148–155.

Keya, N. C. O., F. J. Olsen, and R. Holliday. 1972. Comparison of seedbeds for oversowing a *Chloris gayana–Desmodium uncinatum* mixture in *Hyparrhenia* grasslands. *East Afr. Agr. For. J.* **37**:286–293.

Keya, N. C. O. and D. W. Kalangi. 1973. The seeding and superphosphate rates for the establishment of *Desmodium uncinatum*. *Trop. Grassl.* **7**:319–325.

Kleinschmidt, F. H. 1967. The influence of nitrogen and water on pastures green panic, lucerne and glycine at Lawes, southeastern Queensland. *Aust. J. Exptal. Agr. Anim. Husb.* **7**:441–446.

Landrau, P., G. Samuels, and P. Rodriguez. 1953. Influence of fertilizers, minor elements and soil pH on the growth and protein content of tropical kudzu. *J. Agr. Univ. Puerto Rico* **37**:81–95.

Lee, M. T. and G. L. Wilson. 1972. The calcium and pH components of lime responses in tropical legumes. *Aust. J. Agr. Res.* **23**:257–265.

Lopes, E. S., L. A. C. Lovadini, H. Gargantini, S. Miyasaka, and J. C. Leon. 1971. Capacidade fixadora de nitrogênio de Rhyzobium autóctone associado com soja perenne e siratro, em dois solos do Estado de São Paulo. *Bragantia* **30**:145–154.

Lotero, J., G. Herrera, and L. V. Crowder. 1965. Respuesta de una pradera natural a la aplicación de fertilizantes. *Agr. Tropical (Colombia)* **21**:229–232.

Lotero, J., J. Bernal, and G. Herrera. 1967. Distancia de siembra y aplicación de nitrógeno en pasto elefante. *Rev. Inst. Colomb. Agropec.* **2**:123–133.

Lotero, J., A. Ramírez, and G. Herrera. 1968. Fuentes, dósis y métodos de aplicación de nitrógeno en pasto Elefante. *Rev. Inst. Colomb. Agropec.* **3**:113–121.

Lotero, J., G. Herrera, and A. Ramírez. 1969. Distanciamiento y dósis de nitrógeno en pasto *Axonopus scoparius*. *Rev. Inst. Colomb. Agropec.* **4**:147–157.

Lotero, J., S. Monsalve, A. Ramírez, and F. Villamízar. 1971. Respuestas al encalado de gramineas y leguminosas forrajeras. *Suelos Ecuatoriales* **3**:210–239.

Mannetje, L't., and N. H. Shaw. 1972. Nitrogen fertilizer responses of a *Heteropogon contortus* and a *Paspalum plicatulum* pasture in relation to rainfall in central coastal Queensland. *Aust. J. Exptal. Agr. Anim. Husb.* **12**:28–35.

Mathieu, P. 1962. Activités zootechniques au Burundi de 1952 à 1962. *Boll. Inf. INEAC* **11**:403–438.

McClung, A. C., L. M. M. de Freitas, T. R. Gallo, et al. 1957. Preliminary fertility studies on Campo Cerrado soils of Brazil. *IBEC Res. Inst. Bull.* 13.

McClung, A. C. and L. R. Quinn. 1959. Sulfur and phosphorus response of batatais grass (*Paspalum notatum*). *IBEC Res. Inst. Bull.* 18.

McDowell, L. R., J. H. Conrad, J. E. Thomas, and L. E. Harris. 1974. *Latin American Tables of Feed Composition.* University of Florida, Gainesville.

McDowell, R. E. 1972. *Improvement of Livestock Production in Warm Climates.* Freeman, San Francisco. 709 pp.

McIlroy, R. J. 1972. *An Introduction to Tropical Grassland Husbandry,* 2nd ed. Oxford University Press, London. 160 pp.

Michelin, A., L. A. León, and A. Ramírez. 1974. Uso eficiente de fertilizantes en la producción de pastos en suelos ácidos. *Suelos Ecuatoriales* **6**:265–287.

Michell, T. E., W. W. Bryan, and T. R. Evans. 1972. Budgetary comparison between pangola grass/legume pasture and nitrogen fertilized pangola grass for beef production in the southern Wallum. *Trop. Grassl.* **6**:177–190.

Miller, S. F., L. R. Quinn, and G. O. Mott. Análise econômica de experimentos com forrageiras e gado realizados no Estado de São Paulo. *Pesq. Agrop. Bras.* **5**:101–116.

Minson, D. J. 1971. The nutritive value of tropical pastures. *J. Aust. Inst. Agr. Sci.* **37**:255–263.

Minson, D. J. 1973. Effect of fertilizer nitrogen in digestibility and voluntary intake of *Chloris gayana, Digitaria decumbens* and *Pennisetum clandestinum. Aust. J. Exptal. Agr. Anim. Husb.* **13**:153–157.

Minson, D. J. and M. N. McLeod. 1970. The digestibility of temperate and tropical grasses. Pp. 719–722. In *Proc. 11th Int. Grassl. Congr. (Australia).*

Moody, P. W. 1974. Nutritional problems of northern Australian soils. *Northern Territory Anim. Ind. Agr. Branch Tech. Bull.* 12. 11 pp.

Moore, R. M. (ed.). 1970). *Australian Grasslands.* Australian National University Press, Canberra. 455 pp.

Morales, V. M., P. H. Graham, and R. Cavallo. 1973. Efecto del método de inoculación y encalado en la nodulación de leguminosas en un suelo de Carimagua, Colombia. *Turrialba* **23**:52–55.

Mosse, B., D. S. Hayman, and D. J. Arnold. 1973. Plant responses to vesicular-arbuscular mycorrhiza. 5. Phosphate uptake by three plant species from P-deficient soils labelled with P^{32}. *New Phytol.* **72**:809–815.

Motooka, P. S. et al. 1968. New role for an old jungle. *World Farming* **10**:26–29.

Motooka, P. S. et al. 1969. Pasture establishment in tropical brushlands by aereal herbicide and seeding treatments on Kauai. *Hawaii Agr. Exp. Sta. Tech. Progr. Rept.* **165**:3–18.

Mott, G. O. 1974. Nutrient recycling in pastures. Pp. 323–329. In D. A. Mays (ed.), *Forage Fertilization.* American Society of Agronomy, Madison, Wisc.

Mott, G. O., L. R. Quinn, and W. V. A. Bischoff. 1970. The retention of nitrogen in a soil–plant-animal system in guinea grass (*Panicum maximum*) pastures in Brazil. Pp. 414–416. In *Proc. 11th. Int. Grassl. Congr. (Australia)*.

Norman, M. J. T. 1962. Response of native pasture to nitrogen and phosphate fertilizer at Katherine, N. T. *Aust. J. Exptal. Agr. Anim. Husb.* 2:27–34.

Norman, M. J. T. 1963. The short-term effects of time and frequency of burning on native pastures at Katherine, N. T. *Aust. J. Exptal. Agr. Anim. Husb.* 3:26–29.

Norman, M. J. T. 1966. Katherine Research Station, 1956–65: A review of published work. *CSIRO Div. Land. Res. Tech. Paper* 28.

Norman, M. J. T. and G. A. Stewart. 1964. Investigation on the feeding of beef cattle in the Katherine region. *J. Aust. Inst. Agr. Sci.* 30:39–46.

Norris, D. O. 1967. The intelligent use of inoculants and lime pelleting for tropical pastures. *Trop. Grassl.* 1:107–121.

Norris, D. O. 1970. Nodulation of pasture legumes. Pp. 339–348. In R. P. Moore (ed.), *Australian Grasslands*. Australian National University Press, Canberra.

Norris, D. O. 1971. Seed pelleting to improve nodulation of tropical and subtropical legumes. 1, 2, 3. *Austr. J. Exptal. Agr. Anim. Husb.* 11:194–201, 282–289, 677–683.

Norris, D. O. 1972. Seed pelleting to improve nodulation of tropical and subtropical legumes. 4. *Aust. J. Exptal. Agr. Anim. Husb.* 12:152–158.

Norris, D. O. 1973. Seed pelleting to improve nodulation of tropical and subtropical legumes. 5, 6. *Aust. J. Exptal. Agr. Anim. Husb.* 13:98–101; 700–704.

North Carolina State University. 1973. *Agronomic-Economic Research on Tropical Soils.* Annual Report. Soil Science Department, North Carolina State University, Raleigh. 190 pp.

North Carolina State University. 1974. *Agronomic-Economic Research on Tropical Soils.* Annual Report. Soil Science Department, North Carolina State University, Raleigh. 220 pp.

Oakes, 1967. Effect of nitrogen fertilization and plant spacing on yield and composition of napier grass in the dry tropics. *Trop Agr. (Trinidad)* 44:77–82.

Okorie, I. I. D. H. Hill, and R. J. McIlroy. 1965. The productivity and nutritive value of tropical grass-legume pastures rotationally grazed by N'Dama cattle at Ibadan, Nigeria. *J. Agr. Sci.* 64:235–245.

Olsen, F. J. 1974. Effects of fertilization and cutting management on the yield and botanical composition of permanent pastures. *East Afr. Agr. For. J.* 39:391–396.

Olsen, F. J. 1975. Effect of large applications of nitrogen fertilizer on the productivity and protein content of four tropical grasses in Uganda. *Trop. Agr. (Trinidad)* 49:251–260.

Olsen, F. J. and P. G. Moe. 1971. The effect of phosphate and lime on the establishment, productivity, nodulation and persistence of *Desmodium intortum, Medicago sativa* and *Stylosanthes gracilis. East Afr. For. J.* 37:29–37.

Olsen, F. J. and G. L. Santos. 1975. Effects of lime and fertilizers on natural pastures in Brazil. *Exptal. Agr.* 11:173–176.

Pearson, R. W., F. Abruña, and J. Vicente-Chandler. 1962. Effect of lime and nitrogen applications on the downwards movement of calcium and magnesium in two humid tropical soils of Puerto Rico. *Soil Sci.* 93:77–82.

Pereira, H. C. and V. R. S. Beckley. 1953. Grass establishment on eroded soil in a semi-arid African reserve. *Emp. J. Exptal. Agr.* 21:1–14.

Perez, F. 1970. Efecto del intervalo de corte y fertilización con nitrógeno en la productividad de ocho gramineas. *Rev. Cubana Cienc. Agr.* **4**:137–142.

Plowes, D. C. H. The seasonal variation of crude protein in twenty common veld grasses at Matopos, Southern Rhodesia and related observations. *Rhodesia Agr. J.* **54**:33–55.

Poultney, R. G. 1963. A comparison of direct seeding and undersowing on the establishment of grass and the effect on the cover crop. *East Afr. Agr. For. J.* **29**:26–30.

Quinn, L. R., G. O. Mott, and W. V. A. Bisschoff. 1961. Fertilization of colonial guinea grass pasture and beef production with Zebu steers. *IBEC Res. Inst. Bull.* 24.

Quinn, L. R., G. O. Mott, W. V. A. Bisschoff, and G. L. da Rocha. 1962. Beef production of six tropical grasses. *IBEC Res. Inst. Bull.* 28.

Ramírez, A. and J. Lotero. 1969. Efectos de la frecuencia de aplicación de nitrógeno y la dósis en la fertilidad y propiedades químicas del suelo. *Rev. Inst. Colomb. Agropec.* **4**:227–254.

Ramsay, J. M. and R. Rose-Innes. 1963. Some quantitative observations on the effect of fire on the Guinea savanna vegetation of northern Ghana over a period of eleven years. *Afri. Soils* **8**:41–85.

Rickert, K. G. 1973. Establishment of green panic as influenced by type, amount, and placement of vegetative mulch. *Aust. J. Exptal. Agr. Anim. Husb.* **13**:268–274.

Rivera-Brenes, L., F. J. Marchan, and J. I. Cabrera. 1952. The utilization of grass, legumes and other forage crops for cattle feeding in Puerto Rico. III. Comparison of fertilized guinea grass, para grass, tropical kudzu and guinea grass–tropical kudzu. *J. Agr. Univ. Puerto Rico* **36**:108–114.

Robinson, P. J. and R. K. Jones. 1972. Effect of phosphorus and sulfur fertilization on the growth and distribution of dry matter, nitrogen, phosphorus and sulfur in Townsville Stylo. *Aust. J. Agr. Res.* **23**:633–640.

Rowland, J. W. 1955. The need for fertilizers in crop and ley rotations. *Rhodesia Agr. J.* **52**:171–179.

Russel, D. A., W. J. Free, and D. L. McCune. 1974. Potential for fertilizer usage on tropical forages. Pp. 39–65. In D. A. Mays (ed.), *Forage Fertilization.* American Society of Agronomy, Madison, Wisc.

Russell, J. S. 1966. Plant growth on a low calcium status solodic soil in a subtropical environment. I. Legume species, calcium carbonate, zinc and other minor element interactions. *Aust. J. Agr. Res.* **17**:673–686.

Samuels, G. and P. Landrau. 1952. The effects of fertilizer applications on the yields and nodulation of tropical kudzu. *Soil Sci. Soc. Amer. Proc.* **16**:154–155.

Sanchez, P. A. 1973. Nitrogen fertilization. Pp. 90–125. In P. A. Sanchez (ed.), "A Review of Soils Research in Tropical Latin America." *North Carolina Agr. Exp. Sta. Tech. Bull.* **219**.

Santhirasegaram, K. 1975. Manejo de praderas de praderas de leguminosas y gramíneas en un ecosistema de Selva Lluviosa Tropical en Perú. Pp. 445–466. In E. Bornemisza and A. Alvarado (eds.), *Manejo de Suelos en la America Tropical.* North Carolina State University, Raleigh.

Santhirasegaram, K., V. Morales, L. Pinedo, J. Diez, et al. 1972. *Pasture Development in the Pucallpa Region.* Interim Report. Instituto Veterinario de Investigación para el Trópico y de Altura. Univ. Nacional Mayor de San Marcos, Lima, Perú. 134 pp.

Serrão, E. S. and M. Simão Neto. 1975. The adaptation of tropical forages in the Amazon region. Pp. 31–52. In E. C. Doll and G. O. Mott (eds.), "Tropical Forages in Livestock

Production Systems." *ASA Spec. Publ.* 24. American Society of Agronomy, Madison, Wisc.

Shaw, N. H. and L. Mannetje. 1970. Studies of a speargrass pasture in central Queensland—the effect of fertilizers, stocking rate and oversowing with *Stylosanthes humilis* on beef production and botanical composition. *Trop. Grassl.* 4:43–56.

Shaw, N. H. and M. J. T. Norman. 1970. Tropical and subtropical woodlands and grasslands. Pp. 112–122. In R. M. Moore (ed.), *Australian Grasslands.* Australian National University Press, Canberra.

Shelton, H. M. and L. R. Humphreys. 1971. Effect of variation indensity and phosphate supply on seed production of *Stylosanthes humilis. J. Agr. Sci.* 76:325–328.

Shelton, H. M. and L. R. Humphreys. 1975. Undersowing rice with *Stylosanthes guyanensis. Exptal. Agr.* 11:89–112.

Silva, A. de F. 1963. O kudzu, a planta de três aplicações. *Gaz. Agr. Moçambique* 15:367–369.

Silvey, M. W. and V. W. Carlisle. 1971. Influence of Zn, P, and Ca on yield and chemical composition of hairy indigo in eastern Panama. *Soil Crop Sci. Soc. Fla. Proc.* 31:26–31.

Souto, S. M. and J. Döbereiner. 1969. Toxidés de manganés em quatro leguminosas forrageiras. *Pesq. Agropec. Bras.* 4:129–138.

Souto, M. S. and A. A. Franco. 1972. Sintomatología de deficiência de maconutrientes en *Centrosema pubescens* and *Phaseolus atropurpureus. Pesq. Agropec. Bras.* 7:23–27.

Spain, J. M. 1971. El problema de la acidez en suelos de los Llanos Orientales: posibles soluciones. *Suelos Ecuatoriales* 3:206–209.

Spain, J. M. 1972. El manejo de Oxisoles en el Oriente de Colombia. *IV Congreso Latinoamericano de la Ciencia del Suelo (Maracay)* (in press).

Spain, J. M. 1975. The forage potential of allic soils of the humid lowland tropics of Latin America. Pp. 1–8. In E. C. Doll and G. O. Mott (eds.), "Tropical Forages in Livestock Production Systems." *ASA Spec. Publ.* 24. American Society of Agronomy, Madison, Wisc.

Spain, J. M., C. A. Francis, R. H. Howeler, and F. Calvo. 1975. Diferencias entre especies y variedades de cultivos y pastos tropicales en su tolerancia a la acidez del suelo. Pp. 313–335. In E. Bornemisza and A. Alvarado (eds.), *Manejo de Suelos en la America Tropical.* North Carolina State University, Raleigh.

Stephens, D. 1967. Effects of fertilizers on grazed and cut elephant grass leys at Kawanda Research Station, Uganda. *East Afr. Agr. For. J.* 32:383–392.

Stobbs, T. H. 1965. Beef production from Uganda pastures containing *Stylosanthes gracilis* and *Centrosema pubescens. Proc. 9th Int. Grassl. Congr. (São Paulo)* 2:939–942.

Stobbs, T. H. 1970. The value of *Centrosema pubescens* for increasing animal production and improving soil fertility in northern Uganda. *East Afr. Agr. For. J.* 35:197–202.

Strachan, R. T., F. C. Lambert, and M. Finlay. 1967. A way to establish Townsville Lucerne. *Queensl. Agr. J.* 93:110–112.

Sutmoller, P., A. B. Abreu, J. VanderGrift, and W. G. Sombroek. 1966. Mineral imbalances in cattle in the Amazon Valley. *Royal Trop. Inst. Dept. Agr. Res. Commun.* 53. 85 pp.

Tang, C. N. and P. W. Long, 1970. Study on the nutrition of a tropical pasture legume on lateritic soil. *Taiwan Livestock Res.* 3:98–105.

Teitzel, J. K. 1969a. Responses to phosphorus, copper and potassium on a granite loam on the wet tropical coast of Queensland. *Trop. Grassl.* 3:43–48.

Teitzel, J. K. 1969b. Pastures for the wet tropical coast. 1. *Queensl. Agr. J.* **95**:304–311.

Teitzel, J. K. and R. C. Bruce. 1972a. Pasture fertilizers for the wet tropics. *Queensl. Agr. J.* **98**:13–22.

Teitzel, J. K. and R. C. Bruce. 1972b. Fertility of pasture soils in the wet tropical coast of Queensland. 3. Basaltic soils. *Aust. J. Exptal. Agr. Anim. Husb.* **12**:49–54.

Teitzel, J. K. and R. C. Bruce. 1973. Fertility of pasture soils in the wet tropical coast of Queensland. 5. Mixed alluvial soils. 6. Soils derived from beach sand. *Aust. J. Exptal. Agr. Anim. Husb.* **13**:306–318.

Tennessee Valley Authority. 1973. Nutrition for tropical and subtropical pastures. *TVA Bull.* Y-66. National Fertilizer Development Center. (Bibliography of abstracts).

Tergas, L. E. and W. G. Blue. 1971. Nitrogen and phosphorus in jaragua grass during the dry season in a tropical savanna as affected by nitrogen fertilization. *Agron. J.* **63**:6–9.

Tergas, L. E., W. G. Blue, and J. E. Moore. 1971. Nutritive value of fertilized jaragua grass in the wet–dry Pacific region of Costa Rica. *Trop. Agr.* (*Trinidad*) **48**:1–8.

Tewari, G. P. 1968. Responses of grasses and legumes to fertilizer treatments in Nigeria. *Exptal. Agr.* **4**:87–91.

Thairu, D. M. 1972. The contribution of *Desmodium uncinatum* to the yield of *Setaria sphacelata. East Afr. Agr. For. J.* **37**:215–219.

Tiharuhondi, E. R., F. J. Olsen, and R. J. Musangi, 1973. Application of nitrogen and irrigation to pasture to enhance cattle production during the dry season in Uganda. *East Afr. Agr. For. J.* **38**:383–393.

Uribe, A. and A. Grisales. 1966. Efecto de la fertilización nitrogenada en el pasto pangola. *Cenicafe* (*Colombia*) **17**:99–107.

Vallis, I. 1972. Soil nitrogen changes under continuously grazed grass–legume pastures in subtropical coastal Queensland. *Aust. J. Exptal. Agr. Anim. Husb.* **12**:495–501.

Vallis, I., K. P. Haydock, P. J. Ross, and E. F. Henzell. 1967. Isotopic studies on the uptake of nitrogen by pasture plants. 3. The uptake of small additions of N^{15} labelled fertilizers by rhodes grass and Townsville lucerne. *Aust. J. Agr. Res.* **18**:865–877.

Vallis, I., E. F. Henzell, A. E. Martin, and P. J. Ross. 1973. Isotopic studies on the uptake of nitrogen by pasture plants. 5. N^{15} balance experiments in field microplots. *Aust. J. Agr. Res.* **24**:693–702.

Vallis, I. and R. J. Jones. 1973. Net mineralization of nitrogen in leaves and leaf litter of *Desmodium intortum* and *Phaseolus atropurpureus* mixed with soil. *Soil Biol. Biochem.* **5**:391–398.

Van Rensburg, H. J. 1952. Grass burning experiments on the Msima river stock farm, southern Highlands, Tanganyika. *East Afr. Agr. J.* **17**:119–129.

Vasconcelos, C. N., A. G. Assis, R. M. de Souza, H. A. Villaca, et al. 1974. Valor nutritivo e productividade de cinco leguminosas tropicais em zona de mata, Estado de Minas Gerais. *Rev. Soc. Brasil. Zootec.* **3**:30–53.

Vásquez, R. 1965. Effects of irrigation and nitrogen levels on the yields of guinea grass (*Panicum maximum*), para grass (*Panicum purpurascens*) and guinea grass–kudzu (*Pueraria javanica*) and para grass–kudzu mixtures in the Lajas Valley. *J. Agr. Univ. Puerto Rico* **49**:389–412.

Vásquez, R. A., E. Hess, and M. J. Martinez-Luciano. 1966. Response of native white sorghum to irrigation under different nitrogen-fertility levels and seeding rates in Lajas Valley, Puerto Rico. *J. Agr. Univ. Puerto Rico* **50**:92–112.

Vicente-Chandler, J. 1966. The role of fertilizers in hot and humid tropical pastures. *Soil Crop Sci. Soc. Fla. Proc.* **26**:328–349.

Vicente-Chandler, J. 1967. Intensive pasture production. Pp. 272–295. In K. L. Turk and L. V. Crowder, *Rural Development in Tropical Latin America.* Cornell University, Ithaca, N.Y.

Vicente-Chandler, J. 1974. Fertilization of humid tropical grasslands. Pp. 277–300. In D. A. Mays (ed.), *Forage Fertilization.* American Society of Agronomy, Madison, Wisc.

Vicente-Chandler, J. 1975. Manejo intensivo de pastos y forrajes en Puerto Rico. Pp. 418–444. In E. Bornemisza and A. Alvarado (eds.), *Manejo de Suelos en la America Tropical.* North Carolina State University, Raleigh.

Vicente-Chandler, J. and R. Caro-Costas. 1953. The effect of two heights of cutting and three fertility levels on the yield, protein content and species composition of a tropical kudzu–molases grass pasture. *Agron. J.* **45**:397–400.

Vicente-Chandler, J. and J. Figarella. 1958. Growth characteristics of guinea grass on the semiarid south coast of Puerto Rico and the effect of nitrogen. *J. Agr. Univ. Puerto Rico* **42**:151–160.

Vicente-Chandler, J., S. Silva, and J. Figarella. 1959a. The effect of nitrogen fertilization and frequency of cutting on the yield and composition of three tropical grasses. *Agron. J.* **51**:202–206.

Vicente-Chandler, J., S. Silva, and J. Figarella. 1959b. Effects of nitrogen fertilization and frequency of cutting on the yield and composition of Napier grass in Puerto Rico. *J. Agr. Univ. Puerto Rico* **43**:215–227.

Vicente-Chandler, J., S. Silva, and J. Figarella. 1959c. Effects of nitrogen fertilization and frequency of cutting on the yield and composition of guinea grass in Puerto Rico. *J. Agr. Univ. Puerto Rico* **43**:228–239.

Vicente-Chandler, J., S. Silva, and J. Figarella. 1959d. Effects of nitrogen fertilization and frequency of cutting on the yield and composition of para grass in Puerto Rico. *J. Agr. Univ. Puerto Rico* **43**:240–248.

Vicente-Chandler, J. and R. W. Pearson. 1960. Nitrogen fertilization in hot climate grasses. *Soil Conserv. Mag.* **25**:269–272.

Vicente-Chandler, J. and S. Silva. 1960. The effect of nitrogen fertilization and grass species on soil physical conditions in some tropical pastures. *J. Agr. Univ. Puerto Rico* **44**:77–86.

Vicente-Chandler, J., J. Figarella, and S. Silva. 1961. Effects of nitrogen fertilization and frequency of cutting on the yield and composition of pangola grass in Puerto Rico. *J. Agr. Univ. Puerto Rico* **45**:37–45.

Vicente-Chandler, J. and J. Figarella. 1962. Effects of five nitrogen sources on yield and composition of Napier grass. *J. Agr. Univ. Puerto Rico* **46**:102–106.

Vicente-Chandler, J., S. Silva, and J. Figarella. 1962. Effect of frequency of application on response of Guinea grass to nitrogen fertilization. *J. Agr. Univ. Puerto Rico* **46**:342–349.

Vicente-Chandler, J., R. Caro-Costas, R. W. Pearson, F. Abruña, J. Figarella, and S. Silva. 1964. The intensive management of tropical forages in Puerto Rico. *Univ. Puerto Rico. Bull.* 187 (published in Spanish in 1967 as *Bull.* 202).

Vicente-Chandler, J., E. Rivera, R. Boneta, et al. 1973. The management and utilization of the forage crops of Puerto Rico. *Univ. Puerto Rico Agr. Exp. Sta. Bull.* 116. 90 pp.

Vicente-Chandler, J., F. Abruña, R. Caro-Costas, J. Figarella, S. Silva, and R. W. Pearson. 1974. Intensive grasslands management in the humid tropics of Puerto Rico. *Univ. Puerto Rico Agr. Exp. Sta. Bull.* 223.

Villachica, H., E. Bornemisza, and M. Arca. 1974. Effect of lime and phosphate treatments on yield and macronutrient content of Pangola grass grown on a soil from Pucallpa, Peru. *Agrochimia* 18:344–353.

Villamizar, F. and J. Lotero. 1967. Respuesta del pasto pangola a diferentes fuentes y dósis de nitrógeno. *Rev. Inst. Colomb. Agropec.* 2:57–70.

Walker, B. 1969. Effect of nitrogen fertilizer on natural pastures in western Tanzania. *Exptal. Agr.* 5:215–222.

Wendt, W. B. 1970. Response to pasture species in eastern Uganda to phosphorus, sulfur and potassium. *East Afr. Agr. For. J.* 36:211–219.

Werner, J. C., J. V. S. Pedreira, and E. L. Carelli. 1967. Estudos de parcelamento e niveis de adubação nitrogenada em capím pangola. *Bol. Ind. Anim. (Brazil)* 24:147–154.

Werner, J. C. F. P. Gomes, and E. A. Kalil. 1968. Fertilização nitrogenada e os seus efeitos na produção de forragem. *Bol. Ind. Anim. (Brazil)* 25:151–159.

Werner, J. C. et al. 1968. Comparação de diferentes adubos fosfatados. *Bol. Ind. Anim. (Brazil)* 25:139–149.

Werner, J. C. 1970. Estudo de épocas da adubação nitrogenada em capím Colonião para aumento de produção de forragem nas secas. *Bol. Ind. Anim. (Brazil)* 27–28:361–367.

Werner, J. C. and H. P. Haag. 1972. Nutrição mineral de gramineas forrageiras. *Bol. Ind. Anim. (Brazil)* 29:191–245.

Whitney, A. S. 1970. Effect of harvesting interval, height of cut and nitrogen fertilization on the performance of *Desmodium intortum* mixtures in Hawaii. Pp. 632–636. In *Proc. 11th Int. Grassl. Congr. (Australia)*.

Whitney, A. S. 1974. Growth of kikuyu grass under clipping, 1, 2. *Agron. J.* 66:281–287, 763–767

Whitney, A. S. 1975. Symbiotic and non-symbiotic nitrogen fixation as viewed by an agronomist. Pp. 51–75. In *Proceedings of the Soil and Water Management Workshop*. U. S. Agency for International Development, Washington.

Whitney, A. S., Y. Kanehiro, and G. D. Sherman. 1967. Nitrogen relationships of three tropical forage legumes in pure stands and in grass mixtures. *Agron. J.* 59:47–50.

Whitney, A. S. and R. E. Green. 1969. Pangola grass performance under different levels of nitrogen fertilization in Hawaii. *Agron. J.* 61:577–581.

Williams, C. H. and C. S. Andrew. 1970. Mineral nutrition of pastures. Pp. 321–328. In R. P. Moore (ed.), *Australian Grasslands*. Australian National University Press, Canberra.

Williams, W. A. 1967. The role of the legumes in pasture and soil improvement in the neotropics. *Trop. Agr. (Trinidad)* 44:103–115.

Winks, L. 1973. Townsville stylo research at Swan's Lagoon. *Trop. Grassl.* 7:201–208.

Wilson, J. R. 1975. Influence of temperature and nitrogen on growth, photosynthesis and accumulation of non-structural carbohydrates in a tropical grass, *Panicum maximum* var. *tricholegume. Netherl. J. Agr. Sci.* 23:48–61.

Wilson, J. R. and K. P. Haydock. 1971. The comparative response of tropical and temperate grasses to varying levels of nitrogen and phosphorus nutrition. *Aust. J. Agr. Res.* 22:573–587.

Wollner, H. 1968. Influencia de la fertilización con nitrógeno en el rendimiento de pasto pangola. *Beitr. Trop. Subtrop. Landwirt. Tropenveterinarmed.* 6:27–31.

Wright, J. W. 1971. Townsville stylo for hay in the Peninsula. *Queensl. Agr. J.* 97:473–478.

Younge, O. R., D. L. Plucknett, and P. R. Rotar. 1964. Culture and yield performance of *Desmodium intortum* and *Desmodium canum* in Hawaii. *Hawaii Agr. Exp. Sta. Tech. Bull.* 59.

Younge, O. R. and D. L. Plucknett. 1965. Beef production with heavy phosphorus fertilization in infertile wet lands of Hawaii. *Proc. 9th Int. Grassl. Congr. (São Paulo)* 2:959–963.

INDEX

607